Quantum Theory of Polymers as Solids

Quantum Theory of Polymers as Solids

János J. Ladik
University of Erlangen-Nuremberg
Erlangen, Federal Republic of Germany

Plenum Press • New York and London

Library of Congress Cataloging in Publication Data

Ladik, János, J.
Quantum theory of polymers as solids / János J. Ladik.
p. cm.
Includes bibliographical references and index.
ISBN-13: 978-1-4684-5235-8 e-ISBN-13: 978-1-4684-5233-4
DOI: 10.1007/978-1-4684-5233-4
1. Polymers and polymerization. 2. Quantum chemistry. I. Title.
QD381.L32 1987
547.7′0448–dc19 87-29159
CIP

Softcover reprint of the hardcover 1st edition 1988
A Division of Plenum Publishing Corporation
233 Spring Street, New York, N.Y. 10013

Preface

The goal of this monograph is to summarize the different quantum-mechanical methods developed in the last 20 years to treat the electronic structure of polymers. Owing to the nature of the problem, these methods consist of a mixture of quantum-chemical and solid-state physical techniques. The theory described in Part I treats, besides the Hartree–Fock problem, the electron correlation, and it has also been developed for disordered polymeric systems. Though for obvious reasons the book could not include all the existing calculations, each new method described is illustrated by a few applications, with a discussion of the numerical results obtained. For more details see the Introduction to Part I.

The second part contains the theoretical calculation of different properties of polymers based on the methods systematically introduced in the first part. The properties calculated include the electronic and vibrational spectra of polymers, and the computation of their transport, magnetic, and mechanical properties. In cases where reliable experimental data are available, the theoretical results are compared with them.

The systems for which the illustrative applications have been chosen fall into two main groups: the highly conductive polymers [e.g., $(CH)_x$, $(SN)_x$, and the TCNQ–TTF system], most of which were discovered in the last 10 years, and biopolymers (DNA and proteins). Since the electronic structure of DNA and proteins and their interactions are closely related to the problem of carcinogenesis, the last chapter in Part II also describes different probable solid-state physical mechanisms through which carcinogens bound to DNA can exert long-range effects — effects which may play an important role in the activation of the newly discovered human oncogenes. In a few cases calculations for polymers that are important for the plastics industry are also described [such as the energy band structure of teflon and the mechanical properties of polyethylene $(CH_2)_x$].

I hope that in this way the monograph will be of interest not only to theoretical chemists and theoretical physicists but also to spectroscopists, polymer physical chemists, and biophysicists.

Though the book describes purely theoretical work (which is scattered throughout several hundred papers, to which the appropriate references are given), it presumes only a knowledge of elementary quantum chemistry and linear algebra. The additional mathematical tools applied are described in detail within the appropriate chapters. The few sections that contain more involved mathematical formalisms are denoted by asterisks, and the less mathematically minded reader can omit them with no danger of being unable to understand the remainder of the book.

It is a pleasure for me to acknowledge the cooperation of many colleagues as well as the generous support of different organizations, without which most of the work described here could not have been done. I want first of all to express my deep gratitude to the late Professor Albert Szent-Györgyi, N.L., whose inspiration and enthusiasm contributed very much to the solution of different theoretical problems. I am further very grateful to Professor N. Fiebiger, President of the Friedrich Alexander University at Erlangen–Nürnberg, and to Mr. Kurt Köhler, Chancellor of the same university, whose interest and support were a continual source of strength. I am also very much indebted to the late Professor K. Laki, whose continuous interest was another important source of inspiration.

On the theoretical side I have profited very much from the ongoing cooperation of and discussions with Professors J. Čížek, E. Clementi, J. Delhalle, G. Del Re, W. Forbes, T. A. Hoffmann, M. Lax, F. Martino, and P. G. Mezey. I am grateful to all my present and former colleagues at the Institute for Physical and Theoretical Chemistry, and especially to Professors P. Otto and M. Seel, Dr. W. Förner, Dr. C.-M. Liegener, and Privatdozent Dr. S. Suhai. I am also very grateful to Dr. M. Gies for preparing the figures.

Special thanks are due to Dr. F. Salisbury, President, and Mrs. T. Salisbury, Executive Vice President, of the National Foundation for Cancer Research. Without the continuous and substantial financial help of NFCR, most of the theoretical investigations on biopolymers could not have been performed. I am very much indebted to the Deutsche Forschungsgemeinschaft, the Volkswagen Foundation, the Alexander von Humboldt Foundation, and the Deutscher Akademischer Austauschdienst, whose generous support made it possible to finance large numbers of co-workers and who also provided different Visiting Professorships. Without their help many of the theoretical investigations described herein could not have been performed. Siemens AG and Kraftwerk Union, A.G., as well as IBM Germany and IBM USA (Poughkeepsie–Kingston, New York), all provided substantial amounts of free computer time for the large-scale calculations.

Finally I should like to thank, for their continuous cooperation and help, the members of the Regional Computing Center of the University, and especially its Director, Dr. F. Wolf, for his understanding and for providing as much free computer time as possible for this work and also for helping to obtain time on the computer of the Leibnitz Computing Center of the Technical University of Munich.

János J. Ladik

Erlangen–Waterloo

Contents

Introduction

Polymers play an important role as plastics. Biopolymers, like nucleic acids (DNA and RNA), proteins, polysaccharides, lipids, and so on, have fundamental significance in life processes. In the last decade highly conducting polymers like doped polyacetylene, $(SN)_x$, or the TCNQ-TTF system, with quasi-one-dimensional alternating stacks of the two different types of molecule embedded in a three-dimensional molecular crystal, have become objects of extensive experimental investigations because they are candidates for the discovery of new physical phenomena. Various attempts have also been made at technological application (such as batteries). To understand the different physical and chemical properties of polymers, which in the case of biopolymers also determine their biological functions, one has to have a fair knowledge of their electronic structure.

The last fifteen years have seen considerable progress in the quantum-mechanical treatment of polymers, but the papers that contain the theoretical developments and their applications are scattered over numerous journals. This monograph therefore aims at presenting a unified approach in one volume to the quantum theory of polymers and its applications.

The book presumes only knowledge of elementary quantum mechanics and linear algebra. More involved formalisms (like Green functions) will be explained in the text. Therefore it can be read not only by theoretical physicists and theoretical chemists, but also by experimental physicists and physical chemists.

On the one hand, the parts of a polymer [the Greek word polymer ($\pi\omega\lambda\iota\mu\eta\rho$) means many parts] which are bound together are molecules or fragments of molecules. On the other hand, these units together form quasi-one-dimensional long chains or higher- (two- or three-) dimensional extended systems. Therefore it is evident that for the quantum-theoretical treatment of their electronic structure one must combine molecular physical (quantum-chemical) and solid-state physical methods. In this way the chemical structures of the subunits is not overlooked (as would occur in conventional solid-state physical methods) and also the merging of these two fields has given rise to the Hartree–Fock (HF) crystal-orbital (CO) theory of one- or higher-dimensional periodic polymers.

This book is divided into two parts: the first deals with the fundamental quantum theory of polymers and its use in determining their electronic structure, while the second part treats their different physical properties.

Chapter 1 discusses the HF CO theory of polymers in its *ab initio* form. [The word *ab initio* (which is a standard expression in quantum chemistry) *does not mean* that one has an exact solution of the Schrödinger equation, but it *does mean* that in the fixed-nuclei (Born–Oppenheimer) and Hartree–Fock approximations one takes into account all the electrons of the system and treats all the interaction integrals explicitly (up to a certain number of neighbors in the solid-state physical case).] Further, this chapter shows how to proceed if one has not a simple translation (as in simple solids), but a combined (such as helix) symmetry operation which determines the periodicity of the system. Special attention is also paid here to the problem of how many neighbor interactions must be taken into account for the different types of interaction integrals occurring in the theory in order to obtain reliable results, and how the effects of the nonexplicitly treated integrals can be estimated. This chapter also includes the different orbitals for different spin formalism and concludes with a relativistic Hartree–Fock theory of such extended periodic systems that also contain heavy atoms.

The second chapter examines applications of crystal-orbital theory presented earlier. These applications include polymers widely used in the production of plastics, primarily polyethylene and its fluorine derivatives. Other examples are from the field of highly conducting polymers, such as the different polyacetylenes, $(SN)_x$, and TCNQ and TTF stacks. Applications to nucleotide base stacks and to periodic polynucleotides and periodic polypeptides conclude this part of the book.

The next, short, chapter reviews different semiempirical crystal-orbital theories [π-electron CO theories, namely Hückel and Pariser–Parr–Pople theory, as well as all valence electron CO theories with different degrees of neglect of the so-called differential overlap (CNDO, INDO, MINDO, etc.)]. Applications to highly conducting polymers and to periodic biopolymers are also presented.

Most real polymers like nucleic acids and proteins, but also lightly and mediumly doped polyacetylenes, different copolymers, etc., are aperiodic. Therefore one must apply the theory of disordered systems to described their electronic structure. This rather important but not easy subject is covered in Chapter 4. Since most methods used to investigate this problem apply the Green-function (Green-matrix) formalism, the elements of this theory are presented in the first section of the chapter. Next, Anderson localization of Hartree–Fock orbitals in a disordered system based on the Green-matrix method is discussed. Afterward the negative-factor counting method (based on Dean's negative eigenvalue theorem) is reviewed both in its simple (Hückel) and in its more sophisticated matrix block form. This method, as well as the so-called coherent potential approximation (CPA), the latter again based on the Green-function formalism, uses as input the results of *ab initio* CO calculations performed on periodic chains of the components of the disordered chain. It is noteworthy that the negative-factor counting method is ideally suited to the treatment of quasi-one-dimensional chains and provides for such systems definitely superior results to CPA, but unfortunately cannot be applied to two- or three-dimensional disordered systems. Chapter 4 also contains applications, including a CPA calculation of $(SN)_x$ with hydrogen impurities distributed in a disordered way, and negative-factor counting treatments of the density of states of aperiodic DNA and of disordered polypeptides containing two or three different kinds of amino acids. The last section presents a Hartree–Fock resolvent (Green-matrix) method for the treatment of local perturbations of a periodic chain together with applications.

The next step in the construction of the quantum theory of polymers is the treatment of electron correlation in polymers, discussed in Chapter 5. The correlation energy of a system is defined as the difference between Hartree–Fock energy and the exact energy, which is the eigenvalue of the Schrödinger equation of the system. First, the construction of localized Wannier functions from the delocalized Bloch-type crystal orbitals and — for localization purposes — the optimal choice of their complex phase factor are discussed. Next, correlation in periodic polymers is examined in the framework of many-body perturbation theory and of the coupled-cluster theory using localized excitations between Wannier functions constructed from different energy bands. An alternative method, namely discretization of the energy bands in k-space, applicable also for metallic polymers with a partially filled valence band, is examined later in detail. In this case, no Wannier functions can be constructed because the necessary Fourier transform would also mix in unfilled Bloch orbitals. The application of this method to alternating *trans*-polyacetylene using a good basis and Møller–Plesset second-order perturbation theory resulted in about 75% of the correlation energy. A generalized electronic polaron model was employed to introduce the concept of quasi-particle energy bands, with which it was possible to reduce the Hartree–Fock minimal basis gap of approximately 8.3 eV to about 3.0 eV if the quasi-particle gap (gap with correlation) was computed with a good-quality basis set. The experimental gap is close to 2.0 eV, but this discrepancy can easily be accounted for by the missing part of the correlation and by other physical effects not yet taken into account. This important result presented here shows that, in contrast to the conventional solid-state physical point of view, if one starts from a good-quality Hartree–Fock energy-band structure and takes into account the major part of the correlation, one can obtain not only ground-state properties but also the gap reasonably well.

The treatment of correlation, even in periodic polymers, presents a formidable and as yet only partially solved problem if the unit cell is large. This problem is examined by initially discussing localization techniques in larger molecules, and then presenting the first results of the application of localized orbitals for correlation calculations using many-body perturbation theory and the coupled-cluster approach. It is also shown how localization of the orbitals within a large unit cell can be applied for correlation calculations if this unit cell is repeated in a periodic way in the polymer. This chapter also deals with the correlation in polymers with smaller unit cells (like polyacetylene, polydiacetylene, and polyethylene) and includes a detailed discussion of the results obtained. Finally, some ideas about possibilities of treating correlation in disordered chains are noted.

Theoretical methods for the investigation of interactions between polymer chains are described in Chapter 6. Besides the theoretically clear-cut but, in the case of polymers with larger unit cells, numerically unfeasible, superchain approach, theoretical perturbation methods and the mutually consistent field (MCF) procedure recently developed at Erlangen are reviewed. The first application of the MCF method, which takes into account both the electrostatic part and polarization forces, to polynucleotide–polypeptide interactions (modeling DNA–protein interactions) is presented.

The MCF method applied to generate an effective potential of the water environment surrounding polynucleotide chains is presented in Chapter 7. The respective positions of the water molecules and ions were taken from the results of extensive Monte Carlo simulations performed by Clementi's group. The first results as to how this effective potential caused by the environment affects the band structure of a cytosine stack are also reviewed in this chapter.

The second part of this book treats various physical properties of polymers. Electronic excitations and ionizations in polymers are discussed in Chapter 8. For this purpose the intermediate (charge-transfer) exciton theory is described in some detail and it is shown how one can introduce correlation and screening into the expressions of this theory. A Green-function formalism is developed for the treatment of ionized states in polymers. This chapter also contains applications for the electronic spectra of polyacetylenes, polydiacetylenes, and homopolynucleotides, and concludes with an interpretation of the experimental electronic-absorption spectra and photoelectron spectra of different polymers.

In Chapter 9, the vibrational spectra of linear polymers, including a description how one can obtain force constants and vibrational frequencies by calculating the total electronic energy at different bond distances, are discussed first. The theory is supplemented by methods for the treatment of vibrations in disordered polymers, which are very similar to those used in the investigation of the electronic structure in disordered chains (see Chapter 4). In several cases the theoretically computed and experimental vibrational spectra are compared.

Next, the phonon dispersion curves and phonon wave functions are used to calculate electron–phonon interaction matrix elements. These can serve as input for the study of different transport properties such as mean free path, mobility, and specific conductivity. In the case of periodic polymers, for which the band model is valid, one expects a coherent Bloch-type conduction. Therefore, the electron–phonon interaction matrix elements give the probabilities of the different phonon scattering processes that one can substitute into Bolzmann's transport equation (or into the Kubo formalism), and by obtaining an approximate solution to these equations one can determine the above-mentioned transport properties. On the other hand, in the case of a disordered polymer or of a disorderly doped periodic polymer, these interaction matrix elements determine the primary jump rates between different sites that are the decisive input data for a hopping theory of charge transport. Chapter 9 also contains applications of the theory of transport properties to different highly conducting polymers, to polypeptides, and to nucleotide base stacks. The results are compared with corresponding experimental data.

Different magnetic, electric, and mechanical properties of polymers are discussed in Chapter 10, which opens with the quite straightforward derivation of the Hartree–Fock equations of a periodic polymer in a static magnetic field. The case of the treatment of the interface between two polymers under the influence of different average magnetic fields is worked out in detail using an MCF–Green matrix formalism. In the case of time-dependent magnetic fields, the corresponding theory developed for atoms and molecules (based on the so-called time-dependent coupled Hartree–Fock equations) is sketched for periodic polymers. Applications to some simple polymers and comparison of theoretical results with available experimental data are also presented.

This chapter describes briefly also methods for the calculation of electric polarizabilities, and hyperpolarizabilities which are important in the theory of nonlinear optics.

The third part of Chapter 10 describes the calculation of various mechanical properties (such as bulk modules and shear modules) of some simple polymers based on their band structure but corrected for correlation. In the case of polyethylene and some of its halogenated derivatives the theoretical results are compared with available experimental data.

The last chapter of the book, Chapter 11, examines the probable role of the solid-state physical properties of biopolymers in their biological functions. After a quantum-chemical discussion of the mechanisms leading to point mutations and aging, there is a review of the possible long-range effects of carcinogen binding or of local radiation damage leading to initiation of a malignant transformation of a cell. These different, possible long-range effects are based on the solid-state physical properties of the biopolymers (like the possible formation of a conformational soliton caused by the binding of a carcinogen), which again underlines the importance of the treatment of biopolymers as solids.

I

Quantum Theory of Polymeric Electronic Structure

Chapter 1

Hartree–Fock Crystal-Orbital Theory of Periodic Polymers

1.1. SIMPLE TRANSLATION

1.1.1. Block Diagonalization of the Hamiltonian Matrix

We consider a three-dimensional periodic polymer or molecular crystal containing m orbitals in the elementary cell of one or more atoms. For the sake of simplicity the number of elementary cells in the direction of each crystal axis is taken equal to an odd number: $N_1 = N_2 = N_3 = 2N + 1$. We assume further that there is an interaction between orbitals belonging to different elementary cells. In that case we can describe, in the one-electron approximation, the delocalized crystal orbitals of the polymer with the aid of the linear-combination-of-atomic-orbitals (LCAO) approximation in the form

$$\phi_h^{\mathbf{p}}(\mathbf{r}) = \sum_{\mathbf{q}} \sum_{g=1}^{m} C(\mathbf{p})_{h;\mathbf{q},g} \chi_g^{\mathbf{q}} \tag{1.1}$$

where $\mathbf{p} = (p_1, p_2, p_3)$ and' $\mathbf{q} = (q_1, q_2, q_3)$; integers p_j and q_j $(j = 1, 2, 3)$ run over $-N, \ldots, 0, \ldots, N$ and $\Sigma_{\mathbf{q}}$ is a shorthand notation for $\Sigma_{q_1=-N}^{N} \Sigma_{q_2=-N}^{N} \Sigma_{q_3=-N}^{N}$. Further, $\chi_g = \chi_g(\mathbf{r} - \mathbf{R}_q - \mathbf{r}_{g_A})$ is the gth atomic orbital (AO) (corresponding to the atom with position vector $\mathbf{r}_{g_A}$) in the cell characterized by the vector $\mathbf{R}_q = q_1\mathbf{a}_1 + q_2\mathbf{a}_2 + q_3\mathbf{a}_3$ $(q_j = -N, \ldots, 0, \ldots, N)$, where $\mathbf{a}_1$, $\mathbf{a}_2$, and $\mathbf{a}_3$ are the three basis vectors of the crystal.

If we form the expectation value

$$\frac{\langle \phi_h^{\mathbf{p}} \mid \hat{H}_{\text{eff}} \mid \phi_h^{\mathbf{p}} \rangle}{\langle \phi_h^{\mathbf{p}} \mid \phi_h^{\mathbf{p}} \rangle} = \varepsilon(\mathbf{p})_h \tag{1.2}$$

of the one-electron effective Hamiltonian $\hat{H}_{\text{eff}}$ and perform a Ritz variational procedure for the coefficients in equation (1.1), we obtain in the standard way for the whole polymer the matrix equation

$$\mathbf{HC}(\mathbf{p})_h = \varepsilon(\mathbf{p})_h \mathbf{SC}(\mathbf{p})_h \tag{1.3}$$

The hypermatrix $\mathbf{H}$ of dimension $m(2N+1)^3$ has submatrices $\mathbf{H}_{\mathbf{p},\mathbf{q}}$ of dimension m consisting of interactions between orbitals belonging to the elementary cells characterized by the lattice vectors $\mathbf{R}_\mathbf{p}$ and $\mathbf{R}_\mathbf{q}$. The f,gth element $H_{\mathbf{p},f;\mathbf{q},g}$ of the matrix $\mathbf{H}_{\mathbf{p},\mathbf{q}}$ is then given by

$$H_{\mathbf{p},f;\mathbf{q},g} = \langle \chi_f^{\mathbf{p}} \mid \hat{H}_{\text{eff}} \mid \chi_g^{\mathbf{q}} \rangle \tag{1.4}$$

The overlap matrix $\mathbf{S}$ can be partitioned similarly into blocks with elements

$$S_{\mathbf{p},f;\mathbf{q},g} = \langle \chi_f^{\mathbf{p}} \mid \chi_g^{\mathbf{q}} \rangle \tag{1.5}$$

In consequence of the three-dimensional translational symmetry of the polymer and of the Born–von Kármán periodic boundary conditions, matrices $\mathbf{H}$ and $\mathbf{S}$ are cyclic hypermatrices. For the sake of simplicity we show this for the one-dimensional case; the generalization to two- and three-dimensional cases is straightforward. In the one-dimensional case, if we take into account the translational symmetry, the hypermatrices $\mathbf{H}$ and $\mathbf{S}$ have the form

$$\begin{bmatrix} \boxed{0} & \boxed{1} & \boxed{2} & & & \boxed{2N} \\ \boxed{-1} & \boxed{0} & \boxed{1} & \boxed{2} & & \boxed{2N-1} \\ & & & & & \\ \boxed{-2N} & & & & \boxed{-1} & \boxed{0} \end{bmatrix} \tag{1.6}$$

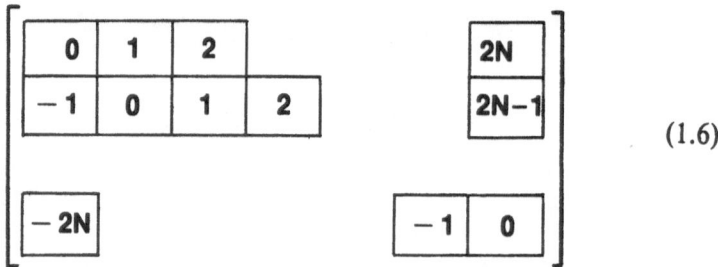

Here, submatrix $\boxed{0}$ denotes interactions within the elementary cell, submatrices $\boxed{1}$ and $\boxed{-1} = \boxed{1^{\text{tr}}}$ correspond to first-neighbor interactions, and so on (all the submatrices have dimension $m \times m$ only). Introduction of the Born–von Kármán periodic boundary conditions yields the relations

$$\boxed{\mathbf{2N}} = \boxed{-\mathbf{1}}, \quad \boxed{\mathbf{2N-1}} = \boxed{-\mathbf{2}}, \qquad \text{and so on}$$

Therefore, hypermatrix (1.6) will become a cyclic hypermatrix. It is well known that one can always find a unitary transformation for a cyclic matrix that diagonalizes it. In the same way there exists a unitary hypermatrix $\mathbf{U}$ with blocks $\mathbf{U}_{p,q}$ ($\mathbf{U}_{p,q}$ is the block of $\mathbf{U}$ corresponding to the pth row of blocks and qth column of blocks) given by

$$\mathbf{U}_{p,q} = \frac{1}{(2N+1)^{1/2}} \exp\left(\frac{2\pi i p q}{2N+1}\right) \mathbf{1} \tag{1.7}$$

which block-diagonalizes the cyclic hypermatrices (the unit matrix $\mathbf{1}$ again has dimension $m \times m$ only). Hence we can write

$$\mathbf{H}' = \mathbf{U}^{+}\mathbf{H}\mathbf{U} = \begin{bmatrix} \Box & & & & \mathbf{0} \\ & \Box & & & \\ & & \Box & & \\ & & & \ddots & \\ \mathbf{0} & & & & \Box \end{bmatrix}, \qquad \mathbf{S}' = \mathbf{U}^{+}\mathbf{S}\mathbf{U} \tag{1.8}$$

where only the diagonal $m \times m$ blocks differ from zero. It is noteworthy that the unitary matrix $\mathbf{U}$ defined by equation (1.7) block-diagonalizes any hypermatrix of dimension $m(2N+1)$ with blocks of dimension m independently of the values of the elements in the submatrices. Therefore, the same matrix $\mathbf{U}$ will block-diagonalize both $\mathbf{H}$ and $\mathbf{S}$.

Returning to the three-dimensional problem, one can show in a similar way[1] that matrices $\mathbf{H}$ and $\mathbf{S}$ are cyclic hypermatrices in this case too and the unitary matrix $\mathbf{U}$ of dimension $m(2N+1)^3$ with $m \times m$ blocks, namely

$$\mathbf{U}_{\mathbf{p},\mathbf{q}} = \frac{1}{(2N+1)^{3/2}} \exp\left(\frac{2\pi i \mathbf{p}\cdot\mathbf{q}}{2N+1}\right) \mathbf{1} \tag{1.9}$$

block-diagonalizes them. Definition (1.9) enables one to easily prove that $\mathbf{U}^{+}\mathbf{U} = \mathbf{I}$, where unit matrix $\mathbf{I}$ has dimension $m(2N+1)^3$ and the $\mathbf{p}$th diagonal blocks of the block-diagonal matrices (1.8) have the form

$$\mathbf{H}'(\mathbf{p}) = \sum_{\mathbf{q}} \exp\left(\frac{2\pi i \mathbf{p}\cdot\mathbf{q}}{2N+1}\right) \mathbf{H}(\mathbf{q}) \tag{1.10}$$

and

$$\mathbf{S}'(\mathbf{p}) = \sum_{\mathbf{q}} \exp\left(\frac{2\pi i \mathbf{p}\cdot\mathbf{q}}{2N+1}\right) \mathbf{S}(\mathbf{q}) \tag{1.11}$$

In these expressions matrices $\mathbf{H}(\mathbf{q})$ and $\mathbf{S}(\mathbf{q})$ are, respectively, submatrices of the original matrices $\mathbf{H}$ and $\mathbf{S}$ before the unitary transformation. Hence in the one-dimensional case $q = 0$ gives the submatrix $\mathbf{0}$ of expression (1.6), $q = 1$ and $q = -1$ refer to blocks $\mathbf{1}$ and $-\mathbf{1}$ (first-neighbor interactions) of expression (1.6), and so on.

If equations (1.3) are multiplied on the left by $\mathbf{U}^+$, and the unit matrix $\mathbf{I}$ in the form $\mathbf{I} = \mathbf{U}\mathbf{U}^+$ is inserted on both sides after $\mathbf{H}$ and $\mathbf{S}$, we obtain

$$\mathbf{U}^+\mathbf{H}\mathbf{U}\mathbf{U}^+\mathbf{C}(\mathbf{p})_h = \varepsilon(\mathbf{p})_h \mathbf{U}^+\mathbf{S}\mathbf{U}\mathbf{U}^+\mathbf{C}(\mathbf{p})_h \tag{1.12}$$

On introducing the notation $\mathbf{U}^+\mathbf{C}(\mathbf{p})_h = \mathbf{D}(\mathbf{p})_h$ and taking into account relationships (1.8), equation (1.12) can be expressed in the form

$$\mathbf{H}'\mathbf{D}(\mathbf{p})_h = \varepsilon(\mathbf{p})_h \mathbf{S}\mathbf{D}(\mathbf{p})_h \tag{1.13}$$

Using the fact that $\mathbf{H}'$ and $\mathbf{S}'$ are block-diagonal, we can decompose equations (1.13) to the much simpler equations

$$\mathbf{H}'(\mathbf{p})\mathbf{d}(\mathbf{p})_h = \varepsilon(\mathbf{p})_h \mathbf{S}'(\mathbf{p})\mathbf{d}(\mathbf{p})_h \tag{1.14}$$

where $\mathbf{H}'(\mathbf{p})$ and $\mathbf{S}'(\mathbf{p})$ are defined by expressions (1.10) and (1.11), respectively.*

If $N \to \infty$ we can consider the new variables

$$k_1 = \frac{2\pi\, p_1}{a_1(2N+1)}, \quad k_2 = \frac{2\pi\, p_2}{a_2(2N+1)}, \quad k_3 \frac{2\pi\, p_3}{a_3(2N+1)} \tag{1.15}$$

as continuous. Since quantities p_j $(j = 1, 2, 3)$ assume the values $-N, \ldots, 0, \ldots, N$, then k_1, k_2, and k_3 will take values between $-\pi/a_1$ and π/a_1, $-\pi/a_2$ and π/a_2, and $-\pi/a_3$ and π/a_3, respectively. By defining the vector

$$\mathbf{k} = k_1\mathbf{b}_1 + k_2\mathbf{b}_2 + k_3\mathbf{b}_3 \tag{1.16}$$

* Actually, the vector $\mathbf{p}$ in equations (1.1) and (1.14) is not the same. However, if one introduces the convention that in the hypervector $\mathbf{c}(\mathbf{p})_h$ its $\mathbf{p}$th segment always corresponds to the reference cell, one can use the same notation for both quantities.

where $\mathbf{b}_1$, $\mathbf{b}_2$, $\mathbf{b}_3$ are the basis vectors of the reciprocal space (by definition, $\mathbf{a}_i\mathbf{b}_j = 2\delta_{ij}$),* we can rewrite equation (1.14) as

$$\mathbf{H}(\mathbf{k})\mathbf{d}(\mathbf{k})_h = \varepsilon(\mathbf{k})_h\mathbf{S}(\mathbf{k})\mathbf{d}(\mathbf{k})_h \qquad (h = 1, 2, \ldots, m) \tag{1.17}$$

where

$$\mathbf{H}(\mathbf{k}) = \sum_{q=-\infty}^{\infty} e^{i\mathbf{k}\cdot\mathbf{R}_q}\mathbf{H}(\mathbf{q}) \tag{1.18}$$

and

$$\mathbf{S}(\mathbf{k}) = \sum_{q=-\infty}^{\infty} e^{i\mathbf{k}\cdot\mathbf{R}_q}\mathbf{S}(\mathbf{q}) \tag{1.19}$$

The solutions of the generalized eigenvalue equations (1.17) for different vectors $\mathbf{k}$ will provide the energy-band structures of the crystal, or in the case of a scalar k, of the linear periodic polymer. The eigenvalues $\varepsilon(\mathbf{k})_h$ corresponding to different values of h provide the different energy bands, and the values of $\varepsilon(\mathbf{k})_h$ corresponding to different vectors $\mathbf{k}$ for a given h give the energy levels within the hth band.†

1.1.2. Elimination of the Overlap Matrix

Equations (1.17) are solved by first eliminating matrix $\mathbf{S}(\mathbf{k})$. This is carried out most simply with the aid of Löwdin's symmetric orthogonalization procedure.[2] If equations (1.17) are multiplied on the left by $\mathbf{S}(\mathbf{k})^{-1/2}$ and $\mathbf{S}(\mathbf{k})^{-1/2}\mathbf{S}(\mathbf{k})^{1/2} = \mathbf{1}$ inserted, we obtain

$$\mathbf{S}(\mathbf{k})^{-1/2}\mathbf{H}(\mathbf{k})\mathbf{S}(\mathbf{k})^{-1/2}\mathbf{S}(\mathbf{k})^{1/2}\mathbf{d}(\mathbf{k})_h = \varepsilon(\mathbf{k})_h\mathbf{S}(\mathbf{k})^{1/2}\mathbf{d}(\mathbf{k})_h \tag{1.20}$$

* The first Brillouin zones of a crystal with given symmetry are usually given in terms of the Cartesian coordinates of $\mathbf{k}$. On the other hand, we can characterize the elementary cells most easily with the aid of the lattice vectors $\mathbf{R}_q$ expressed in terms of unit vectors $\mathbf{a}_1$, $\mathbf{a}_2$, and $\mathbf{a}_3$, which are not necessarily orthogonal. Therefore, we need to transform the rectangular coordinates k_x, k_y, and k_z of $\mathbf{k}$ into the quantities k_j ($j = 1, 2, 3$). It is easy to show that this can be achieved with the aid of the equations

$$k_j = \frac{1}{2\pi}(a_{jx}k_x + a_{jy}k_y + a_{jz}k_z) \qquad (j = 1, 2, 3)$$

where a_{jx}, a_{jy}, and a_{jz} are the rectangular coordinates of the basis vector $\mathbf{a}_j$.

† If $N \to \infty$, or is finite but large, the individual energy levels within a band are closer than k_BT (k_B is the Boltzmann constant) and therefore the level distribution within a band can be regarded as continuous.

With use of the notation

$$\tilde{\mathbf{H}}(\mathbf{k}) = \mathbf{S}(\mathbf{k})^{-1/2}\,\mathbf{H}(\mathbf{k})\mathbf{S}(\mathbf{k})^{-1/2} \tag{1.21}$$

and

$$\mathbf{b}(\mathbf{k})_h = \mathbf{S}(\mathbf{k})^{1/2}\mathbf{d}(\mathbf{k})_h \tag{1.22}$$

equation (1.20) can be written in the form

$$\tilde{\mathbf{H}}(\mathbf{k})\mathbf{b}(\mathbf{k})_h = \varepsilon(\mathbf{k})_h\, \mathbf{b}(\mathbf{k})_h \tag{1.23}$$

The square root and inverse square root of $\mathbf{S}(\mathbf{k})$ are easily calculated via its diagonalization, namely

$$\mathbf{S}_0(\mathbf{k}) = \mathbf{V}(\mathbf{k})^+\mathbf{S}(\mathbf{k})\mathbf{V}(\mathbf{k}) \tag{1.24}$$

where the unitary matrix $\mathbf{V}(\mathbf{k})$ contains as columns the eigenvectors $\mathbf{v}(\mathbf{k})_i$ of the eigenvalue problem

$$\mathbf{S}(\mathbf{k})\mathbf{v}(\mathbf{k})_i = s(\mathbf{k})_i\mathbf{v}(\mathbf{k})_i \tag{1.25}$$

and the diagonal matrix $\mathbf{S}_0(\mathbf{k})$ contains as elements the eigenvalues $s(\mathbf{k})_i$. Equation (1.24) enables function $\mathbf{S}_0(\mathbf{k})$ to be expressed as

$$\mathbf{S}(\mathbf{k}) = \mathbf{V}(\mathbf{k})\mathbf{S}_0(\mathbf{k})\mathbf{V}(\mathbf{k})^+$$

and $\mathbf{S}(\mathbf{k})^{1/2}$ and $\mathbf{S}(\mathbf{k})^{-1/2}$, respectively, as

$$\mathbf{S}(\mathbf{k})^{1/2} = \mathbf{V}(\mathbf{k})\mathbf{S}_0(\mathbf{k})^{1/2}\mathbf{V}(\mathbf{k})^+ \quad \text{and} \quad \mathbf{S}(\mathbf{k})^{-1/2} = \mathbf{V}(\mathbf{k})\mathbf{S}_0(\mathbf{k})^{-1/2}\mathbf{V}(\mathbf{k})^+ \tag{1.26}$$

[The proof of the validity of equations (1.26) can be found elsewhere.[3]] Since all the eigenvalues of $\mathbf{S}(\mathbf{k})$ can be shown to be positive, one can easily calculate quantities $\mathbf{S}_0(\mathbf{k})^{1/2}$ and $\mathbf{S}_0(\mathbf{k})^{-1/2}$.

The simplest way to diagonalize the Hermitian complex matrix $\mathbf{S}(\mathbf{k})$ defined by equation (1.19) is to rewrite its eigenvalue equation (1.25) in the form

$$\begin{aligned}(\mathrm{Re}[\mathbf{S}(\mathbf{k})] + i\,\mathrm{Im}[\mathbf{S}(\mathbf{k})])(\mathrm{Re}[\mathbf{v}(\mathbf{k})_i] + i\,\mathrm{Im}[\mathbf{v}(\mathbf{k})_i]\\ = s(\mathbf{k})_i(\mathrm{Re}[\mathbf{v}(\mathbf{k})_i] + i\,\mathrm{Im}[\mathbf{v}(\mathbf{k})_i])\end{aligned} \tag{1.27}$$

Separation of real and imaginary parts yields

$$\mathrm{Re}[\mathbf{S}(\mathbf{k})]\,\mathrm{Re}[\mathbf{v}(\mathbf{k})_i] - \mathrm{Im}[\mathbf{S}(\mathbf{k})]\,\mathrm{Im}[\mathbf{v}(\mathbf{k})_i] = s(\mathbf{k})_i\,\mathrm{Re}[\mathbf{v}(\mathbf{k})_i] \tag{1.28a}$$

and

$$\mathrm{Im}[\mathbf{S}(\mathbf{k})]\,\mathrm{Re}[\mathbf{v}(\mathbf{k})_i] + \mathrm{Re}[\mathbf{S}(\mathbf{k})]\,\mathrm{Im}[\mathbf{v}(\mathbf{k})_i] = s(\mathbf{k})_i\,\mathrm{Im}[\mathbf{v}(\mathbf{k})_i] \tag{1.28b}$$

The latter two matrix equations can be written as the single matrix eigenvalue equation

$$\begin{pmatrix} \mathrm{Re}[\mathbf{S}(\mathbf{k})] & -\mathrm{Im}[\mathbf{S}(\mathbf{k})] \\ \mathrm{Im}[\mathbf{S}(\mathbf{k})] & \mathrm{Re}[\mathbf{S}(\mathbf{k})] \end{pmatrix} \begin{pmatrix} \mathrm{Re}[\mathbf{v}(\mathbf{k})_i] \\ \mathrm{Im}[\mathbf{v}(\mathbf{k})_i] \end{pmatrix} = s(\mathbf{k})_i \begin{pmatrix} \mathrm{Re}[\mathbf{v}(\mathbf{k})_i] \\ \mathrm{Im}[\mathbf{v}(\mathbf{k})_i] \end{pmatrix} \tag{1.29}$$

of a real matrix of order $2m$, if the order of the Hermitian complex matrix $\mathbf{S}(\mathbf{k})$ is m. Each of the always-real eigenvalues $s(\mathbf{k})_i$ of $\mathbf{S}(\mathbf{k})$ will occur twice in equation (1.29) and, taking into account equation (1.19), we can write

$$\mathrm{Re}[\mathbf{S}(\mathbf{k})] = \sum_{\mathbf{q}=-\infty}^{\infty} \cos(\mathbf{k}\cdot\mathbf{R}_\mathbf{q})\mathbf{S}(\mathbf{q}), \quad \mathrm{Im}[\mathbf{S}(\mathbf{k})] = \sum_{\mathbf{q}=-\infty}^{\infty} \sin(\mathbf{k}\cdot\mathbf{R}_\mathbf{q})\mathbf{S}(\mathbf{q}) \tag{1.30}$$

Similarly, the eigenvalue equation (1.23), namely

$$\tilde{\mathbf{H}}(\mathbf{k})\mathbf{B}(\mathbf{k}) = \mathbf{B}(\mathbf{k})\boldsymbol{\varepsilon}(\mathbf{k}) \tag{1.31}$$

where matrix $\mathbf{B}(\mathbf{k})$ again contains as columns the vectors $\mathbf{b}(\mathbf{k})_h$ and diagonal matrix $\boldsymbol{\varepsilon}(\mathbf{k})$ has as elements eigenvalues $\varepsilon(\mathbf{k})_h$, can be written as

$$\begin{pmatrix} \mathrm{Re}[\tilde{\mathbf{H}}(\mathbf{k})] & -\mathrm{Im}[\tilde{\mathbf{H}}(\mathbf{k})] \\ \mathrm{Im}[\tilde{\mathbf{H}}(\mathbf{k})] & \mathrm{Re}[\tilde{\mathbf{H}}(\mathbf{k})] \end{pmatrix} \begin{pmatrix} \mathbf{B}_1(\mathbf{k}) \\ \mathbf{B}_2(\mathbf{k}) \end{pmatrix} = \begin{pmatrix} \mathbf{B}_1(\mathbf{k}) \\ \mathbf{B}_2(\mathbf{k}) \end{pmatrix} \boldsymbol{\varepsilon}'(\mathbf{k}) \tag{1.32}$$

Here $\mathbf{B}_1(\mathbf{k}) = \mathrm{Re}[\mathbf{B}(\mathbf{k})]$, $\mathbf{B}_2(\mathbf{k}) = \mathrm{Im}[\mathbf{B}(\mathbf{k})]$, and the diagonal matrix $\boldsymbol{\varepsilon}'(\mathbf{k})$ contains each eigenvalue $\varepsilon(\mathbf{k})_h$ twice. After determining matrices $\mathbf{B}_1(\mathbf{k})$ and $\mathbf{B}_2(\mathbf{k})$ on the basis of the relation

$$\mathbf{B}(\mathbf{k}) = \mathbf{B}_1(\mathbf{k}) + i\mathbf{B}_2(\mathbf{k}) = \mathbf{S}(\mathbf{k})^{1/2}\mathbf{D}(\mathbf{k}) \tag{1.33}$$

one can always calculate the real and imaginary parts of the original vectors $\mathbf{d}(\mathbf{k})_h$:

$$\mathbf{D}(\mathbf{k}) = \mathbf{S}(\mathbf{k})^{-1/2}\mathbf{B}(\mathbf{k})$$

or

$$\mathbf{D}_1(\mathbf{k}) + i\mathbf{D}_2(\mathbf{k}) = \{\mathrm{Re}[\mathbf{S}(\mathbf{k})^{-1/2}] + i\,\mathrm{Im}[\mathbf{S}(\mathbf{k})^{-1/2})\}[\mathbf{B}_1(\mathbf{k}) + i\mathbf{B}_2(\mathbf{k})] \tag{1.34}$$

with

$$\mathbf{D}_1(\mathbf{k}) = \mathrm{Re}[\mathbf{S}(\mathbf{k})^{-1/2}]\mathbf{B}_1(\mathbf{k}) - \mathrm{Im}[\mathbf{S}(\mathbf{k})^{-1/2}]\mathbf{B}_2(\mathbf{k})$$

and

$$\mathbf{D}_2(\mathbf{k}) = \mathrm{Re}[\mathbf{S}(\mathbf{k})^{-1/2}]\mathbf{B}_2(\mathbf{k}) + \mathrm{Im}[\mathbf{S}(\mathbf{k})^{-1/2}]\mathbf{B}_1(\mathbf{k})$$

It should be noted that instead of solving equations (1.29) and (1.32), one can solve equations (1.24) and (1.17) also directly using complex arithmetic. At most computing centers where the necessary programs for complex arithmetic are available, this approach is followed. We have included the above description of the problem with real arithmetic only for the use of those for whom complex arithmetic is, for some reason, unfeasible.

1.1.3. Hartree–Fock–Roothaan Crystal-Orbital Formalism

Until now we have not specified matrices $\mathbf{H}(\mathbf{q})$ in equation (1.18). To do this we write the Hamiltonian of the whole polymer (in atomic units) as

$$\hat{H} = \sum_{\mu=1}^{n_e(2N+1)^3} \hat{H}^N(\mu) + \sum_{\mu<\nu}^{n_e(2N+1)^3} \frac{1}{r_{\mu\nu}} \tag{1.35}$$

and

$$\hat{H}^N(\mu) = -\tfrac{1}{2}\Delta_\mu - \sum_{\mathbf{q}} \sum_{\alpha=1}^{M} \frac{Z_\alpha}{|\mathbf{r}_\mu - \mathbf{R}_\alpha^\mathbf{q}|} \tag{1.36}$$

where Greek letters μ and ν refer to the electrons, n_e denotes the number of electrons within the elementary cell, Z_α is the nuclear charge of atom α, $\mathbf{R}_\alpha^\mathbf{q}$ stands for the position vector of the αth nucleus in the cell characterized by $\mathbf{R}_\mathbf{q}$, and M is the number of atoms in the elementary cell. If we take the one-electron orbitals in the LCAO form (1.1), construct with their aid a Slater-determinant many-electron wave function, and perform the same variational procedure on the expectation value $\langle \hat{H} \rangle$ of equation (1.35) (calculated with the above Slater determinant) as Roothaan[4] did for molecules, then we obtain the matrix equation (1.3), where the elements of the hypermatrix $\mathbf{H}$ are defined by relationship (1.4) with $\hat{H}_{\text{eff}}$ replaced by the Fock operator $\hat{F}$:

$$\hat{H}_{\text{eff}} = \hat{F} = \hat{H}^N + \sum_{\mathbf{p}} \sum_{h=1}^{n^*} [2\hat{J}(\mathbf{p}, h) - \hat{K}(\mathbf{p}, h)] \tag{1.37}$$

Here, the operator $\hat{H}^N$ is given by expression (1.36) and the Coulomb and exchange operators $\hat{J}(\mathbf{p}, h)$ and $\hat{K}(\mathbf{p}, h)$ are defined, respectively, as

$$\hat{J}(\mathbf{p}, h, \mu)\phi(\mu) = \left\langle \phi_h^{\mathbf{p}}(\nu) \left| \frac{1}{r_{\mu\nu}} \right| \phi_h^{\mathbf{p}}(\nu) \right\rangle \phi(\mu) \tag{1.38a}$$

and

$$\hat{K}(\mathbf{p}, h, \mu)\phi(\mu) = \left\langle \phi_h^{\mathbf{p}}(\nu) \left| \frac{1}{r_{\mu\nu}} \right| \phi(\nu) \right\rangle \phi_h^{\mathbf{p}}(\mu) \tag{1.38b}$$

and n^* denotes the number of MOs which would be doubly filled by the n_e electrons originating from one elementary cell ($n^* = n_e/2$).

It was seen above that if we take into account the translational symmetry of the system and introduce the Born–von Kármán periodic boundary conditions, our matrix equations (1.3) reduce to relationship (1.17). In the Hartree–Fock–Roothaan case the elements of the Fock matrices $\mathbf{F}(\mathbf{q})$ occurring in the expression

$$\mathbf{F}(\mathbf{k}) = \sum_{\mathbf{q}} e^{i\mathbf{k}\cdot\mathbf{R}_{\mathbf{q}}} \mathbf{F}(\mathbf{q}) \tag{1.39}$$

can be easily obtained, if we substitute into the expression

$$[\mathbf{F}(\mathbf{q})]_{r,s} = \langle \chi_r^{\mathbf{0}} | \hat{F} | \chi_s^{\mathbf{q}} \rangle = \left\langle \chi_r^{\mathbf{0}} \left| \hat{H}^N + \sum_{\mathbf{p}} \sum_{h=1}^{n^*} [2\hat{J}(\mathbf{p}, h) - \hat{K}(\mathbf{p}, h)] \right| \chi_s^{\mathbf{q}} \right\rangle \tag{1.40}$$

expressions (1.38a) and (1.38b) and the LCAO form (1.1) of the crystal orbitals appearing in operators $\hat{J}$ and $\hat{K}$. This approach yields the equation

$$[\mathbf{F}(\mathbf{q})]_{r,s} = \langle \chi_r^{\mathbf{0}} | \hat{H}^N | \chi_s^{\mathbf{q}} \rangle + \sum_{\mathbf{p}} \sum_{h=1}^{n^*} \sum_{\mathbf{q}_1} \sum_{\mathbf{q}_2} \sum_{u=1}^{m} \sum_{v=1}^{m} C(\mathbf{p})^*_{h;\mathbf{q}_1,u} C(\mathbf{p})_{h;\mathbf{q}_2,v}$$
$$\times (2\langle \chi_r^{\mathbf{0}} \chi_u^{\mathbf{q}_1} | \chi_s^{\mathbf{q}} \chi_v^{\mathbf{q}_2} \rangle - \langle \chi_r^{\mathbf{0}} \chi_u^{\mathbf{q}_1} | \chi_v^{\mathbf{q}_2} \chi_s^{\mathbf{q}} \rangle) \tag{1.41}$$

The charge-bond-order matrix of a three-dimensional polymer can be introduced in the form

$$\mathbf{P} = 2 \sum_{\mathbf{p}} \sum_{h=1}^{n^*} \mathbf{C}(\mathbf{p})_h \mathbf{C}(\mathbf{p})_h^+ \tag{1.42}$$

where $\mathbf{C(p)}_h^+$ denotes the adjunct of vector $\mathbf{C(p)}_h$. With the aid of the transformation $\mathbf{C(p)}_h = \mathbf{UD(p)}_h$ [see equation (1.12)] and expression (1.9) for the unitary matrix $\mathbf{U}$, one obtains the subblocks $\mathbf{p}(\mathbf{q}_1, \mathbf{q}_2)$ of $\mathbf{P}$:

$$\mathbf{p}(\mathbf{q}_1, \mathbf{q}_2) = \frac{2}{(2N+1)^3} \sum_{\mathbf{p}} \sum_{h=1}^{n^*} \mathbf{d(p)}_h \mathbf{d(p)}_h^+ \exp\left[\frac{2\pi i \mathbf{p}\cdot(\mathbf{q}_1 - \mathbf{q}_2)}{2N+1}\right] \tag{1.43}$$

The quantity $\mathbf{p}$ can be replaced by the continuous variable $\mathbf{k}$ defined in equations (1.15) and (1.16). Integration over $\mathbf{k}$ instead of summation over $\mathbf{p}$ yields

$$\mathbf{p}(\mathbf{q}_1, \mathbf{q}_2) = \frac{2}{\omega} \int_{\omega} \sum_{h=1}^{n^*} \mathbf{d(k)}_h \mathbf{d(k)}_h^+ \exp[i\mathbf{k}\cdot(\mathbf{R}_{\mathbf{q}_1} - \mathbf{R}_{\mathbf{q}_2})]\, d\mathbf{k} \tag{1.44}$$

where ω is the volume of the first Brillouin zone.

On substituting equation (1.42) into expression (1.41) and using relationship (1.44), one obtains[1]

$$[\mathbf{F}(\mathbf{q})]_{r,s} = \langle \chi_r^{\mathbf{0}} | \hat{H}^N | \chi_s^{\mathbf{q}} \rangle + \sum_{\mathbf{q}_1} \sum_{\mathbf{q}_2} \sum_{u,v} p(\mathbf{q}_1 - \mathbf{q}_2)_{u,v}$$
$$\times (\langle \chi_r^{\mathbf{0}} \chi_u^{\mathbf{q}_1} | \chi_s^{\mathbf{q}} \chi_v^{\mathbf{q}_2} \rangle - \tfrac{1}{2} \langle \chi_r^{\mathbf{0}} \chi_u^{\mathbf{q}_1} | \chi_v^{\mathbf{q}_2} \chi_s^{\mathbf{q}} \rangle) \tag{1.45}$$

where $p(\mathbf{q}_1 - \mathbf{q}_2)_{u,v}$ is the u,vth element of the submatrix

$$\mathbf{p}(\mathbf{q}_1 - \mathbf{q}_2) = \mathbf{P}(\mathbf{q}_1, \mathbf{q}_2) \tag{1.46}$$

which follows immediately from the form (1.44) of $\mathbf{p}(\mathbf{q}_1, \mathbf{q}_2)$. The described *ab initio* SCF LCAO crystal-orbital method was developed independently but simultaneously by Del Re *et al.*[1] and André *et al.*[5]

In the case of a *one-dimensional chain* with translational symmetry all the vectors k, q, q_1, and q_2 become scalars and so equation (1.17) reduces to

$$\mathbf{F}(k)\mathbf{d}(k)_h = \varepsilon(k)_h \mathbf{S}(k)\mathbf{d}(k)_h \tag{1.47}$$

with

$$\mathbf{F}(k) = \sum_{q=-\infty}^{\infty} \exp(ikqa)\mathbf{F}(q) \tag{1.48}$$

and

$$\mathbf{S}(k) = \sum_{q=-\infty}^{\infty} \exp(ikqa)\mathbf{S}(q) \tag{1.49}$$

Here a is the elementary translation,

$$[\mathbf{S}(q)]_{r,s} = \langle \chi_r^0 \mid \chi_s^q \rangle \tag{1.50}$$

and the elements of matrices $\mathbf{F}(q)$ are now defined by

$$[\mathbf{F}(q)]_{r,s} = \left\langle \chi_r^0 \left| -\tfrac{1}{2}\Delta - \sum_{q_1=-\infty}^{\infty} \sum_{\alpha=1}^{M} \frac{Z_\alpha}{|\mathbf{r} - \mathbf{R}_\alpha^{q_1}|} \right| \chi_s^q \right\rangle + \sum_{q_1,q_2=-\infty}^{\infty} \sum_{u,v=1}^{m} p(q_1 - q_2)_{u,v} (\langle \chi_r^0 \chi_u^{q_1} \mid \chi_s^q \chi_v^{q_2} \rangle - \tfrac{1}{2} \langle \chi_r^0 \chi_u^{q_1} \mid \chi_v^{q_2} \chi_s^q \rangle) \tag{1.51}$$

with

$$p(q_1 - q_2)_{u,v} = \frac{a}{2\pi} \int_{-\pi/a}^{\pi/a} 2 \sum_{h=1}^{n^*} d(k)_{h,u}^* d(k)_{h,v} \exp[ika(q_1 - q_2)] dk \tag{1.52}$$

Matrix $\mathbf{S}(k)$ can be eliminated from equation (1.47) in the same way as for the three-dimensional case described above. However, in both the three- and one-dimensional cases it has to be taken into account that vectors $\mathbf{d}(\mathbf{k})_h$ and $\mathbf{d}(k)_h$ [which occur in expressions (1.43) and (1.52), respectively, for the charge-bond-order matrices] must be calculated with the aid of expressions

$$\mathbf{d}(\mathbf{k})_h = \mathbf{S}(\mathbf{k})^{-1/2}\mathbf{b}(\mathbf{k})_h \quad \text{and} \quad \mathbf{d}(k)_h = \mathbf{S}(k)^{-1/2}\mathbf{b}(k)_h \tag{1.53}$$

from vectors $\mathbf{b}(\mathbf{k})_h$ and $\mathbf{b}(k)_h$, respectively [obtained from the solution of eigenvalue problem (1.31) and from its counterpart for the one-dimensional case], at each iteration step. For this reason, and since the dependence of the vector components $\mathbf{d}(\mathbf{k})_{h,u}$ on $\mathbf{k}$ or k is usually not known analytically so that integrations (1.42) and (1.52) must be performed numerically, programming of the described method is not simple.

Andre[6] has developed the first program for the one-dimensional case with finite-neighbor interactions ($q = -N, \ldots, 0, \ldots, N$, where N is a small integer). This program, called POLYMOL, uses a Gaussian basis set and makes allowance for the fact that matrix $\mathbf{F}(k)$ is a Hermitian complex one also in the *ab initio* SCF case, as follows from the relation

$$\mathbf{F}(q) = \mathbf{F}(-q)^{\text{tr}} \tag{1.54}$$

This relation can easily be proven if we take into account the translational symmetry of all the relevant integrals and the relation $\mathbf{p}(q_1 - q_2) = \mathbf{p}(q_2 - q_1)^{tr}$, which follows again from translational symmetry.

Numerous *ab initio* LCAO crystal-orbital programs were subsequently developed for quasi-one-dimensional chains at Erlangen, Namur, Vienna, Budapest, Poughkeepsie (New York), and Kingston (New York) and employed different procedures to treat the finite-neighbor-interaction problem (see Section 1.3 below). The only working, fully *ab initio* (using a nonlocal exchange term and not neglecting three- and four-center integrals) crystal-orbital program for three-dimensional periodic systems has been developed by the group in Turin.[7] However, a description of the potentialities of this program and the results obtained with its help lie outside the scope of this book.

1.2. COMBINED SYMMETRY OPERATION

We can apply the formalism developed in the preceding section also in the case of a combined symmetry operation. To show this let us consider a helix in which we pass from one unit to the next by a translation $\boldsymbol{\tau}$ and simultaneous rotation α. We can then introduce the helix operator

$$\hat{S}(\alpha, \boldsymbol{\tau}) = \hat{D}(\alpha) + \boldsymbol{\tau} \tag{1.55}$$

where $\hat{D}(\alpha)$ denotes the operator of a rotation around the main axis of the helix through angle α.[8] For the sake of simplicity it is further assumed that after n repetitions of the helix operation we obtain the "large" translation $\hat{T}$, namely

$$\hat{S}^n(\alpha, \boldsymbol{\tau}) = \hat{T} \tag{1.56}$$

It should be noted that the following considerations hold also when relationship (1.56) is not satisfied, i.e., $2\pi/\alpha$ is not integral.

We can again introduce the Born–von Kármán periodic boundary conditions in the form

$$\hat{S}^{2N+1} = \hat{1} \tag{1.57}$$

where N is a large integer and measures the number of unit cells.

If $\hat{F}$ is the Fock operator of the helix, then in addition

$$[\hat{S}, \hat{F}] = \hat{0} \tag{1.58}$$

so we can classify the eigenfunctions of $\hat{F}$ according to the one-dimensional representations of the Abelian group $G = \{\hat{S}^m; m = 1, \ldots, 2N+1\}$ (identical with the characters). The kth representation of this group is thus given by

$$\xi_{qk} = \exp\left(\frac{2\pi iqk}{2N+1}\right)$$

This means that the eigenvalue equation

$$\hat{S}^q\psi_k = \exp[2\pi iqk/(2N+1)]\psi_k = \xi_{qk}\psi_k \tag{1.59}$$

must be satisfied, where ψ_k may have again an LCAO form [see equation (1.1)], but k is now defined on the combined symmetry operation. In other words, the hypermatrices **F** and **S** will be again cyclic hypermatrices, if we construct their blocks **F**(q) and **S**(q) with the aid of the combined symmetry operation $\hat{S}$ [i.e., in the matrix elements (1.4) and (1.5) we can pass from the cell characterized by p to the cell characterized by q through repeated application of $\hat{S}$] and introduce periodic boundary conditions. These cyclic hypermatrices can be block-diagonalized again with the aid of the unitary matrix (1.7) and we can introduce again instead of the quantity p the quantity k, which varies continuously according to equation (1.15).

To generate the eigenfunctions ψ_k of $\hat{S}^q$ we can introduce the projection operator $\hat{O}_k$ given by[8]

$$\hat{O}_k = (2N+1)^{-1/2}\sum_{q=1}^{2N+1}\xi_{qk}\hat{S}^{-q} \tag{1.60}$$

which satisfies the relationships $\hat{O}_k^+ = \hat{O}_k$, $\hat{S}^m\hat{O}_k = \xi_{mk}\hat{O}_k$, $\hat{O}_k\hat{O}_l = \delta_{kl}\hat{O}_k$, and $\Sigma_k\hat{O}_k = \hat{1}$. If $\chi_g^{q_1} = \chi_g(\mathbf{r} - \mathbf{R}_{q_1} - \mathbf{r}_{g_A})$ again denotes the gth AO of the q_1th cell, we can generate the generalized LCAO Bloch orbitals of the helix with the aid of the expression

$$\begin{aligned}\psi_{k,g}(\mathbf{r}) &= \hat{O}_k\chi_g(\mathbf{r} - \mathbf{R}_{q_1} - \mathbf{r}_{g_A})\\ &= \hat{O}_k\chi_g(\mathbf{r} - \mathbf{R}_{q_1}^{g_A})\\ &= (2N+1)^{-1/2}\sum_{q=1}^{2N+1}\xi_{qk}\hat{S}^{-q}\chi_g(\mathbf{r} - \mathbf{R}_{q_1}^{g_A}); \quad (\mathbf{R}_{q_1}^{g_A} = \mathbf{R}_{q_1} + \mathbf{r}_{g_A})\end{aligned} \tag{1.61}$$

The same procedure can also be applied if our reference cell includes not only a single AO but a linear combination of them (LCAO MO).

To be able to apply equation (1.61) we require the expression

$$\hat{S}^{-q}\chi_g(\mathbf{r} - \mathbf{R}^{g_A}_{q_1}) = \chi_g[\hat{S}^q(\mathbf{r}) - \mathbf{R}^{g_A}_{q_1}] \tag{1.62}$$

where the right-hand side follows from the well-known relation that $\hat{S}^{-q}$ applied to a function is identical with the transformation of the coordinate system under the inverse operation.[8] On taking into account that

$$\hat{S}^q\mathbf{r} = \hat{D}(q\alpha)\mathbf{r} + q\boldsymbol{\tau}$$

we can further write

$$\chi_g[\hat{S}^q(\mathbf{r}) - \mathbf{R}^{g_A}_{q_1}] = \chi_g[\hat{D}(q\alpha)\mathbf{r} + q\boldsymbol{\tau} - \mathbf{R}^{g_A}_{q_1}] \tag{1.63}$$

The identity[8]

$$\begin{aligned}\hat{D}(q\alpha)\mathbf{r} - \hat{D}(q\alpha)\hat{S}^{-q}(\mathbf{R}^{g_A}_{q_1}) &= \hat{D}(q\alpha)\mathbf{r} - \hat{D}(q\alpha)[\hat{D}(-q\alpha)\mathbf{R}^{g_A}_{q_1} - q\boldsymbol{\tau}]\\ &= \hat{D}(q\alpha)\mathbf{r} - \mathbf{R}^{g_A}_{q_1} + q\boldsymbol{\tau}\end{aligned} \tag{1.64}$$

enables one to write

$$\hat{S}^{-q}\chi_g(\mathbf{r} - \mathbf{R}^{g_A}_{q_1}) = \chi_g\{\hat{D}(q\alpha)[\mathbf{r} - \hat{S}^{-q}(\mathbf{R}^{g_A}_{q_1})]\} \tag{1.65}$$

This means that by applying the helix operator $\hat{S}^{-q}$ to an AO we must (1) repeat the helix operation q times at the position of the nucleus, and (2) rotate the argument of the AO with angle $q\alpha$ around the axis of the helix.

Equation (1.65) enables expression (1.1) to be rewritten for the case of a helical chain in the form

$$\begin{aligned}\phi(k, r)_h = \frac{1}{\sqrt{2N+1}} \sum_{q=-N}^{N} \sum_{q=1}^{m} \exp\left(\frac{2\pi i k q}{2N+1}\right) C(k)_{h,g}\chi_g\\ \times \{\hat{D}(q\alpha)[\mathbf{r} - \hat{S}^{-q}(\mathbf{R}^{g_A}_{q_1})]\}\end{aligned} \tag{1.66}$$

Results (1.65) and (1.66) enable the *ab initio* SCF LCAO CO program for linear chains to be modified and hence applicable also in the case of a combined symmetry operation. This modified program has been applied for the nucleotide base stacks (see Section 2.3 in the next chapter).

Finally, it is noteworthy that this derivation[8] differs from those given by McCubbin[10] and Ukrainski,[11] who start in the usual way with simple translational symmetry and analyze *a posteriori* the effects of other symmetry operations, such as the helix operation.

1.3. METHODS TO TREAT MANY-NEIGHBOR INTERACTIONS

In actual polymer calculations it is necessary to decide how many-neighbor interactions must be taken into account to obtain satisfactory results. This is by no means a trivial question. Experience gained in numerous calculations (see below) indicates that the number of neighbors to be taken into account explicitly is smaller if one performs a band-structure calculation, than if one is interested in the total energy per unit cell. To understand the reason for this we now examine the latter quantity for a linear chain obtained by straightforward generalization of the Hartree–Fock–Roothaan expression[4] for molecules, namely

$$\frac{E}{2N+1} = \frac{1}{2}\sum_q \sum_r \sum_s [h_{r,s}^{0,q} + F_{r,s}(q)]P_{r,s}(q) + \frac{1}{2}\sum_q \sum_\alpha \sum_\beta{}' \frac{Z_\alpha Z_\beta}{|\mathbf{R}_\alpha^q - \mathbf{R}_\beta^0|} \tag{1.67}$$

The prime on the summation sign in the last term means that the term $\beta = \alpha$ is to be excluded for $q = 0$. Here the one-electron matrix element $h_{r,s}^{0,q} = \langle \chi_r^0 | \hat{H}^N | \chi_s^q \rangle$ can be calculated with the aid of the one-dimensional analogue of $\hat{H}^N$ [see equation (1.36)], the Fock matrix elements $F_{r,s}(q)$ are given by expression (1.51) and the generalized charge-bond-order matrix elements [equation (1.52)] if we set $q_2 = 0$.

For an electrically neutral polymer the numerical integration in equation (1.52) must yield the same number of electrons as the total number of positive charges per cell, i.e., the requirement

$$2\sum_q \sum_r \sum_s P_{r,s}(q) S_{r,s}(q) = \sum_q \sum_{\alpha=1}^{M} Z_\alpha \tag{1.68}$$

has to be satisfied.

It can be seen from equation (1.51) that the major difficulties in applying the HF CO method arise when calculating the two-electron integrals whose number is proportional to m^4 for each upper (cell) index triplet (q, q_1, q_2). The method can be applied reasonably only if the three infinite sums in equations (1.48) and (1.51) can be truncated with a relatively small radius around the reference cell. Even in this case only a very small portion of the two-electron integrals has a physically significant value. Therefore, instead of actually calculating m^4 integrals for a (q, q_2, q_3) group, the program developed in Erlangen[12] examines first, on the basis of Mulliken's integral approximation,[13] whether the absolute value of a certain integral is larger than a prescribed threshold. The integral program package developed is based on the Gaussian 76 program system,[14] which is especially efficient if the s- and p-type Gaussians have the same exponent. In this case, for a (s, p_x, p_y, p_z) shell one can decide in one step whether any of the 256 possible integrals gives a significant contribution. For this purpose, a p-type function rotated in the case of each shell into a maximum overlap position and the largest possible integral is tested for significance by Mulliken's approximation. For triplets in which each index is different, one must actually calculate usually far less than 1% of the two-electronic integrals in this way.

Returning to the problem of the cell index truncation, we can see from equations (1.51) and (1.67) that there are six physical quantities to be properly truncated in this procedure:

1. The overlap integrals $S_{r,s}(q)$ defined in equation (1.5) with $p = 0$.
2. The matrix elements of the kinetic energy:

$$t_{r,s}^{0,q} = -\tfrac{1}{2}\langle \chi_r^0(\mathbf{r}) | \Delta_{\mathbf{r}} | \chi_s^q(\mathbf{r}) \rangle \tag{1.69}$$

3. The electron–nucleus interactions:

$$Z_{r,s}^{0,q} = -\sum_{q_1}\sum_{\alpha} \left\langle \chi_r^0(\mathbf{r}) \left| \frac{Z_\alpha}{|\mathbf{r} - \mathbf{R}_{q_1} - \mathbf{R}_\alpha|} \right| \chi_s^q(\mathbf{r}) \right\rangle \tag{1.70}$$

4. The electron–electron Coulomb repulsion:

$$C_{r,s}^{0,q} = 2\sum_{q_1}\sum_{q_2}\sum_{u}\sum_{v} P_{u,v}(q_1 - q_2) \left\langle \chi_r^0(\mathbf{r}_1)\chi_u^{q_1}(\mathbf{r}_2) \left| \frac{1}{r_{12}} \right| \chi_s^q(\mathbf{r}_1)\chi_v^{q_2}(\mathbf{r}_2) \right\rangle \tag{1.71}$$

5. The nonlocal exchange potential:

$$e_{r,s}^{0,q} = -\sum_{q_1}\sum_{q_2}\sum_{u}\sum_{v} P_{u,v}(q_1 - q_2) \left\langle \chi_r^0(1)\chi_s^q(2) \left| \frac{1}{r_2} \right| \chi_u^{q_1}(1)\chi_v^{q_2}(2) \right\rangle \tag{1.72}$$

6. The nucleus–nucleus repulsions:

$$E_{n,n} = \frac{1}{2}\sum_{q}\sum_{\alpha}\sum_{\beta}{}' \frac{Z_\alpha Z_\beta}{|\mathbf{R}_\alpha - \mathbf{R}_q - \mathbf{R}_\beta|} \tag{1.73}$$

The first five quantities contribute not only to the total energy, but also to the Fock matrices, and therefore also play a crucial role in determining the band structure.

For AOs of either Slater or Gaussian type, the overlap and kinetic energy integrals decay exponentially and can be safely truncated after a few neighbors. It is more complicated to truncate the contributions from equations (1.70) and (1.71), both of which behave separately at large distances as $\ln(r)$, that is, they diverge as $N \to \infty$. Their sum acts as an alternating series, only conditionally convergent.* By defining the operators

$$\hat{c}(\mathbf{r}_1) = 2 \sum_{q_1} \sum_{q_2} \sum_{u} \sum_{v} P_{u,v}(q_1 - q_2) \int d\mathbf{r}_2 \frac{\chi_u^{q_1}(\mathbf{r}_2)\chi_v^{q_2}(\mathbf{r}_2)}{|\mathbf{r}_1 - \mathbf{r}_2|} \tag{1.74}$$

and

$$\hat{z}(\mathbf{r}_1) = -\sum_{q_1} \sum_{\alpha} \frac{Z_\alpha}{|\mathbf{r}_1 - \mathbf{R}_{q_1} - \mathbf{R}_\alpha|} \tag{1.75}$$

we obtain the nuclear-attraction and electron-repulsion contributions to each Fock-matrix element in the form

$$\langle \chi_r^0(\mathbf{r}_1) | \hat{c}(\mathbf{r}_1) + \hat{z}(\mathbf{r}_1) | \chi_s^q(\mathbf{r}_1) \rangle \tag{1.76}$$

It follows from these considerations that, for a physically correct truncation of the infinite lattice sums, the following criteria have to be fulfilled:

1. The sum over q_1 in operators $\hat{c}$ and $\hat{z}$ must be truncated with the same radius in such a way that the elements of $\hat{c} + \hat{z}$ have an accuracy of 10^{-6}–10^{-7} au.
2. The sum over q_2 in $\hat{c}$ must extend beyond the sum over q_1 in such a way that all contributions to the electronic charge of the cell q_1 are properly taken into acount. This is necessary, since the electron population of a given cell is built up according to equation (1.68) as a sum over a certain number of neighbors. If one would use the same radius (measured from the reference cell) for the lattice sums over q_2 and q_1, the tail of the electron distribution belonging

* In the expression for the total energy per unit cell of an infinite chain [equation (1.66)] the infinite sum of the nuclear attraction occurs twice [in $h_{r,s}^{0,q}$ and also in $F_{r,s}(q)$] and this is compensated by the positive infinite sums (electron-repulsion and nuclear-repulsion terms). This compensation is not valid even approximately if we truncate the cell summation too early. This causes the total energy per unit cell to be very sensitive in the calculations to the number of neighbors taken into account explicitly.

to the cell at $\mathbf{R}_{q_1}$, but collected from the cells at $\mathbf{R}_{q_1+1}$, $\mathbf{R}_{q_1+2}$, etc., would be neglected. This loss of electronic charge would, of course, violate the electrostatic balance of the crystal.

3. The matrix elements of the nonlocal Hartree–Fock exchange potential stand alone without any balancing counterpart of one-electron integrals. They can therefore be truncated only if their sum has converged with a prescribed accuracy (10^{-6} au as in the case of the other types of integral). The summation radius to be used here is expected to differ for semiconducting and metallic systems, since the long-range tail of the exchange decays exponentially in the former case but follows only a power law for metals.[15] Experience shows, however, no significant difference between the two types of solid in the short and medium range[12] and a too-early cutoff of the exchange leads either to serious convergence difficulties in the self-consistent-field procedure[16] or to nonphysical solutions of the Hartree–Fock equations.
4. In the case of elementary cells containing more than one atom, the truncation must be performed with respect to both the lower (AO) and upper (cell) indices in order to preserve translational and point group symmetry. While analyzing the appearance of spurious gaps in polyene calculations[17] Santry has pointed out that the number of interactions with other atoms included for a given atom in the positive direction must equal the number of interactions in the negative direction. *This requirement can obviously not be satisfied if the whole cell is treated as an entity during the truncation procedure.* Santry demonstrated at the CNDO level that for equidistant polyene (with zero gap at the Hartree–Fock level) the violation of this symmetry produces a gap of 0.3 eV even if 37 unit cells are included in the Coulomb lattice sums. *Ab initio* calculations show the same trend.

For a computationally efficient formulation of these requirements we introduce three different cutoff radii: d_o controlling the (exponential) decay of overlap-type quantities, d_c for the much slower decay of Coulomb-type interactions, and d_e for the exchange terms. The choice of d_o is the easiest, since the order of magnitude of the overlap integrals attains the machine accuracy after a few neighbors. The overlap charge distributions $\chi_r(\mathbf{r})\chi_s(\mathbf{r})$ appear, however, besides kinetic-energy integrals, also in core attractions and electron–electron repulsions. Since the careful balance of the two latter quantities is important for maintaining the crystal electrically neutral, we consider here in more detail the consequence of their truncation.

The AO $\chi_r^0 = \chi_r(\mathbf{r} - \mathbf{R}_\alpha)$ is centered in the reference cell on atom α ($r \in \alpha$). We now calculate the core attraction integral

$$-\left\langle \chi_r^0(\mathbf{r} - \mathbf{R}_\alpha) \left| \frac{Z_\beta}{|\mathbf{r} - \mathbf{R}_{q_1} - \mathbf{R}_\beta|} \right| \chi_r^0(\mathbf{r} - \mathbf{R}_\alpha) \right\rangle \tag{1.77}$$

assuming that the distance between atoms α in the reference cell and β in the q_1th cell is just the Coulomb truncation radius, $|\mathbf{R}_\alpha - \mathbf{R}_{q_1} - \mathbf{R}_\beta| = d_c$. This attraction is compensated in the crystal (besides the corresponding repulsion between the nuclei) by the interelectronic repulsion between the charge distributions $\chi_r^0\chi_r^0$ and $P_{u,v}(q_1 - q_2)\chi_u^{q_1}\chi_v^{q_2}$, summed over all such u for which $u \in \beta$ and over all the cells, to obtain the full charge of atom β according to equation (1.68). As a consequence, when checking the interorbital distances for the intercell integrals of the form $\langle \chi_r^0\chi_u^{q_1} | \chi_r^0\chi_v^{q_2} \rangle$, we must apply d_c for $|\mathbf{R}_\alpha - \mathbf{R}_{q_1} - \mathbf{R}_\beta|$ but $d_c + d_o$ for $|\mathbf{R}_\alpha - \mathbf{R}_{q_2} - \mathbf{R}_\beta|$. In other words, the nth-neighbor approximation always means that for two-electron integrals the neighbors $n + 1$, $n + 2$, etc., are also included. Owing to the long range of the corresponding Coulomb integrals, neglect of these additional neighbors introduces a substantial error in the Fock matrices even for large values of n.

To obtain a more quantitative estimate of the parameters d, one should check first the behavior of the intercellular overlap integrals for a given system and basis set and fixed d_o. It is less straightforward to choose d_e properly. The simplest procedure is to first choose $d_e = d_c$, which overestimates in all cases the range of the exchange integrals. Since only one set of two-electron AO integrals is usually calculated (and is then used for the construction of both Coulomb and exchange contributions to the Fock-matrix elements), this choice involves only a little superfluous effort during the SCF steps. It may be worthwhile, however, to optimize the value of d_e in the region $d_o \leqslant d_e \leqslant d_c$ if, for example, geometry optimization is performed for a system.

Finally, the value of d_c is also system-dependent but qualitative knowledge of the electronic structure of the elementary cell usually helps to estimate the difficulties one encounters in the long-range part of the electrostatic potential. In other words if the earlier-discussed requirement for charge neutrality is fulfilled, the zeroth moment of the electron distribution is exactly compensated by the nuclear charges of a given cell (in the sense of a multipole expansion) so that the long-range part of $\hat{c} + \hat{z}$ starts in the worse case with dipole–dipole interaction, i.e., with an r^{-3}-dependent term. In cases when the unit cell does not possess a permanent dipole moment, the situation is even better. A number of polymers (such as polyethylene and polyacetylenes) have an inversion center in the unit cell, therefore their first long-range electrostatic term is quadrupole–quadrupole interaction, i.e., it decreases at r^{-5}. One must keep in mind, however, that considerations of this kind are valid only for the really long-range region, where charge distributions $\chi_r^0 \chi_s^q$ and $\chi_u^{q_1} \chi_v^{q_2}$ no longer overlap.

Summarizing the above considerations for symmetry-adapted and electrostatically balanced truncation of the infinite lattice sums, the following requirements must be satisfied in actual calculations (using again the convention that indices 0, q, q_1, q_2 refer to cells, and r, s, u, v and α, β refer to orbitals and atoms, respectively).[12]

1. For overlap and kinetic energy integrals

$$|\mathbf{R}_r - (\mathbf{R}_q + \mathbf{R}_s)| \leqslant d_o \tag{1.78}$$

2. For core attraction integrals

$$|\mathbf{R}_r - (\mathbf{R}_q + \mathbf{R}_s)| \leqslant d_o \tag{1.79}$$

$$|\mathbf{R}_r - (\mathbf{R}_q - \mathbf{R}_\alpha)| \leqslant d_c \tag{1.80}$$

$$|\mathbf{R}_q + \mathbf{R}_s - (\mathbf{R}_{q_1} - \mathbf{R}_\alpha)| \leqslant d_c \tag{1.81}$$

3. For electron–electron Couloumb repulsion

$$|\mathbf{R}_r - (\mathbf{R}_q + \mathbf{R}_s)| \leqslant d_o \tag{1.82}$$

$$|\mathbf{R}_{q_1} + \mathbf{R}_u - (\mathbf{R}_{q_2} + \mathbf{R}_v)| \leqslant d_o \tag{1.83}$$

$$|\mathbf{R}_r - (\mathbf{R}_{q_1} + \mathbf{R}_u)| \leqslant d_c \tag{1.84}$$

$$|\mathbf{R}_q + \mathbf{R}_s - (\mathbf{R}_{q_1} + \mathbf{R}_u)| \leqslant d_c \tag{1.85}$$

$$|\mathbf{R}_r + (\mathbf{R}_{q_2} + \mathbf{R}_v)| \leqslant d_o + d_c \tag{1.86}$$

$$|\mathbf{R}_q + \mathbf{R}_s - (\mathbf{R}_{q_2} + \mathbf{R}_v)| \leqslant d_o + d_c \tag{1.87}$$

One should note here that inequalities (1.82) and (1.83) involve distances between the centers of the AOs of the same electron, while in the other four integrals the distances refer to centers of the AOs of different electrons. One finds the same situation in the following requirements:

4. For exchange integrals

$$|\mathbf{R}_r - (\mathbf{R}_{q_1} + \mathbf{R}_u)| \leqslant d_o \tag{1.88}$$

$$|\mathbf{R}_q + \mathbf{R}_s - (\mathbf{R}_{q_2} + \mathbf{R}_v)| \leqslant d_o \tag{1.89}$$

$$|\mathbf{R}_r - (\mathbf{R}_q + \mathbf{R}_s)| \leqslant d_e \tag{1.90}$$

$$|\mathbf{R}_{q_1} + \mathbf{R}_u - (\mathbf{R}_q + \mathbf{R}_s)| \leqslant d_e \tag{1.91}$$

$$|\mathbf{R}_r - (\mathbf{R}_{q_2} + \mathbf{R}_v)| \leqslant d_o + d_e \tag{1.92}$$

$$|\mathbf{R}_q + \mathbf{R}_s - (\mathbf{R}_{q_2} + \mathbf{R}_v)| \leqslant d_o + d_e \tag{1.93}$$

To conclude this discussion we note that after truncating the different integrals as described above, one can account for the long-range parts of the different Coulomb interactions by a multipole expansion. Effective procedures for this approach by applying the multipole expansion within the Fock operator have been worked out by the Namur group. The details can be found in their papers.[(18)]

1.4. Different Orbitals for Different Spin Formalisms

The method of different orbitals for different spins (DODS) is regarded as one possible way of accounting for a part of the correlation. However, in using one single Slater determinant built up from different spatial orbitals for the electrons with spin α and β [the unrestricted Hartree–Fock (UHF) method], the difficulty arises that the many-electron wave function will not be an eigenfunction of the total spin operator $\hat{S}^2$. This difficulty is overcome with the aid of a projection operator by projecting out from the Slater determinant the component with the desired multiplicity $2S + 1$, annihilating all other "contaminating" components. This can be done either after an already performed calculation (spin projection after variation, UHF with annihilation), or, as Löwdin[(19)] has pointed out, one would expect a more negative total energy if the variation is performed with an already spin-projected Slater determinant [spin projection before variation, spin-projected extended Hartree–Fock (EHF) method]. The reason is that a spin-projected Slater determinant is a given linear combination of different Slater determinants. The variation in the expectation value of the Hamiltonian formed with a spin-projected Sater determinant thus provides equations (EHF equations), whose solutions represent the solution of this particular multiconfigurational SCF problem.

The EHF equations for the one-electron orbitals derived in this way[(20)] are rather complicated. Their investigation shows, however, that if the number of electrons $n \rightarrow \infty$ and $n \gg S$, the EHF equations go over into the unprojected UHF equations.[(21)] Therefore, the expectation values of all one- and two-electron operators and all one-electron energies and wave functions will be the same in the projected and unprojected cases.[21] This means that if we wish to apply the DODS method to a molecular crystal or to a polymer, we can use its simple unprojected form.

The unprojected UHF equations obtained from the variation in the expectation value of the Hamiltonian formed with a DODS Slater determinant have the well-known form (for $\tilde{n}^{*\alpha} = \tilde{n}^{*\beta}$)

$$\hat{F}^{\alpha}\phi_i^{\alpha} = \left[\hat{H}^N + \sum_{h=1}^{\tilde{n}^{*\alpha}} (\hat{J}_h^{\alpha} + \hat{J}_h^{\beta} - \hat{K}_h^{\alpha})\right]\phi_i^{\alpha} = \varepsilon_i^{\alpha}\phi_i^{\alpha} \tag{1.94a}$$

and

$$\hat{F}^{\beta}\phi_i^{\beta} = \left[\hat{H}^N + \sum_{h=1}^{\tilde{n}^{*\beta}} (\hat{J}_h^{\alpha} + \hat{J}_h^{\beta} - \hat{K}_h^{\beta})\right]\phi_i^{\beta} = \varepsilon_i^{\beta}\phi_i^{\beta} \tag{1.94b}$$

Here $\tilde{n}^{*\gamma}$ is the number of filled levels with electrons of spin $\gamma(\gamma = \alpha \text{ or } \beta)$ and the Coulomb operators $\hat{J}_l^{\gamma}$ and exchange operators $\hat{K}_l^{\gamma}$ are defined in the usual way by

$$\hat{J}_h^{\gamma}(\mathbf{r}_1)\phi(\mathbf{r}_1) = \int \frac{|\phi_h^{\gamma}(\mathbf{r}_2)|^2}{|\mathbf{r}_1 - \mathbf{r}_2|} d\mathbf{r}_2\phi(\mathbf{r}_1) \qquad (\gamma = \alpha, \beta) \tag{1.95a}$$

and

$$\hat{K}_h^{\gamma}(\mathbf{r}_1)\phi(\mathbf{r}_1) = \int \frac{\phi_h^{\gamma}(\mathbf{r}_2)^*\phi(\mathbf{r}_2)}{|\mathbf{r}_1 - \mathbf{r}_2|} d\mathbf{r}_2\phi_h^{\gamma}(\mathbf{r}_1) \qquad (\gamma = \alpha, \beta) \tag{1.95b}$$

If we introduce a basis $\{\chi\}$ and express all functions ϕ_i^{γ} as linear combinations of the basis functions, namely

$$\phi_h^{\gamma} = \sum_{r=1}^{m} C_{h,r}^{\gamma}\chi_{\rho} \qquad (\gamma = \alpha, \beta) \tag{1.96}$$

we obtain with the help of the Ritz variational procedure the matrix equations

$$\mathbf{F}^{\alpha}\mathbf{C}_h^{\alpha} = \varepsilon_h^{\alpha}\mathbf{S}\mathbf{C}_h^{\alpha} \quad \text{and} \quad \mathbf{F}^{\beta}\mathbf{C}_h^{\beta} = \varepsilon_h^{\beta}\mathbf{S}\mathbf{C}_h^{\beta} \tag{1.97}$$

Here, the elements of matrices $\mathbf{F}^\gamma$ are defined by

$$[\mathbf{F}^\gamma]_{r,s} = \langle \chi_r | \hat{H}^N | \chi_s \rangle + \sum_{u,v=1}^{m} [(p_{u,v}^\alpha + p_{u,v}^\beta)\langle \chi_r\chi_u | \chi_s\chi_v \rangle - p_{u,v}^\gamma \langle \chi_r\chi_u | \chi_v\chi_s \rangle] \quad (\gamma = \alpha, \beta) \tag{1.98}$$

where

$$p_{u,v}^\gamma = \sum_{h=1}^{\bar{n}^{*\gamma}} C_{h,u}^{\gamma *} C_{h,v}^\gamma \qquad (\gamma = \alpha, \beta) \tag{1.99}$$

We now consider a three-dimensional periodic system with $(2N+1)^3$ unit cells and m orbitals within the cell. If we again apply the Born–von Kármán periodic boundary conditions the matrices $\mathbf{F}^\alpha$, $\mathbf{F}^\beta$, and $\mathbf{S}$ become cyclic hypermatrices of order $m(2N+1)^3$. Therefore we can again apply to equations (1.97) the unitary transformation described in Section 1.1. Hence

$$\mathbf{U}^+\mathbf{F}^\alpha\mathbf{U}\mathbf{U}^+\mathbf{C}_h^\alpha = \varepsilon_h^\alpha \mathbf{U}^+\mathbf{S}\mathbf{U}\mathbf{U}^+\mathbf{C}_h^\alpha \tag{1.100a}$$

and

$$\mathbf{U}^+\mathbf{F}^\beta\mathbf{U}\mathbf{U}^+\mathbf{C}_h^\beta = \varepsilon_h^\beta \mathbf{U}^+\mathbf{S}\mathbf{U}\mathbf{U}^+\mathbf{C}_h^\beta \tag{1.100b}$$

or

$$\mathbf{F}^{\alpha\prime}\mathbf{D}_h^\alpha = \varepsilon_h^\alpha \mathbf{S}'\mathbf{D}_h^\alpha \tag{1.101a}$$

and

$$\mathbf{F}^{\beta\prime}\mathbf{D}_h^\beta = \varepsilon_h^\beta \mathbf{S}'\mathbf{D}_h^\beta \tag{1.101b}$$

with $\mathbf{D}_h^\gamma = \mathbf{U}^+\mathbf{C}_h^\gamma$ $(\gamma = \alpha, \beta)$, where the unitary matrix $\mathbf{U}$ was defined by equation (1.9). By the same approach as in Section 1.1, equations (1.101) yield

$$\mathbf{F}^\alpha(\mathbf{k})_h \mathbf{d}^\alpha(\mathbf{k})_h = \varepsilon^\alpha(\mathbf{k})_h \mathbf{S}(\mathbf{k})\mathbf{d}^\alpha(\mathbf{k})_h \tag{1.102a}$$

and

$$\mathbf{F}^\beta(\mathbf{k})_h \mathbf{d}^\beta(\mathbf{k})_h = \varepsilon^\beta(\mathbf{k})_h \mathbf{S}(\mathbf{k})\mathbf{d}^\beta(\mathbf{k})_h \tag{1.102b}$$

as $N \to \infty$. Here matrices $\mathbf{F}^\gamma$, which have now only order m, possess the form[22]

$$\mathbf{F}^\gamma(\mathbf{k}) = \sum_{\mathbf{q}=-\infty}^{\infty} e^{i\mathbf{k}\cdot\mathbf{R}_q}\mathbf{F}^\gamma(\mathbf{q}) \tag{1.103}$$

$$[\mathbf{F}^\gamma(\mathbf{q})]_{r,s} = \langle \chi_r^0 \,|\, \hat{H}^N \,|\, \chi_s^\mathbf{q} \rangle + \sum_{\mathbf{q}_1,\mathbf{q}_2} \sum_{u,v} \{[p(\mathbf{q}_1 - \mathbf{q}_2)_{u,v}^\alpha + p(\mathbf{q}_1 - \mathbf{q}_2)_{u,v}^\beta] \langle \chi_r^0 \chi_u^{\mathbf{q}_1} \,|\, \chi_s^\mathbf{q} \chi_v^{\mathbf{q}_2} \rangle - p(\mathbf{q}_1 - \mathbf{q}_2)_{u,v}^\gamma \langle \chi_r^0 \chi_u^{\mathbf{q}_1} \,|\, \chi_v^{\mathbf{q}_2} \chi_s^\mathbf{q} \rangle\} \quad (\gamma = \alpha, \beta) \qquad (1.104)$$

where $\hat{H}^N$ was defined in equation (1.36), the charge-bond-order matrix elements $p(\mathbf{q}_1 - \mathbf{q}_2)_{u,v}^\gamma$ are given by

$$p(\mathbf{q}_1 - \mathbf{q}_2)_{u,v}^\gamma = \frac{1}{\omega} \int_\omega \sum_{h=1}^{n^{*\gamma}} d^\gamma(\mathbf{k})_{h,u}^* d^\gamma(\mathbf{k})_{h,v} \exp[i\mathbf{k} \cdot (\mathbf{R}_{\mathbf{q}_1} - \mathbf{R}_{\mathbf{q}_2})] d\mathbf{k} \qquad (1.105)$$

and $n^{*\gamma}$ is the number of electrons per unit cell with spin γ [it is easy to see that $\tilde{n}^{*\gamma} = (2N+1)^3 n^{*\gamma}$] and ω is the volume of the first Brillouin zone. Matrix $\mathbf{S}(\mathbf{k})$ in equations (1.102) was defined earlier in equation (1.19).

The total energy of a molecule can be expressed in the UHF case as

$$E = \frac{1}{2} \sum_{r,s} [h_{r,s} p_{r,s} + F_{r,s}^\alpha p_{r,s}^\alpha + F_{r,s}^\beta p_{r,s}^\beta] + \text{NR} \qquad (1.106)$$

where NR stands for the nuclear-repulsion term and $p_{r,s}(\mathbf{q}) = p_{r,s}^\alpha(\mathbf{q})^\alpha + p_{r,s}^\beta(q)^\beta$. For a periodic system expression (1.106) yields in a straightforward way the total energy per unit cell,[22]

$$\frac{E}{(2N+1)^3} = \frac{1}{2} \sum_\mathbf{q} \sum_{r,s} [h_{r,s}^{0,\mathbf{q}} p_{r,s}(\mathbf{q}) + F_{r,s}^\alpha(\mathbf{q}) p_{r,s}^\alpha(\mathbf{q}) + F_{r,s}^\beta(\mathbf{q}) p_{r,s}^\beta(\mathbf{q})] + \text{NR} \qquad (1.107)$$

where NR is defined in equation (1.67). By combining again the long-range parts of $h_{r,\ s}^{0,\mathbf{q}}$, $F_{r,s}^\gamma(\mathbf{q})$ $(\gamma = \alpha, \beta)$ and NR it can be shown[23] that they will cancel, thus giving a convergent result for the total energy per unit cell.

The one-dimensional form of the described *ab initio* DODS SCF LCAO crystal-orbital scheme can be obtained in exactly the same way from the above-described three-dimensional expressions, as in Section 1.1 for the conventional Hartree–Fock scheme [see equation (1.54)].

The aforementioned *ab initio* DODS scheme so far possesses very few applications. Some general considerations based on the total electronic density[24] (for which a criterion similar to Mott's criterion in the case of metal–insulator transition[25] could be formulated) and experience based on different calculations seem to indicate that if $n^{*\alpha} = n^{*\beta}$ (closed-shell ground state) at equilibrium bond distances in most cases of an *ab initio*[26] or of an all-valence electron calculation,[27] one cannot expect a splitting of the levels belonging to electrons with spin α and β. (Some counterexamples exist if there is a very-low-lying triplet excited state extremely close to the singlet ground state. This is, however, practically never the case for polymers.) The same result was obtained by Calais and Sperber[28] in their Li crystal calculation using the alternant MO (AMO) method, which is a special case of the DODS SCF LCAO scheme. On the other hand, if $n^{*\alpha} \neq n^{*\beta}$ general experience shows that one does obtain a splitting and with it a lowering of the total energy. Only after application of the DODS CO method to more polymers can it be decided whether no lowering of the total energy can really be achieved if $n^{*\alpha} = n^{*\beta}$. On the other hand, the method will certainly give a splitting of the bands of electrons with spin α and β, if $n^{*\alpha} \neq n^{*\beta}$.

A further shortcoming of the DODS crystal-orbital method is that it takes the correlation within the unit cell (intracell correlation) into account only partially, and completely neglects the electron correlation between different cells (intercell correlation). If the interaction between the different unit cells is not very strong (as in periodic DNA stacks, or in other stacked systems), this intercell correlation might be unimportant. On the other hand, if the interactions between cells are large (as in a polyacetylene chain) the long-range intercell correlations can be very important. One possibility of taking them into account in the DODS scheme is to multiply the LCAO Bloch orbitals Φ_h^α and Φ_h^β by different standing waves η_h^α and η_h^β which possess wavelengths several times the elementary translation (suitably chosen for a given polymer) and different phases for electrons with spin α and β.[29] In this way determined total population of electrons with spin α will be large in a given cell and that of electrons with spin β will be small, while in the next two neighbor cells (in the case of a linear chain) the reverse situation will occur, and so on. Of course, only numerical calculations performed on some polymers can decide whether the described simple idea is really useful for taking intercell correlations into account.

In applying equation (1.107) to the one-dimensional case we substitute only the vector $\mathbf{q}$ by q to obtain the total energy per unit cell of a linear chain in the DODS case. By comparing the total energies per unit cell obtained by the one-dimensional counterpart of equation (1.107) and by expression (1.67), one can study the improvement in energy of a linear chain provided by the DODS method for the $n^{*\alpha} \neq n^{*\beta}$ case and (if there is a splitting) also for the $n^{*\alpha} = n^{*\beta}$ case.

Finally, we note that the DODS CO method is also most useful if one wishes to study the spin distribution (spin density wave) in the $n^{*\alpha} \neq n^{*\beta}$ case.

1.5. RELATIVISTIC FORMULATION*

1.5.1. Introductory Remarks

There is increasing interest in the relativistic treatment of atoms,[30] molecules,[31] and solids.[32] A relativistic Hartree–Fock scheme [Hartree–Fock–Dirac (HFD) method] based on the variation in the total energy obtained with a single Slater determinant (in which the one-electron orbitals are four-component Dirac spinors), using a Dirac-type Hamiltonian for each electron and including Coulomb interaction, was developed some time ago.[33] For the remaining interaction terms the first-order perturbation of the Breit interaction operator[34] reduced to large components (Pauli approximation) is usually taken into account (see, however, the work of Mann and Johnson[35]).

In the case of nonrelativistic calculations, the Hartree–Fock approximation provides the best one-electron orbitals from the standpoint of minimizing the total energy. In a similar way, the relativistic Hartree–Fock (HFD) equations must be solved in order to find the best relativistic total energy in the one-electron approximation. These equations can be used as starting orbitals for the relativistic treatment of a many-electron system. It should be mentioned also that *only* if the nonrelativistic or relativistic Hartree–Fock problem is solved can one define the nonrelativistic or relativistic correlation energy, respectively. This was the motivation for developing a formalism to solve, within any desired degree of accuracy, the relativistic Hartree–Fock problem in the general case (atoms, molecules, and crystals).

In the case of atoms and molecules with central symmetry, in which case a one-center expansion is feasible,[30,36] these equations are solved numerically. In the case of molecules with no central symmetry and in the case of crystals, this procedure is not possible and one must expand the molecular orbitals (MOs) or crystal orbitals (COs) as a linear combination of some basis functions (LCAO expansion). This was first conducted for molecules by Malli and Oreg[37] and applications can be found for diatomic or linear molecules,[38,39] the only exception being the H_2CO molecule, which was treated by Aoyama *et al.*[40]

* This section assumes knowledge of elementary relativistic quantum mechanics. If the reader is unfamiliar with it, this section can be omitted without hindering an understanding of the subsequent material.

Here we shall rederive the HFD equations in an LCAO form (relativistic Hartree–Fock–Roothaan equations) in the general case, i.e., using the different coefficients for all four components of the atomic spinors (basis functions), taking no advantage of the molecular symmetry, and applying the full relativistic molecular or crystal Fock operator with the nonlocal exchange term. Such an approach will make the generalization to the case of crystals straightforward.

It is noteworthy that by developing this formalism we did not first think at all of heavy metals, which have been treated successfully in recent years with the aid of relativistic versions of the APW or KKR[32] methods, but rather more complicated solids in which, besides a few heavy atoms, there are several light atoms in the unit cell (such as AsF_5, SbF_5, U_3O_8, RbF, and KF crystals). In such cases there are also high-density regions (regions of heavy atoms), low-density regions (regions between atoms and regions containing light atoms), and correspondingly more extended regions with larger density gradients than in heavy metals. For such systems, experience gained in the nonrelativistic case indicates that density functional methods (or their approximations, such as the KKR method) or methods based on a plane-wave (OPW) or modified-plane-wave (like the APW method) description do not work well. One can expect a similar situation to arise also in the relativistic case. Therefore, in order to treat these systems relativistically, one must return to the computationally more difficult, but clearly defined Hartree–Fock–Dirac approximation. After obtaining the energy bands of a crystal in the HFD approximation, there is an advantage in that one can proceed systematically to include the same types of corrections (corrections for the excited states to decrease the gap, short- and long-range correlation corrections) as in the nonrelativistic case.[41]

1.5.2. Derivation of the Relativistic Hartree–Fock–Roothaan Equations for Molecules and Crystals

We shall assume here that we have a crystal of $2N + 1$ unit cells in each direction and m orbitals in the unit cell, and further that we have four basis sets $\{\chi^{(t)}\}$ $(t = 1, 2, 3, 4)$. With their help any four-component relativistic crystal (Bloch) orbital $\tilde{\Psi}_i^{\mathrm{CO}}$ can be expressed in the form

$$\tilde{\Psi}_i^{\mathrm{CO}} = \begin{bmatrix} \tilde{\psi}_i^{(1)} \\ \tilde{\psi}_i^{(2)} \\ \tilde{\psi}_i^{(3)} \\ \tilde{\psi}_i^{(4)} \end{bmatrix} = \sum_{\mathbf{q}=-N}^{N} \sum_{g=1}^{m} \begin{bmatrix} C_{i,\mathbf{q},g}^{(1)} \chi_{\mathbf{q},g}^{(1)} \\ C_{i,\mathbf{q},g}^{(2)} \chi_{\mathbf{q},g}^{(2)} \\ C_{i,\mathbf{q},g}^{(3)} \chi_{\mathbf{q},g}^{(3)} \\ C_{i,\mathbf{q},g}^{(4)} \chi_{\mathbf{q},g}^{(4)} \end{bmatrix} e^{-iW_i t} \tag{1.108}$$

where $\mathbf{q}$ again denotes the unit cell, and $\chi_{\mathbf{q},g}^{(t)} = \chi^{(t)}(\mathbf{r} - \mathbf{R}_\mathbf{q} - \mathbf{g}_A)_g$ denotes the gth basis functions of the tth set centered at atom g_A (to which the orbital g belongs) in the unit cell $\mathbf{R}_\mathbf{q}$. Finally, W_i is the relativistic one-electron energy belonging to the CO $\tilde{\Psi}_i^{\text{CO}}$.

One should note that a four-component relativistic AO always contains only two different radial-dependent parts (large and small components). The four different components are then formed with the help of different combinations of these two radial functions with the well-known angular-dependent parts of the wave functions [see, for instance, equations (54), (99), (105), and (107) in the book by Mott and Sneddon[42]]. Already in the $l = 0$, $j = s = \frac{1}{2}$ case ($1s$ electron) four different components are obtained in this way for the relativistic AO. Therefore, if one wishes to perform an LCAO procedure, the basis functions must possess four different components which have to be varied independently. This is the reason for ansatz (1.108).

Furthermore, one could most probably use only two different basis sets, one for the large and one for the small component as a function of r, because the angular part of the functions is fixed in the case of an H atom. On the other hand, to maintain the possibility of basis functions not centered at the atomic nuclei (such as function centers at the midpoints of chemical bonds, or floating Gaussians), we prefer to work with four different basis sets. In a practical case the four basis sets could be equal, $\{\chi^{(1)}\} = \{\chi^{(2)}\} = \{\chi^{(3)}\} = \{\chi^{(4)}\}$, or one could use only one basis $\{\chi^{(1)}\}$ for the large components and another $\{\chi^{(2)}\}$ for the small ones.

A Slater determinant could be built up in the usual way with the aid of the COs $\tilde{\Psi}_i^{\text{CO}}$, and the expectation value of the approximate relativistic total Hamiltonian constructed in the form

$$\hat{H}_{\text{rel}} = \sum_{l=1}^{M} \hat{H}^{\text{D}}(l) + \sum_{k<l} 1/r_{kl} \tag{1.109}$$

where $\hat{H}^{\text{D}}$ is the Dirac operator of the lth electron and $M = n(2N+1)^3$ (n is the number of electrons per unit cell), with this Slater determinant. The variation of this expectation value with allowance for the auxiliary conditions $\langle \tilde{\Psi}_i^{\text{CO}} | \tilde{\Psi}_j^{\text{CO}} \rangle = \delta_{ij}$ and only positive of values W_i (in the relativistic sense) to avoid electronic states with negative energies yields in the usual way the relativistic Hartree–Fock equations of the system (a molecule, or, in the case of translational symmetry, a crystal)

$$\hat{F}^{\text{rel}} \tilde{\Psi}_i^{\text{CO}} = \left\{ \hat{H}^{\text{D}} + \sum_{\mathbf{p}=1}^{(2N+1)^3} \sum_{h'=1}^{n^*} [\hat{J}(\mathbf{p}, h') - \hat{K}(\mathbf{p}, h')] \right\} \tilde{\Psi}_i^{\text{CO}} = W_i \tilde{\Psi}_i^{\text{CO}} \tag{1.110}$$

Here n^* is the number of (in the restricted Hartree–Fock case, doubly filled) bands and $\hat{J}$ and $\hat{K}$ are again the Coulomb and exchange operators, respectively, formed now with the aid of the four-component relativistic HF COs instead of the nonrelativistic one-component HF COs. In deriving equation (1.110) the time-dependent factor $e^{-iW_i t}$ of $\tilde{\Psi}_i^{CO}$ has been eliminated by differentiating with respect to time contained in the original Dirac operator $\hat{H}^D$,[42]

$$-i\left[\frac{1}{i}\frac{\partial}{ic\partial t}\tilde{\Psi}^{CO}(\mathbf{r},t)\right] = -i\left\{\frac{1}{i}\frac{\partial}{ic\partial t}[\Psi^{CO}(\mathbf{r})\exp(-iW_i t)]\right\}$$

$$= \frac{1}{c}W_i\Psi^{CO}(\mathbf{r})\exp[-iW_i t) \qquad (1.111)$$

The scalar potential of the crystal is

$$\Phi = -\sum_{\alpha=1}^{M}\sum_{\mathbf{p}=1}^{(2N+1)^3}\frac{Z_\alpha}{|\mathbf{r}-\mathbf{R}_\alpha^{\mathbf{p}}|} \qquad (1.112)$$

where M is again the number of atomic nuclei in the unit cell, Z_α is the nuclear charge of the αth nucleus, and $\mathbf{R}_\alpha^{\mathbf{p}}$ is the position vector of the αth nucleus in the unit cell characterized by vector $\mathbf{p}$. We assume further that there is no external magnetic field, so that $\mathbf{A} = \mathbf{0}$. The detailed form of $\hat{H}^D$ (in atomic units)[42] then allows equation (1.110) to be written in the form

$$\left\{\frac{c}{i}\sum_{j=1}^{3}\boldsymbol{\alpha}_j\frac{\partial}{\partial x_j} + \Phi + \boldsymbol{\alpha}_4 m_0 c + \sum_{\mathbf{p}'}\sum_{h'}[\hat{J}(\mathbf{p}',h') - \hat{K}(\mathbf{p}',h')]\right\}\Psi_i^{CO}$$

$$= \frac{c}{i}\begin{bmatrix}\frac{\partial}{\partial x}\psi^{(4)}(\mathbf{p},h)\\ \frac{\partial}{\partial x}\psi^{(3)}(\mathbf{p},h)\\ \frac{\partial}{\partial x}\psi^{(2)}(\mathbf{p},h)\\ \frac{\partial}{\partial x}\psi^{(1)}(\mathbf{p},h)\end{bmatrix} + c\begin{bmatrix}-\frac{\partial}{\partial y}\psi^{(4)}(\mathbf{p},h)\\ \frac{\partial}{\partial y}\psi^{(3)}(\mathbf{p},h)\\ -\frac{\partial}{\partial y}\psi^{(2)}(\mathbf{p},h)\\ \frac{\partial}{\partial y}\psi^{(1)}(\mathbf{p},h)\end{bmatrix} + \frac{c}{i}\begin{bmatrix}\frac{\partial}{\partial z}\psi^{(3)}(\mathbf{p},h)\\ -\frac{\partial}{\partial z}\psi^{(4)}(\mathbf{p},h)\\ \frac{\partial}{\partial z}\psi^{(1)}(\mathbf{p},h)\\ -\frac{\partial}{\partial z}\psi^{(2)}(\mathbf{p},h)\end{bmatrix}$$

(continued)

$$
+ m_0c^2 \begin{bmatrix} \psi^{(1)}(\mathbf{p},h) \\ \psi^{(2)}(\mathbf{p},h) \\ -\psi^{(3)}(\mathbf{p},h) \\ -\psi^{(4)}(\mathbf{p},h) \end{bmatrix} + \Phi \begin{bmatrix} \psi^{(1)}(\mathbf{p},h) \\ \psi^{(2)}(\mathbf{p},h) \\ \psi^{(3)}(\mathbf{p},h) \\ \psi^{(4)}(\mathbf{p},h) \end{bmatrix}
$$

$$
+ \sum_{\mathbf{p}',h'} \Big\langle [\psi^{(1)}(\mathbf{p}',h';\mathbf{r}_2)\psi^{(2)}(\mathbf{p}',h';\mathbf{r}_2)\psi^{(3)}(\mathbf{p}',h';\mathbf{r}_2)\psi^{(4)}(\mathbf{p}',h';\mathbf{r}_2)]
$$

$$
\times \left| \frac{1}{r_n}(1-\hat{P}_{12}) \right| \begin{bmatrix} \psi^{(1)}(\mathbf{p}',h';\mathbf{r}_2) \\ \psi^{(2)}(\mathbf{p}',h';\mathbf{r}_2) \\ \psi^{(3)}(\mathbf{p}',h';\mathbf{r}_2) \\ \psi^{(4)}(\mathbf{p}',h';\mathbf{r}_2) \end{bmatrix} \Big\rangle_2 \begin{bmatrix} \psi^{(1)}(\mathbf{p},h;\mathbf{r}_1) \\ \psi^{(2)}(\mathbf{p},h;\mathbf{r}_1) \\ \psi^{(3)}(\mathbf{p},h;\mathbf{r}_1) \\ \psi^{(4)}(\mathbf{p},h;\mathbf{r}_1) \end{bmatrix} = W(\mathbf{p},h) \begin{bmatrix} \psi^{(1)}(\mathbf{p},h) \\ \psi^{(2)}(\mathbf{p},h) \\ \psi^{(3)}(\mathbf{p},h) \\ \psi^{(4)}(\mathbf{p},h) \end{bmatrix} \tag{1.113}
$$

Here, the general index i [$i = 1, \ldots, M$; $M = n(2N+1)^3$] of the COs has been replaced by double indexing (for future convenience), where h denotes the band indices and $\mathbf{p}$ the levels inside the bands. The subscript 2 at the bracket indicates that integration must be performed over the coordinates of the second electron while $\hat{P}_{12}$ is the exchange operator. The detailed form of the relativistic Hartree–Fock equation for a crystal involves the definitions of the Dirac matrices in the form

$$
\boldsymbol{\alpha}_1 = \begin{bmatrix} 0 & 0 & 0 & 1 \\ 0 & 0 & 1 & 0 \\ 0 & 1 & 0 & 0 \\ 1 & 0 & 0 & 0 \end{bmatrix}, \quad \boldsymbol{\alpha}_2 = \begin{bmatrix} 0 & 0 & 0 & -i \\ 0 & 0 & i & 0 \\ 0 & -i & 0 & 0 \\ i & 0 & 0 & 0 \end{bmatrix}
$$

$$
\boldsymbol{\alpha}_3 = \begin{bmatrix} 0 & 0 & 1 & 0 \\ 0 & 0 & 0 & -1 \\ 1 & 0 & 0 & 0 \\ 0 & -1 & 0 & 0 \end{bmatrix}, \quad \boldsymbol{\alpha}_4 = \begin{bmatrix} 1 & 0 & 0 & 0 \\ 0 & 1 & 0 & 0 \\ 0 & 0 & -1 & 0 \\ 0 & 0 & 0 & -1 \end{bmatrix} \tag{1.114}
$$

If the LCAO forms of the four components of $\boldsymbol{\Psi}^{\mathrm{CO}}(\mathbf{p},h)$ are introduced [see equation (1.108)] one obtains the matrix equation

$$\mathbf{F}^{\text{rel}}\begin{bmatrix} \mathbf{C}^{(1)}(\mathbf{p},h) \\ \mathbf{C}^{(2)}(\mathbf{p},h) \\ \mathbf{C}^{(3)}(\mathbf{p},h) \\ \mathbf{C}^{(4)}(\mathbf{p},h) \end{bmatrix} = W(\mathbf{p},h)\mathbf{S}^{\text{rel}}\begin{bmatrix} \mathbf{C}^{(1)}(\mathbf{p},h) \\ \mathbf{C}^{(2)}(\mathbf{p},h) \\ \mathbf{C}^{(3)}(\mathbf{p},h) \\ \mathbf{C}^{(4)}(\mathbf{p},h) \end{bmatrix} \tag{1.115}$$

In complete analogy to the derivation of the nonrelativistic Hartree–Fock–Roothaan equations,[4] the generalized overlap matrix $\mathbf{S}^{\text{rel}}$ is here given by

$$\mathbf{S}^{\text{rel}} = \begin{bmatrix} \mathbf{S}^{(1,1)} & & & \mathbf{0} \\ & \mathbf{S}^{(2,2)} & & \\ & & \mathbf{S}^{(3,3)} & \\ \mathbf{0} & & & \mathbf{S}^{(4,4)} \end{bmatrix} \tag{1.116}$$

where, for instance, the diagonal block $\mathbf{S}^{(3,3)}$ contains only overlap integrals between basis functions belonging to the third basis set:

$$\mathbf{S}^{(3,3)}_{\mathbf{q}',g';\mathbf{q},g} = \langle \chi^{(3)}_{\mathbf{q}',g'} \mid \chi^{(3)}_{\mathbf{q},g} \rangle \tag{1.117}$$

$\chi^{(3)}_{\mathbf{q}',g'}$ denotes the g'th basis function in the unit cell $\mathbf{q}'$ belonging to the third set.

The left-hand side of equation (1.115) can be expressed in detail by substituting into equation (1.113) the LCAO form [equation (1.108)] of the COs $\Psi^{(t)}$. This leads to the following matrix equations (again in complete analogy to the nonrelativistic derivation[4]):

$$\begin{bmatrix} \mathbf{F}_1^{(1,4)} & & & \mathbf{0} \\ & \mathbf{F}_1^{(2,3)} & & \\ & & \mathbf{F}_1^{(3,2)} & \\ \mathbf{0} & & & \mathbf{F}_1^{(4,1)} \end{bmatrix} \begin{bmatrix} \mathbf{c}^{(4)}(\mathbf{p},h) \\ \mathbf{c}^{(3)}(\mathbf{p},h) \\ \mathbf{c}^{(2)}(\mathbf{p},h) \\ \mathbf{c}^{(1)}(\mathbf{p},h) \end{bmatrix}$$

$$+ \begin{bmatrix} \mathbf{F}_2^{(1,4)} & & & \\ & \mathbf{F}_2^{(2,3)} & & \\ & & \mathbf{F}_2^{(3,2)} & \\ \mathbf{0} & & & \mathbf{F}_2^{(4,1)} \end{bmatrix} \begin{bmatrix} -\mathbf{c}^{(4)}(\mathbf{p},h) \\ \mathbf{c}^{(3)}(\mathbf{p},h) \\ -\mathbf{c}^{(2)}(\mathbf{p},h) \\ \mathbf{c}^{(1)}(\mathbf{p},h) \end{bmatrix}$$

(continued)

$$+\begin{bmatrix} \mathbf{F}_3^{(1,3)} & & & \mathbf{0} \\ & \mathbf{F}_3^{(2,4)} & & \\ & & \mathbf{F}_3^{(3,1)} & \\ \mathbf{0} & & & \mathbf{F}_3^{(4,2)} \end{bmatrix} \begin{bmatrix} \mathbf{c}^{(3)}(\mathbf{p},h) \\ -\mathbf{c}^{(4)}(\mathbf{p},h) \\ \mathbf{c}^{(1)}(\mathbf{p},h) \\ -\mathbf{c}^{(2)}(\mathbf{p},h) \end{bmatrix}$$

$$+\begin{bmatrix} \mathbf{F}_4^{(1,1)} & & & \mathbf{0} \\ & \mathbf{F}_4^{(2,2)} & & \\ & & \mathbf{F}_4^{(3,3)} & \\ \mathbf{0} & & & \mathbf{F}_4^{(4,4)} \end{bmatrix} \begin{bmatrix} \mathbf{c}^{(1)}(\mathbf{p},h) \\ \mathbf{c}^{(2)}(\mathbf{p},h) \\ -\mathbf{c}^{(3)}(\mathbf{p},h) \\ -\mathbf{c}^{(4)}(\mathbf{p},h) \end{bmatrix}$$

$$+\begin{bmatrix} \mathbf{F}_5^{(1,1)} & & & \mathbf{0} \\ & \mathbf{F}_5^{(2,2)} & & \\ & & \mathbf{F}_5^{(3,3)} & \\ \mathbf{0} & & & \mathbf{F}_5^{(4,4)} \end{bmatrix} \begin{bmatrix} \mathbf{c}^{(1)}(\mathbf{p},h) \\ \mathbf{c}^{(2)}(\mathbf{p},h) \\ \mathbf{c}^{(3)}(\mathbf{p},h) \\ \mathbf{c}^{(4)}(\mathbf{p},h) \end{bmatrix}$$

$$= W(\mathbf{p},h) \begin{bmatrix} \mathbf{S}^{(1,1)} & & & \mathbf{0} \\ & \mathbf{S}^{(2,2)} & & \\ & & \mathbf{S}^{(3,3)} & \\ \mathbf{0} & & & \mathbf{S}^{(4,4)} \end{bmatrix} \begin{bmatrix} \mathbf{c}^{(1)}(\mathbf{p},h) \\ \mathbf{c}^{(2)}(\mathbf{p},h) \\ \mathbf{c}^{(3)}(\mathbf{p},h) \\ \mathbf{c}^{(4)}(\mathbf{p},h) \end{bmatrix} \tag{1.118}$$

The blocks of the relativistic Fock matrix $\mathbf{F}^{\text{rel}}$ are here defined in terms of their elements as

$$(\mathbf{F}_1^{(t',t)})_{\mathbf{q}',g';\mathbf{q},g} = \frac{c}{i}\left\langle \chi_{\mathbf{q}',g'}^{(t')} \middle| \frac{\partial}{\partial x} \chi_{\mathbf{q},g}^{(t)} \right\rangle \tag{1.119a}$$

$$(\mathbf{F}_2^{(t',t)})_{\mathbf{q}',g';\mathbf{q},g} = c\left\langle \chi_{\mathbf{q}',g'}^{(t')} \left| \frac{\partial}{\partial y} \chi_{\mathbf{q},g}^{(t)} \right.\right\rangle \tag{1.119b}$$

$$(\mathbf{F}_3^{(t',t)})_{\mathbf{q}',g';\mathbf{q},g} = \frac{c}{i}\left\langle \chi_{\mathbf{q}',g'}^{(t')} \left| \frac{\partial}{\partial z} \chi_{\mathbf{q},g}^{(t)} \right.\right\rangle \tag{1.119c}$$

$$(\mathbf{F}_4^{(t,t)})_{\mathbf{q}',g';\mathbf{q},g} = m_0c^2\langle \chi_{\mathbf{q}',g'}^{(t)} \mid \chi_{\mathbf{q},g}^{(t)} \rangle = m_0c^2(\mathbf{S}^{(t,t)})_{\mathbf{q}',g';\mathbf{q},g} \tag{1.119d}$$

$$(\mathbf{F}_5^{(t,t)})_{\mathbf{q}',g';\mathbf{q},g} = \left\langle \chi_{\mathbf{q}',g'}^{(t)} \mid \Phi + \sum_{\mathbf{p}'} \sum_{h'} [\hat{J}(\mathbf{p}', h') - \hat{K}(\mathbf{p}',h')] \mid \chi_{\mathbf{q},g}^{(t)} \right\rangle \tag{1.119e}$$

$$(t', t = 1, 2, 3, 4)$$

The operators $\hat{J}(p', h')$ and $\hat{K}(p', h')$ were written down in more detail in equation (1.113) and their LCAO form in the crystal case will be given subsequently.

The hypermatrix equation (1.118) can be expressed alternatively by the following four matrix equations:

$$(\mathbf{F}_1^{(1,4)} - \mathbf{F}_2^{(1,4)})\mathbf{C}^{(4)}(\mathbf{p}, h) + \mathbf{F}_3^{(1,3)}\mathbf{C}^{(3)}(\mathbf{p}, h) + (\mathbf{F}_4^{(1,1)} + \mathbf{F}_5^{(1,1)})\mathbf{C}^{(1)}(\mathbf{p}, h)$$
$$= W(\mathbf{p}, h)\mathbf{S}^{(1,1)}\mathbf{C}^{(1)}(\mathbf{p}, h) \tag{1.120a}$$

$$(\mathbf{F}_1^{(2,3)} + \mathbf{F}_2^{(2,3)})\mathbf{C}^{(3)}(\mathbf{p}, h) - \mathbf{F}_3^{(2,4)}\mathbf{C}^{(4)}(\mathbf{p}, h) + (\mathbf{F}_4^{(2,2)} + \mathbf{F}_5^{(2,2)})\mathbf{C}^{(2)}(\mathbf{p}, h)$$
$$= W(\mathbf{p}, h)\mathbf{S}^{(2,2)}\mathbf{C}^{(2)}(\mathbf{p}, h) \tag{1.120b}$$

$$(\mathbf{F}_1^{(3,2)} - \mathbf{F}_2^{(3,2)})\mathbf{C}^{(2)}(\mathbf{p}, h) + \mathbf{F}_3^{(3,1)}\mathbf{C}^{(1)}(\mathbf{p}, h) + (\mathbf{F}_4^{(3,3)} - \mathbf{F}_5^{(3,3)})\mathbf{C}^{(3)}(\mathbf{p}, h)$$
$$= W(\mathbf{p}, h)\mathbf{S}^{(3,3)}\mathbf{C}^{(3)}(\mathbf{p}, h) \tag{1.120c}$$

$$(\mathbf{F}_1^{(4,1)} + \mathbf{F}_2^{(4,1)})\mathbf{C}^{(1)}(\mathbf{p}, h) - \mathbf{F}_3^{(4,2)}\mathbf{C}^{(2)}(\mathbf{p}, h) + (\mathbf{F}_4^{(4,4)} - \mathbf{F}_5^{(4,4)})\mathbf{C}^{(4)}(\mathbf{p}, h)$$
$$= W(\mathbf{p}, h)\mathbf{S}^{(4,4)}\mathbf{C}^{(4)}(\mathbf{p}, h) \tag{1.120d}$$

It is easy to show that all matrices $\mathbf{F}_l^{(t',t)}$ $(l = 1, 2, 3, 4, 5)$ of order $m(2N+1)^3 \times m(2N+1)^3$ are cyclic hypermatrices if one takes into account the translational symmetry of the crystal and introduces periodic boundary conditions. Therefore, one can apply to all the terms in all four equations a unitary transformation which block-diagonalizes all the matrices by multiplying from the left all the equations by $\mathbf{U}^+$ and introducing everywhere the unit matrix $\mathbf{I} = \mathbf{U}\mathbf{U}^+$ between matrices $\mathbf{F}^{(t',t)}$ and eigenvectors $\mathbf{C}^{(t)}$.[1] Here we can write

$$\mathbf{U}^+\mathbf{F}_l^{(t',t)}\mathbf{U}\mathbf{U}^+\mathbf{C}^{(t)} = \mathbf{F}_l^{(t',t)\mathrm{BD}}\mathbf{D}^{(t)} \tag{1.121}$$

where $\mathbf{F}_l^{(t',t)\mathrm{BD}} = \mathbf{U}^+\mathbf{F}_l^{(t',t)}\mathbf{U}$ and $\mathbf{D}^{(t)} = \mathbf{U}^+\mathbf{C}^{(t)}$. The form of matrix $\mathbf{U}$ was given in equation (1.9) in Section 1.1.

In all four equations (1.120) we have only block-diagonal matrices after the unitary transformation, so we can easily separate these equations according to their blocks.[1] Assuming further that $N \to \infty$, in which case we can replace the discrete variables $p_i = -N, -N+1, \ldots, 0, \ldots, N$ ($i = 1, 2, 3$), which specify the blocks, by the continuous variables $k_i = 2\pi p_i / a_i(2N+1)$, we arrive at the following relativistic Hartree–Fock–Roothaan equations governing a crystal (again in complete analogy to the corresponding nonrelativistic derivation[1]):

$$[\mathbf{F}_1^{(1,4)}(\mathbf{k}) - \mathbf{F}_2^{(1,4)}(\mathbf{k})]\mathbf{d}^{(4)}(\mathbf{k})_h + \mathbf{F}_3^{(1,3)}(\mathbf{k})\mathbf{d}^{(3)}(\mathbf{k})_h + [\mathbf{F}_4^{(1,1)}(\mathbf{k}) + \mathbf{F}_5^{(1,1)}(\mathbf{k})]\mathbf{d}^{(1)}(\mathbf{k})_h$$
$$= W(\mathbf{k})_h \mathbf{S}^{(1,1)}(\mathbf{k})\mathbf{d}^{(1)}(\mathbf{k})_h \qquad (1.122a)$$

$$[\mathbf{F}_1^{(2,3)}(\mathbf{k}) + \mathbf{F}_2^{(2,3)}(\mathbf{k})]\mathbf{d}^{(3)}(\mathbf{k})_h - \mathbf{F}_3^{(2,4)}(\mathbf{k})\mathbf{d}^{(4)}(\mathbf{k})_h + [\mathbf{F}_4^{(2,2)}(\mathbf{k}) + \mathbf{F}_5^{(2,2)}(\mathbf{k})]\mathbf{d}^{(2)}(\mathbf{k})_h$$
$$= W(\mathbf{k})_h \mathbf{S}^{(2,2)}(\mathbf{k})\mathbf{d}^{(2)}(\mathbf{k})_h \qquad (1.122b)$$

$$[\mathbf{F}_1^{(3,2)}(\mathbf{k}) - \mathbf{F}_2^{(3,2)}(\mathbf{k})]\mathbf{d}^{(2)}(\mathbf{k})_h + \mathbf{F}_3^{(3,1)}(\mathbf{k})\mathbf{d}^{(1)}(\mathbf{k})_h + [\mathbf{F}_4^{(3,3)}(\mathbf{k}) - \mathbf{F}_5^{(3,3)}(\mathbf{k})]\mathbf{d}^{(3)}(\mathbf{k})_h$$
$$= W(\mathbf{k})_h \mathbf{S}^{(3,3)}(\mathbf{k})\mathbf{d}^{(3)}(\mathbf{k})_h \qquad (1.122c)$$

$$[\mathbf{F}_1^{(4,1)}(\mathbf{k}) - \mathbf{F}_2^{(4,1)}(\mathbf{k})]\mathbf{d}^{(1)}(\mathbf{k})_h - \mathbf{F}_3^{(4,2)}(\mathbf{k})\mathbf{d}^{(4)}(\mathbf{k})_h + [\mathbf{F}_4^{(4,4)}(\mathbf{k}) - \mathbf{F}_5^{(4,4)}(\mathbf{k})]\mathbf{d}^{(4)}(\mathbf{k})_h$$
$$= W(\mathbf{k})_h \mathbf{S}^{(4,4)}(\mathbf{k})\mathbf{d}^{(4)}(\mathbf{k})_h \qquad (1.122d)$$

In these equations all the $m \times m$ matrices $\mathbf{F}^{(t',t)}(\mathbf{k})$ and the overlap matrices $\mathbf{S}^{(t',t)}(\mathbf{k})$ have the form of a Fourier transform,[1] namely

$$\mathbf{F}_1^{(t',t)}(\mathbf{k}) = \sum_{\mathbf{q}} \mathbf{F}_1^{(t',t)}(\mathbf{q}) \exp(i\mathbf{k}\mathbf{R}_{\mathbf{q}}) \qquad (1.123a)$$

$$\mathbf{S}^{(t',t)}(\mathbf{k}) = \sum_{\mathbf{q}} \mathbf{S}^{(t',t)}(\mathbf{q}) \exp(i\mathbf{k}\mathbf{R}_{\mathbf{q}}) \qquad (1.123b)$$

Matrices $\mathbf{S}^{(t',t)}(\mathbf{q})$ and $\mathbf{F}_1^{(t',t)}(\mathbf{q})$ have already been defined in detail in terms of their elements [see equations (1.117) and (1.119)], with the exception of matrices $\mathbf{F}_5^{(t',t)}(\mathbf{q})$.

The form of matrices $\mathbf{F}_5^{(t',t)}(\mathbf{q})$ is obtained by substituting in operators $\hat{J}$ and $\hat{K}$ in equation (1.113) the LCAO form of the COs $\Psi^{(t)}(\mathbf{p}', h')$. After defining the relativistic generalized charge-bond-order matrix elements

$$P(\mathbf{q}_1 - \mathbf{q}_2)_{u,v}^{(j)} = \frac{1}{\omega} \sum_{h'=1}^{n^*} \int_{\omega} d(\mathbf{k})_{h',u}^{(j)*} d(\mathbf{k})_{h',v}^{(j)} \exp[i\mathbf{k}(\mathbf{R}_{\mathbf{q}_1} - \mathbf{R}_{\mathbf{q}_2})] d\mathbf{k}$$
$$(j = 1, 2, 3, 4) \qquad (1.124)$$

(again, in complete analogy with the corresponding nonrelativistic definitions[1]), a straightforward derivation leads to the expression

$$[\mathbf{F}_5^{(t,t)}(\mathbf{q})]_{g',g} = \langle \chi_{\mathbf{0},g'}^{(t)} \,|\, \Phi \,|\, \chi_{\mathbf{q},g}^{(t)} \rangle + \sum_{\mathbf{q}_1,\mathbf{q}_2} \sum_{u,v} \sum_{j=1}^{4} P(\mathbf{q}_1 - \mathbf{q}_2)_{u,v}^{(j)}$$

$$\times [\langle \chi_{\mathbf{0},g'}^{(t)}(1)\chi_{\mathbf{q}_1,u}^{(j)}(2) \left| \frac{1}{r_{12}}(1 - \hat{P}_{12}) \right| \,|\, \chi_{\mathbf{q},g}^{(t)}(1)\chi_{\mathbf{q}_2,v}^{(j)}(2) \rangle] \qquad (1.125)$$

In these latter two expressions ω denotes the volume of the first Brillouin zone, n^* the number of filled bands (in the closed-shell case), Φ was given in equation (1.112) for a crystal, and all other quantities were defined earlier.

Instead of solving the four coupled matrix equations (1.122) we can rewrite them as a hypermatrix equation. To this end we use the relation

$$\tilde{\mathbf{d}}^{(1)} \equiv \begin{bmatrix} \mathbf{d}^{(4)}(\mathbf{k})_h \\ \mathbf{d}^{(3)}(\mathbf{k})_h \\ \mathbf{d}^{(2)}(\mathbf{k})_h \\ \mathbf{d}^{(1)}(\mathbf{k})_h \end{bmatrix} = \boldsymbol{\alpha}_1 \begin{bmatrix} \mathbf{d}^{(1)}(\mathbf{k})_h \\ \mathbf{d}^{(2)}(\mathbf{k})_h \\ \mathbf{d}^{(3)}(\mathbf{k})_h \\ \mathbf{d}^{(4)}(\mathbf{k})_h \end{bmatrix} \equiv \boldsymbol{\alpha}_1 \tilde{\mathbf{d}}^{(5)} \qquad (1.126)$$

and similar relations for the other three hypervectors $\tilde{\mathbf{d}}^{(2)}$, $\tilde{\mathbf{d}}^{(3)}$, and $\tilde{\mathbf{d}}^{(4)}$:

$$\tilde{\mathbf{d}}^{(2)} = \frac{1}{i}\boldsymbol{\alpha}_2 \tilde{\mathbf{d}}^{(5)}, \quad \tilde{\mathbf{d}}^{(3)} = \boldsymbol{\alpha}_3 \tilde{\mathbf{d}}^{(5)}, \quad \text{and} \quad \tilde{\mathbf{d}}^{(4)} = \boldsymbol{\alpha}_4 \tilde{\mathbf{d}}^{(5)}$$

With the aid of these equations we can condense the matrix equations (1.122) (by performing matrix multiplications and adding corresponding block matrices) into the generalized eigenvalue equation

$$\mathbf{F}^{\text{rel}}(\mathbf{k})\tilde{\mathbf{d}}^{(5)}(\mathbf{k})$$

$$= \begin{bmatrix} \mathbf{F}^{(1,1)}(\mathbf{k}) + \mathbf{F}_5^{(1,1)}(\mathbf{k}) & \mathbf{0} & \mathbf{F}_3^{(1,3)}(\mathbf{k}) & \mathbf{F}_1^{(1,4)}(\mathbf{k}) - \mathbf{F}_2^{(1,4)}(\mathbf{k}) \\ \mathbf{0} & \mathbf{F}_4^{(2,2)}(\mathbf{k}) + \mathbf{F}_5^{(2,2)}(\mathbf{k}) & \mathbf{F}_1^{(2,3)}(\mathbf{k}) + \mathbf{F}_2^{(2,3)}(\mathbf{k}) & -\mathbf{F}_3^{(2,4)}(\mathbf{k}) \\ \mathbf{F}_3^{(3,1)}(\mathbf{k}) & \mathbf{F}_1^{(3,2)}(\mathbf{k}) - \mathbf{F}_2^{(3,2)}(\mathbf{k}) & \mathbf{F}_4^{(3,2)}(\mathbf{k}) - \mathbf{F}_5^{(3,3)}(\mathbf{k}) & \mathbf{0} \\ \mathbf{F}_1^{(4,1)}(\mathbf{k}) + \mathbf{F}_2^{(4,1)}(\mathbf{k}) & -\mathbf{F}_3^{(4,2)}(\mathbf{k}) & \mathbf{0} & \mathbf{F}_4^{(4,4)}(\mathbf{k}) - \mathbf{F}_5^{(4,4)}(\mathbf{k}) \end{bmatrix} \tilde{\mathbf{d}}^{(5)} \qquad (1.127)$$

$$= W(\mathbf{k})_h \begin{bmatrix} \mathbf{S}^{(1,1)}(\mathbf{k}) & & & \mathbf{0} \\ & \mathbf{S}^{(2,2)}(\mathbf{k}) & & \\ & & \mathbf{S}^{(3,3)}(\mathbf{k}) & \\ \mathbf{0} & & & \mathbf{S}^{(4,4)}(\mathbf{k}) \end{bmatrix} \begin{bmatrix} \mathbf{d}^{(1)}(\mathbf{k})_h \\ \mathbf{d}^{(2)}(\mathbf{k})_h \\ \mathbf{d}^{(3)}(\mathbf{k})_h \\ \mathbf{d}^{(4)}(\mathbf{k})_h \end{bmatrix}$$

It is easy to show with the aid of the relationship $\mathbf{F}_3^{(1,3)} = -(\mathbf{F}_3^{(3,1)})^{tr} = (\mathbf{F}_3^{(3,1)})^{+}$, and of similar relationships in the cases of other corresponding off-diagonal matrix blocks (which can all be demonstrated via integration by parts of their elements), such as

$$\left\langle \chi_{0,g'}^{(t')} \left| \frac{\partial \chi_{q,g}^{(t)}}{\partial x_i} \right.\right\rangle = -\left\langle \chi_{q,g}^{(t)} \left| \frac{\partial \chi_{0,g'}^{(t')}}{\partial x_i} \right.\right\rangle \qquad (i = 1, 2, 3)$$

that $\mathbf{F}^{rel}$ is an Hermitian matrix. Hence, after a Löwdin symmetric orthogonalization procedure[2] applied for a crystal,[1]

$$\mathbf{S}^{-1/2}(\mathbf{k})\mathbf{F}^{rel}(\mathbf{k})\mathbf{S}^{-1/2}(\mathbf{k})\mathbf{S}^{+1/2}(\mathbf{k})\tilde{\mathbf{d}}^{(5)}(\mathbf{k})_h = \tilde{\mathbf{F}}^{rel}(\mathbf{k})\mathbf{b}(\mathbf{k})_h$$

$$= W(\mathbf{k})_h \begin{bmatrix} \mathbf{b}^{(1)}(\mathbf{k})_h \\ \mathbf{b}^{(2)}(\mathbf{k})_h \\ \mathbf{b}^{(3)}(\mathbf{k})_h \\ \mathbf{b}^{(4)}(\mathbf{k})_h \end{bmatrix} \qquad (1.128)$$

one obtains real (relativistic) one-electron energies $W(\mathbf{k})_h$. In this way equations (1.127) and (1.128) together with definitions (1.117), (1.119), (1.124), and (1.125) fully define the relativistic Hartree–Fock–Roothaan (relativistic *ab initio* SCF LCAO) equations with various coefficients for all the functions belonging to the four different basis sets [see equation (1.118) for a crystal].

The case of a molecule (or an extended system with no translational symmetry) is a special case of this formalism when one does not use a cell index, only an orbital index. The form of all the equations remains in this case the same, except that one must omit the **k**-dependence, and in the definition of the charge-bond-order matrix elements [see equation (1.124)] one does not integrate over **k**.

1.5.3. Concluding Remarks

The above relativistic *ab initio* SCF LCAO CO method (further details can be found elsewhere[43]) has the shortcoming that, in the final equations (1.127) and (1.128), the states denoted by h are not specified. To do this, as in the case of atoms and molecules, one must classify the states (by applying advanced group-theoretical methods) not only in the usual way but also according to j (the internal quantum number). This classification must be carried out before performing actual calculations on solids by developing the appropriate programs (similar to the cases of atoms and molecules).

Such a program must also incorporate an appropriate choice for the *four basis sets*. One possibility is to start with relativistic hydrogen-like functions. Since these functions consist essentially of a product of a confluent hypergeometric function with a nonintegral power of r, the calculation of matrix elements (1.119) and especially (1.125) (four-center integrals) may cause serious difficulties in performing a computation.

A more realistic way of solving this problem would most probably be to start with the AOs obtained from an atomic relativistic Hartree–Fock calculation and try to fit the numerical output of this procedure with a linear combination of Gaussians. One can hope that in this way the integral packages of the nonrelativistic *ab initio* molecular programs could be used. In order to attempt this procedure a relativistic Hartree–Fock treatment of the carbon atom was performed (using the Desclaux program[31,32]) and its output (for the filled $1s_{1/2}$, $2s_{1/2}$, $2p_{1/2}$ states) fitted by a nonlinear least-squares procedure using a linear combination of Gaussians. It was found that if one wishes to fit, with acceptable accuracy, the radial part of a relativistic AO, about 30 Gaussians are needed to cover the whole range of r from 0 to 40 atomic units, which are required in the case of relativistic functions.[44]

To overcome this difficulty the range of r has been divided into four regions (see Table 1.1), identical for both the large and small components of the radial part of all three AOs ($1s_{1/2}$, $2s_{1/2}$, $2p_{1/2}$). In this way it was always possible to fit a component of the radial part of a relativistic carbon orbital in a given region with three Gaussians (see Tables 1.2, 1.3, and 1.4).[43] This means that the nonrelativistic molecular integral packages using Gaussian basis functions could be used without any change if one takes for each component of each function the linear combination of the corresponding three Gaussian functions. Further, one should observe

TABLE 1.1. Regions of the Radial Coordinate Chosen for the Primitive Gaussians in the Approximation of the Numerical Relativistic HF AOs of the Carbon Atom[a]

Region	r_{min} (au)	r_{max} (au)
1	3.150×10^{-5}	8.124×10^{-4}
2	8.540×10^{-4}	2.203×10^{-2}
3	2.316×10^{-2}	5.972×10^{-1}
4	6.278×10^{-1}	4.187×10^{1}

[a] The regions are the same for all six relativistic AOs.

TABLE 1.2. Linear Coefficients and Orbital Exponents of the Gaussian Expansion of the Numerical Output for the Radial Part of the $1s_{1/2}$ Relativistic Carbon AO

	Large component		Small component	
Region	Exponent	Coefficient	Exponent	Coefficient
1	0.709246×10^{10}	0.0174281	0.115250×10^{11}	−0.003167
	0.663154×10^{9}	0.1342063	0.109683×10^{10}	−0.144682
	0.668763×10^{8}	1.0000000	0.119233×10^{9}	−1.000000
2	0.547988×10^{7}	0.009344	0.115413×10^{8}	−0.002981
	0.454239×10^{6}	0.096842	0.109590×10^{7}	−0.053046
	0.346166×10^{5}	1.000000	0.113213×10^{6}	−1.000000
3	0.128783×10^{4}	0.214370	0.917609×10^{4}	−0.007226
	-0.458103×10^{2}	0.681347	0.922993×10^{3}	−0.193120
	-0.137436×10^{2}	1.000000	0.842388×10^{2}	−1.000000
4	0.159355×10^{1}	0.046071	0.125850×10^{2}	−0.090276
	0.172433×10^{1}	0.370211	0.300573×10^{1}	−0.124683
	0.390551×10^{0}	1.000000	0.598104×10^{0}	−1.000000

TABLE 1.3. Linear Coefficients and Orbital Exponents of the Gaussian Expansion of the Numerical Output for the Radial Part of the $2s_{1/2}$ Relativistic Carbon AO

	Large component		Small component	
Region	Exponent	Coefficient	Exponent	Coefficient
1	0.854596×10^{10}	−0.043226	0.130785×10^{11}	0.009131
	0.832298×10^{9}	−0.163678	0.126895×10^{10}	0.081642
	0.879549×10^{8}	−1.000000	0.140308×10^{9}	1.000000
2	0.808547×10^{7}	−0.017281	0.138757×10^{8}	0.026423
	0.713038×10^{6}	−0.097700	0.135426×10^{7}	0.258674
	0.663103×10^{5}	−1.000000	0.144756×10^{6}	1.000000
3	0.912482×10^{4}	−0.187896	0.170530×10^{5}	0.462386
	0.928049×10^{3}	−0.552046	0.129597×10^{4}	0.487012
	0.305059×10^{2}	−1.000000	0.118313×10^{3}	1.000000
4	0.253196×10^{1}	−0.08596	0.161403×10^{2}	0.453691
	0.120361×10^{0}	0.643077	0.183581×10^{1}	0.613625
	0.121074×10^{0}	1.000000	0.308486×10^{0}	1.000000

TABLE 1.4. Linear Coefficients and Orbital Exponents of the Gaussian Expansion of the Numerical Output for the Radial Part of the $2p_{1/2}$ Relativistic Carbon AO

	Large component		Small component	
Region	Exponent	Coefficient	Exponent	Coefficient
1	0.174033×10^{11}	0.001762	0.131826×10^{11}	0.003274
	0.173246×10^{10}	0.018436	0.133796×10^{10}	0.078761
	0.191477×10^{9}	1.000000	0.149506×10^{9}	1.000000
2	0.184518×10^{8}	0.016723	0.149050×10^{8}	0.001416
	0.170380×10^{7}	0.158463	0.146798×10^{7}	0.326620
	0.166997×10^{6}	1.000000	0.158568×10^{6}	1.000000
3	0.121435×10^{5}	0.115279	0.136523×10^{5}	0.146574
	0.988808×10^{3}	0.464636	0.144676×10^{4}	0.270990
	0.501331×10^{2}	1.000000	0.128094×10^{3}	1.000000
4	0.182731×10^{1}	0.199345	0.144363×10^{2}	0.183627
	0.139551×10^{0}	0.877747	0.236021×10^{1}	0.217206
	0.887712×10^{-1}	1.000000	0.304874×10^{0}	1.000000

that the orbital exponents of the corresponding large and small components (in contrast to the H atom problem[(42)]) are not equal (see Tables 1.2–1.4). This is obviously caused by the deviation of the simple central Coulomb potential in the H atom from the effective Hartree–Fock potential in the case of atoms with more than one electron.

The relativistic orbitals of other atoms can most probably be expanded in Gaussians in a similar way (thus maintaining a low number of Gaussians for a given region of r), so it should not be very difficult to develop *ab initio* relativistic SCF LCAO programs (at least in the closed-shell case[(45)]) for molecules with arbitrary symmetry and for crystals. The investigation of these problems, especially from the viewpoint of an optimal choice of orbital exponents (with the help of a nonlinear fitting program) and of the contraction coefficients, as well as the development of the corresponding programs, is being carried out by different research groups.

Since the eigenvalues of $\hat{H}_{\text{rel}}$ in equation (1.109) are not bounded from below, in the course of analytical solutions of the relativistic Hartree–Fock equations one usually obtains an admixture of *negative energy states* in addition to the states with positive energies

$$\hat{F}\,|\phi_i^+\rangle = \varepsilon_i\,|\phi_i^+\rangle$$
$$|\phi_i^+\rangle = \sum_p c_{p,i}\,|\varphi_p\rangle + \sum_n c_{n,i}\,|\varphi_n\rangle \tag{1.129}$$

where $|\varphi_p\rangle$ are the positive and $|\varphi_n\rangle$ the negative energy four-component eigenspinors which satisfy the HFD equation. The admixture of these negative-energy states can reduce the lowest positive-energy level well below the "true" ground state, a situation frequently observed in atomic or molecular calculations.[45]

This is not the case in numerical HFD calculations,[30,46,47] where the boundary conditions of the eigenspinors effectively eliminate the admixture of the negative-energy states, and thus variational instability does not arise.

To avoid this difficulty one can introduce in the spinors $\langle\phi_i^+|$ the further constraint

$$\langle\phi_i^+|\varphi_n\rangle = 0 \qquad \text{for all } i \text{ and } n \tag{1.130}$$

After some algebra[48] this leads to the expression

$$\hat{P}\hat{F}\hat{P}\,|\phi_i^+\rangle = \varepsilon_i^+\,|\phi_i^+\rangle \tag{1.131}$$

where the projector operator $\hat{P}$ is defined as

$$\hat{P} = \sum_p |\varphi_p^{\mathrm{D}}\rangle\langle\varphi_p^{\mathrm{D}}| \tag{1.132}$$

Several authors (such as Datta[49]) have suggested that in equation (1.132), spinors satisfying the Dirac equation in the case of the hydrogen atom should be replaced by spinors which are eigenfunctions of the DHF equations. This procedure, however, would again admix to the positive-energy solutions also negative-energy states (due to their occurrence in the Coulomb and exchange operators in $\hat{F}_{\mathrm{rel}}$[50]).

A possible good compromise would be to start with the form of equation (1.132) for $\hat{P}$ and introduce at the beginning also Dirac spinors φ_p^{D} in the form of an appropriate linear combination of Gaussians in the expressions [see equation (1.113)] for the Coulomb and exchange operators, $\hat{J}(\mathbf{p}, h)$ and $\hat{K}(\mathbf{p}, h)$, respectively. (More precisely, in the case of a molecule or a solid one should start both in the case of $\hat{P}$ and in the cases of $\hat{J}$ and $\hat{K}$ with a linear combination of Dirac spinors.) This would secure that, at the first iteration, one would obtain purely positive-energy solutions. The next step would be to reconstruct both the operators $\hat{J}$ and $\hat{K}$ and also the projection operator $\hat{P}$ with the help of the approximate (but purely positive-energy) $|\phi_i^{HF+}\rangle$ solutions, and the iterations (including the iteration of $\hat{P}$) can be repeated until self-consistency.

This problem of course requires further, more detailed research. First, with the help of the $\delta^2 = 0$ criterion of Thouless and of Čižek and Paldus,[51] the stability of the purely positive-energy solutions reached in this way should be carefully analyzed. Further, the role of the size of the basis sets that influences also the quality of the operator $\hat{P}$ should be investigated.

Most recently Grant[52] has found that if one takes into account the finite size of the atomic nucleus, the postulated admixture of negative energy states $|\phi_n\rangle$ does not occur and therefore one does not have to project out the positive energy states $|\phi_p\rangle$.

REFERENCES

1. G. Del Re, J. Ladik, and G. Biczó, *Phys. Rev.* **155**, 997 (1967).
2. P.-O. Löwdin, *J. Chem. Phys.* **18**, 365 (1950).
3. J. Ladik, *Quantenchemie*, p. 124, Enke Verlag, Stuttgart (1973) (in German).
4. C. C. J. Roothaan, *Rev. Mod. Phys.* **23**, 69 (1951).
5. J.-M. André, L. Gouverneur, and G. Leroy, *Int. J. Quantum Chem.* **1**, 427, 451 (1967).
6. J.-M. André, *J. Chem. Phys.* **50**, 1536 (1969).
7. C. Pisani and R. Dovesi, *Int. J. Quantum Chem.* **17**, 501 (1980); R. Dovesi, C. Pisani, and C. Rovetti, *Int. J. Quantum Chem.* **17**, 517 (1980).
8. A. Blumen and C. Merkel, *Phys. Status Solidi B* **83**, 425 (1977); J. Ladik, in: *Excited States in Quantum Chemistry* (C. A. Nicolaides and D. R. Beck, eds.), p. 495, D. Reidel Publ. Co., Dordrecht–Boston–New York (1979).
9. See, for instance, M. Hammermesh, *Group Theory and its Application to Physical Problems*, p. 80, Addison-Wesley Publ. Co., Reading, Mass. (1964).
10. W. L. McCubbin, in: *Electronic Structure of Polymers and Molecular Crystals* (J.-M. André and J. Ladik, eds.), p. 171, Plenum Press, London–New York (1975).
11. I. I. Ukrainski, *Theor. Chem. Acta (Berlin)* **38**, 139 (1975).
12. S. Suhai, *J. Chem. Phys.* (unpublished result).
13. R. S. Mulliken, *J. Chem. Phys.* **23**, 1833 (1955).
14. J. B. Binkley, R. A. Whiteside, P. C. Hariharan, R. Seeger, and J. A. Pople, QCPE No. 368, Indiana University, Bloomington, Indiana (1976).
15. L. Piela, J.-M. André, J. G. Fripiat, and J. Delhalle, *Chem. Phys. Lett.* **77**, 143 (1981); H. J. Monkhorst and M. Kertész, *Phys. Rev.* **21**, 3105 (1981).
16. S. Suhai, P. S. Bagus, and J. Ladik, *Chem. Phys.* **68**, 467 (1982).
17. S. F. O'Shea and D. P. Santry, *Chem. Phys. Lett.* **25**, 164 (1974).
18. L. Piela and J. Delhalle, *Int. J. Quantum Chem.* **13**, 605 (1978); L. Piela, in: *Recent Advances in the Quantum Theory of Polymers* (J.-M. André, J.-L. Brédas, J. Delhalle, J. Ladik, G. Leroy, and C. Moser, eds.), Lecture Notes in Physics No. 113,

p. 104, Springer-Verlag, Berlin–Heidelberg–New York (1980); J.-L. Brédas, J.-M. André, and J. Delhalle, *Chem. Phys.* **45**, 109 (1980).
19. P.-O. Löwdin, *Phys. Rev.* **97**, 1474, 1490, 1509 (1955).
20. F. Martino and J. Ladik, *J. Chem. Phys.* **52**, 2262 (1970); I. Meyer, J. Ladik, and G. Biczó, *Int. J. Quantum Chem.* **7**, 583 (1973).
21. F. Martino and J. Ladik, *Phys. Rev. A* **3**, 862 (1971).
22. J. Ladik, in: *Electronic Structure of Polymers and Molecular Crystals* (J.-M. André and J. Ladik, eds.), p. 23, Plenum Press, London–New York (1975).
23. K. F. Berggreen and F. Martino, *Phys. Rev.* **184**, 484 (1969); F. E. Harris and H. J. Monkhorst, *Chem. Phys. Lett.* **23**, 1026 (1969).
24. F. Martino and J. Ladik (unpublished).
25. N. F. Mott and E. A. Davis, *Electronic Process in Non-Crystalline Materials*, Ch. V, p. 121, Clarendon Press, Oxford (1971).
26. J. Ladik and P. Otto, *Chem. Phys. Lett.* **31**, 83 (1975).
27. A. Bartha and J. Ladik (unpublished).
28. J. L. Calais and G. Sperber, *Int. J. Quantum Chem.* **7**, 501 (1973).
29. J. Ladik (unpublished).
30. See, for instance, A. Rosen and I. Lindgren, *Phys. Rev.* **176**, 114 (1968); I. P. Grant, *Adv. Phys.* **19**, 747 (1970); J. P. Desclaux, *The Status of Relativistic Calculations for Atoms and Molecules*, Physica Scripta, Proceedings of the Nobel Symposium on Many Body Theory of Atomic Systems, Göteborg (1979) and references cited therein: I. P. Grant, in: *Relativistic Effects in Atoms, Molecules and Solids* (G. L. Malli, ed.), pp. 73, 89, 101, Plenum Press, London–New York (1983).
31. J. Ladik, *Acta Acad. Sci. Hung.* **10**, 271 (1959); J. Ladik, *Acta Acad. Sci. Hung.* **13**, 123 (1961); W. C. Mackrodt, *Mol. Phys.* **18**, 697 (1970); J. P. Desclaux and P. Pyykkö, *Chem. Phys. Lett.* **29**, 534 (1974); J. P. Desclaux, *Chem. Phys.* **34**, 261 (1978) and references cited therein; as well as papers in the book cited at the end of Reference 30.
32. See, for instance, T. L. Loucks, *Augmented Plane Wave Method*, Benjamin Inc., New York (1967); D. D. Koelling and A. J. Freeman, in: *Plutonium and Other Actinides* (J. Blank and R. Lindner, eds.), p. 291, North-Holland Publ. Co., Amsterdam (1976); as well as papers in the book cited at the end of Reference 30.
33. See, for instance, T. P. Das, *Relativistic Quantum Mechanics of Electrons*, Harper and Row, New York (1973).
34. G. Breit, *Phys. Rev.* **34**, 553 (1929); H. Bethe and E. E. Salpeter, in: *Encyclopedia of Physics* (S. Flügge, ed.), Vol. 355, p. 140, Springer-Verlag, Berlin–New York–Heidelberg (1957).
35. J. B. Mann and W. R. Johnson, *Phys. Rev. A* **4**, 41 (1971).
36. J. P. Desclaux and P. Pyykkö, *Chem. Phys. Lett.* **39**, 300 (1976); P. Pyykkö and J. P. Desclaux, *Chem. Phys. Lett.* **43**, 545 (1976); P. Pyykkö and J. P. Desclaux, *Chem. Phys. Lett.* **50**, 503 (1977).
37. G. Malli and J. Oreg, *J. Chem. Phys.* **63**, 830 (1975).
38. A. Rosen, *Int. J. Quantum Chem.* **13**, 509 (1978).
39. J. Oreg and G. Malli, *J. Chem. Phys.* **61**, 4349 (1975); J. Oreg and G. Malli, *J. Chem. Phys.* **65**, 1755 (1976); H. T. Toivoren and P. Pyykkö, *Int. J. Quantum Chem.* **11**, 695 (1977).
40. B. Aoyama, H. Yamakowa, and O. Matsuoka, *J. Chem. Phys.* **73**, 1329 (1980).
41. See, for instance, T. C. Collins, in: *Electronic Structure of Polymers and Molecular Crystals* (J. M. André and J. Ladik, eds.), p. 389, Plenum Press, London–New York (1975).

42. See, for instance, N. F. Mott and G. Sneddon, *Quantum Mechanics and its Applications*, Chapt. XI, Clarendon Press, Oxford (1948).
43. J. Ladik, J. Čížek, and P. K. Mukherjee, in: *Relativistic Effects in Atoms, Molecules and Solids* (G. L. Malli, ed.), p. 305, Plenum Press, London–New York (1983).
44. H. Ruder, S. Suhai, and J. Ladik (unpublished).
45. Y. K. Kim, *Phys. Rev.* **154**, 17 (1967); T. Kagawa, *Phys, Rev. A* **12**, 2245 (1975); G. Malli, *Chem. Phys. Lett.* **68**, 529 (1979); V. Bonifrei and S. Huzinaga, *J. Chem. Phys.* **60**, 2779 (1974).
46. S. J. Rose, I. P. Grant, and N. C. Pyper, *J. Phys. B* **11**, 1711 (1978).
47. P. Pyykkö, in: *Adv. Quant. Chem.* (P.-O. Löwdin, ed.), p. 353, Academic Press, London–New York (1978).
48. E. R. Davidson, *Chem. Phys. Lett.* **21**, 565 (1973); Y. Ishikawa and G. Malli, *Chem. Phys. Lett.* **74**, 568 (1978).
49. S. N. Datta, *Chem. Phys. Lett.* **74**, 568 (1980).
50. G. E. Brown and O. G. Ravenhall, *Proc. R. Soc. London, Ser. A* **208**, 552 (1951); M. H. Mittleman, *Phys. Rev. A* **4**, 893 (1971).
51. D. J. Thouless, *The Quantum Mechanics of Many Body Systems*, Academic Press, New York (1961); J. Čížek and J. Paldus, *J. Chem. Phys.* **47**, 3976 (1967); J. Paldus and J. Čížek, *Phys. Rev. A* **2**, 2268 (1970).
52. I. P. Grant (personal communication).

Chapter 2

Examples of *Ab Initio* Calculations on Quasi-One-Dimensional Polymers

This chapter presents some examples of the application of the *ab initio* crystal-orbital method described in Chapter 1. Though these applications range from the field of plastics (polyethylene and its fluoro derivatives) through highly conducting polymers [polyacetylenes and polydiacetylenes, $(SN)_x$, TCNQ and TTF stacks] to biopolymers (homopolynucleotides and homopolypeptides), they are only illustrative. No attempt has been made to review the numerous other applications performed by the Namur group and by other researchers, as this would increase unduly the size of this book.

2.1. SOME POLYMERS USED FOR THE PRODUCTION OF PLASTICS: POLYETHYLENE AND ITS FLUORO DERIVATIVES

2.1.1. The Energy-Band Structure of Polyethylene

After a number of semiempirical (extended Hückel,[1] CNDO/2,[2] INDO,[3] and MINDO[3]) calculations* in 1970, André[5] performed the first *ab initio* band-structure computation of an infinite polyethylene chain. Since this polymer is mass-produced in the plastics industry, the theoretical study of its properties (especially of its mechanical properties, see Chapter 10) is of great practical interest, which explains the large

* We give as reference for the semiempirical calculations only the first paper dealing with each method. For more detailed references see Table II in the review paper of André.[4]

number of subsequent *ab initio* calculations (for references see Table II of André[4]).

In a recent paper, Suhai[6] performed detailed calculations on this chain using the minimal STO-3G[7] as well as the split valence 6-31G[8] and the latter basis with polarization functions (a set of *d* functions on C and a set of *p* functions on H).[9] He optimized the geometry of the polyethylene chain in the case of the three basis sets.[6] He also found, in agreement with previous authors,[4,5] that the chain is planar, has a zigzag structure, and the quality of the basis set left the internal geometry of the CH_2 group practically unchanged but considerably influenced the other geometrical parameters. The same is true for the case when the C–C bond distance was stretched by applying different values of $\varepsilon = \Delta l / l_{equ}$, where Δl is the change in bond length with equilibrium value l_{equ}. The calculations aimed at determining the Young modulus theoretically (see Chapter 10) and were performed for $\varepsilon = 0.02, 0.04, 0.06, 0.10, 0.15$, and 0.20. The STO/3G geometry of the CH_2 group was maintained for all elongations and basis-set parameters but $R_{CC_{equ}}$ and the CCC bond angle was reoptimized for all cases (basis sets and elongations).[6]

The infinite lattice sums must be truncated in such a way that both the balance between the repulsive and attractive terms and the translational symmetry are preserved.[10] As the elementary cell of polyethylene (CH_2–CH_2) has no dipole moment, the purely electrostatic interactions converge very rapidly (the leading term is the quadrupole–quadrupole interaction) and the long-range tail of the lattice sums is determined by the exchange potential. However, an early cutoff of this contribution leads, even in an otherwise corrrect truncation scheme, to numerical instabilities.[11] In fact, depending on the basis set applied, no physically reasonably Hartree–Fock solutions can be obtained for polyethylene (PE) until the fourth or fifth interacting neighbors are included and, for a proper convergence, eight neighbors were needed in these calculations.

Table 2.1 gives some ground-state properties of polyethylene (in its equilibrium geometry) obtained by Suhai[6] with the aid of his best (6-31G**) basis and by optimizing the geometry. This table shows that, though the width of the valence band agrees quite well with experiment, all other quantities (especially the theoretical gap, which is 4.5 eV larger than the experimental value) exhibit rather poor agreement. One should not forget, however, that in this calculation no correlation effects, which in the case of semiempirical methods are hidden by the choice of approximation of the integrals and parameters, were taken into account. In Chapter 5, after discussing how to correct a Hartree–Fock band structure for the major part of correlation using a generalized electronic polaron model, we shall see that this corrected (quasi-particle) band structure of polyethylene agrees much better with experimental data.

TABLE 2.1. Different Theoretical[6] and Experimental Quantities Characterizing the Electronic Structure of Polyethylene in Its Ground State

	Hartree–Fock[a]	Experimental[b]
Ionization potential[c]	10.2	7.6–8.8
Valence bandwidth	6.2	6.0
Electron affinity[c]	3.2	0.6–1.2
Conduction bandwidth	4.3	—
Fundamental gap	13.4	8.8

[a] Obtained with a 6-31G** basis set.
[b] See the paper of Less and Wilson.[12]
[c] Applying Koopmans' theorem.[13]

2.1.2. Band Structures of Fluorinated Polyethylenes

After a number of semiempirical crystal-orbital (CO) calculations (extended Hückel CO[14] and CNDO/2 CO calculations[15]; for further references see Table III of André[4]) Otto *et al.*[16] performed the first *ab initio* band-structure calculations on different fluorinated polyethylenes. The calculations were conducted for all six different polyfluoroethylenes, namely $(CFH–CH_2)_x$, $(CF_2–CH_2)_x$, $(CFH–CFH)_{x\,anti}$, $(CFH–CFH)_{syn}$, $(CF_2–CFH)_x$, and $(CF_2)_x$, which can be obtained from polyethylene through fluorine substitution.

For this purpose, a new computer program for the case of a linear periodic macromolecule has been written on the basis of the one- and two-electron integral packages of the molecular program IBMOL, which also takes into account the necessary rotation of the basis functions in the case of a combined symmetry operation in the polymer (translation along the helix axis and rotation around it). In several cases different-neighbor interactions have been included together with a proper electrostatically balanced cutoff[10] to investigate the convergence of the total energy per unit cell with respect to the number of neighbor interactions explicitly taken into account.

To obtain information on how the electronic and energetic properties of the macromolecules depend on the choice of basis set, the band-structure calculations were performed using four different atomic basis sets. On the one hand, the minimal basis sets STO-3G[7] and Clementi's basis set $7s3p$ (denoted by C-$7s3p$) were used, and, on the other, the valence-split set 4-31G (contracted to a $3s2p$ basis[8]) and Clementi's double-zeta basis set $10s/5p$ (denoted by C-$10s5p$) contracted to a $4s2p$ basis.

The values 1.540 Å, 1.375 Å, and 1.09 Å, respectively, were selected for the bond lengths of the C–C, C–F, and C–H bonds, assuming tetrahedral bond angles. Optimization of the bond lengths in the case of the basis set C-7s3p leads to unreasonably large values for the C–F bond.

Table 2.2 presents theoretical results for poly(ethylene) and all its fluoro derivatives that can be formed by replacing step-by-step the hydrogen atoms by fluorine atoms. A planar zigzag configuration of the carbon backbone chains has been assumed for these calculations. In the case of poly(1,2-difluorethylene) we can distinguish between two isomers. In the so-called syn form all the fluorine atoms are placed on the same side of a plane, which is defined by the carbon atoms, while in the anti-isomer they are alternating. Table 2.2 lists, for the different applied basis sets, the upper and lower limits and the width of the valence and conduction bands. The last column contains the Hartree–Fock gap.

It may be more interesting to deduce some general trends from Table 2.2 without discussing the individual numbers in too much detail. Comparing the results obtained with the basis set C-7s3p, we observe a decrease in the width of the valence band ranging from 0.5 eV for polytetrafluoroethylene (PTFE) up to 1.5 eV for syn-poly (1.2-difluoroethylene) with respect to poly(ethylene). This is due to the shift of the lower band limit to higher energies in the case of the second compound and to a shift of the upper limit in the opposite direction. In PTFE, however, both bounds are shifted remarkably to more negative energies. The energy of the conduction band in the fluorinated compounds is decreased, compared with polyethylene, but the changes in bandwidth do not show a regular behavior. The last quantity is smaller for all polymers, compared with polyethylene, but the changes in bandwidth do not show a regular behavior. The last quantity is smaller for all polymers, except for PTFE and antipoly(1,2-difluoroethylene). Again, in the case of PTFE the changes are more dramatic than for the partially fluorinated polymers (energy shift of the conduction band about 6.7 eV compared with polyethylene). The gap (22.317 eV for PE) is gradually reduced by increasing the number of fluorine atoms to 17.806 eV for PTFE.

The results (Table 2.2) using the basis set C-10s5p show that the valence and conduction bands are shifted to somewhat higher and considerably lower energies, respectively, in comparison with the minimal basis-set calculations. The widths of the valence bands increase by about 1 eV, while the same quantity for the conduction band shows no regular behavior. The gap is now reduced by about 4–7 eV. Again, PTFE deviates from these trends, showing almost no change in value for the gap, compared with the C-7s3p result.

It is noteworthy that the position of the bands is changed by a large shift toward higher energies in the case of the basis-set STO-3G basis.

TABLE 2.2. Band Characteristics for Polyethylene and Its Six Fluoro Derivates Using Different Basis Sets[16]

System		Valence band			Conduction band			
Poly-	Basis set	$\mathcal{E}_{min}$ (eV)	$\mathcal{E}_{max}$ (eV)	Width (eV)	$\mathcal{E}_{min}$ (eV)	$\mathcal{E}_{max}$ (eV)	Width (eV)	Gap (eV)
(CHH)	STO-3G	− 14.052	− 9.703	4.349	14.248	18.779	4.531	23.951
(CHH)	C-7*s*3*p*	− 16.094	− 12.775	3.319	9.542	14.614	5.072	22.317
(CHH)	4-31G	− 14.286	− 12.524	1.762	4.612	10.306	5.694	17.136
(CHH)	C-10*s*5*p*	− 16.606	− 11.116[a]	5.490[b]	2.809[c]	6.609	3.800[d]	13.925[e]
(CFH-CHH)	C-7*s*3*p*	− 14.944	− 12.936	2.008	7.237	11.642	4.405	20.173
(CFH-CHH)	C-10*s*5*p*	− 12.681	− 9.565	3.116	3.608	7.543	3.935	13.172
(CFF-CHH)	C-7*s*3*p*	− 15.868	− 13.540	2.328	5.945	9.780	3.935	19.485
(CFF-CHH)	C-10*s*5*p*	− 14.187	− 11.074	3.113	3.458	8.434	4.974	14.532
(CFH-CHF) anti	C-7*s*3*p*	− 15.742	− 13.402	2.340	5.037	10.727	5.690	18.439
(CFH-CHF) anti	C-10*s*5*p*	− 14.589	− 10.964	3.625	3.050	7.039	3.989	14.015
(CFH-CHF) syn	C-7*s*3*p*	− 15.228	− 13.357	1.871	6.266	10.561	4.295	19.623
(CFF-CFH)	C-7*s*3*p*	− 15.971	− 14.034	1.937	4.187	8.908	4.721	18.221
(CFF)	STO-3G	− 13.277	− 10.309	2.968	8.812	13.573	4.761	19.121
(CFF)	C-7*s*3*p*	− 17.854	− 14.968	2.889	2.838	8.358	5.520	17.806
(CFF)	4-31G	− 17.035	− 13.472	3.563	3.849	9.465	5.616	17.321
(CFF)	C-10*s*5*p*	− 18.372	− 14.676	3.696	2.426	7.967	5.541	17.102

[a] The corresponding value obtained with the 6-31G** basis and geometry optimization is − 10.2 eV; see Table 2.1 and work of Suhai.[6]
[b] With 6-31G** basis, 6.2 eV.
[c] With 6-31G** basis, 3.2 eV.
[d] With 6-31G** basis, 4.3 eV.
[e] With 6-31G** basis, 13.4 eV.

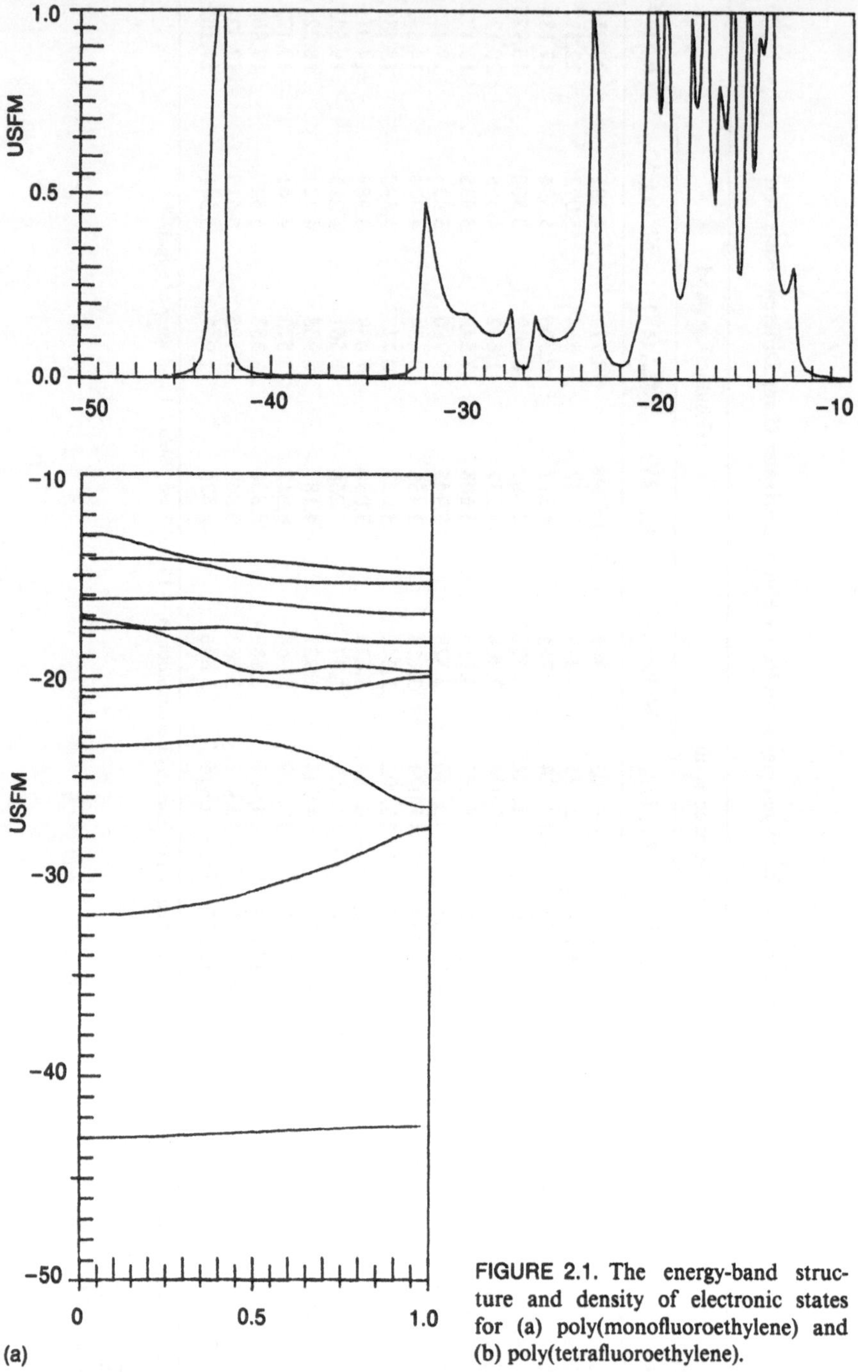

FIGURE 2.1. The energy-band structure and density of electronic states for (a) poly(monofluoroethylene) and (b) poly(tetrafluoroethylene).

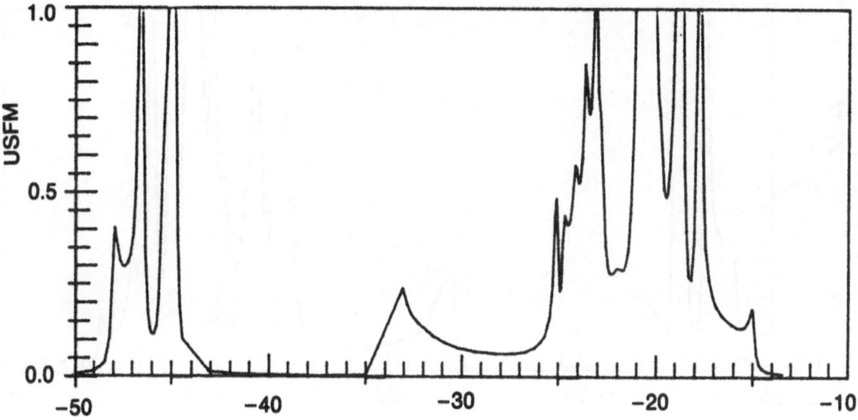

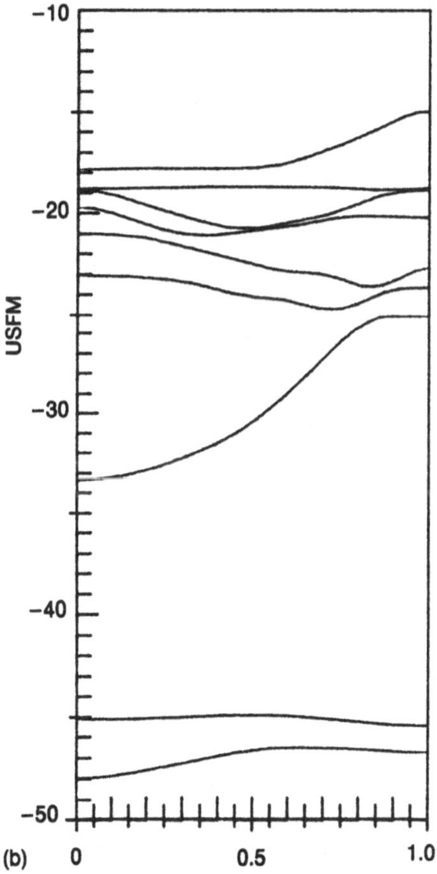

FIGURE 2.1 (*continued*)

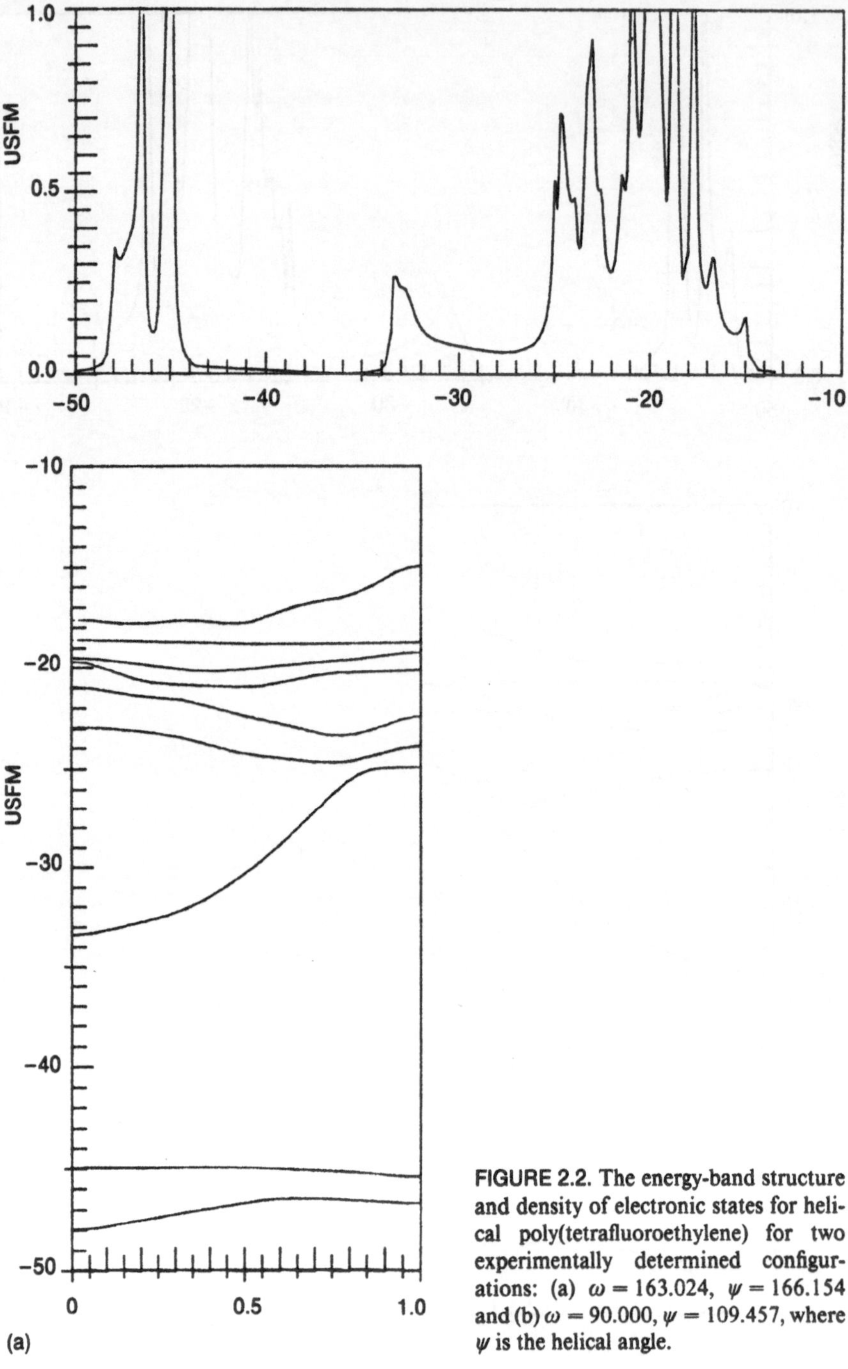

FIGURE 2.2. The energy-band structure and density of electronic states for helical poly(tetrafluoroethylene) for two experimentally determined configurations: (a) $\omega = 163.024$, $\psi = 166.154$ and (b) $\omega = 90.000$, $\psi = 109.457$, where ψ is the helical angle.

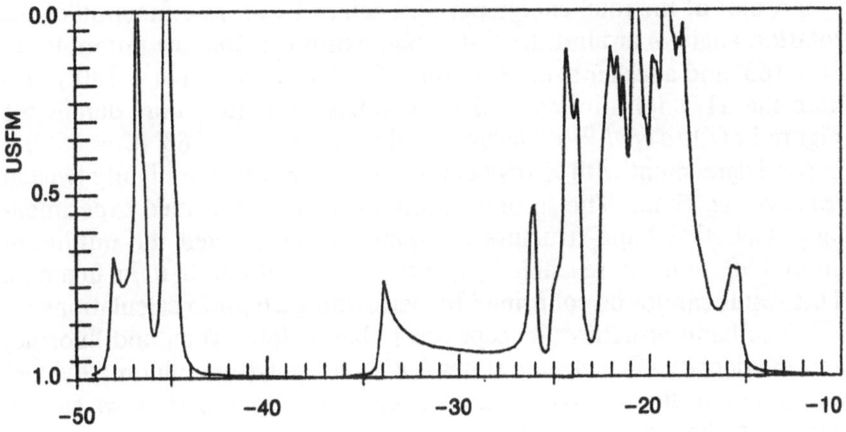

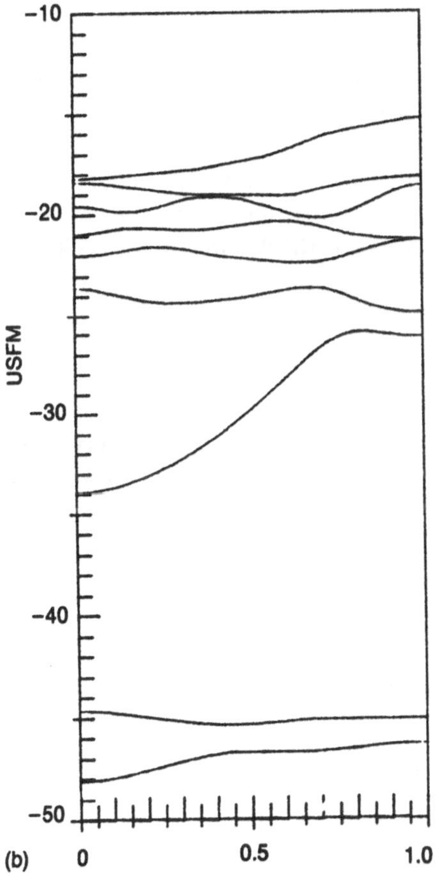

(b) FIGURE 2.2 (*continued*)

A plot of the total energy per unit cell of PTFE as a function of the rotation angle ω around the C–C bond exhibits a shallow minimum at $\omega = 163°$ and another at $\omega = 90°$ ($\omega = 0°$ corresponds to $\omega = 180°$) with both the STO-3G minimal and C-$7s3p$ basis sets (for more details see Figure 1 of Otto *et al.*[16]). The first minimum near $\omega = 163°$ ($\psi = 166°$) is in good agreement with X-ray results, corresponding to a slightly twisted *trans* zigzag chain. The second minimum is not confirmed experimentally. CNDO/2 band-structure calculations[15] predicted the minimum around 90° to be more stable by about -0.2 kcal/mol than the first one. This result cannot be confirmed by performing *ab initio* calculations.

The band structures (except the $1s$ bands for carbon and fluorine) and the density of electronic states (DOS) for poly(monofluoroethylene) and poly(tetrafluoroethylene) are shown in Figure 2.1. Here we should note that the density of states at a given energy $\mathscr{E}_0$ in a quasi-one-dimensional polymer is defined as the number of states in the band per unit energy interval enclosing $\mathscr{E}_0$,

$$D(\mathscr{E}_0) = \left(\frac{dN(\mathscr{E})}{d\mathscr{E}}\right)_{\mathscr{E}=\mathscr{E}_0} \tag{2.1}$$

For each nondegenerate band there are two electrons with opposite spin for each allowed k value. We have seen in Chapter 1 that $-\pi/a < k < \pi/a$, so if the number of states in k space is N (the number of unit cells) we can write

$$dN = \frac{1}{\pi}\, dk \tag{2.2}$$

(The factor 2 in the denominator disappears because we have two electrons on each level belonging to a certain vector **k**.) Substitution of expression (2.2) into equation (2.1) yields

$$D(\mathscr{E}_0) = \frac{1}{\pi}\left(\frac{dk}{d\mathscr{E}(k)}\right)_{\mathscr{E}=\mathscr{E}_0} = \frac{1}{\pi}\left(\frac{d\varepsilon(k)}{dk}\right)^{-1}_{\mathscr{E}=\mathscr{E}_0} \tag{2.3}$$

In the general case, if we have many bands then we can write

$$D(\mathscr{E}_0) = \frac{1}{\pi}\sum_{h=1}^{n^*}\left(\frac{d\mathscr{E}_h(k)}{dk}\right)^{-1}_{\mathscr{E}=\mathscr{E}_0} \tag{2.4}$$

where n^* is again the number of partially or completely filled bands.

The numerical calculation of the DOSs, especially if at certain points (such as the band edges of one-dimensional systems) $d\mathscr{E}_h(k)/dk = 0$, is not always trivial and sometimes requires rather sophisticated techniques. The details can be found in Delhalle's paper.[17]

Returning to Figure 2.1, one should note that these curves are based on results computed with the help of the C-$7s3p$ minimal basis set. In Figure 2.2, the same information is presented for those two PTFE helices (again using the C-$7s3p$ basis set) for which we found a minimum of the total energy per unit cell as a function of ω ($\omega = 90°$, $\omega = 163°$).

The experimental ESCA spectrum of PTFE[18] possesses four main peaks, three of intermediate intensity centered at approximately 20, 25, and 31 eV (peaks A, B, C, respectively), and one of high intensity at 41 eV (peak D). Since we have not included the contributions of the different atomic orbitals (transition probabilities) to the total transition cross-sections to establish the theoretical ESCA curves, we are not yet able to discuss the relative intensities of the peaks, but only their positions.

The increase in the number of fluorine atoms is correlated with a shift of about 1 eV per F atom to lower energies for the peak D, while the other peaks do not change appreciably. The spectrum predicted from the calculated band structure for PTFE is in good agreement with experiment, underlining once more that calculations of periodic polymers are quite successful in interpreting ESCA spectra. This agreement will be more comprehensive when allowance is made for the dependence of the ionization cross section of the crystal orbitals on their atomic constituents (intensity calculations), and correlation effects are also taken into account. Furthermore, better-resolved experimental spectra are probably also needed to check the detailed theoretical information obtained for relatively more complex polymers.

The DOS results for the two PTFE helices shown in Figure 2.2a and b are very similar to the *trans*-PTFE spectra with respect to the energetic position of the four peaks.

2.2. HIGHLY CONDUCTING POLYMERS: $(CH)_x$, $(SN)_x$, TNCQ AND TTF STACKS

2.2.1. Hartree–Fock Calculations on *cis*- and *trans*-Polyacetylenes (Polyenes)

The electronic structure of polyacetylenes [PA; called also polyenes, $(CH)_x$, $(C_2H_2)_x$] has been a subject of interest for several decades in theoretical chemistry owing to the central role of the polyene backbone in various organic compounds. These investigations have been considerably stimulated in recent years by the fascinating solid-state physical properties of different doped polyacetylene crystals. It has been shown that by doping with electron acceptors (bromine, iodine, AsF_5) or donors (lithium, sodium) the electrical properties can be varied over a wide range.[18,19] The specific conductivity of films of $(CH)_x$ varies, for instance, over twelve orders of magnitude from insulator ($\sigma \sim 10^{-9}\ \Omega^{-1}\ cm^{-1}$) to highly conducting metallic polymers ($\sigma \sim 10^3\ \Omega^{-1}\ cm^{-1}$).[20–22] Furthermore, compensation and junction formation have been demonstrated on various n- and p-type samples.[23] Partial orientation of the polymer fibers results in highly anisotropic electrical[21] and optical properties.[24] Optical-absorption studies suggest for most cases a direct band-gap semiconductor with a very anisotropic band structure. For certain dopant concentrations in AsF_5-doped polyacetylene, however, some qualitative changes in the electrical and optical properties[25,26] indicate a semiconductor-to-metal transition with conductivities in excess of 2000 $\Omega^{-1}\ cm^{-1}$.

There are two possible isomers of pure polyacetylene (the *cis* and *trans* forms) that show characteristic differences in their physical and chemical properties. The *cis*-isomer is thermodynamically unstable in pure form. Its doped form shows, however, electrical conductivities that exceed in some cases those of the *trans*-modification.[19] Early experimental observations suggested a bond-length-alternating molecular structure for *trans*-polyene[27] and this model was used also as a possible explanation of the observed energy gap in long polyenes.[28] In fact, for the *cis*-polyene there are two further possible structures with bond-length alternation, the *cis*-transoid and *trans*-cisoid forms. Since highly oriented polymer samples of pure polyacetylenes are still unavailable, it has not yet been possible to directly determine the structural parameters from X-ray diffraction alone.[29] Although Raman spectroscopic investigations suggest[30] a *cis*-transoid structure for the *cis*-isomer, the relative stability of the various polyene structures still presents an open problem (especially in their doped form). Only recently have Fincher *et al.*[29] been able to combine X-ray results with independent information on the basis structure of single chains to determine the dimerization distortion of PA.

Most of the theoretical work on polyenes has concentrated on the gap problem,[31–48] and these papers can be divided roughly into two groups. The authors in the first group based their investigations on a one-electron picture (band theory)[31–46] while the (smaller) second group (such as Harris and Falicov[47] and Ovchinnikov *et al.*[48]) stressed the importance of many-electron (correlation) effects. It has been shown that, in fact, the two kinds of effects can be present simultaneously.[49]

In Section 5.5, we shall return in detail to the gap problem of alternating *trans*-PA. We point out here only that the gap obtained from every *ab initio* Hartree–Fock calculation is usually too large (though it is essentially decreasing with increase in the basis set). Besides geometry optimization and correct treatment of the many-neighbor interactions in the Hartree–Fock case (see Section 1.3) it is also absolutely necessary to include correlation effects to obtain reasonable agreement with experiment.[48].

André and Leroy[39] have performed the first *ab initio* calculation for the energy-band structure of PA. This calculation was followed by the work of Kertész *et al.*[43,44] and of Karpfen and Petkov.[45] Suhai,[50] using also a minimal (STO-3G) basis, has applied the method described in Section 1.3 for the treatment of many-neighbor interactions in his calculation on five different PA models. Models I and II are the two *trans*-isomers with regular and alternating backbone structures, respectively (Figure 2.3). The three proposed *cis*-isomers, namely the regular (Model III), the *trans*-cisoid (Model IV), and the *cis*-transoid structures (Model V), are shown in Figure 2.4. The bond-length and bond-angle values of these models have been proposed by Baughman *et al.* on the basis of X-ray diffraction studies combined with a crystal-packing analysis.[29] The translational elementary cell consists of two CH units in the case of the *trans*-isomers and four CH units for the *cis*-isomers. Owing to the presence of an additional symmetry, however, the actual asymmetry unit reduces to one CH group in the case of the regular models and to one C_2H_2 group for the alternating *cis*-isomers. Accordingly, the band structures show the corresponding degeneracies at the high-symmetry points of the Brillouin zone. A full PA infinite-chain calculation including all

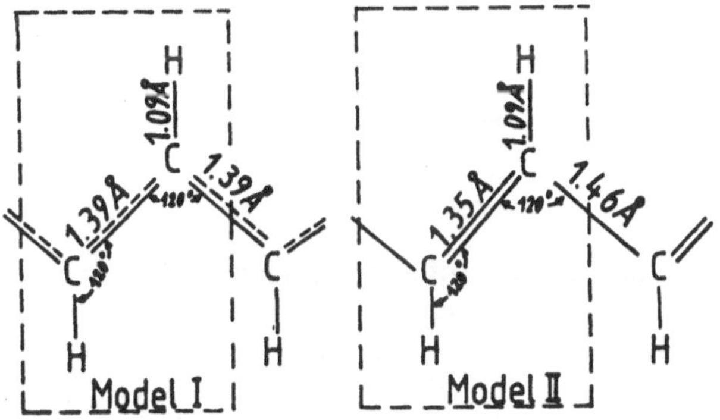

FIGURE 2.3. Regular (Model I) and alternating (Model II) *trans*-polyacetylene structures. The unit cells are surrounded by broken lines.

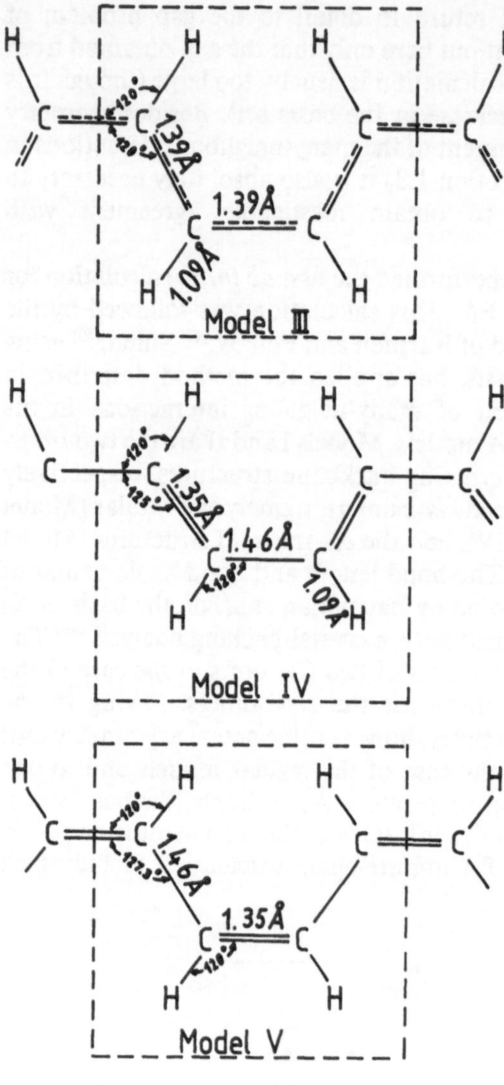

FIGURE 2.4. *cis*-polyacetylene structures: regular (Model III), *trans*-cisoid (Model IV), and *cis*-transoid (Model V). The unit cells are surrounded by broken lines.

interactions within a region of approximately 140 Å required about 20 min on a CDC Cyber 176 computer.

Subsequent CO calculations on PAs with more extended basis sets[50] have shown that even with the rather good 6-31G** basis (with the exception of the gap, of the total energy per unit cell, and of the bond alternation $\Delta R = R_1 - R_2'$; see Table 2.3 for *trans*-PA) most ground-state

TABLE 2.3. Optimized Single–Double Bond Length Difference ΔR, Total Energy per Unit Cell, and the Fundamental Gap in *trans*-PA in the Hartree–Fock Approximation Using Different Basis Sets[49]

Basis set	ΔR_{HF} (Å)	$E_{HF}{}^a$ (au)	$\Delta E_{HF}{}^b$ (eV)
STO-3G	0.16	− 75.947	8.28
4-31G	0.11	− 76.777	
6-31G	0.11	− 76.859	4.93
6-31G*	0.10	− 76.887	
6-31G**	0.10	− 76.893	4.43

[a] Total HF energy per unit cell (C_2H_2 unit).
[b] The HF gap.

properties of the PAs do not change fundamentally. In connection with Table 2.3 it should be noted that for each basis set a full geometry optimization was performed and the trend for all three quantities shown in the table becomes even stronger if one also takes correlation effects into account.[49]

Returning to the discussion of the STO-3G results, researchers found[50] that among the PAs, the alternating *trans*-PA has about the same energy per C_2H_2 unit as the *cis*-transoid and *trans*-cisoid structures and these structures have very much the same energy. The energy differences between these structures (~ 0.1 and ~ 2.5 kcal/mol, respectively) are too small to allow one to predict theoretically the most stable structure without much more sophisticated calculations (good basis and correlation with geometry optimization) and without taking interchain interactions into account. On the other hand, these three structures are in agreement with experiment — considerably more stable than the regular *trans*- and *cis*-PA structures.

If one examines the dependence of the different physical properties on the number of neighbor interactions explicitly taken into account, one finds that the position of the valence and conduction bands (or, if one recalls Koopmans' theorem, the ionization potential and electron affinity needed for the interpretation of ESCA spectra), their widths (which have values between 6 and 10 eV), as well as the gap and the charge on the carbon atom become practically constant after a few (5–7) neighbor interactions. On the other hand, this is not the case for the total energy per unit cell where one must account for a larger number of neighbors (the reasons for this were discussed in Section 1.3). For further details we refer the reader to the paper of Suhai.[54]

TABLE 2.4. Total Energy per C_2H_2 Unit (E_{tot}), Relative Energy Measured from the Equidistant Symmetry-Adapted State (ΔE), the Single-Particle Gap (ΔE_g), and Off-Diagonal π-Electron Density Matrix (Bond Order) Elements ($P_{\pi,\pi}$) Obtained for Various HF Solutions of *trans*-PA[51]

HF solution	$E_{tot}(C_2H_2)$ (au)	ΔE (kcal/mol)	ΔE_g (eV)	$P^0_{\pi,\pi}$	$P^+_{\pi,\pi}$
Equidistant symmetry-adapted (RHF)	−75.936	0	0	0.474	0.474
Equidistant off-diagonal CDW (RHF)	−75.939	−2.09	4.44	0.290	0.644
Bond alternating (RHF)	−75.947	−7.13	8.28	0.153	0.735
Equidistant SDW (UHF)	−75.966	−18.91	13.44	0.364[a]	0.364[a]

[a] Sum of the contributions of the α and β electrons.

In a later paper[51] the Peierls distortion (bond alternation) of *trans*-PA was investigated in detail both at the Hartree–Fock level and also by including correlation (see Chapter 5). The calculations were performed with an STO-3G basis (the qualitative conclusions reached did not change in the case of better basis sets) but optimizing the geometry. The calculations were executed for (1) equidistant symmetry-adapted *trans*-PA using the restricted Hartree–Fock (RHF) method, (2) equidistant *trans*-PA with an off-diagonal charge-density wave using again the RHF method, (3) *trans*-PA with bond alternation using the RHF method, and finally (4) equidistant *trans*-PA with a spin-density wave (SDW) applying the unrestricted Hartree–Fock (UHF) or, in other words, different orbitals for different spin CO methods (see Section 1.4). The most important results are summarized in Table 2.4.

The table shows that the symmetry-adapted (metallic) solution of the equidistant chain turned out to be both singlet- and triplet-unstable.[52]* We found both diagonal and off-diagonal charge-density waves (CDWs) showing an atomic charge and a bond-order alternation, respectively. The diagonal CDW state, however, is still singlet-unstable (respresenting a saddle point on the energy hypersurface[53]) and easily goes over to the singlet-stable off-diagonal CDW. The bond orders $P^0_{\pi,\pi}$ and $P^+_{\pi,\pi}$ connecting the π orbitals of the atoms C_1 and C_2 in the elementary cell and those

* A molecule or linear chain is singlet unstable if there is a diagonal or off-diagonal CDW solution which has a more negative total energy than the symmetry adapted solution, but the orbitals preserve their double occupancy by two electrons with opposite spins.[52] A system has a triplet instability if there is a SDW, namely the different orbitals for different spin solutions are more negative than the RHF one, which assumes double occupancy of the energy levels.[52]

of C_2 and C_1 between neighboring cells, respectively, show substantial alternation (0.290–644), though the energy difference is only ≈ 2 kcal/mol as compared to the symmetry-adapted state.

From this picture it is obvious that the nuclear framework has to relax in order to adopt the same spatial symmetry as the density matrix of the CDW state. The subsequent optimization with different bond lengths R_1 and R_2 leads to the bond-alternating structure (Peierls distortion), which is singlet-stable and properly symmetry-adapted. We also observe that the major part of the energetic stabilization of the bond-alternating state ($\Delta E \approx 7$ kcal/mol) originates from the relaxation of the nuclear framework, since the CDW state lowers the energy only by approximately 2 kcal/mol.

If we also let the spin symmetry (UHF calculation) relax, we arrive at an SDE solution with a substantial atomic spin density of $0.878e$. The corresponding energy lowering ($\Delta E \approx 19$ kcal/mol) is, of course, on another scale than the previous ones since this state also includes correlation energy (though only a very small fraction of it). It can be seen from Table 2.4 that all symmetry-breaking solutions introduce a gap in the single-particle spectrum (ΔE_g) whose value is, however, unphysically large in the case of the SDW state (the value of about 8.3 eV obtained for the bond-alternating structure will reduce to the correct order of magnitude if basis set and correlation effects are taken into account; see Chapter 5).

The minimal basis set provides, of course, only qualitatively correct results. Therefore, both double-ζ-type extended basis sets (4-31G and 6-31G[53]) and five d-type polarization functions on carbon (6-31G*) as well as three p functions on hydrogen (6-31G**)[54] were applied. The results obtained with these three different basis sets were shown earlier in Table 2.3.

Three aspects of these results seem to be interesting: (1) the improvement of the atomic basis reduces the long bond and increases the short bond almost equally, leading to an extrapolated HF value of the bond alternation of $\Delta R \approx 0.1$ Å; (2) the energy barrier separating the semiconducting and metallic phases reduces at the same time from approximately 7 kcal/mol to approximately 3.5 kcal/mol; and (3) the lattice period does not change substantially and is in good agreement with the experimental values of 2.47–2.48 Å[55] and 2.46 Å.[56]

2.2.2. The Energy-Band Structure of $(SN)_x$

The interesting physical properties of the inorganic polymer $(SN)_x$ (polysulfur nitride) as a highly anisotropic metal at higher temperatures and as a superconductor below 0.26 K are experimentally rather well documented.[57] In spite of a considerable number of theoretical calculations,[58] however, the description of the electronic structure in this material is by no means complete. To contribute to the better understanding of the properties of $(SN)_x$ an *ab initio* SCF LCAO Hartree–Fock crystal-orbital calculation was performed for two different models of one-dimensional $(SN)_x$ chains.[59]

The method used constructs the one-electron Bloch orbitals as a linear combination of a double-ζ (split-shell) type atomic basis set, i.e., 10 and 18 contracted atomic orbitals are applied on atoms N and S, respectively. These atomic orbitals themselves are taken as contractions of Gaussian lobe functions using 25 and 46 primitive functions for atoms N and S, respectively. The exponents of the Gaussians as well as the contraction coefficients were taken from the paper of Roos *et al.*[60] Based on previous test calculations, we have taken into account interactions up to the fourth neighboring SN unit to be able to stabilize the band structures to three decimals in electron volts. All multicenter two-electron integrals, including the correct Hartree–Fock exchange, have been calculated up to the threshold value of 10^{-8} au.

Since important details concerning the true structure of $(SN)_x$ are still a matter of controversy, we performed the calculations for linear zigzag chains of sulfur nitride using both structural data sets already published. In the first configuration obtained by electron diffraction analysis[61] there is one SN unit per elementary cell along the chain with alternating SN and NS bond distances of 1.73 and 1.55 Å, respectively, having a bond angle of 113.5°. X-ray diffraction measurements, on the other hand, resulted in a more complex structure with the unit cell containing two SN molecules.[62] The bond lengths and bond angles in this model are alternately 1.63 Å, 1.59 Å and 120°, 106°, respectively. Though our calculations prefer this latter structure, since the binding energy of the polymer per SN units is 0.32 eV lower in this case, the significance of this result is considerably reduced by the fact that interchain interactions may be quite different in the two structural models. The differences in the shortest interchain atomic distances are about 0.3–0.4 Å. It is evident, therefore, that only proper multidimensional investigations may help to clarify this important problem. In this respect it should be noted that the already-performed 3D orthogonalized-plane-wave (OPW) calculation[63] is — in the author's opinion — not very relevant, as the nature of this method most probably causes the chemical details of the $(SN)_x$ system to be washed away.

The band structures obtained for the two systems show no significant differences in essential details. Since the unit cell is quasi-doubled there are twice as many bands in the second case while the Brillouin zone is half as small. The bands of this more complicated system, however, do not differ significantly from those that would be obtained by folding over the bands of the simpler chain in the middle of the zone corresponding to a unit cell doubled in size. We shall therefore discuss some physical properties of the $(SN)_x$ polymer on the basis of the band structure obtained for the more probable structure.[62]

In consequence of the internal symmetry of the unit cell each of the original 56 bands is degenerate at the zone boundary, thus resulting in 28 distinct bands of which 11 are completely filled and one is half-filled. The physically most interesting metallic band is a π band (its Bloch function is composed of the $2p_z$ AO of N and the $2p_z$ AO of S, both of π symmetry) with a width of 4.12 eV ($E_{min} = -9.85$ eV, $E_{max} = -5.734$ eV) and the Fermi level lies at $E_F = -7.821$ eV in this calculation. There is unfortunately no direct experimental evidence concerning the width of this band in $(SN)_x$. Therefore, we calculated some quantities from the band structure, which can be better related to experiment. For the effective electronic mass at the Fermi level we obtained $1.7m_e$ contrasted with the experimental value of $2m_e$ deduced by Grant *et al.*[64] from the analysis of polarized reflectivity spectra of single crystalline $(SN)_x$. The density of states is calculated to be 0.14/(eV spin molecule) at the Fermi level; this value is in quite good agreement with the value of 0.18/(eV spin molecule) found by Greene *et al.*[65] from the contribution of the linear temperature-dependent term to the specific heat.

It is worth noting here that a similar *ab initio* calculation using a minimal atomic basis set,[66] i.e., five contracted orbitals on atom N and nine on atom S, resulted in a conduction bandwidth of about 10 eV with a corresponding effective mass of $0.72m_e$ and density of states of 0.06/(eV spin molecule), both values taken at the Fermi level. Comparing the results of the two calculations, on the one hand, and the experimental values, on the other, it is clear that, in accordance with general experience, the properties of the conduction band are strongly dependent on the size of the basis set applied.

We note finally that according to a Mulliken-type population analysis[67] of the wave function, a charge of $0.402e$ is transferred in the ground state from S to N, which observation seems to agree with those of Mengel *et al.*[68] who concluded from their X-ray photoemission (XPS) measurements that an extra charge of 0.3–$0.4e$ is present on the nitrogen.

In this connection we think that the rather good agreement between the experimental and theoretical values obtained for the effective mass, density of states, and charge distribution by our double-ζ *ab initio* calculation may encourage the use of this method for the further investigation of interchain interactions in $(SN)_x$ that seems to be inevitable for a better understanding of the unusual transport properties of this material.[57] Work along these lines is in progress in our laboratory.

2.2.3. *Ab Initio* Calculation of Infinite TCNQ and TTF Stacks

It is well known that the molecular crystal built up of alternating TCNQ (7,7′, 8,8′-tetracyanoquinodimethane; see Figure 2.5a) and TTF (2,2′-bis-1,3-tetratiofulvalene; see Figure 2.5b) stacks exhibit high conductivity in the direction of the stacks.[69] Owing to its interesting physical properties, the TCNQ–TTF system has been the subject of a large number of experimental investigations,[70] which have also shown that the high conductivity is primarily due to a charge transfer of 0.6*e* between a TCNQ–TTF molecule pair, TCNQ being the electron acceptor and TTF the donor.

Ab initio SCF LCAO CO calculations on the infinite neutral TCNQ and TTF stacks were performed[71] using a TCNQ (see Figure 2.5a) or TTF (Figure 2.5b) molecule as unit cell. These *ab initio* Hartree–Fock band structures can serve as a starting point for further improvements, such as the treatment of charge transfer between the stacks and interac-

(a)

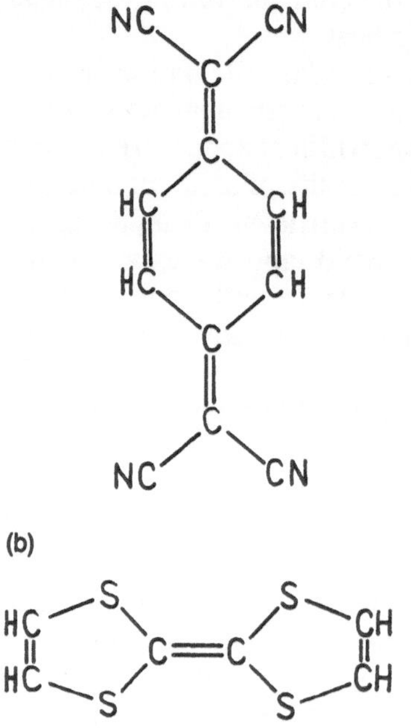

FIGURE 2.5. The chemical formulas of (a) TCNQ and (b) TTF.

tions between them, and for the calculation of differrent correlation effects. This work is in progress at our laboratory.

The calculations were performed in the second-neighbor interaction approximation and, for the stacked chains, the geometry found in the mixed TCNQ–TTF crystal(72) (3.18 Å interplane distance in the TCNQ stack and 3.47 Å in the TTF stack) was applied. The GAUSSIAN 74 program generalized to infinite chains with periodic boundary conditions was used by applying the STO-3G(73) basis set. To attain self-consistency, about 30 iteration steps were needed to fulfill simultaneously the applied SCF criteria:

$$|P(q)_{r,s}^{(n+1)} - P(q)_{r,s}^{(n)}| \leqslant 5 \times 10^{-5} \qquad (q = 0, 1, 2) \tag{2.5}$$

The numerical integrations necessary to calculate the generalized bond orders $P(q)_{r,s}$ $(q = 0, 1, 2)$ involved nine different values of k in the interval 0–π/a (according to our previous experience, nine different values of k already give a consistent band structure).

Table 2.5 presents the positions and widths of the valence and conduction bands of poly(TNCQ) and poly(TTF) obtained with the help of the above-described procedure. The most striking result in the table, as in the case of the previous semiempirical band-structure calculations,[74] is that the valence band of poly(TTF) [from which the charge transfer (CT) occurs] is comparatively broad (~ 0.3 eV) and the conduction band of poly(TCNQ) (to which the charge is transferred) is broad (~ 1.2 eV), while both the valence band of poly(TCNQ) and the conduction band of poly(TTF) (which do not take part in the CI process) have widths less than 0.1 eV. One should also note that the positions of the conduction band of poly(TCNQ) and the valence band of poly(TTF), respectively,

TABLE 2.5. Valence and Conduction Bands of Poly(TCNQ) and Poly(TTF) and Their Widths (in eV)

	$\mathscr{E}^{MO}$	$\mathscr{E}^{CO}_{min}$	$\mathscr{E}^{CO}_{max}$	$\delta\mathscr{E}$
Valence band				
Poly(TCNQ)	-6.84	$-7.25(\pi/a)^a$	$-7.16(0)$	0.10
Poly(TTF)	-3.77	$-3.80(\pi/a)$	$-3.50(0)$	0.30
Conduction band				
Poly(TCNQ)	-0.44	$0.49(0)$	$0.69(\pi/a)$	1.18
Poly(TTF)	8.47	$8.51(\pi/a)$	$8.59(0)$	0.08

[a] In parentheses the corresponding ka values.

favor still more the CT process rather than the LEMO and HOMO, respectively, of the corresponding single molecule,

$$\mathscr{E}^{\mathrm{LEMO}}_{\mathrm{TCNQ}} - \mathscr{E}^{\mathrm{HOMO}}_{\mathrm{TTF}} = 4.21 \text{ eV}$$

$$\mathscr{E}^{\mathrm{cond}}_{\mathrm{min,TCNQ}} - \mathscr{E}^{\mathrm{val}}_{\mathrm{max,TTF}} = 3.01 \text{ eV}$$

This may throw some light on the fact that neither theoretical (*ab initio* SCF LCAO MO) calculations on a TCNQ–TTF molecule pair with realistic structural data modeling the situation in the mixed crystal[(75)] nor measurements on aqueous solutions containing both TCNQ and TTF molecules as solutes[(76)] indicated CT, which seems to be a purely solid-state physical effect in the mixed crystal ($\sim 0.6e$ per TCNQ–TTF pair).[(77)]

Comparison of our present result with the previous MINDO/3 ones (see the paper by Singh and Ladik[(74)]) shows that the bandwidth in the case of the valence band of poly(TTF) is only about 55% of the MINDO/3 value, while the conduction band of poly(TCNQ) became more than twice as broad as in the MINDO case. The application of a better basis set may change the bandwidths somewhat and correlation effects will most probably decrease them. One expects, however, that the following qualitative features will remain the same:

1. The conduction and valence bands of both systems are π bands with opposite dispersions.
2. Bands that participate in the CT process are relatively broad, and those valence and conduction bands which play no role in CT are very narrow.
3. The position of the conduction band of poly(TCNQ) and that of the valence band of poly(TTF) favor still more CT than the corresponding LEMO and HOMO levels.

2.3. PERIODIC BIOPOLYMERS: HOMOPOLYNUCLEOTIDES AND HOMOPOLYPEPTIDES

2.3.1. Homopolynucleotides

Owing to the central role of DNA in biochemistry and biophysics, the computation of the electronic structure of periodic polymers constructed from nucleotide bases, base pairs, and nucleotides by applying the *ab initio* Hartree–Fock crystal-orbital method has attracted much interest. These investigations aim at a better knowledge of the physical properties and chemical behavior of these macromolecules and the development with time of a realistic model for DNA. Recently, the results of *ab initio* energy-band-structure calculations for the four nucleotide base stacks, for the sugar–phosphate chain, and for polycytidine, which contains one cytosine–sugar–phosphate unit in the elementary cell,[78] have been reported, using the truncated STO-3G basis set. In subsequent calculations[79] a somewhat larger minimal basis set (containing a larger number of primitive Gaussians) has been used to assess the basis-set dependence of the previous results. It is noted, however, that even with the more extended basis set the computations are far from the Hartree–Fock limit. Also, the basis set used later is the same as that adopted in recent molecular-orbital computations[80] of large fragments of B- and Z-DNA, thus allowing for additional comparisons.

The computations have been obtained with a new computer program[81] for SCF–LCAO crystal orbitals specially adapted to chain polymers with helical (rotational-translational) symmetry. The new program makes use of the one- and two-electron integral packages of a previously published molecular program.[82] The new program also contains all the necessary routines to rotate the basis functions, thus saving computational time when "local symmetry"[82] is present (see Section 1.2).

Second-neighbor interactions (namely $q = -2, -1, 0, 1, 2$) have been included with a correct, electrostatistically balanced cutoff.[83] This means that the two-electron repulsion integrals have been neglected in a way that balances the neglect of the electron–nuclear attraction integrals. Therefore, some of the two-electron integrals containing centers three units apart from one other had to be retained.

For the structural parameters of the helix and the geometry of the molecules in the unit cell, the experimental data of B-DNA[84] have been added with the aid of a special computer program[85] making use of previously reported bond lengths and bond angles for the hydrogen atoms in the four bases[86] and in the sugar unit.[87] In the study of the polynucleotides, a Na^+ ion has been attached to each phosphate group; in this way these chains attain electroneutrality. The position of these added counterions was assumed to correspond to the one determined from quantum-mechanical computations on the interaction of one Na^+ ion and one sugar–phosphate fragment.[88] This assumption, however, is only an approximation, since the Na^+ ion in a chain of periodic polynucleotides interacts not only with one sugar–phosphate unit but with the entire DNA chain.[89]

In all calculations the minimal basis set (4s functions for hydrogen atoms, $7s$ and $3p$ functions for the second-row elements, and $9s$ and $6p$ primitive Gaussian functions for the P atoms) has been applied. While in the case of the four base stacks a starting density matrix for the elementary cell, which is built up from the wave vectors resulting from atomic calculations, is sufficient to lead to convergence in the SCF procedure, at least a reasonable guess for the interaction density matrix elements between the reference cell and the first-neighbor cells is required for the more complicated DNA models. For the construction of the starting $\mathbf{P}(0)$ and $\mathbf{P}(1)$ matrices (defined as the charge-bond-order matrices within the cell and between two neighboring cells, respectively) the results of a MINDO/3 molecular computation of the system containing a double cell of the chain turned out to be suitable. After reaching convergence, by taking only first-neighbor interactions into account, the additional interactions from the cells further away have been switched on. In most cases 25–30 iterations were necessary to achieve self-consistency using the criteria

$$\Delta P_{rs} = |P(q)_{rs}^{n} - P(q)_{rs}^{n-1}| \leq 10^{-5}, \qquad \text{where } q = 0, 1, \text{ or } 2$$

Based on experience it is expected that computations with extended basis sets might be aggravated by convergence problems.

The largest DNA model studied was poly(adenylic acid) with 125 basis functions; it took 42 h to compute the two-electron integrals and 70 min for each iteration step on an IBM/370–3033 computer. Even if the recent and faster algorithm for the computation of the two-electron integrals[90] would be included in the computer program,[86] it seems clear from the above computer times that a more accurate description than second-neighbor interactions will remain a rather difficult computational task.

Table 2.6 contains the characteristics of the highest filled and lowest unfilled bands of the four nucleotide base stacks, while Table 2.7 gives the same information for the double-stranded poly(base pairs) poly(A–T) and poly(G–C). The third and fourth columns report the highest and lowest value for each band, the fifth column gives the bandwidths, and the sixth column contains the energy of the corresponding molecular orbital obtained via Hartree–Fock computations of the molecular system representing the unit cell. Finally, the last column lists the value of the gap. The correspondence between the individual molecular levels and the bands is always unambiguous. Although the symmetry is broken in the periodic chains as a result of the stacked arrangement of the units, one can still define quasi-π-type bands, which are located mainly around the Fermi level.

TABLE 2.6. Limits and Widths of the Highest Filled and Lowest Unfilled Bands of the Four Nucleotide Base Stacks (in eV)[79]a

System	Type of band[b]	Band minimum	Band maximum	Bandwidth	MO	Gap
PolyC	$n^* + 1^b$	1.70	2.52	0.82	1.64	10.44
	n^*	−9.53	−8.74	0.79	−9.95	
PolyT	$n^* + 1$	1.23	1.54	0.31	1.59	11.95
	n^*	−11.25	−10.72	0.53	−10.99	
PolyA	$n^* + 1$	2.44	3.18	0.76	1.92	12.12
	n^*	−10.12	−9.68	0.44	−10.37	
PolyG	$n^* + 1$	2.57	3.28	0.71	2.43	11.27
	n^*	−9.47	−8.70	0.77	−9.49	

[a] The locations of the corresponding MOs are given, for comparison.
[b] n^* and $n^* + 1$ denote the valence and conduction bands, respectively.

TABLE 2.7. Characteristics of the Two Highest Filled and Lowest Unfilled Bands of the Two Poly(base pairs), Poly(A–T) and Poly(G–C), and of the Three Polynucleotides, Poly(adenylic acid), Polythymidine[79] and Polycytidine[78] (in ev)[a]

System	Type of band[b]	Band minimum	Band maximum	Bandwidth	MO	Gap
Poly(A–T)	$n^* + 1^b$	1.72	2.01	0.30	1.54	11.69
	n^*	−10.52	−9.97	0.52	−10.14	
Poly(G–C)	$n^* + 1$	1.69	2.41	0.72	1.61	11.79
	n^*	−10.94	−10.10	0.84	−9.29	
Poly(ASP)	$n^* + 1$	2.65	2.93	0.28		12.16
	n^*	−9.82	−9.51	0.31		
Poly(TSP)	$n^* + 1$	1.99	2.31	0.32		11.79
	n^*	−10.10	−9.80	0.30		
Poly(CSP)[c]	$n^* + 1$	6.55	7.38	0.83		10.91
	n^*	−5.19	−4.36	0.83		

[a] The locations of MOs of the base pair are given for comparison.
[b] n^* and $n^* + 1$ denote the valence and conduction bands, respectively.
[c] See the work of Ladik and Suhai.[78]

It can be seen from Table 2.6 that the physically most important valence and conduction bands in the later *ab initio* calculations are much broader (0.435–0.789 and 0.245–0.820 eV, respectively) than those obtained by application of different semiempirical crystal-orbital methods.[83,91] With the simple PPP–CO approximation, the corresponding values for the highest filled bands are 0.218–0.299 eV and those for the conduction band are about 0.109 eV. Energy-band-structure calculations for the base stacks taking into account the effect of the other valence electrons with the aid of the CNDO/2 CO method give again broader bands (valence bandwidths of 0.136–0.490 eV and conduction bandwidths of 0.109–0.245 eV), while the MINDO/2 CO results indicate somewhat less-broad bands (valence bandwidths of 0.027–0.299 eV and conduction bandwidths of 0.027–0.163 eV). For futher details on the semiempirical crystal-orbital calculations see also Chapter 3.

Comparing our results in Table 2.6 with previously reported *ab initio* crystal-orbital calculations[78] for which the STO-3G minimal atomic basis set had been applied, we find that the bandwidths of both the valence and the conduction bands are in good agreement. The positions of these bands and of the molecular orbitals are shifted by a constant amount, about 4.30 eV, toward deeper energies owing to the more flexible minimal atomic basis set used in these computations. Therefore both calculations give a more than 10 eV gap between the highest occupied and lowest unoccupied bands in all cases.

On comparing the widths of the conduction and valence bands of the stacks of the four nucleotide bases (values between 0.3 and 0.8 eV; see Table 2.6) with those of the TCNQ and TTF stacks (1.2 eV for the conduction band of the TCNQ stack and 0.3 eV for the valence band of the TTF stack; see Table 2.5), one would expect the electron mobilities to be rather similar in both kinds of stacks. It was noted before in Section 2.2.2 that the TCNQ–TTF system is highly conducting owing to the internal charge transfer between the TTF and TCNQ molecules. Therefore, if one could generate free charge carriers in the nucleotide base stacks, one would expect a similarly large conductivity.

The fact that the gap values obtained in these calculations (see Table 2.6) are always above 10 eV, seems to exclude the possibility of intrinsic conductivity in these stacks. On the other hand, we shall see that in the case of a whole nucleotide (base + sugar + phosphate) the question is still open owing to the internal charge shift that occurs. (More details are given below.)

Attempts were made at the Max Planck Institute for Solid State Physics in Stuttgart to dope periodic polynucleotides (polycytidine and polyguanidine) with the usual acceptors used to dope polyacetylene (Br_2,

J_2, AsF_5).[93] These initial attemps failed (there was no conductivity change along the whole frequency range that we investigated) because these inorganic dopants most probably reacted with the K^+ ions or the phosphate groups of the homopolynucleotides and so never reached the base-stack regions. At several American Universities well-defined periodic homopolynucleotide samples are now being produced and purified using the HPLC techniques, and will be doped by such organic electron acceptors as may partially intercalate to the base stacks. One can only hope that after such preparation nucleotide base stacks can be successfully doped and in this way, free-charge carriers generated in them that would provide high conductivities in periodic polynucleotides.

Little is available in the literature pertaining to our energy-band structures for the more complex periodic DNA models reported in Table 2.7. For poly(A–T) and poly(G–C), the features of the bands, their positions, and their bandwidths can be related to those of the corresponding bases from which they originate. Equivalent findings have been obtained also with the aid of the semiempirical PPP and CNDO/2 CO calculations.[95]

As regards the one-electron level patterns (obtained from semiempirical methods[95] and *ab initio* calculations[96]), earlier comparison between the G–C and A–T base pairs with the single bases shows that in all cases the one-electron levels of nucleotide base pairs can be obtained — to a good approximation — by superposition of the corresponding levels of the constituent molecules. On applying this procedure to the calculated conduction and valence bands of the poly(base pairs), energy-band structures have been proposed for both poly(A–T) and poly(G–C).[97] In the case of poly(A–T) the previous estimate[97] and our present results both show an overlap of the valence band with the second highest filled ($n^* - 1$) band. Therefore, in this case the superposition of the band structures of the constituents becomes questionable. In the last row of Table 2.7 (headed "Gap") the smallest possible excitation energy of the poly(basc pairs) is given. In poly(A–T) the highest filled and lowest unfilled levels originate from poly A and poly T, respectively, and therefore the first transition is of the interbase type, in agreement with the conclusion drawn from the band structure estimated by superposition. In contrast to the proposed intrabase-type excitation in poly(G–C), our results indicate again an interbase-type transition from guanine to cytosine.

With the help of Mulliken's population analysis, the amount of charge shifted from the sugar–phosphate chain to the adenine or thymine chain has been computed using the results of the poly(ASP) and poly(TSP) chain calculations given in Table 2.7. The amounts of 0.212

and 0.190e per molecule pair are transferred to adenine and thymine, respectively. This is in agreement with the result of a previous calculation[78] for poly(CSP) (in this computation a hydrogen atom was attached to each phosphate group, hence the polymer chain was neutral) in which a charge transfer of 0.187e from the sugar–phosphate unit to cytosine was found. Recently, the internal electronic charge shift in B-DNA has been the subject of detailed computational investigations.[80] The different molecular fragments were chosen in such a way that each base is situated in a field similar to the one experienced in the macromolecule; the fragments consist of one phosphate group, two sugar residues, and the base bound to one of the sugar units (SPS-base). In addition, the possible unit-to-unit charge shift was taken into account by studying an even larger fragment of B-DNA composed of three sugar units, two phosphate groups, and two bases, G and C. These molecular computations yielded a charge shift from the sugar units to the bases, confirming the results of our crystal-orbital computations. A quantitative comparison, however, shows a decrease of about 20% for poly(ASP) and poly(TSP) relative to the corresponding values obtained for the molecular models (SPS-A and SPS-T, respectively). The above difference may be due to the unit-to-unit interactions, which are considered up to second neighbors in the periodic macromolecule. Another reason for the above discrepancy is that the interacting bases are different in the two computations. The main point, however, is confirmed: namely different methods yield a notable charge shift from the sugar to the base. These results raise a question as to the validity of those electrostatic potential studies — frequently published in the last few years — where this charge shift is neglected.

The charge shift of approximately 0.2e from the sugar to the base-stack part of DNA poses the question whether periodic DNA (homopolynucleotides) cannot be intrinsic conductors. For such a situation it would be necessary to conduct the iterations in the SCF procedure so that charge 0.2e is removed from the sugar–phosphate-type $(n^* - 1)$th band [both the n^*th and $(n^* - 1)$th bands are base-stack-type bands] and put into the $(n^* + 1)$th base-stack-type band. If one looks at the large gaps in Table 2.7 it seems that, though the Hartree–Fock total energy is not just the sum of the one-electron energies, in this way one would never obtain a more negative total energy than filling up the bands in the usual way, in which case DNA is an insulator. It should be recalled, however, that by using a large basis set and taking into account the major part (75%) of correlation in alternating *trans*-polyacetylene the original gap of 8.2 eV could be reduced to about 3 eV[49] (the experimental value is about 2 eV; see Section 2.2.1). Therefore one could imagine that, eventually, when it will also be possible to perform (with still larger computers) similarly precise calculations (4-31G** + MP/2) for the nucleotide base stacks, the minimal basis gap over 10 eV will be reduced to approximately 5 eV. Actually, a recent valence double-ζ + second-order perturbation-theory calculation of a cytosine stack has reduced the minimal basis gap of 10.4 eV to 8 eV[98] (since no polarization functions were included in the basis, one obtained only about a 2.5 eV decrease of the gap, but with a more accurate basis the correlation effects would also decrease the gap more substantially; see Chapter 5).

If the gap of the nucleotide base stacks is really about 5 eV, the question of whether periodic (homo)polynucleotides could become intrinsic conductors because of the 0.2e charge shift between the sugar and base molecules seems to be legitimate, and will only be answered through even more accurate computations than are now possible.

2.3.2. Homopolypeptides

Ab initio SCF LCAO band structures[99] have been computed for eight different periodic homopolypeptides: polyglycine, polyalanine, polyhistidine, polyserine, polycysteine, polyasparagine, polyleucine, and polyisoleucine. Clementi's $7s/3p \rightarrow 2s/1p$ minimal basis was applied to these calculations.[100] For the geometry of the polypeptide chain, the parallel-chain β-pleated sheet conformation was taken using the atomic coordinates of Pauling and Corey,[101] and the computations were performed taking into account second-neighbor interactions (in the simplest case of polyglycine the band structure was also corrected for long-range interactions and, in a third case, a double-ζ basis applied; see Table 2.8).

Table 2.8 gives the highest filled and lowest unfilled energy bands, their widths, the fundamental gap, and the total energy per unit cell of these periodic polypeptides obtained from the *ab initio* calculations. The chemical formulas of the eight calculated polypeptides are given in Figure 2.6.

If can be seen from Table 2.8 that the position of the bands, and especially the widths of the valence and conduction bands, are strongly influenced by the side chains (values between 0.02 eV and 1.4 eV). In the case of polyglycine the results agree qualitatively with those of previous MINDO/2 CO calculations.[102] It is interesting also to observe that the double-ζ calculation provided substantially broader bands than the minimal-basis computation.

The gap varies between about 16 and 11 eV. These values are far too large to allow intrinsic semiconductivity in proteins (however, according to the discussion in Section 2.3.1, the use of a better basis set and the

TABLE 2.8. Energy Band Structures of Periodic Polypeptides[a]

System	Band[e]	$\mathcal{E}_{min}$	$\mathcal{E}_{max}$	$\delta\mathcal{E}$	Gap	E_{tot}/N (au)
Polyglycine[b]	n^*+1	3.591	4.166	0.575	15.777	−206.061237
	n^*	−12.579	−12.186	0.393		
Polyglycine[c]	n^*+1	3.627	4.159	0.532	15.816	−206.051241
	n^*	−12.545	−12.189	0.356		
Polyglycine[d]	n^*+1	3.633	4.124	0.491	13.465	−206.622290
	n^*	−10.671	−9.832	0.839		
Polyalanine	n^*+1	3.722	4.395	0.673	15.776	−244.929313
	n^*	−12.288	−12.054	0.234		
Polyhistidine	n^*+1	3.222	3.619	0.397	13.507	−467.418315
	n^*	−10.301	−10.285	0.016		
Polyserine	n^*+1	4.265	4.996	0.731	15.366	−319.529368
	n^*	−11.176	−11.101	0.075		
Polycysteine	n^*+1	4.100	4.789	0.689	14.736	−641.666535
	n^*	−10.686	−10.636	0.050		
Polyasparagine	n^*+1	3.988	4.587	0.599	15.572	−412.110392
	n^*	−11.704	−11.584	0.120		
Polyleucine	n^*+1	5.606	6.863	1.257	10.824	−361.162814
	n^*	−5.241	−5.118	0.123		
Polyisoleucine	n^*+1	5.898	7.266	1.367	11.877	−361.205390
	n^*	−6.131	−5.979	0.152		

[a] $7s/3p \rightarrow 2s/1p$ basis set; second-neighbor interactions; all quantities with the exception of the total energy per unit cell are in eV.
[b] Second-neighbor interactions + long-range interactions.
[c] Second-neigbor interactions.
[d] Second-neighbor interactions; double-ζ basis set $9s/5p \rightarrow 4s/2p$.
[e] n^* denotes the valence band, n^*+1 the conduction band.

inclusion of correlation effects would decrease the gap considerably), but are most probably still not enough to allow conductivity due to thermal excitations of the electrons into the conduction band.

In this respect the finding of Szent-Györgyi[(103)] that unsaturated ketones or dicarbonyls act as electron acceptors against proteins is of great importance. According to Szent-Györgyi, for instance, methylglyoxal molecules can act as electron acceptors against a peptide bond, and in this way they can produce positive holes in the valence band of a protein thus transforming it from an insulator to a conductor. *Ab initio* SCF MO supermolecule calculations of a formamide and a glyoxal molecule (the former served as a model for the peptide groups in a protein) have shown that if the molecular planes are parallel at an intermolecular distance of 1.80 Å, the charge shift from formamide to glyoxal is 0.2e in the ground state of the system.[(104)]

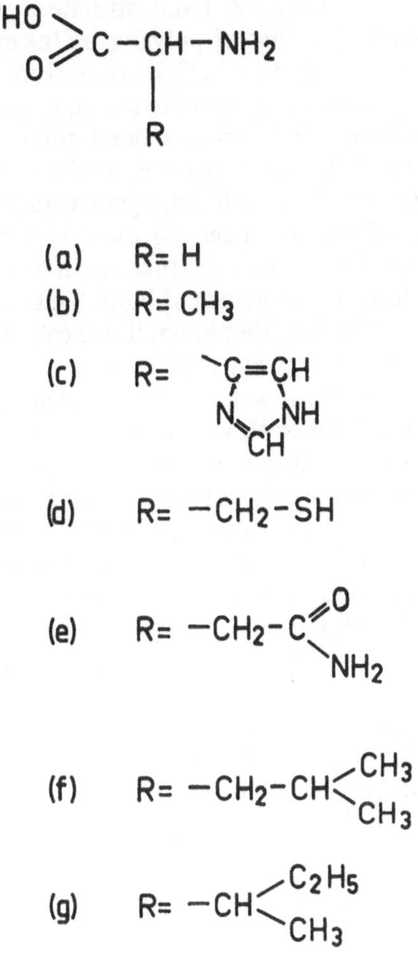

FIGURE 2.6. The chemical formulas of (a) glycine, (b) alanine, (c) histidine, (d) cysteine, (e) asparagine, (f) leucine, and (g) isoleucine.

The results of *ab initio* band-structure calculations of six different polyglycine helices have recently been reported by Otto *et al.*[105] As expected, the widths of the conduction and valence bands of these helices depend strongly on their structural (geometrical) parameters, but the gap is in all cases still larger than those given in Table 2.8. For further details we refer to Tables I and II of Otto *et al.*

The next step in the investigations of the *ab initio* band structure of polyglycine, namely its treatment as a two-dimensional periodic system (periodicity along the polypeptide chain and perpendicularly through the hydrogen bonds), is in progress.

Though the band structure of polyalanine (Table 2.8) is rather similar to that of polyglycine (as expected, because an alanine residue differs only in one CH_3 group from a glycine molecule), a calculation of an alternating glycine-alanine chain, poly(glyala),[99] resulted in the splitting of the valence band into two bands with a gap of approximately 0.2 eV between them. A similar situation has been found in the case of the conduction band, again with a gap (of 0.1 eV width) between the resulting two new narrower bands. These results alone indicate that aperiodicity (see Chapter 3) can be expected to have a crucial effect on the electronic structure of biopolymers.

On the other hand, if one calculates not a periodic poly(glyala) chain but a 1:1 gy-ala chain with a random sequence of subunits, the DOS curve shows substantial broadening of the regions of allowed energies (see Chapter 4). Very recent results obtained for the DOS of random four-component (gly, ser, cys, and his) polypeptide chains show a very large broadening effect (4.0 eV broad valence- and conduction-band regions) with only two small gaps in both regions. These surprising results indicate that it is very probable that even if all the states in native proteins are localized because of the strong disorder (see Chapter 4), in consequence of charge shift (e.g., between the phosphate groups of a polynucleotide chain and the side chains of the peptide groups in a polypeptide chain) free charge carriers are generated in the valence- or conduction-band regions, and one can expect at least a rather strong hopping-type conduction in proteins. These theoretical results seem to verify Szent-Györgyi's 45-year-old prediction.[106]

REFERENCES

1. W. L. McCubbin and R. Hanner, *Chem. Phys. Lett.* **2**, 230 (1968).
2. K. Morokuma, *Chem. Phys. Lett.* **6**, 186 (1970).
3. D. L. Beveridge, I. Jano, and J. Ladik, *J. Chem. Phys.* **56**, 4744 (1972).
4. J.-M. André, in: *Electronic Structure of Polymers and Molecular Crystals* (J.-M. André and J. Ladik, eds.), p. 1, Plenum Press, London–New York (1975).
5. J.-M. André, *Comput. Phys. Commun.* **1**, 39 (1970).
6. S. Suhai, *J. Polym. Sci., Polym. Phys. Ed.* **21**, 1341 (1983); S. Suhai, in: *Quantum Chemistry of Polymers; Solid State Aspects* (J. Ladik and J.-M. André, eds.), p. 101, D. Reidel Publ. Co., Dordrecht–Boston, 1984.
7. W. J. Hehre, R. F. Stewart, and J. A. Pople, *J. Chem. Phys.* **51**, 2657 (1969).
8. R. Ditchfield, J. W. Hehre, and J. A. Pople, *J. Chem. Phys.* **54**, 724 (1971).
9. R. A. Binkley, P. C. Hariharan, R. Seeger, R. A. Whiteside, and J. A. Pople, Gaussian 76, *QCPE* No. 368, Bloomington, Indiana (1968).
10. S. Suhai, *J. Chem. Phys.* **73**, 3843 (1980).
11. S. Suhai, P. S. Bagus, and J. Ladik, *Chem. Phys. Lett.* **68**, 467 (1982).

12. K. J. Less and E. G. Wilson, *J. Phys. C.* **6**, 3110 (1973).
13. T. Koopmans', *Physica* **1**, 104 (1933).
14. W. L. McCubbin, *Chem. Phys. Lett.* **8**, 507 (1971); J.-M. André and J. Delhalle, *Chem. Phys. Lett.* **17**, 145 (1972); J Delhalle, S. Delhalle, and J.-M. André, *Bull. Soc. Chim. Belq.* **83**, 107 (1974); J. Delhalle, *J. Chem. Phys.* **5**, 306 (1974).
15. K. Morokuma, *J. Chem. Phys.* **54**, 962 (1971).
16. P. Otto, J. Ladik, and W. Förner, *Chem. Phys.* **95**, 365 (1985).
17. J. Delhalle, in: *Electronic Structure of Polymers and Molecular Crystals* (J.-M. André and J. Ladik, eds.), p. 53, Penum Press, London–New York (1975).
18. H. Shirakawa, E. J. Louis, A. G. MacDiarmid, C. K. Heeger, and A. J. Heeger, *J. Chem. Soc., Chem. Commun.* 578 (1977).
19. C. K. Chiang, M. A. Drug, S. C. Gau, A. J. Heeger, H. Shirakawa, E. J. Louis, A. G. MacDiarmid, and Y. W. Park, *J. Am. Chem. Soc.* **100**, 1013 (1978).
20. C. K. Chiang, C. R. Fincher, Jr., Y. W. Park, A. J. Heeger, H. Shirakawa, E. J. Louis, S. C. Gau, and A. G. MacDiarmid, *Phys. Rev. Lett.* **39**, 1098 (1977).
21. Y. W. Park, M. A. Drug, C. K. Chiang, A. J. Heeger, A. G. MacDiarmid, H. Shirakawa, and S. Ikeda, *J. Polym. Sci., Polym. Chem. Ed.* **17**, 195 (1979).
22. C. K. Chiang, Y. W. Park, A. J. Heeger, H. Shirakawa, E. J. Louis, and A. G. MacDiarmid, *J. Chem. Phys.* **69**, 5098 (1978).
23. C. K. Chaing, S. C. Gau, C. R. Fincher, Jr., Y. W. Park, A. G. MacDiarmid, and A. J. Heeger, *Appl. Phys. Lett.* **33**, 181 (1978).
24. C. R. Fincher, Jr., D. L. Peebles, A. J. Heeger, M. A. Drug, Y. Matsumara, and A. G. DacDiarmid, *Solid State Commun.* **27**, 489 (1978).
25. C. R. Fincher, Jr., M. Ozaki, A. J. Heeger, and A. C. MacDiarmid, *Phys. Rev. B* **19**, 4140 (1979).
26. Y. W. Park, A. Denenstein, C. K. Chiang, A. J. Heeger, and A. G. MacDiarmid, *Solid State Commun.* **29**, 747 (1979).
27. L. G. S. Brooker, *J. Am. Chem. Soc.* **73**, 1087, 5332 (1951).
28. H. Kuhn, *Helv. Chim. Acta* **31**, 1441 (1948).
29. R. H. Baughman, S. L. Hsu, G. P. Pez, and A. J. Signorelli, *J. Chem. Phys.* **68**, 5405 (1978); A. J. Heeger, *Comments Solid State Phys.* **10**, 53 (1981); G. Wegner, *Angew. Chem.* **93**, 352 (1981); C. R. Fincher, Jr., C.-E. Chen, A. J. Heeger, W. G. MacDiarmid, and J. B. Hastings, *Phys. Rev. Lett.* **48**, 100 (1982).
30. T. Ito, H. Shirakawa, and S. Ikeda, *J. Polym. Sci., Polym. Chem. Ed.* **13**, 1943 (1975).
31. J. E. Lennard-Jones, *Proc. R. Soc. London, Ser. A* **158**, 280 (1937).
32. C. A. Coulson, *Proc. R. Soc. London, Ser. A* **164**, 383 (1938); **169**, 413 (1939).
33. H. Labhart, *J. Chem. Phys.* **27**, 957 (1957).
34. H. C. Longuet-Higgins and L. Salem, *Proc. R. Soc. London, Ser. A* **251**, 172 (1959).
35. J. A. Pople and S. H. Walmsley, *Trans. Faraday Soc.* **58**, 441 (1962).
36. M. J. S. Dewar and G. J. Gleicher, *J. Am. Chem. Soc.* **87**, 692 (1965).
37. R. Hoffman, *Tetrahedron* **22**, 521 (1966).
38. W. Kutzelnigg, *Theor. Chim. Acta (Berlin)* **4**, 417 (1966).
39. J.-M. André and G. Leroy, *Theor. Chem. Acta (Berlin)* **9**, 123 (1967).
40. S. F. O'Shea and D. P. Santry, *J. Chem. Phys.* **54**, 2667 (1971).
41. D. L. Beveridge, I. Jano, and J. Ladik, *J. Chem. Phys.* **56**, 4744 (1972).
42. J.-M. André and G. Leroy, *Int. J. Quantum Chem.* **5**, 557 (1971).
43. M. Kertész, J. Koller, and A. Ažman, *J. Chem. Phys.* **67**, 1180 (1977).

44. M. Kertész, J. Koller, and A. Ažman, *J. Chem. Soc., Chem. Commun.* 575 (1978).
45. A. Karpfen and J. Petkov, *Theor. Chim. Acta (Berlin)* **53**, 65 (1979).
46. P. Grant and I. P. Batra, *Solid State Commun.* **29**, 225 (1979).
47. R. A. Harris and L. M. Falicov, *J. Chem. Phys.* **51**, 5034 (1969).
48. A. A. Ovchinnikov, I. I. Ukrainski, and G. V. Kventsel, *Usp. Fiz. Nauk* **108**, 81 (1973); *Sov. Phys. Usp.* **15**, 575 (1979).
49. S. Suhai, *Phys. Rev. B* **27**, 3506 (1983).
50. S. Suhai, *J. Chem. Phys.* **73**, 3843 (1980).
51. S. Suhai, *Chem. Phys. Lett.* **96**, 619 (1983).
52. J. Čížek and J. Paldus, *J. Chem. Phys.* **47**, 3976 (1967); J. Padus and J. Čížek, *Phys. Rev. A* **3**, 2268 (1970).
53. W. J. Hehre, R. F. Stewart, and J. A. Pople, *J. Chem. Phys.* **51**, 2657 (1969); R. Ditchfield, W. J. Hehre, and J. A. Pople, *J. Chem. Phys.* **54**, 726 (1971).
54. J. S. Binkley, R. A. Whiteside, P. C. Hariharan, R. Seeger, J. A. Pople, W. J. Hehre, and M. D. Newton, Gaussian 76, *QCPE* No. 368, Bloomington, Indiana (1968).
55. R. H. Baughman, S. I. Hsu, L. R. Anderson, G. P. Pez, and A. J. Signoreli, in: *Molecular Metals* (W. Hatfield, ed.), p. 189, Plenum Press, New York–London (1979).
56. C. R. Fincher, Jr., C.-E. Chen, A. J. Heeger, A. G. MacDiarmid, and J. B. Hastings, *Phys. Rev. Lett.* **48**, 100 (1982).
57. H. P. Geserich and L. Pintschovius, *Adv. Solid State Phys.* **16**, 65 (1976).
58. H. Kamimura, A. J. Grant, F. Levy, A. D. Yoffe, and G. D. Pitt, *Solid State Commun.* **17**, 49 (1976); D. E. Parry and J. M. Thomas, *J. Phys. C* **8**, L45 (1975); V. T. Rajan and L. M. Falicov, *Phys. Rev. B* **12**, 1240 (1975); A. Zunger, *Chem. Phys.* **63**, 4854 (1975); C. Merkel and J. Ladik, *Phys. Lett.* **56A**, 395 (1976); S. Suhai and M. Kertész, *J. Phys. C* **9**, L347 (1976); W. E. Rudge and P. M. Grant, *Phys. Rev. Lett.* **35**, 1799 (1975).
59. S. Suhai and J. Ladik, *Solid State Commun*, **22**, 227 (1977).
60. B. Roos and P. Siegbahn, *Theor. Chim. Acta (Berlin)* **17**, 209 (1970).
61. M. Boudeulle and P. Michelle, *Acta Crystallogr., Sect. A* **28**, S199 (1972).
62. C. M. Mikulski, P. J. Russo, M. S. Saran, A. G. MacDiarmid, A. F. Garito, and A. J. Heeger, *J. Am. Chem. Soc.* **97**, 6358 (1975).
63. W. E. Rudge and P. M. Grant, *Phys. Rev. Lett.* **35**, 1799 (1975).
64. P. M. Grant, R. L. Greene, and G. B. Street, *Phys. Rev. Lett.* **35**, 1740 (1975).
65. R. L. Greene, P. M. Grant, and G. B. Street, *Phys. Rev. Lett.* **34**, 89 (1975).
66. M. Kertész, J. Koller, A. Ažman, and S. Suhai, *Phys. Lett.* **55A**, 107 (1975).
67. R. S. Mulliken, *J. Chem. Phys.* **23**, 1833,1841 (1955).
68. P. Mengel, P. M. Grant, W. E. Rudge, and B. H. Schechtman, *Phys. Rev. Lett.* **35**, 1803 (1975).
69. L. B. Coleman, M. J. Cohen, D. J. Sandman, F. G. Yamagishi, A. G. Garito, and A. J. Heeger, *Solid State Commun.* **12**, 1125 (1973).
70. See the following review papers: I. F. Schegolev, *Phys. Status Solidi* **12**, 9 (1972); H. R. Zeller, *Adv. Solid State Phys.* **13**, 31 (1973); Z. G. Soos, *Ann. Rev. Phys. Chem.* **25**, 12 (1974); A. J. Heeger and A. F. Garito, in: *Low-dimensional Cooperative Phenomena* (H. J. Keller, ed.), p. 89, Plenum Press, New York–London (1975); G. A. Thomas, D. E. Schafer, F. Wudl, P. M. Horn, D. Rimai, J. W. Cook, D. A. Glocker, M. J. Skove, C. W. Chu, R. P. Groff, J. L. Gillson, R. C. Wheland, L. R. Melby, M. B. Salamon, R. A. Craven, G. De Pasquali, A. N. Bloch, D. O. Cowan, V. V. Walatka, R. E. Pyle, R. Gemmer, T. O. Poehler,

G. R. Johnson, M. G. Miles, J. D. Wilson, J. P. Ferraris, T. F. Finnegan, R. J. Warmack, V. F. Raaen and D. Jerome, *Phys. Rev. B* **11**, 5105 (1976); M. J. Cohen, L. B. Coleman, A. F. Garito, and A. J. Heeger, *Phys. Rev. B* **13**, 5111 (1976).
71. S. Suhai and J. Ladik, *Phys. Lett.* **77A**, 25 (1980).
72. T. J. Kistenmacher, T. E. Phillips, and D. O. Cowan, *Acta Crystallogr.* **33**, 76 (1974).
73. W. J. Hehre, R. F. Stewart, and J. A. Pople, *J. Chem. Phys.* **18**, 932 (1965).
74. A. Karpfen, J. Ladik, G. Stollhoff, and P. Fulde, *Chem. Phys.* **8**, 215 (1975); J. Ladik, *Int. J. Quantum Chem.* **S9**, 1563 (1975); R. D. Singh and J. Ladik, *Phys. Lett.* **65A**, 264 (1975).
75. F. Cavallone and E. Clementi, *J. Chem. Phys.* **63**, 4304 (1975).
76. J. André, Lecture at the Cecam Symposium on Quantum Theory of Polymers, Namur (1979).
77. F. Donoyer, R. Comes, A. F. Garito, and A. J. Heeger, *Phys. Rev. Lett.* **35**, 445 (1975).
78. J. Ladik and S. Suhai, *Int. J. Quantum Chem., Quantum Biol. Symp.* **7**, 181 (1980).
79. P. Otto, E. Clementi, and J. Ladik, *J. Chem. Phys.* **78**, 454 (1980).
80. E. Clementi and G. Corongiu, *Int. J. Quantum Chem. Quantum Biol. Symp.* **9**, 213 (1982). See also IBM Research Report POK-09, March 3, 1982.
81. P. Otto and E. Clementi (unpublished results).
82. See, for example, C. Castiglioni, D. Ortolera, and E. Clementi, *Comput. Phys. Commun.* **19**, 337 (1980).
83. J. Ladkik and S. Suhai, in: *Theoretical Chemistry* (C. Thomson, ed.), Vol. 4, p. 49, 'Specialists' Report, Royal Society of Chemistry, London (1981).
84. R. Fieldman, in: *Atlas of Macromolecular Structure on Microfiche* (*AMSOM*), Document 13.2.1.1.1, National Institutes of Health, Bethesda, Washington (1976).
85. An adapted version of the program REFINE originally written by J. Hermans, D. R. Ferro, J. E. McQueen, and S. C. Wang, in: *Environmental Effects on Molecular Structure and Properties* (B. Pullman, ed.), p. 459, Reidel, Dordrecht (1976), has been used.
86. E. Cementi, J.-M. André, C. André, C. Klint, and D. Hahn, *Acta Phys. Acad. Sci. Hung.* **27**, 493 (1969).
87. G. Corongiu and E. Clementi, *Gazz. Chim. Ital.* **108**, 273 (1978).
88. E. Clementi, in: *Lecture Notes in Chemistry*, Vol. 19, Springer-Verlag, New York (1980).
89. E. Clementi and G. Corongiu, *Biopolymers* **20**, 2427 (1981); **21**, 763 (1982).
90. E. Clementi, G. Corongiu, M. Gratarola, P. Habitz, C. Lupo, P. Otto, and D. Vercauteren, *Int. J. Quantum Chem., Symp.* **16**, 409 (1982). See also IBM Research Report POK-03, September 23, 1982.
91. J. Ladik, *Int. J. Quantum Chem.* **4**, 307 (1971).
92. J. Ladik, in: *Advances in Quantum Chemistry* (P.-O. Löwdin, ed.), Vol. 7, p. 377, Academic Press, New York (1973).
93. J. Ladik, *Int. J. Quantum Chem., Quantum Biol. Symp.* **1**, 651 (1974).
94. S. Roth and K. Dransfeld (personal communication).
95. T. A. Hoffman and J. Ladik, *Adv. Chem. Phys.* **7**, 184 (1964); R. Rein and J. Ladik, *J. Chem. Phys.* **40**, 2466 (1964).
96. E. Clementi, J. Mehl, and W. von Niessen, *J. Chem. Phys.* **54**, 508 (1971).
97. J. Ladik, S. Suhai, P. Otto, and T. C. Collins, *Int. J. Quantum Chem., Quantum Biol. Symp.* **4**, 55 (1977); J. J. Ladik and S. Suhai, in: *Molecular Interactions* (H. Ratajczak and W. J. Orville-Thomas, eds.), p. 151, Wiley, New York (1980).

98. S. SUHAI, *Int. J. Quantum Chem., Quantum Biol. Symp.* **11** (in press).
99. P. OTTO, A. K. BAKHSHI, M. SEEL, AND J. LADIK (unpublished).
100. E. CLEMENTI (unpublished results).
101. L. PAULING AND R. B. COREY, *Proc. Natl. Acad. Sci. U.S.A.* **39**, 253 (1953).
102. S. SUHAI AND J. LADIK, *Theor. Chim. Acta (Berl.)* **28**, 67 (1974); S. SUHAI, *Theor. Chim. Acta (Berl.)* **34**, 157 (1974); S. SUHAI AND J. LADIK, *Acta Chim. Hung. Acad. Sci.* **82**, 67 (1974).
103. A. SZENT-GYÖRGYI, *Int. J. Quantum Chem., Quantum Biol. Symp.* **3**, 45 (1976); A. SZENT-GYÖRGYI, *Bioenergetics* **4**, 535 (1973); A. SZENT-GYÖRGYI, *Electronic Biology and Cancer*, Marcel Dekker, New York (1976).
104. P. OTTO, S. SUHAI AND J. LADIK, *Int. J. Quantum Chem., Quantum Biol. Symp.* **4**, 451 (1977).
105. P. OTTO, E. CLEMENTI, J. LADIK, AND F. MARTINO, *J. Chem. Phys.* **80**, 5294 (1984).
106. A. SZENT-GYÖRGYI, *Nature* **148**, 157 (1941); *Science* **93**, 609 (1941).

Chapter 3

Semiempirical Band-Structure Calculations

Different semiempirical crystal-orbital (CO) methods have been used to calculate the energy-band structures of one-dimensional periodic chains with larger unit cells (such as periodic DNA or protein models, TCNQ and TTF stacks) or of two-dimensional periodic protein models. The π-electron structure of the different periodic DNA stacks and of the H–N–C = O· · ·H–N–C = O· · · chain in polypeptides has been calculated with the aid of the Pariser–Parr–Pople (PPP) CO method for one-dimensional systems, while for the sugar–phosphate chain of DNA [poly(SP)], for the homopolynucleotides, and for the one-dimensional polypeptide and polyglycine chains, the CNDO/2 and MINDO/2 methods for one-dimensional systems have been applied. Finally, in the case of the two-dimensional polyformamide and polyglycine systems the two-dimensional versions of the all-valence-electron CO methods discussed earlier have been used. Here we shall show in detail how to obtain expressions pertaining to the PPP CO method for linear chains starting from the corresponding *ab initio* equations. In the case of the other semiempirical CO methods only the final expressions will be given.

3.1. SEMIEMPIRICAL CRYSTAL-ORBITAL METHODS

3.1.1. The Pariser–Parr–Pople Crystal-Orbital Method

We saw in Chapter 1 that in the *ab initio* case the band structure of a linear chain can be obtained from the matrix equation (1.47)

$$\mathbf{F}(k)\mathbf{d}(k)_h = \varepsilon(k)_h\mathbf{S}(k)\mathbf{d}(k)_h \tag{3.1}$$

where $\mathbf{F}(k)$ and $\mathbf{S}(k)$ were defined by Fourier transforms (1.48) and (1.49), respectively. The elements of matrices $\mathbf{F}(q)$ were defined as [see equation (1.51)]

$$[\mathbf{F}(q)]_{r,s} = \langle \chi_r^0(1)\hat{H}^{\mathrm{N}} \mid \chi_s^q(1)\rangle + \sum_{u,v=1}^{m} \sum_{q_1,q_2=-\infty}^{+\infty} P(q_1-q_2)_{u,v}$$
$$\times [\langle \chi_r^0(1)\chi_u^{q_1}(2) \mid \chi_s^q(1)\chi_v^{q_2}(2)\rangle - \tfrac{1}{2}\langle \chi_r^0(1)\chi_u^{q_1}(2) \mid \chi_v^{q_2}(1)\chi_s^q(2)\rangle] \tag{3.2}$$

with

$$\hat{H}^{\mathrm{N}} = -\tfrac{1}{2}\Delta_1 + \sum_{q_1=-\infty}^{+\infty} \sum_{\alpha=1}^{M} \frac{Z_\alpha}{|\mathbf{r}_1 - \mathbf{R}_\alpha^{q_1}|}$$

Here again χ_s^q denotes the sth AO in the qth cell, m is the number of orbitals, M the number of nuclei in the unit cell, Z_α the charge (in atomic units) of the αth nucleus, $\mathbf{r}_1$ the position vector of the electron and $\mathbf{R}_\alpha^{q_1}$ that of the αth nucleus in the q_1th cell, and finally, a is the elementary translation. Further, the four-center integral

$$\langle \chi_r^0(1)\chi_u^{q_1}(2) \mid \chi_s^q(1) \mid \chi_v^{q_2}(2)\rangle$$
$$\equiv \int \chi_r^0(\mathbf{r}_1)^*\chi_u^{q_1}(\mathbf{r}_2)^* \frac{1}{r_{12}} \chi_s^q(\mathbf{r}_1)\chi_v^{q_2}(\mathbf{r}_2)d\mathbf{r}_1 d\mathbf{r}_2 \tag{3.3}$$

and the matrix elements of the generalized charge-bond-order matrix are defined as

$$P(q_1-q_2)_{u,v} = (a/2\pi) \int_{-\pi/a}^{\pi/a} 2 \sum_{h=1}^{n_f} d(k)_{h,u}^* d(k)_{n,v} \exp[ik(q_1-q_2)a]dk \tag{3.4}$$

where n_f denotes the number of filled bands. Finally, the elements of the overlap matrices $\mathbf{S}(q)$ are defined again by the equation

$$[\mathbf{S}(q)]_{r,s} = \langle \chi_r^0 \mid \chi_s^q\rangle \equiv \int \chi_r^0(\mathbf{r}_1)^*\chi_s^q(\mathbf{r}_1)d\mathbf{r}_1 \tag{3.5}$$

The PPP CO method[1] is derived by taking into account explicitly, as in the case of molecules, only the π electrons of the system. Putting $\mathbf{S}(k) = \mathbf{1}$, equation (3.1) reduces to the matrix eigenvalue equation

$$\mathbf{H}(k)\tilde{\mathbf{d}}(k)_h = \tilde{\varepsilon}(k)_h\tilde{\mathbf{d}}(k)_h \tag{3.6}$$

Following the PPP method[2] in the case of molecules, we introduce systematically the neglect of differential overlap and, taking into account only first-neighbor interactions, the elements of matrices $\mathbf{F}(q)$ in expression (3.2) take the forms[3]

$$[\mathbf{H}(0)]_{r,r} = \langle \chi_r^0(1) \,|\, \hat{H}^{\text{eff}} \,|\, \chi_r^0(1)\rangle + \sum_{u=1}^{m_\pi} \sum_{q_1=-1}^{+1} P(0)_{u,u} \langle \chi_r^0(1)\chi_u^{q_1}(2) \,|\, \chi_r^0(1)\chi_u^{q_1}(2)\rangle - \tfrac{1}{2}P(0)_{r,r}\langle \chi_r^0(1)\chi_r^0(2) \,|\, \chi_r^0(1)\chi_r^0(2)\rangle \quad (3.7a)$$

$$[\mathbf{H}(q)]_{r,s} = \langle \chi_r^0(1) \,|\, \hat{H}^{\text{eff}} \,|\, \chi_s^q(1)\rangle - \tfrac{1}{2}P(-q)_{r,s}\langle \chi_r^0(1)\chi_s^q(2) \,|\, \chi_r^0(1)\chi_s^q(2)\rangle$$
$$(q = -1, 0, 1; \text{ if } q = 0 \text{ then } r \neq s) \quad (3.7b)$$

where χ_r^0 now denotes the rth π orbital in the reference cell and

$$\hat{H}^{\text{eff}} = \hat{H}^{\text{N}} + \sum_{q_1=-1}^{+1} \sum_{\alpha=1}^{m} V_{\text{el}\alpha}^{\text{eff}q_1} \quad (3.8)$$

Here $V_{\text{el}\alpha}^{\text{eff}q_1}$ is the potential of the inner-shell and σ electrons of the αth atom with a π orbital in the cell characterized by q_1, $\hat{H}^{\text{N}}$ is given by the expression after equation (3.2), but now q_1 runs only from -1 to $+1$ and m_π is the number of π orbitals in the unit cell. In terms of the usual approximations and notations of the PPP MO method[2] generalized for our case,[1,3] namely

$$\langle \chi_r^0\chi_s^{q_1} \,|\, \chi_r^0\chi_s^{q_1}\rangle = \gamma(q_1)_{r,s} \quad (3.9a)$$

$$\langle \chi_r^0\chi_r^0 \,|\, \chi_r^0\chi_r^0\rangle = I_r - E_r \quad (3.9b)$$

$$\langle \chi_r^0 \,|\, \hat{H}^{\text{eff}} \,|\, \chi_r^0\rangle = -I_r - \sum_{q_1=-1}^{+1} \sum_{s=1}^{m_\pi} Z_s\gamma(q_1)_{r,s} \quad (3.9c)$$

$$\langle \chi_r^0 \,|\, \hat{H}^{\text{eff}} \,|\, \chi_s^q\rangle = \beta(q)_{r,s} \qquad (s \neq r \text{ if } q = 0) \quad (3.9d)$$

we can write

$$[\mathbf{H}(0)]_{r,r} = -I_r + \tfrac{1}{2}P(0)_{r,r}(I_r - E_r) + \sum_{s=1}^{m_\pi} [P(0)_{s,s} - Z_s]\gamma(0)_{r,s} + \sum_{s=1}^{m_\pi} [P(0)_{s,s} - Z_s][\gamma(1)_{r,s} + \gamma(-1)_{r,s}] \quad (3.10a)$$

$$[\mathbf{H}(q)]_{r,s} = \beta(q)_{r,s} - \tfrac{1}{2}P(-q)_{r,s}\gamma(q)_{r,s} \quad (\text{if } q = 0 \text{ then } r \neq s) \quad (3.10b)$$

Here I_r and E_r are the ionization potential and electron affinity, respectively, of the rth atom in its appropriate valence state; their values can be taken from Hinze and Jaffé[4] and the generalized charge-bond orders have been defined previously by expression (3.4). The Coulomb integrals $\gamma(q)_{r,s}$ have been approximated in the actual calculations with the aid of the expressions

$$\gamma(q)_{r,s} = \frac{1}{a_{r,s} + R(q)_{r,s}}, \quad \frac{1}{a_{r,s}} = \tfrac{1}{2}[I_r + I_s - E_r - E_s] \tag{3.11}$$

given by Mataga and Nishimoto,[5] where $R(q)_{r,s}$ is the distance between atom r in the reference cell and atom s in the qth cell. For the core integrals $\beta(0)_{r,s}$ between atoms within the same molecule the usual PPP values[2] have been used, while the intercell core integrals $\beta(1)_{r,s}$ have been taken proportional to the corresponding overlap integrals.[3] In calculating and storing the relevant quantities the relations

$$\mathbf{R}(-1) = \mathbf{R}(1)^{\text{tr}}, \quad \boldsymbol{\gamma}(-1) = \boldsymbol{\gamma}(1)^{\text{tr}}, \quad \mathbf{P}(-1) = \mathbf{P}(1)^{\text{tr}}, \quad \boldsymbol{\beta}(-1) = \boldsymbol{\beta}(1)^{\text{tr}}$$

have been used, so it was not necessary to calculate them separately for $q = -1$.

Since the k-dependence of the eigenvector-components $d(k)_{h,r}$ is not known analytically, integrations (3.4) were performed numerically at each iteration step to obtain the elements of matrices $\mathbf{P}(q)$. This involved solving the eigenvalue problem of the Hermitian complex matrix

$$\mathbf{H}(k) = \mathbf{H}(0) + \mathbf{H}(1)e^{ika} + \mathbf{H}(-1)e^{-ika} \tag{3.12}$$

at each iteration step by different values of k (usually, seven k values saturated the band structure in this sense). The SCF criteria used in the calculations of periodic DNA models were

$$|P(0)_{r,s}^{(n+1)} - P(0)_{r,s}^{(n)}| \leqslant 10^{-4} \quad \text{and} \quad |P(1)_{r,s}^{(n+1)} - P(1)_{r,s}^{(n)}| \leqslant 10^{-4} \tag{3.13}$$

simultaneously for all r and s.

Finally, it should be mentioned that if one completely neglects electron–electron interaction terms and introduces an effective Hamiltonian $\hat{H}_{\text{eff}}$, which includes not only the kinetic-energy term but the potential field both of the nuclei and of *all* the other electrons, one obtains expressions for the simple tight-binding (Hückel) crystal-orbital theory for the π electrons. In this formalism, still useful for qualitative orientation, equation (3.6) with definition (1.48) of $\mathbf{H}(k)$ still remains valid, however the elements of $\mathbf{H}(0)$ and $\mathbf{H}(1)$ will be substituted by Hückel parameters[6]

$$[\mathbf{H}(0)]_{r,r} = \alpha_r, \quad [\mathbf{H}(0)]_{r,s} = \beta_{r,s}(0),$$

$$[\mathbf{H}(q)]_{r,s} = \beta_{r,s}(q) \qquad (q \neq 0) \tag{3.14}$$

In the latter relationships, the usual molecular values can be assumed while quantities $\beta_{r,s}(q)$ can be estimated on the basis of the corresponding overlap integrals[7]

$$\frac{S(0)_{r,s}}{S(q)_{r,s}} = \frac{\beta_{r,s}(0)}{\beta_{r,s}(q)}$$

3.1.2. Semiempirical All-Valence Electron Crystal-Orbital Schemes

In the case of the CNDO/2 CO scheme for the valence electrons of one-dimensional periodic systems we can write equation (3.6), with expression (3.12) for $\mathbf{H}(k)$, if we apply again the first-neighbor-interaction approximation. In this case the expressions for the elements of matrices $\mathbf{H}(q)$ that can be derived again from the corresponding *ab initio* equations by introducing the simplifications and approximations of the CNDO/2 method,[8] will be[9]

$$[\mathbf{H}(0)]_{r,r} = U_r + [P(0)_{\mathrm{A,A}} - \tfrac{1}{2}P(0)_{r,r}]\gamma(0)_{\mathrm{A,A}} + \sum_{\mathrm{B}} [P(0)_{\mathrm{B,B}} - Z_{\mathrm{B}}]\gamma(0)_{\mathrm{A,B}}$$

$$+ \sum_{\mathrm{B}} [P(0)_{\mathrm{B,B}} - Z_{\mathrm{B}}][\gamma(1)_{\mathrm{A,B}} + \gamma(-1)_{\mathrm{A,B}}] \quad \begin{matrix}(\mathrm{B} \neq \mathrm{A}) \\ (r \in \mathrm{A})\end{matrix} \tag{3.14a}$$

and

$$[\mathbf{H}(q)]_{r,s} = \beta^0_{\mathrm{A,B}} S(q)_{r,s} - \tfrac{1}{2}P(q)_{r,s}\gamma(q)_{\mathrm{A,B}}$$

$$(\text{if } q = 0, r \neq s; r \in \mathrm{A}, S \in \mathrm{B}) \tag{3.14b}$$

Here

$$P(0)_{\mathrm{A,A}} = \sum_{r \in \mathrm{A}} P(0)_{r,r} \tag{3.15}$$

with expression (3.4) for $P(q)_{r,s}$, a generalization of the integral expression occurring in the CNDO method,

$$\gamma(q)_{A,B} = \int |\chi_A^0(1)|^2 \frac{1}{r_{12}} |\chi_B^q(2)|^2 dV_1 dV_2 \quad (q = -1, 0, 1) \qquad (3.16)$$

where χ_A^0 is an appropriate valence s orbital centered on atom A in the reference cell. Further, the parameter $\beta_{A,B}^0 = \frac{1}{2}(\beta_A^0 + \beta_B^0)$, where constants β_A^0 and β_B^0 are given numerically for the elements of the first two rows of the periodic table,[10] Z_B is the core charge of atom B, and the core integrals $U_r = \langle \chi_r^0 | \hat{H}_{A,C}^{\text{eff}} | \chi_r^0 \rangle$ can be calculated with the aid of the valence-state ionization potentials and electron affinities of atom A ($r \in A$) and of its positive and negative ions. Finally $\hat{H}_{A,C}^{\text{eff}} = -\frac{1}{2}\Delta + V_{A,C}^{\text{eff}}$, where $V_{A,C}^{\text{eff}}$ is the effective potential of the atomic core (nucleus and inner-shell electrons) of A in the reference cell.

In the two-dimensional case equation (3.6) becomes

$$\mathbf{H}(k_1, k_2)\mathbf{d}(k_1, k_2)_h = \varepsilon(k_1, k_2)_h \mathbf{d}(k_1, k_2)_h \qquad (3.17)$$

On introducing the second-neighbor-interaction approximation, we can write

$$\begin{aligned}\mathbf{H}(k_1, k_2) = {} & \mathbf{H}(0, 0) + \mathbf{H}(1, 0)\exp(ik_1a) + \mathbf{H}(-1, 0)\exp(-ik_1a) \\ & + \mathbf{H}(0, 1)\exp(ik_2a_2) + \mathbf{H}(0, -1)\exp(-ik_2a_2) \\ & + \mathbf{H}(1, 1)\exp[i(k_1a_1 + k_2a_2)] + \mathbf{H}(1, -1)\exp[i(k_1a_1 - k_2a_2)] \\ & + \mathbf{H}(-1, 1)\exp[-i(k_1a_1 - k_2a_2)] \\ & + \mathbf{H}(-1, -1)\exp[-i(k_1a_1 + k_2a_2)] \\ = {} & \mathbf{H}(0, 0) + \tilde{\mathbf{H}}^{1\text{nb}} + \tilde{\mathbf{H}}^{2\text{nb}} \end{aligned} \qquad (3.18)$$

where $\tilde{\mathbf{H}}^{1\text{nb}}$ denotes the sum of the second, third, fourth, and fifth terms and $\tilde{\mathbf{H}}^{2\text{nb}}$ the sum of the sixth, seventh, eighth, and ninth terms in the latter expression.

If the simplifications and parametrizations characteristic of the CNDO/2 method[8] are again introduced, then the elements of matrices $\mathbf{H}(q_1, q_2)$ in expression (3.18) reduce to

$$\begin{aligned}[\mathbf{H}(0, 0)]_{r,r} = {} & U_r + [P(0, 0)_{A,A} - \tfrac{1}{2}P(0, 0)_{r,r}]\gamma(0, 0)_{A,A} \\ & + \sum_{B \neq A} [P(0, 0)_{B,B} - Z_B]\gamma(0, 0)_{A,B} \\ & + \sum_{B} \left\{ [P(0, 0)_{B,B} - Z_B] \sum_{q_1=-1}^{+1} \sum_{q_2=-1}^{+1} \gamma(q_1, q_2)_{A,B} \right\}_{(r \in A)} \end{aligned} \qquad (3.19a)$$

and

$$[\mathbf{H}(q_1, q_2)]_{r,s} = \beta^0_{A,B} S(q_1, q_2)_{r,s} - \tfrac{1}{2} P(q_1, q_2)_{r,s} \gamma(q_1, q_2)_{A,B} \quad (3.19b)$$

Now

$$P(0, 0)_{A,A} = \sum_{r \in A} P(0, 0)_{r,r} \quad (3.20a)$$

and

$$P(q_1, q_2)_{r,s} = \frac{1}{\omega_{2D}} \omega_{2D} \int \sum_{h=1}^{n_t} d^*(k_1, k_2)_{h,r} d(k_1, k_2)_{h,s}$$

$$\times \exp[i(k_1 q_1 a_1 + k_2 q_2 a_2)] dk_1 dk_2 \quad (3.20b)$$

Here a_1 and a_2 are the elementary translations in the two different directions, ω_{2D} is the first Brillouin zone of the two-dimensional lattice, and also

$$\gamma(q_1, q_2)_{A,B} = \int |\chi_A^{0,0}(1)|^2 \frac{1}{r_{12}} |\chi_B^{q_1,q_2}(2)|^2 dV_1 dV_2 \quad (3.21)$$

where $\chi_B^{q_1,q_2}$ is the appropriate valence s Slater orbital centered on atom B in the cell characterized by

$$\mathbf{q} = q_1 \mathbf{a}_1 + \mathbf{q}_2 a_2$$

The CNDO/2 parametrization can be replaced by the elements of the matrices $\mathbf{H}(q_1, q_2)$ also in the MINDO/2 parametrization (3.11). Hence[12]

$$[\mathbf{H}(0, 0)]_{r,r} = U_r + \tfrac{1}{2} P(0, 0)_{r,r} \langle \chi_r^0 \chi_r^0 \mid \chi_r^0 \chi_r^0 \rangle$$

$$+ \sum_{\substack{r \in A \\ \sigma \neq A}} P(0, 0)_{\sigma,\sigma} [\langle \chi_r^0 \chi_r^0 \mid \chi_r^0 \chi_r^0 \rangle - \tfrac{1}{2} \langle \chi_r^0 \chi_\sigma^0 \mid \chi_r^0 \chi_\sigma^0 \rangle]$$

$$+ \tilde{\sum_{B}} \quad (r \in A) \quad (3.22a)$$

$$[\mathbf{H}(0, 0)]_{r,s} = P(0, 0)_{r,s} [\tfrac{3}{2} \langle \chi_r^0 \chi_s^0 \mid \chi_r^0 \chi_s^0 \rangle - \tfrac{1}{2} \langle \chi_r^0 \chi_r^0 \mid \chi_s^0 \chi_s^0 \rangle]$$

$$(r \neq s;\ r, s \in A) \quad (3.22b)$$

and

$$[\mathbf{H}(q_1, q_2)]_{r,s} = \Omega_{A,B}(I_r + I_s)S(q_1, q_2)_{r,s} - \tfrac{1}{2}P(q_1, q_2)_{r,s}\gamma(q_1, q_2)_{A,B}$$

$$(r \in A; \text{ if } q_1 = q_2 = 0 \text{ then } s \in B) \tag{3.22c}$$

Here the term $\tilde{\Sigma}_B$ in (3.22a) replaces all the expressions summed over B in equation (3.19a); for the one-center core integrals U_r, the valence-state ionization potential I_r, and parameters $\Omega_{A,B}$ appropriate values have been given by Baird and Dewar.[11] All the one-center two-electron integrals

$$\langle \chi_r^0 \chi_s^0 \,|\, \chi_r^0 \chi_s^0 \rangle \quad \text{and} \quad \langle \chi_r^0 \chi_r^0 \,|\, \chi_s^0 \chi_s^0 \rangle$$

have been calculated with the aid of appropriate valence-electron Slater orbitals, while for the two-center Coulomb integrals $\gamma(q_1, q_2)^{A,B}$ the Ohno–Klopman expression[13]

$$\gamma(q_1, q_2)_{A,B} = \frac{1}{[R(q_1, q_2)_{A,B}^2 + a_{A,B}^2]^{1/2}} \tag{3.23a}$$

with

$$\frac{1}{a_{A,B}} = \tfrac{1}{2}(I_A + I_B - E_A - E_B) \tag{3.23b}$$

has been used. Here I_A and E_A are the valence-state s-electron ionization potential and electron affinity of atom A, respectively. All other quantities were defined earlier.

The corresponding formulas for the MINDO/2 CO method for one-dimensional periodic systems in the first-neighbor-interaction approximation can be obtained very easily from the above expressions by substituting equations (3.22) into equation (3.12) and suppressing the variable q_2 throughout.[14]

In the actual all-valence electron calculations the SCF criteria (3.13) were again used and convergence reached in 12–20 iteration steps, the MINDO/2 CO methods requiring a smaller number of iteration steps.

It should be noted that using an INDO[15] rather than a CNDO or MINDO parameterization, one can easily also write the matrix elements for an INDO crystal-orbital theory. This has been done; for the sake of brevity we refer the reader to the original paper[16] for the details.

Finally, within the crystal-orbital formalism for the valence electrons one can also use the extended Hückel method of Roald Hoffmann[17] that leads to extended Hückel band structures of quasi-one-dimensional periodic chains.[18] Unfortunately, we cannot here enter into the details of this method (neither in its simple nor in its iterative form).

3.2. APPLICATIONS TO HIGHLY CONDUCTING POLYMERS AND BIOPOLYMERS

In this section we cannot, of course, review the numerous semiempirical band-structure calculations available in the literature, but we shall concentrate on a few illustrative examples.

3.2.1. *Trans*-Polyacetylene

An INDO and MINDO/2 CO calculation was performed in the case of *trans* equidistant *polyacetylene*.[16] Figure 3.1 shows the geometry employed in this calculation. Only the planar *trans* extended form of the polymer was considered, with all bond angles taken as 120°. The electronic energy band structure of the polymer calculated by the INDO method is given in Figure 3.2, juxtaposed with the calculated orbital energy levels of the isolated elementary cell. Considerable band overlap is evident, so that the highest occupied and lowest unoccupied bands are of mixed σ and π character. The calculated gap between the highest occupied band and lowest unoccupied band is 10.14 eV. [It was noted in Chapter 2 that the *ab initio* minimal basis (STO-3G) gap is 8.3 eV.] The $\varepsilon(k)$ curves for each band are given on the right-hand side of the diagram. The calculated charge distribution is recorded in Figure 3.1, and reflects accurately the low polarity expected for the C–H bond.

For the MINDO/2 calculations, the C–H bond length was increased to 1.19 Å as recommended by Dewar *et al.*[19] for molecular calculations.

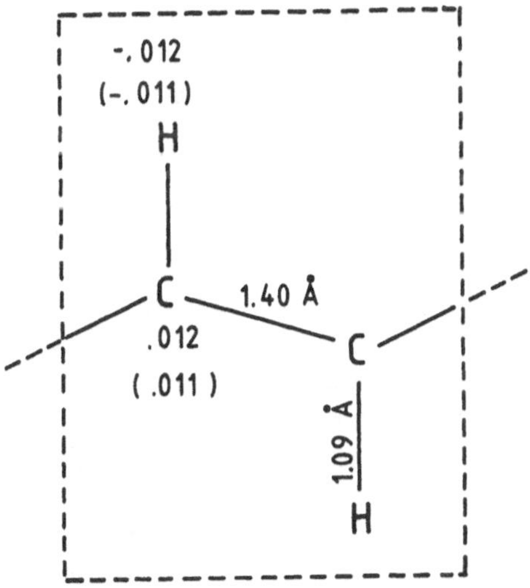

FIGURE 3.1. The unit-cell geometry and calculated net atomic charges for polyacetylene using the INDO and MINDO/2 methods. MINDO/2 charge values are in parentheses.

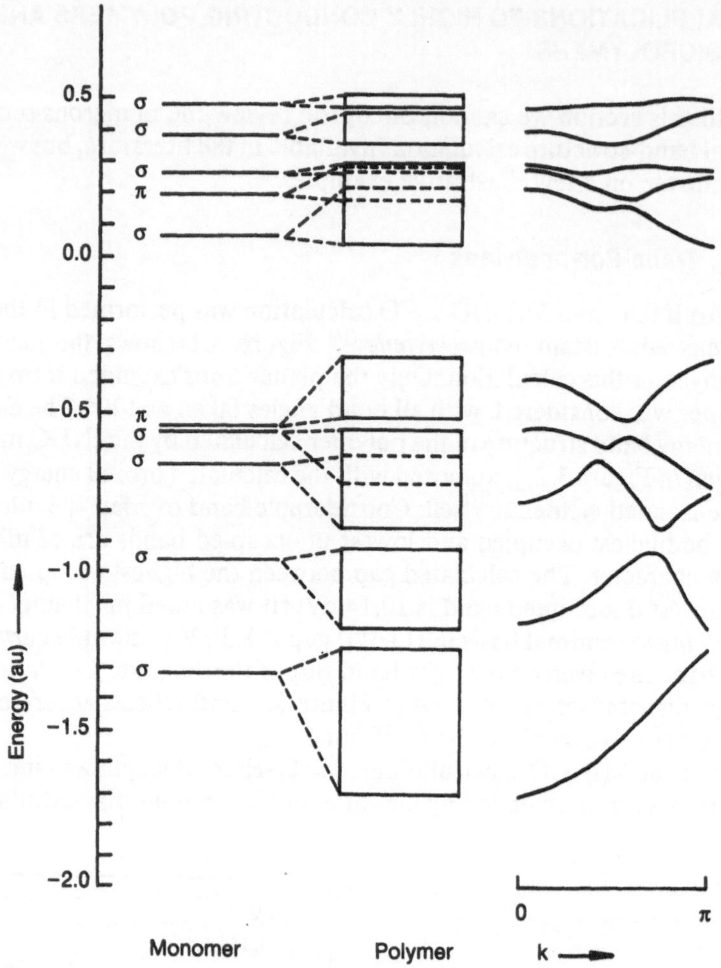

FIGURE 3.2. Energy-band structure of polyacetylene computed with the aid of the INDO CO method.

The calculated band structure was similar qualitatively to the INDO results, but differed in quantitative detail. In particular, the calculated band gap was 5.97 eV, substantially lower than the INDO value.[16] The MINDO/2 calculated net atomic charges are recorded in parentheses in Figure 3.1 and show C–H with slightly less charge separation than the INDO values.

The INDO calculated band structure and the corresponding CNDO calculation can be compared by inspecting our Figure 3.2 with Figure 3 of

O'Shea and Santry.[20] The results, as expected, are very similar, considering the slight difference in geometrical parameters. We note that for overlapping bands in our methodology, the σ/π nature of a given band may change as k varies, and resolution into pure bands is necessary for direct comparison with their calculations. This is readily accomplished by inspection of the matrices $\mathbf{C}(k)$.

3.2.2. TCNQ and TTF Stacks

It is well known that the mixed three-dimensional molecular crystal built up from alternating 7,7′,8,8′-tetracyanoquinodimethane (TCNQ) and tetrathiofulvalene (TTF) stacks exhibits metallic conductivity owing to charge transfer of about $0.6e$ from a TTF molecule to a TCNQ unit. For the band-structure calculations of the two stacks we have applied the CNDO/2 method, and in the case of the TCNQ stack also the MINDO/2 method.[21] (No MINDO/2 calculations could be conducted for the TTF chains owing to the lack of a parametrization scheme for sulfur at that time.) All band-structure calculations were performed in the first-neighbor-interaction approximation. This approximation is well justified in view of the large distances between neighboring molecules in the chain.

The following SCF criteria were applied during the course of calculations:

$$|P(0)_{r,s}^{(n+1)} - P(0)_{r,s}^{(n)}| < 5 \times 10^{-4} \text{ and } |P(1)_{r,s}^{(n+1)} - P(1)_{r,s}^{(n)}| < 5 \times 10^{-4}$$

where $P(q)_{r,s}^{(n)}$ denotes the r, s element of matrix $P(q)$ ($q = 0, 1$) obtained at the nth iteration step. In order to fulfill these criteria simultaneously 16–30 iteration steps were required, whereby a convergence-acceleration procedure was used.

In order to calculate the elements of matrices $\mathbf{P}(q)$, a numerical integration over k in the interval $(0, \pi/a)$ is required. This was performed using Simpson's rule.

For each TCNQ and TTF chain the eigenvalue problem of a complex Hermitian matrix of order 68 and 64, respectively, had to be solved 80–150 times. The calculations were accelerated by diagonalizing the Hermitian complex matrices with the aid of a complex matrix eigenvalue program based on the QR algorithm.[22] The resulting computing time was only 6–9 min for one chain on the IBM 360/91 computer.

Table 3.1 shows the band structures obtained for neutral TCNQ stacks with $a = 3.17$ Å. We also tested the case $a = 3.45$ Å, which is the stacking distance in pure TCNQ crystals. Table 3.1 also contains MINDO/2 results for poly(TCNQ) at $a = 3.17$ Å.

TABLE 3.1. Energy Bands of Poly(TCNQ) (in eV) Calculated in the CNDO/2 CO Approximation

	Poly(TCNQ) ($a = 3.17$ Å)				Poly(TCNQ) ($a = 3.45$ Å)		
	ε_{MO}	ε_{min} (k_{min})	ε_{max} (k_{max})	$\delta\varepsilon$	ε_{min} (k_{min})	ε_{max} (k_{max})	$\delta\varepsilon$
Second lowest unfilled band	3.56	3.12 (0)[a] − 1.28 (0)	3.24 (π/a) − 1.21 (π/a)	0.12 0.07	3.19 (0)	3.25 (π/a)	0.06
Conduction band	− 1.23	− 2.21 (0) − 3.50 (0)	− 0.83 (π/a) − 2.87 (π/a)	1.38 0.63	− 1.91 (0)	− 1.14 (π/a)	0.77
Valence band	− 10.06	− 10.43 (π/a) − 9.81 (π/a)	− 10.34 (0) − 9.75 (0)	0.09 0.06	− 10.37 (π/a)	− 10.34 (0)	0.03
Second highest filled band	− 13.90	− 14.34 (π/a) − 10.97 (π/a)	− 14.17 (0) − 10.93 (0)	0.17 0.04	− 14.26 (π/a)	− 14.18 (0)	0.08
Lowest filled band	− 53.42	− 55.98 (0) − 44.81 (0)	− 52.01 (π/a) − 41.12 (π/a)	3.97 3.69	− 54.98 (0)	52.63 (π/a)	2.35

[a] The second set of numbers represents MINDO/2 CO results.

TABLE 3.2. Energy Bands of Poly(TTF) (in eV) Calculated in the CNDO/2 CO Approximation

	Poly(TTF) ($a = 3.47$ Å)				Poly(TTF) ($a = 3.62$ Å)		
	ε_{MO}	ε_{min} (k_{min})	ε_{max} (k_{max})	$\delta\varepsilon$	ε_{min} (k_{min})	ε_{max} (k_{max})	$\delta\varepsilon$
Second lowest unfilled band	0.76	0.86 ($\pi/4a$)	0.93 (π/a)	0.07	0.85 ($\pi/4a$)	0.89 (π/a)	0.04
Conduction band	0.55	0.53 (π/a)	0.63 ($\pi/4a$)	0.10	0.56 (π/a)	0.63 ($\pi/4a$)	0.07
Valence band	− 9.61	− 10.03 (π/a)	− 9.03 (0)	1.00	− 9.92 (π/a)	− 9.16 (0)	0.76
Second highest filled band	− 11.43	− 11.38 ($\pi/4a$)	− 11.37 (π/a)	0.01	− 11.38 (0)	− 11.37 (π/a)	0.01
Lowest filled band	− 44.65	− 45.67 (0)	− 43.50 (π/a)	2.17	− 45.38 (0)	− 43.78 (π/a)	1.60

The geometry of the TCNQ molecule was taken from work of Long *et al.*[23] (see Figure 3.3). The relative positions of the different molecules were chosen as they were found in TCNQ–TTF crystals.

Within the CNDO/2 scheme we found a very broad π-like conduction band (1.4 eV) compared with a (also π-like) valence band of about 0.1 eV width. The dependence of bandwidth on stacking distance can be seen in Table 3.1.

Similar results are found within the MINDO/2 approximation scheme. The width of the conduction band of the TCNQ chain (0.63 eV) is again much larger than that of the valence band (0.06 eV), both bands again being π-like. However, it is one-half that found by the CNDO/2 method.

Table 3.2 shows the band-structure results within the CNDO/2 scheme for the neutral TTF stack with $a = 3.47$ Å. This corresponds to the stacking distance within the TTF–TCNQ crystal. We have also treated the case $a = 3.62$ Å, which corresponds to the pure TTF crystal. The geometry of the molecule was taken from work of Cooper *et al.*[24] (see Figure 3.4).

For the neutral stack we obtain a broad π-type valence band (1.0 eV) and a narrower σ-type conduction band (0.1 eV).

In Table 3.3 the charge distributions in the stack obtained for a TCNQ and TTF molecule, respectively, are shown. They clearly indicate strong charge transfer (CT) between the two molecules.

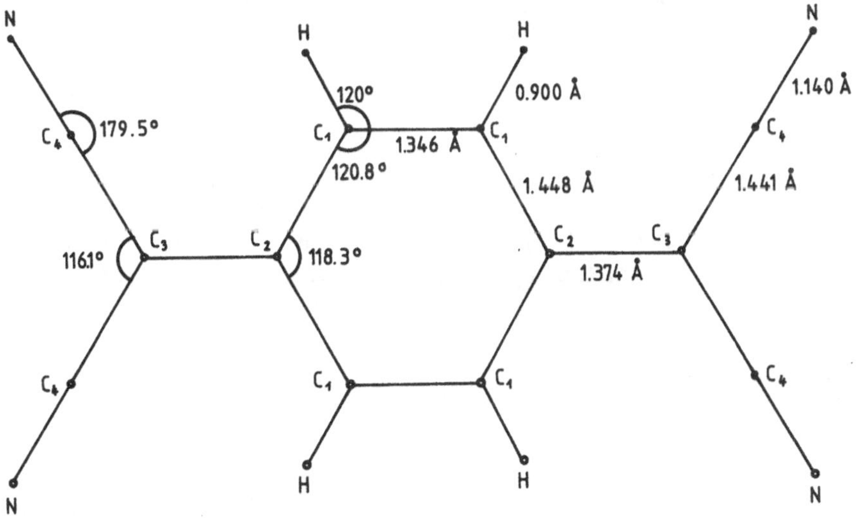

FIGURE 3.3. The molecular geometry of TNCQ (after Lang *et al.*[23]).

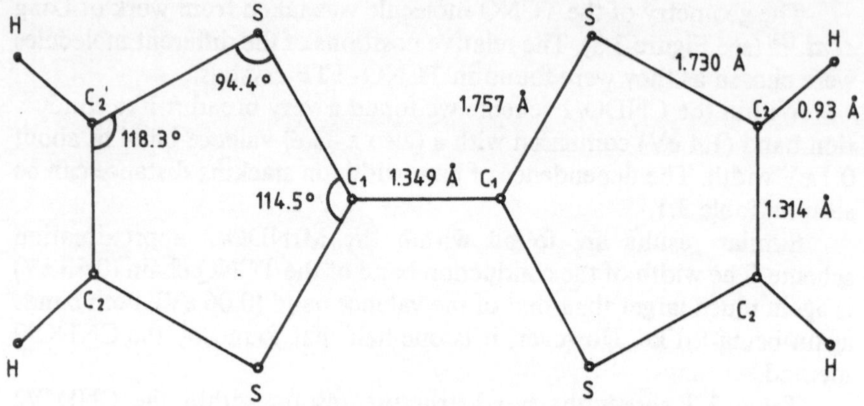

FIGURE 3.4. The molecular geometry of TTF (after Cooper *et al.*[24]).

TABLE 3.3. Net Charges in the Unit Cell of Poly(TCNQ) and Poly(TTF) Chains Obtained in the CNDO/2 CO Approximation

	Poly(TCNQ)			Poly(TTF)	
Atom	$a = 3.17$ Å	$a = 3.45$ Å	Atom	$a = 3.47$ Å	$a = 3.62$ Å
C_1	−0.015	−0.014	C_1	0.050	0.050
C_2	0.056	0.056	S	−0.068	−0.068
C_3	0.016	0.017	C_2	0.005	0.005
C_4	0.089	0.089	H	0.038	0.038
N	−0.140	−0.140			
H	0.030	0.030			

The most striking result of these calculations is that the conduction band of the TCNQ stack (to which CT occurs) and the valence band of TTF (from which CT takes place) are, in the CNDO/2 case, one order of magnitude larger than the physically insignificant valence band of TCNQ and conduction band of TTF, respectively. The same is also true in the MINDO/2 CO calculation of the TCNQ stack, though now the width of the bands is about one half the CNDO/2 values. If we scale down the CNDO/2 widths of the valence band of the TTF stack by this factor 2, we obtain approximately 0.5 eV, which is in good agreement with the experimental result of 0.4–0.5 eV.[25]

With respect to the shape of the conduction bands, the following results are evident: The valence band of poly(TTF) is cosine-like as well as the conduction band of poly(TCNQ) with the minimum at $k = 0$ for TCNQ and $k = \pi/a$ for TTF. A fit of the five different values of k to the expansion $E(k) = \Sigma_{\nu=0}^{4} \alpha_\nu \cos(\nu ka)$ showed that α_2/α_1 equals about 0.01. The situation was found to differ for the valence band of TCNQ and the conduction band of TTF. These bands did not show a simple cosine-like behavior, but rather α_2/α_1 was found to be 0.3. Higher-order coefficents α_ν were found to be negligible. Stronger deviations from a cosine-like behavior can be found for lower-lying bands.

3.2.3. Periodic DNA and Protein Models

The geometry used in the periodic DNA models was developed by Spencer[26] and Langridge *et al.*,[27] while for the two-dimensional polyglycine network a simplified model geometry (with 90° valence angles) and the parallel-chain pleated-sheet conformation of β polyglycine[28] has been used.

The band structures of the periodic DNA models* obtained with the aid of the PPP CO method can be divided into two groups. The five homopolynucleotides (also including poly U)[3] and the two poly (base pairs) [poly(A–T) and poly(G–C)][29] possess rather broad bands (the widths of the valence bands are 0.2–0.3 eV, those of the conduction bands about 0.1 eV, and the widths of the lowest filled bands are in most cases approximately 1 eV). On the other hand, for the more complicated periodic DNA models, where we have two different base pairs in the unit cell, such as poly($^{\text{A-T}}_{\text{G-C}}$),[30] the bands are very narrow (the widths of the valence bands are usually 0.01–0.03 eV, those of the conduction bands about 0.01 eV, and the widths of the lowest filled bands are approximately 0.1 eV). The CNDO/2 CO calculations of periodic base stacks[31] and of the sugar–phosphate chain of DNA[32] again give rather broad bands [valence bandwidths between 0.15 and 0.50 eV for the base stacks and 0.05 eV for poly(SP), and conduction bandwidths 0.1–0.25 eV for both kinds of system], while the MINDO/2 CO results for the base stacks indicate somewhat less-broad bands (widths of valence bands 0.03–0.30 eV and of conduction bands 0.03–0.17 eV). Further details of the PPP and all-valence-electron CO calculations of different periodic DNA models can be found in original papers[3,29–32] as well as in a review paper.[33]

On the basis of perturbation theory it is easy to understand that if the

* The number of necessary iteration steps was 5–7 for periodic DNA models with smaller unit cells, while in cases of a large elementary cell (such as two different interacting base pairs forming the unit cell) 10 or 11 iterations were required to fulfill the criteria (3.13).

bases or base pairs are repeated [periodic base stacks of the poly(A–T), poly(G–C) systems], splitting of the levels of the constituent single systems is much larger in the polymer than if different subsystems are alternating in it (as is the case in all periodic DNA models for which narrow bands have been obtained).

We can conclude on the basis of the band structures obtained that for the complicated narrow-band periodic DNA models, a delocalized, Bloch-type description of charge transport is not very correct and therefore the application of the hopping model seems more realistic. On the other hand, the Bloch picture is justified in the cases of periodic DNA models with broad bands. These conclusions are also supported by the calculation of the mean free path and mobility values of the different periodic DNA models starting from their band structures[34] (see Chapter 9 for more details).

The width of the forbidden band between the valence and conduction bands was found from the PPP CO calculation of the poly($\substack{\text{A-T}\\\text{G-C}}$) system to be about 6.1 eV.[30] On the other hand, the UV absorption maximum of DNA lies at approximately 4.5 eV. Therefore, the PPP CO method has given a seemingly too high gap. The reason for this difficulty, and a detailed discussion can be found elsewhere,[29] is that in any closed-shell Hartree–Fock scheme (as also in the PPP scheme) the excitation energy is not just the difference of the corresponding one-electron energies, but

$$^1\Delta E_{i\to j} = \varepsilon_j(k) - \varepsilon_i(k) - J_{i,j}(k) + 2K_{i,j}(k) \tag{3.24}$$

Here the $J_{i,j}(k)$ Coulomb and $K_{i,j}(k)$ exchange integrals disappear if the number of cells $N \to \infty$.[29] (If we take into account all neighbor interactions they disappear as $\ln N/N$.[35]) Thus equation (3.24) reduces to the difference of one-electron energies, giving too high results. The physical explanation of the problem is that, using equation (3.24), we treat the motion of the excited electron and the remaining hole as completely uncorrelated, which is not correct. [The correct treatment of the gap will be discussed in detail in Chapter 5 and the calculation of excited (exciton) states in Chapter 8.]

As a first step in investigating the effect of impurities on the electronic structure of DNA, the band structure of poly(G–C) has been computed in the PPP CO approximation assuming that one of two water molecules is bound by hydrogen bonds to the NH_2 group of each C molecule or/and to the $C=O$ group of each G molecule.[36] According to the results obtained for these systems the additional π-orbital of the H_2O

molecules produced an extra π-band between the lowest filled bands, while the other band remained practically unchanged.[36]

The next step was to recalculate the band structures of poy(G–C) and poly(A–T) in the presence of Mg^{2+} ions again using the PPP CO method modified suitably to account for the presence of charged ions (details can be found elsewhere[36,37]). (The ability of divalent metal ions, especially Mg^{2+} ions, to react with a variety of electron-donor sites of the polynucleotide chains was demonstrated experimentally.[38]) The calculations were performed for the poly(G–C) and poly(A–T) systems with all possible types of Mg^{2+} attachments to the heterocyclic bases (including both in the plane of the base pairs and out-of-plane positions), taking one Mg^{2+} per unit cell. According to the results obtained for these model calculations the presence of the Mg^{2+} ions drastically changes the band structures: the bands become generally broader by a factor of 2–3 and also their positions change considerably. This causes great changes in the band gap, for instance, if the Mg^{2+} ion is attached to the NH_2 nitrogen of C in the G–C base pair it decreases from about 6.0 eV to 2.0 eV.

If the interactions in the case of one-dimensional polyformamide chains are taken along the hydrogen bonds, the widths of the valence and conduction bands are about 0.2 and 0.3 eV, respectively, in the CNDO/2 case, while the MINDO/2 CO calculations provide smaller widths ($\sim$0.1 eV and $\sim$0.05 eV, respectively).[12] On the other hand, if the interactions are taken along the main chain, the corresponding bandwidths are much larger (the width of the valence band is given by $\delta\varepsilon_v \approx 2.6$ eV and the width of the conduction band $\delta\varepsilon_c \approx 0.6$ eV in the CNDO/2 CO case, while the corresponding MINDO/2 CO results are $\delta\varepsilon_v \approx 0.09$ eV and $\delta\varepsilon_c \approx 2.5$ eV). Finally, taking into account both types of interaction, the CNDO/2 CO results for the two-dimensional polyformamide network are $\delta\varepsilon_v \approx 2.6$ eV and $\delta\varepsilon_c \approx 0.8$ eV, while the MINDO/2 CO results are $\delta\varepsilon_v \approx 2.0$ eV and $\delta\varepsilon_c \approx 1.6$ eV, indicating that periodic protein models must really be treated as two-dimensional networks.[12]

In a subsequent calculation of a two-dimensional polyglycine model network, in which the valence angles of a glycine molecule were distorted to rectangular so as to be able to place only one glycine molecule in the unit cell, it was demonstrated that the band structures are essentially different (for obvious geometrical reasons) if we take into account not only first-neighbor interactions but also the effect of second neighbors.[12] Here, for $\delta\varepsilon_v \sim 2.1$ eV $\rightarrow \sim 1.5$ eV and for $\delta\varepsilon_c \sim 2.6$ eV $\rightarrow \sim 3.1$ eV in the CNDO/2 case; for $\delta\varepsilon_v \sim 2.7$ eV $\rightarrow 1.8$ eV and for $\delta\varepsilon_c \sim 3.0$ eV $\rightarrow \sim 3.5$ eV if we go from first-neighbor to second-neighbor interactions.

Finally, the MINDO/2 CO calculation of the two-dimensional parallel-chain pleated-sheet β-polyglycine network (two glycine molecules in the unit cell) using second-neighbor interactions has resulted in $\delta\varepsilon_v \approx 1.2$ eV, $\delta\varepsilon_c \approx 1.7$ eV, and a forbidden bandwidth of about 4.8 eV,[39] showing that the choice of the geometry considerably influences the band structure. Further, it should be noted that while in the cases of the polyformamide and of the model polyglycine network we had planar structures and so could define σ- and π-electron bands (in both cases the valence band was a π band and the conduction band a σ band), this classification becomes impossible in the case of the parallel-chain pleated-sheet conformation of β-polyglycine because it is not planar.

REFERENCES

1. T. E. Peacock and R. McWeeny, *Proc. R. Soc. London* **74**, 385 (1959); J. Ladik, *Acta Phys. Acad. Sci. Hung.* **18**, 185 (1965).
2. R. Pariser and R. G. Parr, *J. Chem. Phys.* **21**, 466, 707 (1953); J. A. Pople, *Trans. Faraday Soc.* **49**, 375 (1953).
3. J. Ladik, D. K. Rai, and K. Appel, *J. Mol. Spectrosc.* **27**, 72 (1968); J. Ladik, in: *Electronic Structure of Polymers and Molecular Crystals* (J.-M. André and J. Ladik, eds.), p. 663, Plenum Press, London–New York (1975).
4. J. Hinze and J. Jaffé, *J. Am. Chem. Soc.* **84**, 540 (1962).
5. N. Mataga and K. Nishimoto, *Z. Phys. Chem.* **13**, 140 (1977).
6. J. Ladik, *Acta, Phys. Acad. Sci. Hung.* **18**, 173 (1965).
7. J. Koutecký and R. Zahradnik, *Collect. Czech. Chem. Commun.* **25**, 811 (1960); J. Ladik and K. Appel, *J. Chem. Phys.* **40**, 2470 (1964).
8. J. A. Pople, D. P. Santry, and G. A. Segal, *J. Chem. Phys.* **43**, 129 (1965); J. A. Pople and G. A. Segal, *J. Chem. Phys.* **43**, 136 (1965).
9. J. Ladik and G. Biczó, *Acta Chim. Acad. Sci. Hung.* **64**, 397 (1971); H. Fujita and A. Imamura, *J. Chem. Phys.* **53**, 4555 (1970); K. Morokuma, *J. Chem. Phys.* **54**, 962 (1971).
10. J. A. Pople and D. C. Beveridge, *Approximate Molecular Orbital Theory*, McGraw-Hill Publ. Co., New York (1970).
11. N. C. Baird and M. J. S. Dewar, *J. Chem. Phys.* **54**, 1262 (1967).
12. S. Suhai and J. Ladik, *Theor. Chem. Acta* **28**, 27 (1972); S. Suhai and J. Ladik, *Acta Chim. Hung.* **82**, 67 (1974).
13. K. Ohno, *Theor. Chim. Acta* **2**, 219 (1964); G. Klopman, *J. Am. Chem. Soc.* **87**, 3300 (1965).
19. S. Suhai and J. Ladik, *Int. J. Quantum Chem.* **7**, 547 (1973).
15. J. A. Pople, D. L. Beveridge, and P. A. Dobosh, *J. Chem. Phys., Suppl.* **47**, 2026 (1967); R. N. Dixon, *Mol. Phys.* **12**, 83 (1967).
16. D. L. Beveridge, I. Jano, and J. Ladik, *J. Chem. Phys.* **56**, 4744 (1972).
17. R. Hoffmann, *J. Chem. Phys.* **39**, 1397 (1963).
18. See, for instance, J.-M. André, in: *Electronic Structure of Polymers and Molecular Crystals* (J.-M. André and J. Ladik, eds.), p. 1, Plenum Press, London–New York (1975).
19. M. J. S. Dewar and E. Haselbach, *J. Am. Chem. Soc.* **92**, 590 (1970).

20. S. O'Shea and D. P. Santry, *J. Chem. Phys.* **54**, 2667 (1971).
21. A. Karpfen, J. Ladik, G. Stollhoff, and P. Fulde, *Chem. Phys.* **8**, 215 (1975).
22. See, for instance, J. H. Wilkinson, *The Algebraic Eigenvalue Problem*, p. 515, Clarendon Press, Oxford (1965).
23. R. E. Long, R. A. Sparks, and K. N. Trueblood, *Acta Crystallogr.* **18**, 932 (1965).
24. W. F. Cooper, N. C. Hemy, J. W. Edmonds, A. Nagel, F. Wudl, and P. Coppens, *Chem. Commun.* 889 (1971).
25. D. Allender, J. W. Bray, and J. Bardeen, *Phys. Rev. B* **9**, 119 (1974); A. A. Bright, A. F. Garito, and A. J. Heeger, *Solid State Commun.* **13**, 943 (1973).
26. M. Spencer, *Acta Crystallogr.* **12**, 59, 66 (1959).
27. R. Langridge, J. Marvin, W. Seeds, H. R. Wilson, C. W. Hooper, M. H. F. Wilkins, and L. D. Hamilton, *J. Mol. Biol.* **2**, 38 (1960).
28. L. Pauling and R. B. Corey, *Proc. Natl. Aci. Sci. U.S.A.* **39**, 253 (1953).
29. J. Avery, J. Packer, J. Ladik, and G. Biczó, *J. Mol. Spectrosc.* **29**, 194 (1969).
30. B. F. Rozsnyai, F. Martino, and J. Ladik, *J. Chem. Phys.* **52**, 5708 (1970).
31. S. Suhai and J. Ladik, *Int. J. Quantum Chem.* **7**, 547 (1973).
32. S. Suhai, *Biopolymers*, **13**, 1739 (1974).
33. J. Ladik, in: *Advances in Quantum Chemistry* (P.-O. Löwdin, ed.), p. 397, Academic Press, New York–London (1973).
34. S. Suhai, *J. Chem. Phys.* **57**, 5599 (1972).
35. A. Bierman and J. Ladik (unpublished).
36. B. F. Rozsnyai and J. Ladik, *J. Chem. Phys.* **52**, 5711 (1970).
37. B. F. Rozsnyai and J. Ladik, *J. Chem. Phys.* **53**, 4325 (1970).
38. G. L. Eichhorn and Y. A. Shin, *J. Am. Chem. Soc.* **90**, 7323 (1970); P. Cheng, D. Honbo, and J. Rozsnyai, *Biochemistry* **1**, 239 (1969).
39. J. Josse, A. D. Kaiser, and A. Kornberg, *J. Biol. Chem.* **236**, 864 (1961); see also a recent statistical analysis of longer nucleic acid sequences: J. F. Gentleman, M. A. Shadbolt-Forbes, J. W. Hawkins, J. Ladik, and W. F. Forbes, *Mathematical Scientist* **9**, 125 (1984).
39. S. Suhai, *Theor. Chim. Acta* **34**, 157 (1974).

Chapter 4

The Treatment of Aperiodicity in Polymers

4.1. ELEMENTARY GREEN FUNCTION THEORY

We introduce the elementary theory of Green functions following Raines[1] by looking for the solution of a simple inhomogeneous differential equation. That is, we do not introduce second quantization and diagrammatic techniques (as more sophisticated treatments of Green functions do[2] at this point).

4.1.1. Solution of Inhomogeneous Differential Equation by Means of Green Functions

We consider the differential equation

$$(\hat{L} - \lambda)u(\mathbf{r}) = f(\mathbf{r}) \tag{4.1}$$

where $\hat{L}$ is a linear differential operator, λ a given constant, and $f(\mathbf{r})$ a given function. Equation (4.1) is inhomogeneous because $u(\mathbf{r})$ does not appear as a factor in the term on the right-hand side. It is clear that, provided $f(\mathbf{r})$ is not identically zero, any multiple of a solution is not itself a solution. The equation

$$(\hat{L} - \lambda)\psi(\mathbf{r}) = 0 \tag{4.2}$$

on the other hand, is homogeneous, and any multiple of a solution is itself a solution.

We wish to solve equation (4.1) within some region of space V, subject to given boundary conditions. However, although the equation is inhomogeneous, we shall assume that the boundary conditions are homogeneous: that is to say, any multiple of $u(\mathbf{r})$ satisfies the same boundary conditions as $u(\mathbf{r})$. For example, $u(\mathbf{r}) = 0$ over the boundary surface of V is a homogeneous boundary condition, and so are the periodic boundary conditions familiar in the theory of metals.

We shall now suppose that $\psi_n(\mathbf{r})$ is an eigenfunction of equation (4.2) corresponding to the eigenvalue λ_n, namely

$$(\hat{L} - \lambda_n)\psi_n(\mathbf{r}) = 0 \tag{4.3}$$

assuming the same boundary conditions pertaining to equation (4.1), and we expand $u(\mathbf{r})$ and $f(\mathbf{r})$ in terms of the complete set of $\psi_n(\mathbf{r})$. Thus

$$u(\mathbf{r}) = \sum_n a_n\psi_n(\mathbf{r}) \quad \text{and} \quad f(\mathbf{r}) = \sum_n b_n\psi_n(\mathbf{r}) \tag{4.4}$$

where it is assumed for simplicity that the eigenvalues λ_n form a discrete set. Equation (4.1) then becomes

$$\sum_n a_n(\lambda_n - \lambda)\psi_n(\mathbf{r}) = \sum_n b_n\psi_n(\mathbf{r}) \tag{4.5}$$

giving

$$a_n = \frac{b_n}{\lambda_n - \lambda} \tag{4.6}$$

Now, if ψ_n is normalized within V so that

$$\int |\psi_n|^2 d\mathbf{r} = 1 \tag{4.7}$$

we obtain

$$b_n = \int \psi_n^*(\mathbf{r})\, f(\mathbf{r}) d\mathbf{r} \tag{4.8}$$

the integrals being taken throughout V. Hence

$$\begin{aligned} u(\mathbf{r}) &= \sum_n \frac{\psi_n(\mathbf{r}) \int \psi_n^*(\mathbf{r}')\, f(\mathbf{r}') d\mathbf{r}'}{\lambda_n - \lambda} \\ &= -\int G(\mathbf{r}, \mathbf{r}'; \lambda)\, f(\mathbf{r}') d\mathbf{r}' \end{aligned} \tag{4.9}$$

where

$$G(\mathbf{r}, \mathbf{r}'; \lambda) = \sum_n \frac{\psi_n(\mathbf{r})\psi_n^*(\mathbf{r}')}{\lambda - \lambda_n} \tag{4.10}$$

which is called the Green function for the problem.

In order to find the differential equation satisfied by the Green function, we set

$$f(\mathbf{r}) = -\delta(\mathbf{r} - \mathbf{r}_0)$$

Then equation (4.9) yields

$$u(\mathbf{r}) = \int G(\mathbf{r}, \mathbf{r}'; \lambda)\delta(\mathbf{r}' - \mathbf{r}_0)d\mathbf{r}' = G(\mathbf{r}, \mathbf{r}_0; \lambda) \tag{4.11}$$

from which we obtain that $G(\mathbf{r}, \mathbf{r}'; \lambda)$ satisfies the equation

$$(\hat{L} - \lambda)G(\mathbf{r}, \mathbf{r}'; \lambda) = -\delta(\mathbf{r} - \mathbf{r}') \tag{4.12}$$

subject to the same boundary conditions as $u(\mathbf{r})$.

On the assumption that quantities λ_n are real (as is the case when $\hat{L}$ is a Hermitian operator) and that λ is real, it follows from equation (4.10) that

$$G(\mathbf{r}, \mathbf{r}'; \lambda) = [G(\mathbf{r}', \mathbf{r}; \lambda)]^* \tag{4.13}$$

A difficulty occurs if λ is equal to one of the eigenvalues, say λ_m, of $\hat{L}$. In this case, equation (4.5) has the form

$$\sum_n a_n(\lambda_n - \lambda_m)\psi_n(\mathbf{r}) = \sum_n b_n\psi_n(\mathbf{r}) \tag{4.14}$$

which can be so only if $b_m = 0$. In other words, equation (4.1) has a solution in this case only if

$$b_m = \int \psi_m^*(\mathbf{r})\, f(\mathbf{r})d\mathbf{r} = 0 \tag{4.15}$$

A useful application of the Green function method is the conversion of a homogeneous differential equation into an integral equation. For example, the equation

$$[\hat{L} + V(\mathbf{r}) - \lambda]u(\mathbf{r}) = 0$$

or

$$(\hat{L} - \lambda)u(\mathbf{r}) = -V(\mathbf{r})u(\mathbf{r}) \tag{4.16}$$

is homogeneous. However, by simply applying the procedure described above with $-V(\mathbf{r})u(\mathbf{r})$ substituted for $f(\mathbf{r})$, equation (4.9) becomes

$$u(\mathbf{r}) = \int G(\mathbf{r}, \mathbf{r}'; \lambda)V(\mathbf{r}')u(\mathbf{r}')d\mathbf{r}' \tag{4.17}$$

with $G(\mathbf{r}, \mathbf{r}'; \lambda)$ again given by equation (4.10) and satisfying equation (4.12). Equation (4.17) is an integral equation for $u(\mathbf{r})$.

Although here we have specified three spatial dimensions, clearly the procedure is the same for any number of dimensions.

4.1.2. Application of Green Functions to the Solution of the Time-Independent Schrödinger Equation

The use of Green functions is now widespread in atomic and molecular physics, solid-state theory, nuclear physics, and the theory of many-particle systems in general. At this point we consider the simple problem of the solution of the one-electron Schrödinger equation by Green functions.

We consider a single electron moving in an electrostatic field with potential energy $V(\mathbf{r})$. The time-independent Schrödinger equation is then

$$-\tfrac{1}{2}\nabla^2\psi(\mathbf{r}) + V(\mathbf{r})\psi(\mathbf{r}) = E\psi(\mathbf{r}) \tag{4.18}$$

which we shall write in the form

$$(\hat{H}_0 - E)\psi(\mathbf{r}) = -V(\mathbf{r})\psi(\mathbf{r}) \tag{4.19}$$

where

$$\hat{H}_0 = -\frac{1}{2m}\nabla^2 \tag{4.20}$$

Since we shall be mainly concerned with applications to the theory of solids (particularly to polymers), periodic boundary conditions will be applied over a large cube, say, of side L and volume $V = L^3$.

According to the method described in the previous section, a solution of equation (4.19) is given by the integral equation

$$\psi(\mathbf{r}) = \int G(\mathbf{r}, \mathbf{r}'; E)V(\mathbf{r}')\psi(\mathbf{r}')d\mathbf{r}' \tag{4.21}$$

where $G(\mathbf{r}, \mathbf{r}'; E)$ is the Green function for the problem and satisfies the equation

$$(\hat{H}_0 - E)G(\mathbf{r}, \mathbf{r}'; E) = -\delta(\mathbf{r} - \mathbf{r}') \tag{4.22}$$

subject to the same periodic boundary conditions as $\psi(\mathbf{r})$. The integral in equation (4.21) is taken throughout the volume V.

That function (4.21) is indeed a solution of equation (4.19) can be verified immediately by substitution and the use of equation (4.22). In this way we obtain

$$\begin{aligned}(\hat{H}_0 - E)\psi(\mathbf{r}) &= (\hat{H}_0 - E)\int G(\mathbf{r}, \mathbf{r}'; E)V(\mathbf{r}')\psi(\mathbf{r}')d\mathbf{r}' \\ &= \int (\hat{H}_0 - E)G(\mathbf{r}, \mathbf{r}'; E)V(\mathbf{r}')\psi(\mathbf{r}')d\mathbf{r}' \\ &= -\int \delta(\mathbf{r} - \mathbf{r}')V(\mathbf{r}')\psi(\mathbf{r}')d\mathbf{r}' \\ &= -V(\mathbf{r})\psi(\mathbf{r}) \end{aligned} \tag{4.23}$$

If $\psi_k(\mathbf{r})$ is an eigenfunction, and ε_k the corresponding eigenvalue, of equation (4.19) with the right-hand side put equal to zero, then

$$(\hat{H}_0 - \varepsilon_{\mathbf{k}})\psi_{\mathbf{k}}(\mathbf{r}) = 0 \tag{4.24}$$

subject to the same periodic boundary conditions as before. The normalized eigenfunctions of this latter equation are

$$\psi_{\mathbf{k}}(\mathbf{r}) = \frac{1}{\sqrt{V}}\exp(i\mathbf{k}\cdot\mathbf{r}) \tag{4.25}$$

with eigenvalues

$$\varepsilon_{\mathbf{k}} = k^2/2m \tag{4.26}$$

In terms of $\psi_{\mathbf{k}}(\mathbf{r})$ the Green function is given by

$$G(\mathbf{r}, \mathbf{r}'; E) = \sum_{\mathbf{k}} \frac{\psi_{\mathbf{k}}(\mathbf{r})\psi_{\mathbf{k}}^*(\mathbf{r}')}{E - \varepsilon_{\mathbf{k}}} \tag{4.27}$$

[cf. equation (4.10)]. This again may be verified by substitution in equation (4.22), since

$$(\hat{H}_0 - E)G(\mathbf{r}, \mathbf{r}'; E) = \sum_{\mathbf{k}} \frac{\psi_{\mathbf{k}}^*(\mathbf{r}')}{E - \varepsilon_{\mathbf{k}}} (\hat{H}_0 - E)\psi_{\mathbf{k}}(\mathbf{r})$$

$$= -\sum_{k} \psi_{\mathbf{k}}^*(\mathbf{r}')\psi_{\mathbf{k}}(\mathbf{r})$$

$$= -\delta(\mathbf{r} - \mathbf{r}') \tag{4.28}$$

by the closure or completeness property of the eigenfunctions.

Using equation (4.25), we find that

$$G(\mathbf{r}, \mathbf{r}'; E) = \frac{1}{V} \sum_{\mathbf{k}} \frac{\exp[i\mathbf{k} \cdot (\mathbf{r} - \mathbf{r}')]}{E - \varepsilon_{\mathbf{k}}} \tag{4.29}$$

and, assuming that the eigenvalues form a quasi-continuum, we may substitute an integral for the sum, according to the usual prescription

$$\sum_{\mathbf{k}} f(\mathbf{k}) \rightarrow \frac{V}{8\pi^3} \int f(\mathbf{k}) d\mathbf{k} \tag{4.30}$$

so that

$$G(\mathbf{r}, \mathbf{r}'; E) = \frac{1}{8\pi^3} \int \frac{\exp[i\mathbf{k} \cdot (\mathbf{r} - \mathbf{r}')]}{E - \varepsilon_{\mathbf{k}}} d\mathbf{k} \tag{4.31}$$

the integral being taken over the whole **k**-space.

One can form the Fourier transform of the Green function (4.31), i.e.,

$$G(\mathbf{k}; E) = 8\pi^3 \int G(\mathbf{r}, \mathbf{r}'; E) \exp[-i\mathbf{k} \cdot (\mathbf{r} - \mathbf{r}')] d\mathbf{r}\, d\mathbf{r}' \tag{4.32}$$

which results in the expression[2]

$$G(\mathbf{k}, E) = \frac{1}{E - \varepsilon_{\mathbf{k}}} \tag{4.33}$$

If we recall the theorem

$$f(\hat{L})\psi_n = f(\lambda_n)\psi_n \tag{4.34}$$

if

$$\hat{L}\psi_n = \lambda_n \psi_n \tag{4.35}$$

and take into account

$$\langle \psi_{\mathbf{k}} | \hat{H}_0 | \psi_{\mathbf{k}} \rangle = \varepsilon_{\mathbf{k}} \tag{4.36}$$

the Green operator $\hat{G}$ can be expressed in the form

$$\hat{G} = (E - \hat{H}_0)^{-1} \tag{4.37}$$

namely, substitution of this expression into the relationship

$$\begin{aligned} \langle \psi_{\mathbf{k}} | \hat{G}_0 | \psi_{\mathbf{k}} \rangle &= \langle \psi_{\mathbf{k}} | (E - \hat{H}_0)^{-1} | \psi_{\mathbf{k}} \rangle \\ &= \left\langle \psi_{\mathbf{k}} \left| \frac{1}{E - \varepsilon_{\mathbf{k}}} \right| \psi_{\mathbf{k}} \right\rangle = \frac{1}{E - \varepsilon_{\mathbf{k}}} \langle \psi_{\mathbf{k}} | \psi_{\mathbf{k}} \rangle = \frac{1}{E - \varepsilon_{\mathbf{k}}} \end{aligned} \tag{4.38}$$

yields back equation (4.33).

4.1.3. Simple Derivation of the Dyson Equation

If it is assumed that we work again with a nonorthogonal basis set $\{\chi_r\}$, then the unperturbed problem assumes the form

$$\mathbf{H}_0\mathbf{c} = E\mathbf{S}^0\mathbf{c} \tag{4.39}$$

The Green matrix of this system can be given, by analogy with expression (4.37), as

$$\mathbf{G}^0 = (E\mathbf{S}^0 - \mathbf{H}_0)^{-1} \tag{4.40}$$

Hence for the perturbed system with $\mathbf{H} = \mathbf{H}^0 + \mathbf{V}$

$$\mathbf{H}\mathbf{c} = (\mathbf{H}_0 + \mathbf{V})\mathbf{c} = E(\mathbf{S}^0 + \mathbf{S}')\mathbf{c} \tag{4.41}$$

where $\mathbf{S}'$ is the change in the overlap matrix due to the perturbation. Then the Green matrix of the perturbed system can be formulated as

$$\mathbf{G} = [E(\mathbf{S}^0 + \mathbf{S}') - (\mathbf{H}_0 + \mathbf{V})]^{-1} = [E\mathbf{S}^0 - \mathbf{H}_0 - (\mathbf{V} - E\mathbf{S}')]^{-1} \tag{4.42}$$

We recall the identity

$$(\mathbf{A} - \mathbf{B})^{-1} = \mathbf{A}^{-1} + \mathbf{A}^{-1}\mathbf{B}(\mathbf{A} - \mathbf{B})^{-1} \tag{4.45}$$

the validity of which is evident in the case of scalar quantities a and b:

$$\frac{1}{a-b} = \frac{1}{a} + \frac{b}{a(a-b)}$$

If matrix $\mathbf{A}$ is set equal to $E\mathbf{S}^0 - \mathbf{H}^0$ and $\mathbf{V} - E\mathbf{S}' = \mathbf{B}$, we obtain immediately

$$\mathbf{G} = [E\mathbf{S}^0 - \mathbf{H}^0]^{-1} + [E\mathbf{S}^0 - \mathbf{H}^0]^{-1}[\mathbf{V} - E\mathbf{S}'][E\mathbf{S}^0 - \mathbf{H}^0 - (\mathbf{V} - E\mathbf{S}')]^{-1} \tag{4.44}$$

The expression (4.40) finally yields the Dyson equation from (4.44) in the form

$$\mathbf{G} = \mathbf{G}^0 + \mathbf{G}^0 \triangle \mathbf{G}, \qquad \triangle = \mathbf{V} - E\mathbf{S}' \tag{4.45}$$

which provides a connection between the one-particle Green matrices of the unperturbed and perturbed systems ($\mathbf{G}^0$ and $\mathbf{G}$), respectively.

4.2. DEMONSTRATION OF THE EFFECTS OF APERIODICITY ON THE ELECTRONIC STRUCTURE OF POLYMERS

Most polymers are in themselves disordered (aperiodic). The nucleic acids and proteins are disordered quasi-one-dimensional (1D) systems built up of 4 or 20 different subunits, respectively, in a nonrepetitive way. Many copolymers, which are important as plastics, are aperiodic too. Even if we start with a periodic polymer (like polyacetylene) and dope it with electron acceptors or donors, the system becomes disordered due to the doping. Similarly, if an inhomogeneous magnetic field with a nonzero component in the chain's direction acts on a periodic polymer, the chain becomes aperiodic. Therefore, the methods used in solid-state physics to treat three-dimensional (3D) disordered systems must be applied (after proper modification) to aperiodic polymers. The reason that Green functions were introduced in the previous section (in a very elementary way) is that most of these methods [such as the coherent potential approximation (CPA), see Section 4.3; or methods to investigate the localization properties of wave function in a disordered chain, see Section 4.5] apply this formalism. There is, however, another method [the negative-factor counting technique (NFC), that we have developed further from its original form, see Section 4.4], which is especially suited to treat disordered quasi-1D systems. However, before applying these more sophisticated methods to disordered chains, we shall demonstrate in two simple cases (two- and three-component polypeptide chains and a ring of H atoms with a disordered charge distribution) the effect of aperiodicity on the energy-level distribution and on the localization properties of the wave functions.

4.2.1. Effect of Side-Chain Disorder on the Electronic Structure of Proteins

The importance of aperiodicity on the electronic structure of proteins can be demonstrated by taking a polypeptide chain composed of three different amino-acid residues: glycine, alanine, and serine (see Figure 4.1).

In the calculations of the electronic structure of these polypeptides,[3] we applied the all-valence electron MINDO crystal-orbital method in its MINDO/3 parametrization.[4] The calculations involved atomic coordinates corresponding to the antiparallel-β-pleated-sheet structure of Pauling and Corey[5]; hence we had a combined symmetry operation in the case of poly(gly): each gly unit was obtained in this polymer from its neighboring one by a translation along the helical axis through 3.25 Å and by a simultaneous 180° rotation around it. In order to preserve the cyclic property of the Fock hypermatrix of the whole polymer (which is

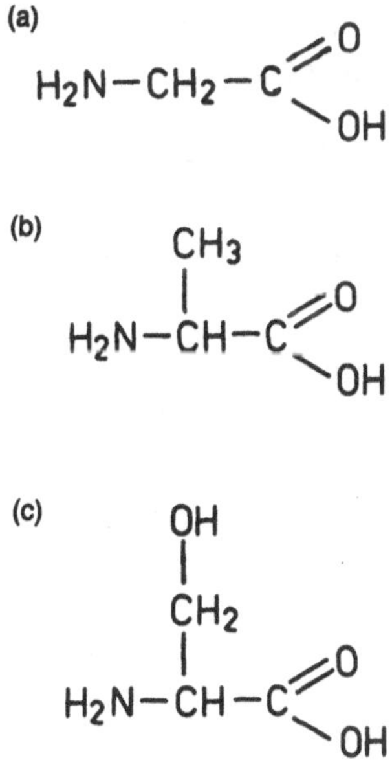

FIGURE 4.1. The chemical formulas of (a) glycine, (b) alanine, and (c) serine.

necessary for its block diagonalization[6]) we had to rotate the local atomic coordinate systems in different elementary cells together with the unit cell. (A similar operation is also necessary in the case of an α-helical structure.) For poly(gly-ala), poly(gly-ser), and poly(ala-ser) no rotation of the local coordinate systems is necessary, since there is a pure translation with $a = 6.5$ Å in these polymers. In the more complicated poly(gly-gly-ala), poly(gly-gly-ser), and poly(gly-ala-ser) the translation with 9.75 Å had to be combined again with a rotation of 180°.

In the case of β-poly(gly) the matrix elements of the Fock operator must be calculated up to the third-neighboring peptide unit to be able to stabilize the energy levels for five digits in electron volts (an accuracy completely sufficient to calculate reliable density-of-state diagrams). In the more complicated polypeptide models one should cut off carefully the intercell interactions individually for each peptide unit with the same interaction radius to avoid inconsistencies.

The density of electronic states (DOS) defined for the band n by

$$\rho_n(E) = \frac{1}{\pi}\left(\frac{dk}{d\varepsilon_n(k)}\right)_{\varepsilon_n = E} \tag{4.46}$$

were calculated from the resulting energy bands; see also equation (2.1). For this purpose we performed a quadratic interpolation for each section of the $\varepsilon_n(k)$ curves, calculated the energy values for 1500 intermediate k points, and obtained finally the functions $\rho_n(E)$ by a sample counting. For plotting, they were rescaled to make the highest peak for each band of equal height.

Table 4.1 presents the minima and maxima of the valence and conduction bands (n^* and $n^* + 1$, respectively) in pure poly(gly),

TABLE 4.1. Energy Parameters of the Valence and Conduction Bands in Pure Poly(gly), Poly(ala), and Poly(ser)[a]

System	Type of band[b]	Band minimum	Band maximum	Bandwidth
Poly(gly)	$n^* + 1$	0.2755	1.1040	0.8285
	n^*	−9.9196	−9.4242	0.4954
Poly(ala)	$n^* + 1$	0.4723	1.2454	0.7731
	n^*	−9.6414	−9.1552	0.4862
Poly(ser)	$n^* + 1$	0.1970	1.1754	0.9784
	n^*	−9.7440	−9.2480	0.4960

[a] All quantities are in eV.
[b] n^* and $n^* + 1$ denote the valence and conduction bands, respectively.

poly(ala), and poly(ser). It can be seen that the corresponding bandwidths differ by only a few percent and the band positions are shifted by 0.2–0.3 eV owing to the different potentials in the side chains. However, these band shifts exert a profound effect on the electronic structure of the composite polymers. It should be noted here that these bandwidths are in good agreement with the results of recent *ab initio* calculations on poly(gly) and poly(ala) with an STO-3G basis set.[7]

In Figure 4.2 we present the density-of-states curve for the valence band of poly(gly). The curves for the conduction band and for the corresponding bands of the other two homopolymers are very similar apart from their positions on the energy scale (which can be seen from Table 4.1), therefore they are not shown there. Figures 4.3 and 4.4 show the density of states for the valence bands in the poly(gly-ala) and poly(gly-ser) mixed periodic polymers, respectively. It can be seen that a gap develops in the energy region of -9.4 eV in poly(gly-ala) with a width of $\Delta E_{gap} = 0.1520$ eV [cf. the band limits for poly(gly-ala) in Table 4.2, in poly(gly-ser) with $\Delta E_{gap} = 0.1447$ eV, and in poly(ala-ser) with $\Delta E_{gap} = 0.0274$ eV]. Similar curves were obtained also for the conduction bands of these polymers. The resulting gaps can be extracted from the corresponding columns of Table 4.2. It is very interesting to see what happens if we diminish the "impurity concentration." Instead of the

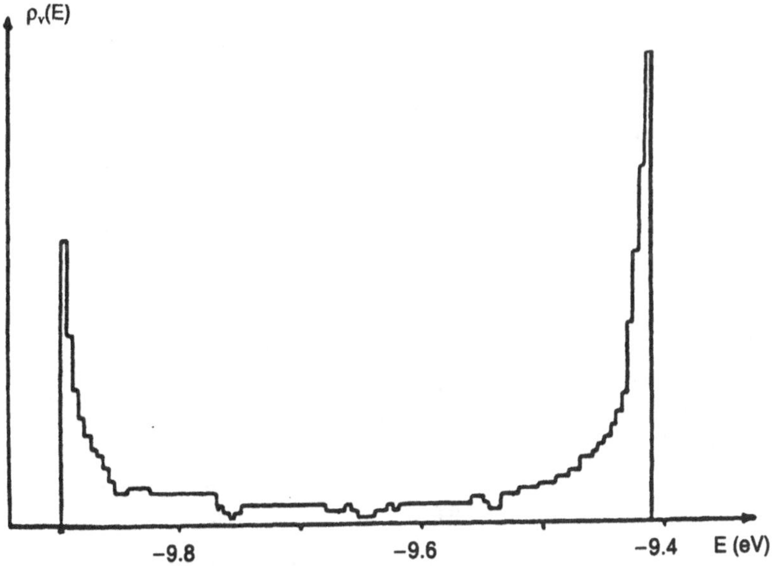

FIGURE 4.2. Density-of-state curve for the valence band of poly(gly).

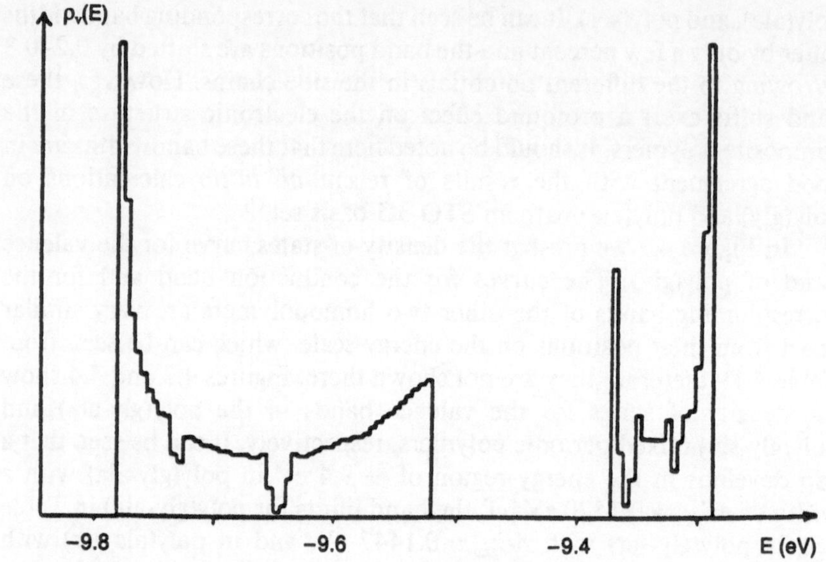

FIGURE 4.3. Density-of-state curve for the valence-band region of the poly(gly-ala) periodic mixed polymer.

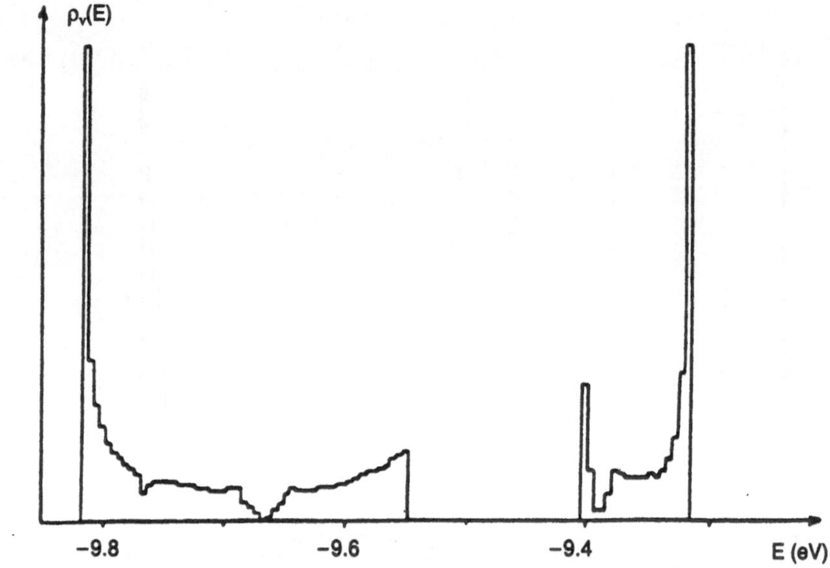

FIGURE 4.4. Density-of-state curve for the valence-band region of the poly(gly-ser) mixed polymer.

TABLE 4.2. Energy Parameters of Four Bands in the Poly(gly-ala), Poly(gly-ser), and Poly(ala-ser) Mixed Polymers Lying in the Energy Region of the Conduction and Valence Bands of the Pure Systems[a]

System	Type of band[b]	Band minimum	Band maximum	Bandwidth
Poly(gly-ala)	n^*+2	0.8117	1.1834	0.3717
	n^*+1	0.3743	0.7608	0.3865
	n^*	−9.3713	−9.2853	0.0860
	n^*-1	−9.7781	−9.5233	0.2548
Poly(gly-ser)	n^*+2	0.7898	1.1661	0.3763
	n^*+1	0.2404	0.6131	0.3727
	n^*	−9.4068	−9.3141	0.0927
	n^*-1	−9.8220	−9.5515	0.2705
Poly(ala-ser)	n^*+2	0.8054	1.2499	0.4445
	n^*+1	0.3760	0.7200	0.3440
	n^*	−9.3205	−9.1963	0.1242
	n^*-1	−9.6819	−9.3479	0.3340

[a] All quantities are in eV.
[b] n^* and n^*+1 denote the valence and conduction bands, respectively.

previous 1:1 mixed polymers we can study 1:2 mixing, if we substitute the H atom of only each third glycine residue by a $-CH_3$ or $-CH_2-OH$ group, respectively. The resulting density-of-state curves for the valence bands of these poly(gly-gly-ala) and poly(gly-gly-ser) polymers are shown in Figures 4.5 and 4.6, respectively. It can be seen from these curves that the developing gaps [0.1117 and 0.0834 eV for poly(gly-gly-ala) and 0.0504 and 0.1097 eV for poly(gly-gly-ser), respectively] are smaller than those for the 1:1 mixed systems. On the basis of these results the (physically plausible) statement that the width of the evolving gap is a monotonic function of the content of "foreign" side-chain groups seems to be justified. It is not surprising, on the other hand, that in the 1:2 mixed systems there are two gaps. This is a consequence of the fact that the two peptide groups with side chains $R_1 = H$ and $R_2 = H$ see a different potential in the crystal owing to the presence of the "impurity groups" $R_3 = CH_3$ or $R_3 = CH_2-OH$, respectively. The situation is very similar also in the case of the conduction bands of these systems. To save space we do not show here the corresponding curves but the resulting gaps can be obtained from Table 4.3.

For the case of the still more complicated polypeptide poly(gly-ala-ser) Figures 4.7 and 4.8 show the computed density-of-state curves for the valence- and conduction-band regions, respectively, of this polymer.

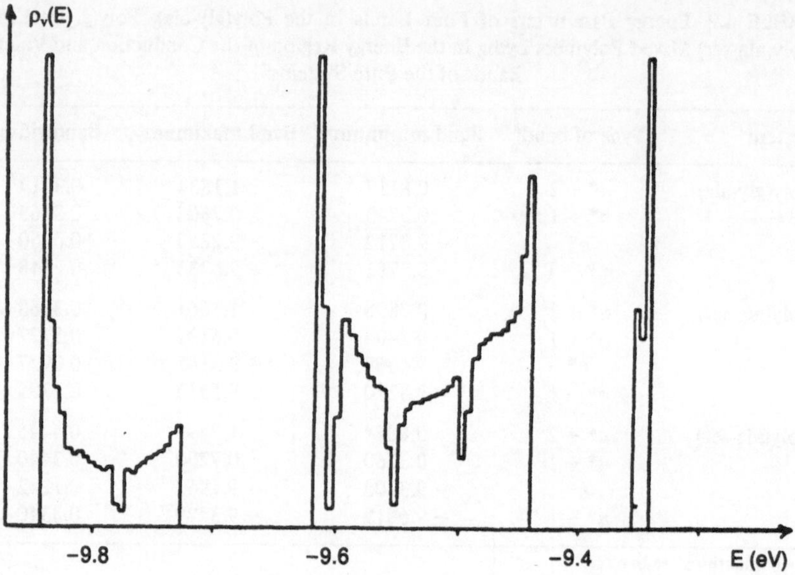

FIGURE 4.5. Density-of-state curve for the valence-band region of the poly(gly-gly-ala) polymer.

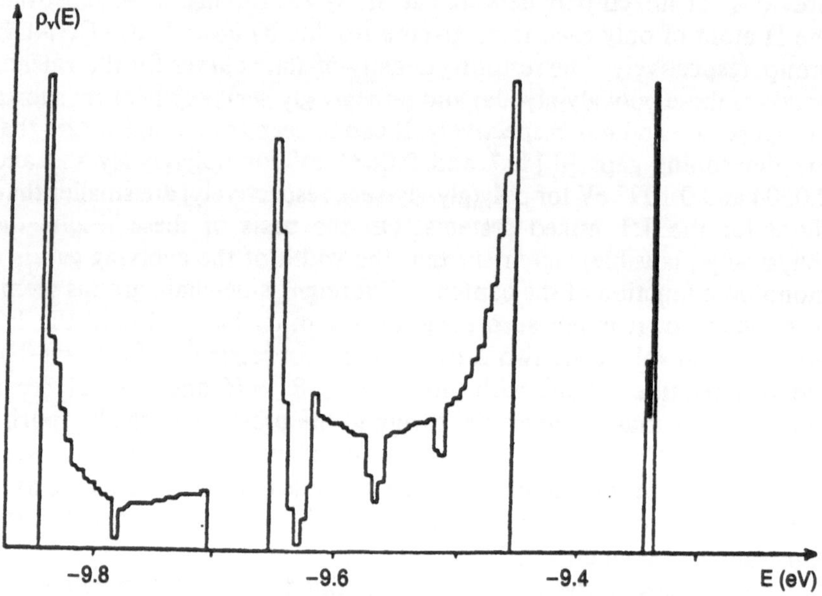

FIGURE 4.6. Density-of-state curve for the valence-band region of the poly(gly-gly-ser) polymer.

It can be seen that owing to the presence of three different side groups in the elementary cell both the valence and conduction bands (as well as other bands not shown here) are split into three new bands separated by gaps of $\Delta E_{gap} = 0.1337$ eV and $\Delta E_{gap} = 0.0931$ eV in the case of the valence-band and $\Delta E_{gap} = 0.1375$ and $\Delta E_{gap} = 0.0471$ for the conduction-band region, respectively.

The most important message to be gotten from all these results is that — contrary to usual chemical intuition — even a small chemical (substitutional) disorder (substituting one H atom in gly by a methyl group in alanine and by an oxy-methyl group in serine) has a strong effect on the electronic structure causing the appearance of new gaps with nonnegligible widths. This clearly demonstrates that the treatment of disorder in aperiodic polymers is of profound importance if we wish to describe their electronic structure and to interpret their different physical properties.

TABLE 4.3. Energy Parameters of Six Bands in the Poly(gly-gly-ala), Poly(gly-gly-ser), and Poly(gly-ala-ser) Mixed Polymers Lying in the Energy Region of the Conduction and Valence Bands of the Pure Systems[a]

System	Type of band[b]	Band minimum	Band maximum	Bandwidth
Poly(gly-gly-ala)	$n^* + 3$	1.0112	1.1568	0.1456
	$n^* + 2$	0.6125	0.9557	0.3432
	$n^* + 1$	0.3657	0.5369	0.1712
	n^*	−9.3547	−9.3424	0.0123
	$n^* - 1$	−9.6164	−9.4381	0.1783
	$n^* - 2$	−9.8459	−9.7281	0.1178
Poly(gly-gly-ser)	$n^* + 3$	0.9708	1.1446	0.1738
	$n^* + 2$	0.5566	0.9323	0.3757
	$n^* + 1$	0.2724	0.4417	0.1693
	n^*	−9.3636	−9.3577	0.0059
	$n^* - 1$	−9.6706	−9.4733	0.1973
	$n^* - 2$	−9.8631	−9.7210	0.1421
Poly(gly-ala-ser)	$n^* + 3$	0.9933	1.8870	0.1954
	$n^* + 2$	0.6071	0.9462	0.3391
	$n^* + 1$	0.3347	0.4696	0.1349
	n^*	−9.2669	−9.2650	0.0019
	$n^* - 1$	−9.5311	−9.3600	0.1711
	$n^* - 2$	−9.7833	−9.6648	0.1185

[a] All quantities are in eV.
[b] n^* and $n^* + 1$ denote the valence and conduction bands, respectively.

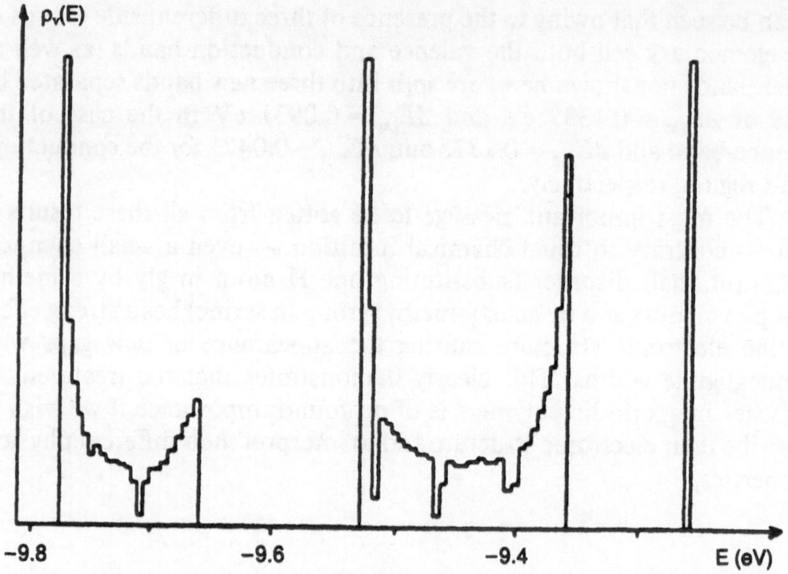

FIGURE 4.7. Density-of-state curve for the valence-band region of poly(gly-ala-ser).

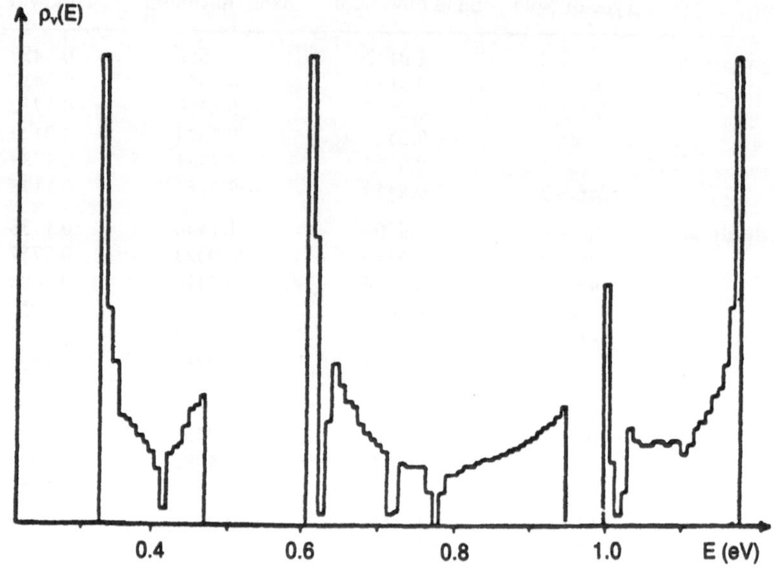

FIGURE 4.8. Density-of-state curve for the conduction-band region of poly(gly-ala-ser).

4.2.2. Localization of Wave Functions in a Disordered Hydrogen Ring

Since the pioneering work of Anderson[8] and Mott[9] it is known that the wave functions of disordered systems are localized. The degree of this so-called Anderson localization increases with increasing disorder. Anderson's original argument was based on a simple model Hamiltonian. In a number of cases different authors[10] carried out numerical studies on disordered systems still using different simplified model Hamiltonians.

In a recent investigation[11] *ab initio* calculations were performed for a ring of equidistant H atoms but varying the nuclear charge Z_i of the ith H atom according to a Gaussian distribution about an average $\langle Z \rangle = 1$. (In a previous study of a disordered H ring the internuclear distance was varied randomly.[12])

The method of solution for the eigenfunctions in the model is that of spin-restricted Hartree–Fock calculations for clusters of atoms treated as giant macromolecules. Thus all-neighbor interactions are taken into account. Molecular orbitals (MOs) for the macromolecule are formed as a linear combination of atomic orbitals (LCAOs). The self-consistent-field (SCF) solutions for the MOs were calculated by means of the GAUSSIAN 70 program,[13] using the STO-3G (minimal) basis set of Gaussian orbitals. Calculations on clusters of 40, 50, and 60 atoms were repeated as tests of our statistics. On the basis of these tests, it would seem that increasing cluster size even further would not materially affect the results in an essential manner. All results shown below are for clusters of 50 atoms.

Although the model is still oversimplified compared to actual systems of experimental interest, it does contain two aspects lacking in previous numerical work: the calculations succeed in demonstrating localization in systems involving clusters of realistic atoms and all-neighbor interactions. Most probably, the quasi-one-dimensional nature (in k space) of the model may introduce peculiarities that will not hold for three-dimensional systems. On the other hand, the results indicate general trends that should also prevail for the more elaborate systems. In this regard we should stress that the clusters of atoms are in three dimensions. The many-center one- and two-electron integrals over AOs, all of which are computed for the SCF procedure, are of course computed in three dimensions. Thus all theories applying to strictly mathematically one-dimensional systems should presumably not apply to this model (with the exception of an infinite number of H atoms in the ring).

If the AO centered on the ith atom is denoted by ϕ_i, then the k_nth eigenfunction can be expanded in the form

$$\psi_{k_n} \approx \sum_j a_j(k_n)\phi_j \tag{4.47}$$

For a perfect ring of H atoms the symmetry implies that (as shown in Chapter 1)

$$a_j(k_n) \approx \exp(ik_n j) \tag{4.48}$$

where $k_n = 2\pi n/N$ and n is an integer such that $-N/2 < n \leqslant N/2$. The orbital of lowest energy is given by $n = 0$.

The graphs that follow show some typical results for rings of 50 hydrogen atoms in which the coefficients $a_j(k_n)$ are plotted as a function of atomic position j. The curves were involved by treating j as a continuous variable. Thus the plot for the lowest occupied state for the ordered system (see Figure 4.9) is a straight line. For localized eigenstates there are regions in which coefficients a_j are sharply peaked, and other regions where they go to zero. This is contrasted with the periodic system in which the coefficients essentially behave in a sinusoidal fashion around the ring (since we are taking a_j to be real). For simplicity, we show graphs for the case of cellular compositional disorder described above, in which the Gaussian standard deviation from the average $Z = 1$ is denoted by σ.

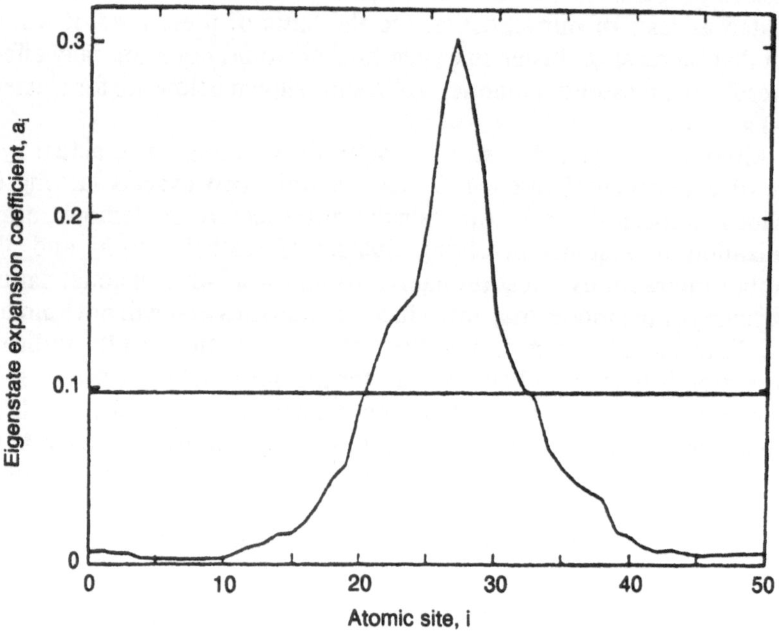

FIGURE 4.9. Smoothed lowest occupied eigenstate for the ordered (straight line) and disordered ($\sigma = 0.05$) hydrogen ring.

Results for structural disorder are very similar. We should stress that the results presented in the figures are those from one illustrative run with a single (computer-generated) Gaussian-distribution nuclear charge. These results are, however, completely illustrative of the many distributions actually calculated for each value of σ. That is, the degree of localization for any given state in the band is determined by n (see below), but the location itself of the state depends on the particular nuclear-charge distribution. It should not be presumed that this location is simply determined by the location of the deepest well, however. Normalization considerations ensure that the localized states distribute themselves all around this ring for any given distribution of charges.

The general trend of the results may be described as follows. Even for very small values of σ the sudden localization of the states at the extremities of the band is quite dramatic. An example is shown in Figure 4.9, which shows a_i for the lowest eigenstate when $\sigma = 0.0$ (ordered ring for which a_i is constant) and when $\sigma = 0.05$. Clearly, the states at the very bottom of the band become very sharply localized immediately. A more quantitative measure of whether a state is localized or not is required, particularly when one is interested in locating something like a "mobility edge" (i.e., the energy separating the localized from the extended states). for this purpose one can calculate a kind of probability centroid for the eigenstates. We place the ring of hydrogen atoms on a unit circle and calculate

$$\bar{X}(k) = \sum_j a_j^2(k) X_j \Big/ \sum_j a_j^2$$

with $\bar{Y}$ similarly defined. Then $R^2(k) = \bar{X}^2(k) + \bar{Y}^2(k)$, where $0 \leqslant R^2 \leqslant 1$ is a measure of localization, with $R = 0$ for perfect delocalization and $R = 1$ for complete localization. The transition in R from $R = 0$ is very sharp. Using this criterion we show the localized states on the energy dispersion curve for the case $\sigma = 0.05$ in Figure 4.10. The dispersion curve has been smoothed to represent an infinite system.

The already sharply localized states at the extremities become even more localized as σ is increased. An illustration of this is given in Figure 4.11, which shows the state at the bottom of the band for $\sigma = 0.10$. This state is about one-third the breadth of the similar state for $\sigma = 0.05$, and is now localized over just a few atoms, and is zero elsewhere. (We note that the atoms at which the state is localized differ from Figure 4.9 because a different random distribution has been taken.)

As σ is increased, the number of localized states also increases. The region of localization at the top and bottom of the band as shown in Figure 4.10 spreads toward to Fermi energy.

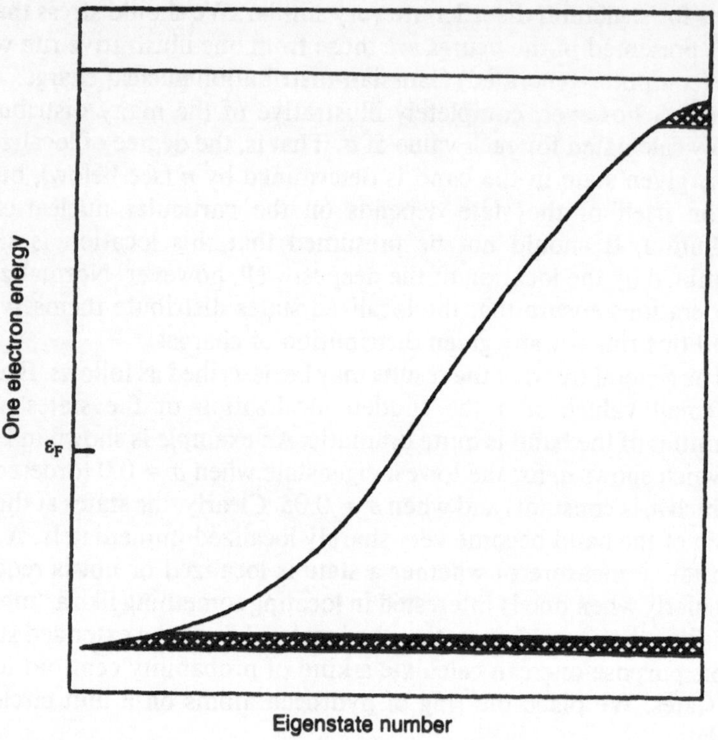

FIGURE 4.10. Smoothed energy dispersion for $\sigma = 0.05$; cross-hatching indicates localized states.

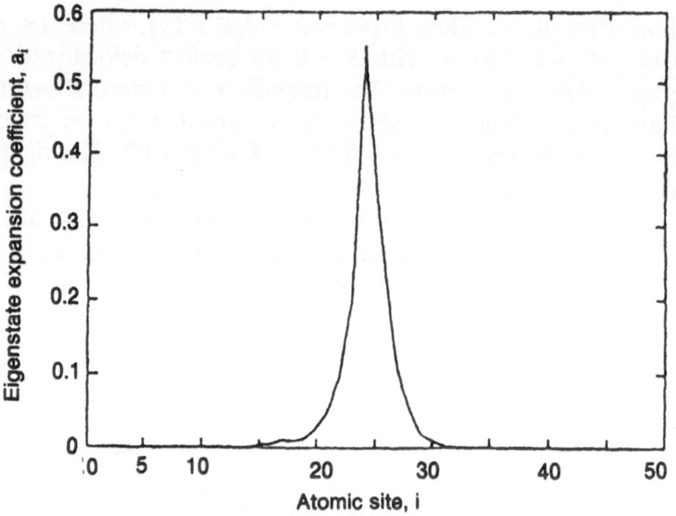

FIGURE 4.11. Lowest occupied eigenstate for $\sigma = 0.10$.

We found that the states at the top of the band are less sharply localized than those at the bottom for equal degrees of disorder. This is probably because these states have on the order of N nodes and are thus inherently more difficult to localize. Figure 4.12 shows just the envelope of the state at the top of the band for $\sigma = 0.10$.

Certain aspects of our results appear to be characteristic of the Hartree–Fock SCF procedure as applied to systems capable of showing charge-density alternation. In such systems a lower total energy may be achieved by lowering the electron–electron Coulomb interaction energy by alternating charges and/or bonds.[14] When this happens a gap appears in the density of states around the Fermi level. In our calculations it appears that the introduction of disorder in the rings favored such a rearrangement of charge, and accompanying this rearrangement a gap broadening, as a function of increasing disorder, appeared at the Fermi level. It should be stressed that it is evident from the charge-bond-order matrices that it is the electron–electron interaction, and not some simple alternation of nuclear charge, that is responsible for the gap. As expected, the states at the edges of the gap, occupied and unoccupied, were also localized. (We observe such a gap in solutions for the ordered chains as

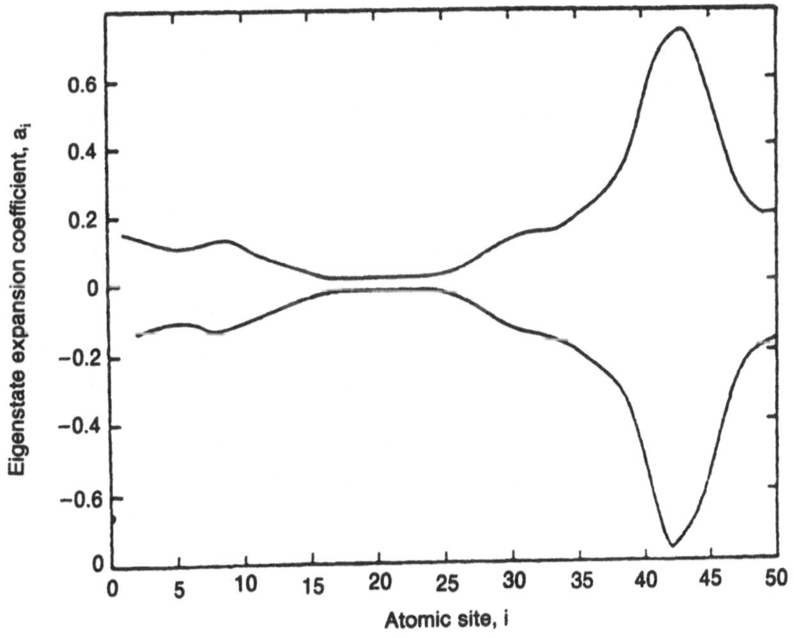

FIGURE 4.12. Smoothed envelope of the highest unoccupied eigenstates for $\sigma = 0.10$.

well, and also in the ordered rings, if symmetry-breaking solutions are allowed, in agreement with other results.[14])

Another characteristic of our SCF procedure as applied to clusters of hydrogen atoms is that of increasing difficulty of convergence as disorder is increased. More sophisticated convergence techniques will in general overcome this problem, but for the present we did not obtain convergent solutions in the case when all the states in the band are localized and thus did not calculate a value for a critical disorder, σ_c.

After demonstrating the effect of disorder on the energy-band structure of a chain and on the localization properties of the corresponding wave functions, we shall describe in the subsequent sections several methods to determine the energy-level distribution (density of states) and wave functions of an aperiodic chain.

4.3. SINGLE-SITE COHERENT POTENTIAL APPROXIMATION AND ITS APPLICATION TO $(SN)_x$ WITH HYDROGEN IMPURITIES

The Coherent Potential Approximation (CPA) of the electronic structure of a disordered system is based on the idea that the disordered system should be replaced by an effective medium. This medium has the property that the statistical average of the fluctuation from its potential (the so-called coherent potential) should be equal to zero.* In other words, each site but one is replaced by an unknown coherent potential. One then puts at the reference site (see Figure 4.13) an A or a B component (in the simplest case of a binary disordered system) with respective probabilities $1-f$ and f. After that one solves the problem of this single impurity imbedded in the effective medium characterized by the coherent potential, which is determined by the requirement that the average scattering (fluctuation) from the reference site is also zero.

If one assumes that no two subsequent scatterings can happen on the same site, one speaks about single-site CPA.[16] On the other hand, if one relaxes this condition and allows two or more subsequent scatterings on the same site (before a scattering happens at another site) one obtains the scheme of cluster CPA, which is more relevant if extended molecules like the nucleotide bases in DNA rather than single atoms or ions are sitting at the lattice sites. Since, however, the cluster CPA method requires a

* We shall see that the coherent potential defined in this way is different from the simple potential of a disordered system obtained by averaging the potentials of the components according to the composition of the system. This latter, so-called virtual crystal approximation only gives acceptable results if the different chemical components of the system are rather similar.[15]

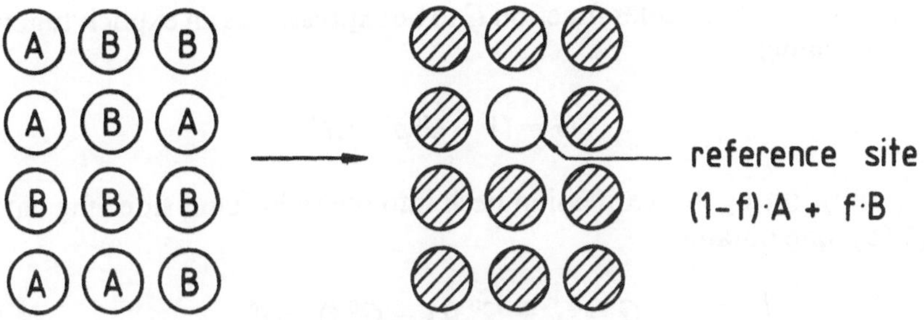

FIGURE 4.13. Replacement of a binary disordered system with components A and B by an effective medium (denoted by hatched circles). At the reference site we can find component A with probability $1-f$ or component B with probability f.

rather elaborate formalism, we shall give here the mathematical formulation only of the simpler single-site CPA for a binary system in the one-band (such as the valence-band) approximation.

4.3.1. Derivation of the Single-Site CPA Equation

To start the mathematical formulation, we replace the Dyson equation by [in the single-site one-orbital-per-unit-cell approximation all quantities occurring in equation (4.45) become scalar]

$$G = G^0 + G^0 \Delta G \tag{4.49}$$

where

$$G^0(\mathbf{k}, E) = [E - \varepsilon^{A}(\mathbf{k})]^{-1} \tag{4.50}$$

if A is the host system. Integration of this equation over k yields

$$G_0(E) = \Omega^{-1} \sum_{\mathbf{k}} [E - \varepsilon^{A}(\mathbf{k})]^{-1} \tag{4.51}$$

where Ω is the volume of the first Brillouin zone. The deviation Δ occurring in expression (4.49) can be defined (again, only in the single-site diagonal perturbation case[16,17]) in its Fourier-transform form as

$$\Delta(\mathbf{k}) = \varepsilon^{B}(\mathbf{k}) - \varepsilon^{A}(\mathbf{k}) \tag{4.52}$$

The scattering matrix element T[16] can be introduced in expression (4.49) through the equation

$$G = G^0 + G^0 T G^0 \tag{4.53}$$

Equation (4.49) enables function G to be expressed as an explicit function of G^0, namely

$$G = (1 - G^0\Delta)^{-1}G^0$$

On substituting this expression back into the right-hand side of equation (4.49), one obtains

$$G = G_0 + G^0\Delta(1 - G^0\Delta)^{-1}G^0 \tag{4.54}$$

hence comparison of equations (4.54) and (4.53) yields the following expression for T:

$$T = \Delta(1 - G^0\Delta)^{-1} \tag{4.55}$$

The introduction of an effective medium enables the Dyson equation to be rewritten in the form

$$G_e = G^0 + G^0\Sigma G_e \tag{4.56}$$

where G_e is the Green function of the effective medium and the self-energy Σ is defined through equation (4.56). The quantity G^0 can now be expressed as a function of G_e from equation (4.56):

$$G^0 = G_e(1 + \Sigma G_e)^{-1} \tag{4.57}$$

substitution of this expression for G^0 back into the original Dyson equation (4.49) yields

$$G = G^0(1 + \Delta G) = G_e(1 + \Sigma G_e)^{-1}(1 + \Delta G) \tag{4.58}$$

multiplying this equation from the left by $(1 + \Sigma G_e)G_e^{-1}$, one has

$$(1 + \Sigma G_e)G_e^{-1}G = 1 + \Delta G$$

and multiplying this latter equation again from the left by G_e gives

$$G + G_e\Sigma G = G_e + G_e\Delta G$$

or

$$G = G_e + G_e(\Delta - \Sigma)G = G_e + G_e\Delta_e G \tag{4.59}$$

with

$$\Delta_e = \Delta - \Sigma \tag{4.60}$$

At sites A, Δ of course vanishes and therefore $\Delta_e = -\Sigma$, while at sites B, $\Delta_e = \Delta - \Sigma$.

With the help of equation (4.55), substitution of Δ by Δ_e and G^0 by G_e together with definition (4.60) allows one to form the average of the scattering matrix element T:

$$\langle T \rangle = (1-f)(-\Sigma)[1 + G_e\Sigma]^{-1} + f(\Delta - \Sigma)[1 - G_e(\Delta - \Sigma)]^{-1} \tag{4.61}$$

From the condition $\langle T \rangle = 0$, namely the average fluctuation from the effective medium should be zero, one can derive finally the single-site CPA equation (further details are reported elsewhere[16]):

$$\Sigma(\mathbf{k}, E) = f\Delta(\mathbf{k})[1 + (\Sigma(\mathbf{k}, E) - \Delta(\mathbf{k})G_e(E)]^{-1} \tag{4.62}$$

It is easy to show from equation (4.56) that

$$G_e(\mathbf{k}, E) = [E - \varepsilon^A(\mathbf{k}) - \Sigma(\mathbf{k}, E)]^{-1} \tag{4.63}$$

or

$$G_e(E) = \Omega^{-1} \sum_{\mathbf{k}} G_e(\mathbf{k}, E) \tag{4.64}$$

It should be noted that if the density-of-state curves calculated for the (periodic) constituent systems have the same shape but are shifted by a constant energy value, one can approximate $\Sigma(\mathbf{k}, E)$ by a constant Σ value[16] and then, by comparing equations (4.50) and (4.64), one can use the equation

$$G_e(E) = G^0(E - \Sigma) \tag{4.65}$$

In many cases of alloys the density-of-state curves of the constituents are really just shifted by a constant value and therefore many calculations in this field employ equation (4.65). Unfortunately, this is not the case for aperiodic polymers, where the DOS curves usually have a quite different shape. This will be shown in the case of the chains of

$$(\mathrm{SN})_x \quad \text{and} \quad \begin{bmatrix} \mathrm{SN} \\ | \\ \mathrm{H} \end{bmatrix}_x$$

Therefore, for aperiodic polymers equation (4.65) generally cannot be used and one must work with the **k**- and energy-dependent $\Sigma(\mathbf{k}, E)$ self-energies.

To develop the theory further we recall the expressions for the spectral density[2],

$$A(\mathbf{k}, E) = -\pi^{-1} \operatorname{Im} G_e(\mathbf{k}, E + i0^+) \tag{4.66}$$

and for the density of states[2],

$$\rho(E) = \Omega^{-1} \sum_{\mathbf{k}} A(\mathbf{k}, E) = -\pi^{-1} \operatorname{Im} G_e(E + i0^+) \tag{4.67}$$

Further, we need the Kramers–Kronig relation[2] for the Green function of the effective medium,

$$G_e(E) = \int_{-\infty}^{\infty} \frac{\rho_e(E')}{E - E'} \, dE' \tag{4.68}$$

In order to calculate the spectral density from expression (4.66) we require the well-known relationship for the ratio of two complex numbers:

$$\frac{\alpha_1 + i\beta_1}{\alpha_2 + i\beta_2} = \frac{\alpha_1\alpha_2 + \beta_1\beta_2}{\alpha_2^2 + \beta_2^2} + i\,\frac{\alpha_2\beta_1 - \alpha_1\beta_2}{\alpha_2^2 + \beta_2^2} \tag{4.69}$$

Application of this to equation (4.63) (that is, $\alpha_1 = 1$ and $\beta_1 = 0$) yields

$$\operatorname{Im} G_e(\mathbf{k}, E) = \frac{\operatorname{Im} \Sigma(\mathbf{k}, E)}{[E - \varepsilon^{A}(\mathbf{k}) - \operatorname{Re} \Sigma(\mathbf{k}, E)]^2 + [\operatorname{Im} \Sigma(\mathbf{k}, E)]^2} \tag{4.70}$$

Calculation with a **k**- and E-dependent self-energy, reported in the literature in only a very few cases, involves first starting from relation (4.68) with the zeroth iteration for $\rho_e(E')^{(0)}$, the virtual crystal value $\rho_{vc}(E)$, which can be obtained from the energy bands of A and B:

$$\varepsilon_{vc}(\mathbf{k}) = (1 - f)\varepsilon^{A}(\mathbf{k}) + f\varepsilon^{B}(\mathbf{k}) \tag{4.71}$$

to determine from expression (4.71) the DOS $\rho_{vc}(E)$. One can apply the method of Delhalle.[18] On putting $\rho_{vc}(E)$ into expression (4.68) one obtains, using for instance the method of Kirkpatrick *et al.*,[19] an already complex starting value for $G_e(E)^{(0)}$. Substitution of this value together with $\Sigma(k)^0 = f\Delta(k)$ [again the virtual crystal aproximation for $\Sigma(k)$] into the CPA equation (4.62) leads to a new and also complex self-energy $\Sigma(\mathbf{k}, E)^{(1)}$. Insertion of this latter quantity into equation (4.70), and the resulting expression for Im $G_e(\mathbf{k}, E)^{(1)}$ into equations (4.66) and (4.67), enables one to calculate the density of states $\rho(E)^{(1)}$ in the next iteration. By substituting back into relation (4.68) the procedure can be repeated until self-consistency is reached.[20]

4.3.2. Application of the Single-Site CPA Method to $(SN)_x$ with Hydrogen Impurities

At IBM (San Jose), 4- to 8-mol% hydrogen was found in $(SN)_x$.[21] The position of the hydrogen impurities is unknown, but the H atoms most probably bind to the N atoms. In this way, they change the hybridization state of the N atoms and the number of π electrons in the partially filled band of $(SN)_x$ (in a –S=N– unit there are three π electrons while in a –S–N–H– unit there are four).

To determine theoretically the shift of the Fermi level and the change in the density of states when $(SN)_x$ is modified, the coherent potential approximation with a k-dependent self-energy (see Section 4.3.1) was applied to calculate the density of states of the partially filled band of the mixed chain.[20] For this calculation, the available *ab initio* band structures of the pure $(SN)_x$ and pure $(SNH)_x$ chains were first applied as input.

Figure 4.14 shows the density-of-state curves of these two periodic chains.[20] The shape of the curves is completely different, which makes it necessary to apply the SCF procedure with $\Sigma(\mathbf{k}, E)$ described in Section 4.3.1.

The condition

$$\max_{(i)} |\rho_e^{(n)}(E_i) - \rho_e^{(n-1)}(E_i)| \leq 10^{-3} \quad (\text{au mol})^{-1} \tag{4.72}$$

was chosen as SCF criterion. Iteration steps were needed to reach self-consistency between 24 ($f = 0.3$) and 70 ($f = 0.5$). The corresponding computing times on a CYBER 172 were 100 and 200 s, respectively.

The other details of the numerical procedure are the following:

1. In building up a mesh for the energy range in which the density of states is calculated, the mesh points were not chosen equidistant but were made more dense in the range of $\rho^B(E) \neq 0$ and in regions with narrow peaks. Finally, 93 points were chosen in the energy range of about 10 eV and the same number of k points ($0 \leq k_i \leq \pi/a$) for which $\Sigma(k, E)$ was calculated in every iteration step. The corresponding 93 points of

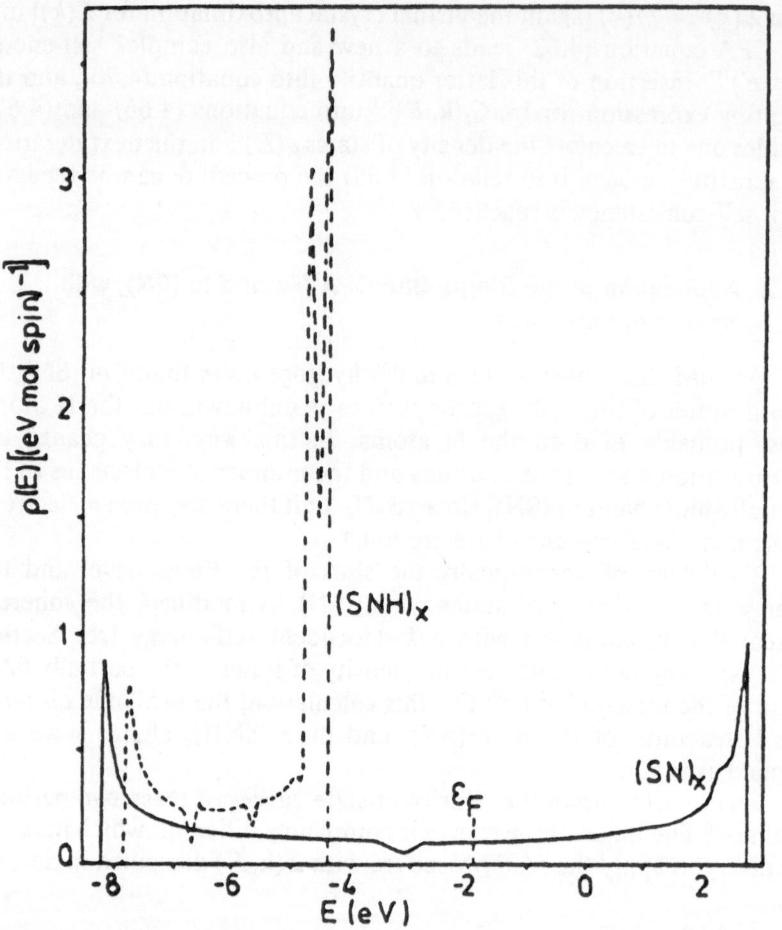

FIGURE 4.14. Density-of-state curves of pure $(SN)_x$ and $(SNH)_x$ chains (in eV mole $spin^{-1}$ units) (after Suhai[23]).

$\Delta(k)$ were calcuated by means of a parabolic interpolation fit of the nine points of the $\varepsilon^A(k)$ and $\varepsilon^B(k)$ curves.

2. The starting values for $\Sigma(k, E)$ were calculated with the virtual-crystal values for $\Sigma^{(0)}$ and $G_e^{(0)}(E)$, as mentioned in Section 4.3.1. These starting values yielded faster convergence than the other guesses obtained with $\Sigma^{(0)} = 0$ and $G_e^{(0)} = G^0$.

3. For $f\Delta(k) \approx 0$ and $\Sigma(k, E) \approx 0$, with allowance for equation (4.70) and the relation[22]

$$\delta(x) = \lim_{\alpha \to 0} \frac{1}{\pi} \frac{\alpha}{x^2 + \alpha^2} \tag{4.73}$$

one obtains sharp δ peaks in the spectral density (4.66). This situation was handled for discrete points E_1 in the numerical integration by employing the normalization condition for$A(\mathbf{k}, E)$ in equation (4.66) at every iteration step. The procedure was checked with $f = 10^{-5}$ for which the spectral density exhibits practically the δ structure of the pure system A.

It was checked by numerical integration that the total area under the curves in Figures 4.14–4.16 is 2.0, thus leading to the proper number of electrons per unit cell for a single band (taking into account both spin directions). since in pure $(SN)_x$ the valence band is only half occupied (one electron per unit cell in the valence band), the position of the Fermi level ε_F was determined by numerical integration from the condition

$$\int_{\varepsilon_{min}}^{\varepsilon_F} \rho_e(e)dE = 1 + f \tag{4.74}$$

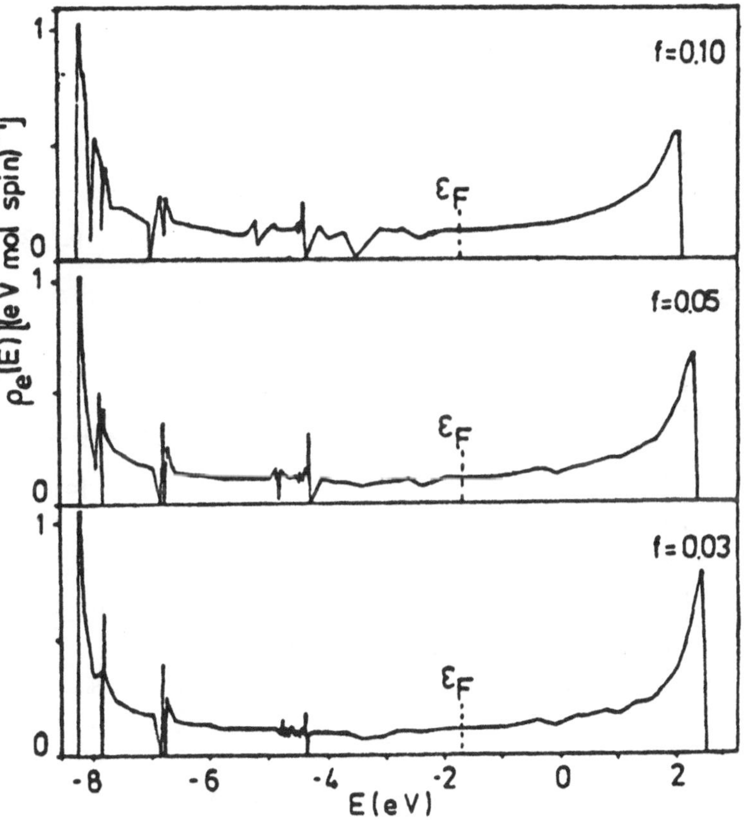

FIGURE 4.15. Density-of-state curves of the $(SN)_x$ and $(SNH)_x$ mixed system obtained in the CPA with $f = 0.03$, 0.05, and 0.10.

Values of f were taken between 0.03 and 0.1 and between 0.2 and 0.5, in order to see the effect of possible hydrogen doping of $(SN)_x$.

Figures 4.15 and 4.16 show the density-of-state curves of the mixed systems obtained with the help of the described method, with $f =$ 0.03, 0.05, and 0.10, and 0.20, 0.30, and 0.50, respectively.

The most surprising result is the complicated structure of the impurity band with spikes and dips where the density of states goes to zero. This structure is well known from computer experiments for linear chains[23] and from cluster CPA calculations,[24] but could not be obtained with the simple CPA using a k-independent Σ. To exclude a numerical effect the number and position of the mesh points for both E and k

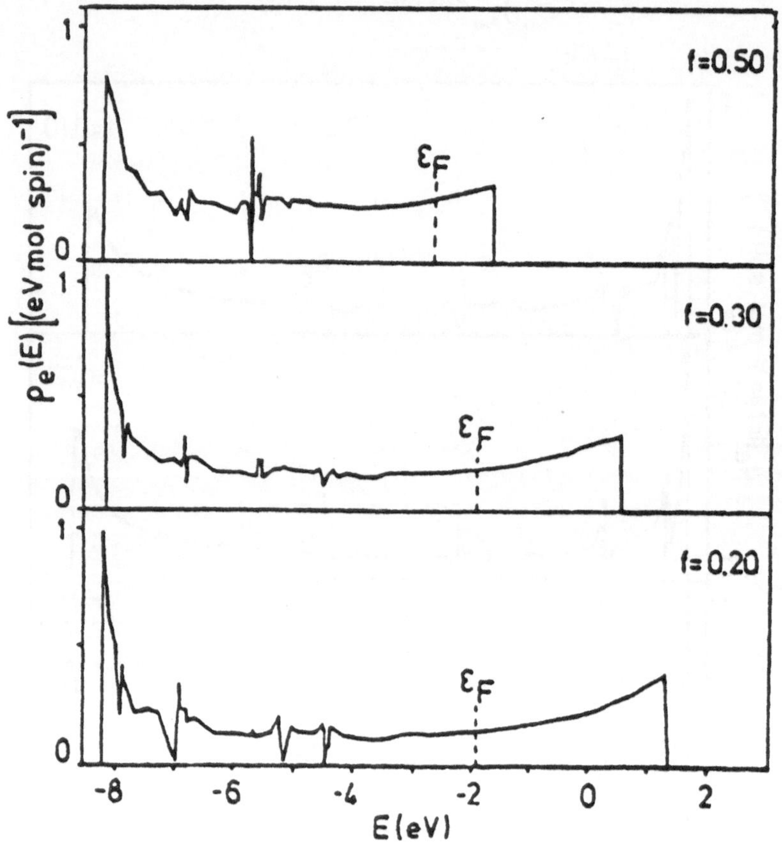

FIGURE 4.16. Density-of-state curves of the $(SN)_x$ and $(SNH)_x$ mixed system obtained in the CPA with $f = 0.2$, 0.3, and 0.5.

independently were changed and a denser mesh introduced at the points where $\rho_e(E)$ were chosen. Further attempts were made to damp out the oscillations in the first iteration steps. For $f \geq 10^{-2}$, however, these spikes and dips always recurred when self-consistency was reached. The method therefore seems to be a simple and rapid way to overcome the nonphysical features of the usual CPA, such as a k-independent self-energy and an impurity band with no structure.

It can be seen from Figure 4.14 that the $(SNH)_x$ periodic chain in an STO-3G basis has, as expected, a much narrower valence band then $(SN)_x$, with corresponding large peaks in the density of states. The Fermi level of $(SN)_x$ lies at -1.90 eV, while the upper limit of the completely filled valence band of $(SNH)_x$ is at -4.38 eV. One would expect, on the basis of the rather large differences in the density-of-state curves of the two systems that in the mixed system even a small percentage of $(SNH)_x$ would have a comparatively large influence on the density-of-state curve of pure $(SN)_x$.

This expectation is fulfilled, as can be seen from Figure 4.15, where the density-of-state curves obained in the CPA approximation for the mixed system with 10% (or less) hydrogen are shown. Even at very low hydrogen concentration ($f = 0.03$) new peaks in the density-of-state curve start to develop in the region between -4.4 and -8.0 eV. At higher f values (see Figure 4.16) peaks of $(SN)_x$ and $(SNH)_x$ in the region between -7 and -8 eV fuse to one broader peak. On the other hand, owing to the very high but extremely narrow peaks of $(SNH)_x$ between -4.8 and -4.4 eV new peaks develop but are shifted to the region around -5.6 eV (see especially the curve corresponding to $f = 0.50$). In this case one sees clearly demonstrated the fact that the CPA method, especially at higher impurity concentration, gives essentially different results than the simple virtual crystal approximation.

The position of the Fermi level of the mixed system is not a sensitive function of f at low concentrations ($\varepsilon_F = -1.9$ eV for $f = 0.00$, $\varepsilon_F = -1.6$ eV for $0.03 \leq f \leq 0.1$, $\varepsilon_F = -1.9$ eV for $f = 0.2$). At higher concentrations, of course, its position shifts toward lower energies ($\varepsilon_F = -2.6$ eV at $f = 0.50$). The density of states at the Fermi level increases monotonically with increasing f [$\rho_e(\varepsilon_F) = 0.10, 0.12, 0.13, 0.18,$ and 0.26 at $f = 0.00, 0.05, 0.10, 0.30,$ and 0.50, respectively].

The calculated very narrow spikes and dips in the density of states due to the hydrogen impurities could, in principle, be detected by ultraviolet-photoemission spectroscopy (UPS) or X-ray-photoelectron spectroscopy (XPS). However, due to a combination of poor XPS resolution and incomplete knowledge of exactly how to treat the background corrections near ε_F [known from the XPS determination of the valence-band structure of $(SN)_x$[25]] it is not certain if such experimental measurements would show the structure in the valence band. On the other hand, if such narrow peaks were found in XPS data, they could probably be attributed to hydrogen impurities as the calculations show.

According to the BCS theory the transition temperature of superconductivity T_c depends exponentially on the electronic density of states at the Fermi level,[26]

$$T_c = 1.14\theta_D \exp[-1/\rho(\varepsilon_F)V] \tag{4.75}$$

where θ_D is the Debye temperature and V is the effective BCS electron–electron interaction parameter, which can be estimated from measurements of the electric resistivity. Early experiments using alloys of molybdenum and niobium[27] over the whole concentration range from pure Nb to pure Mo showed that T_c varies by a factor of over 500, while the electron–phonon coupling constant $\rho(\varepsilon_F)V$ changes only by a factor of less than 2.

Therefore, assuming an approximate constancy of V over the concentration range[27,28] (or estimating it from electric-resistivity data), the densities of state and the positions of the Fermi level calculated for different concentrations and different impurities (H, Br_2, I_2, ICl) can give important indications of the variations of T_c. From the fact that $\rho_e(\varepsilon_F)$ increases monotonically with the H impurity concentration (f) one would expect, on the basis of equation (4.75), that via doping $(SN)_x$ by hydrogen one could increase the transition temperature T_c of superconductivity. As far as the author knows, such experiments have not yet been performed.

4.4. THE NEGATIVE FACTOR COUNTING (NFC) TECHNIQUE AND ITS APPLICATION TO APERIODIC DNA AND PROTEINS

4.4.1. The Negative Factor Counting Technique in Its One Band (Simple Tight-Binding Form)

Dean[29] developed a simple method (based on his so-called negative eigenvalue theorem) for determining the distribution of eigenvalues (density of states) in order to calculate the vibrational spectra of disordered systems. This method cannot be used — as we shall see — for simple topological reasons in the case of the electronic states of two- and three-dimensional solids (and therefore was generally not applied in solid-state physics), but gives rather accurate DOS in quasi-one-dimensional disordered systems.

Dean's method was first applied to electronic-structure calculations by Pumpernick *et al.*[30] for a bundle of short disordered alloy chains (quasi-three-dimensional systems), and by Seel[31] to long disordered polypeptide chains.

Suppose we have a linear chain of N units and take into account in the simple tight-binding (Hückel) approximation only one orbital per unit cell. (In this way we obtain the DOS belonging, for instance, to the valence bands of a binary or multicomponent system if the positions of these bands are not very different.) Then the Hückel determinant will be tridiagonal if only first-neighbor interactions are considered, namely

$$|\mathbf{H}(\lambda)| = \begin{vmatrix} \alpha_1-\lambda & \beta_2 & 0 & 0 & \cdots & 0 \\ \beta_2 & \alpha_2-\lambda & \beta_3 & 0 & & \\ 0 & \beta_3 & \alpha_3-\lambda & \beta_4 & & \\ \vdots & & & & & \\ 0 & & & & & \beta_N \alpha_N-\lambda \end{vmatrix} = 0 \quad (4.76)$$

Here α_i and β_{ij} are the usual Hückel parameters and λ denotes the unknown root (energy eigenvalue) of the determinant. In a disordered chain the values of α_i and β_{ij} are different from each other. The determinant $|\mathbf{H}(\lambda)|$ can be easily transformed into a didiagonal form with the help of successive Gaussian eliminations. To achieve this we must subtract from the second row of the determinant the first row multiplied by $\beta_2/(\alpha_1-\lambda)$. This will eliminate the element β_2. Continuing this procedure in a similar way we can obtain zeros for all the elements of the lower diagonal of the originally tridiagonal determinant. Therefore the determinant

$$|\mathbf{H}(\lambda)| = \prod_{i=1}^{N} (\lambda_i - \lambda) \quad (4.77)$$

can be written in the form

$$|\mathbf{H}(\lambda)| = \prod_{i=1}^{N} \varepsilon_i(\lambda) \quad (4.78)$$

(the value of a didiagonal determinant is just the product of its diagonal elements), where the diagonal elements of the determinant are given by the simple recursion relation

$$\varepsilon_i(\lambda) = \alpha_1 - \lambda - \beta_i^2/\varepsilon_{i-1}(\lambda), \qquad i = 1, 2, 3, \cdots, N \qquad (4.79a)$$

$$\varepsilon_1(\lambda) = \alpha_1 - \lambda \qquad (4.79b)$$

determined by the Gaussian elimination procedure. Comparison of equations (4.77) and (4.78) clearly shows that for a given value of λ, the number of eigenvalues less than λ ($\lambda_i < \lambda$) must equal the number of negative factors $\varepsilon_i(\lambda)$ in equation (4.78).[29] [The calculation of the eigenvalues λ_i for a long chain ($N = 10^4$ or 10^3) is difficult but the computation of factors $\varepsilon_i(\lambda)$ with the help of equations (4.79a) and (4.79b) is very rapid.] By giving λ different values throughout the spectrum and taking the difference of the number of negative quantities $\varepsilon_i(\lambda)$ belonging to consecutive values of λ, one can obtain a histogram for the distribution of eigenvalues (density of states) of **H** to any desired accuracy.

In actual calculations, one has to compute the band structure for each component of the disordered chain assuming that it is repeated periodically. Then the values α_i (diagonal elements of **H**) can be determined from the positions of the bands of the components (the middle point or weighted middle points of the bands) and the off-diagonal elements β_i from the widths of the bands. It should be pointed out again that the NFC in this simple form gives only the level distribution belonging to one band (such as the valence or conduction band) of the disordered chain.

The (NFC) method gives much more detail of the density-of-state curves than the CPA procedure (this will be demonstrated in the case of aperiodic DNA in Section 4.4.3). Therefore, it would really be a breakthrough in the theory of disordered systems if it could also be applied (especially in its *ab initio* form; see below) to 2D and 3D disordered systems. Unfortunately, for obvious topological reasons, in these cases there is no way of expressing the Hückel matrix in a tridiagonal form. (Attempts to use matrices with four or six indices instead of two indices were of no help in solving this problem.[32]) If one applies the first-neighbor-interaction approximation in two or three dimensions, one obtains in the Hückel matrix, besides the diagonal, four or six off-diagonals, respectively, containing nonzero elements. Using the Givens–Housholder routine for diagonalizing matrices of order 100 and 200 (describing a 2D disordered system with 10×10 and 20×10 sites, respectively), we found[33] that to bring these five-diagonal matrices into tridiagonal form took about one-third of the time necessary to completely diagonalize the matrices. To treat only 100 chains, each of 100 units (which still does not provide a very good description of a 2D disordered system), would thus require one-third of the time to diagonalize a matrix of order 10^4. Obviously this lies outside the reach of present computational possibilities. We also plan to try other diagonalization routines, but it seems improbable that an essential reduction in computer time can be attained. Therefore, in the case of 2D and 3D disordered systems, one must still use the other well-known methods.[34]

4.4.2. The NFC Method in Its *Ab Initio* (Matrix-Block) Form

In the case of disordered quasi-one-dimensional systems the NFC method can also be applied to the case of an arbitrary number of orbitals per site either in an *ab initio* form[35] or in a semiempirical, for instance, extended Hückel form.[36] In the *ab initio* case one has the secular determinant instead of a tridiagonal in a triblock-diagonal form

$$\mathbf{M}(\lambda) = |\mathbf{F} - \lambda \mathbf{S}|$$

$$= \begin{vmatrix} \mathbf{A}_1 - \lambda \mathbf{S}_1 & \mathbf{B}_2 - \lambda \mathbf{Q}_2 & 0 & & \\ \mathbf{B}_2^{\mathrm{T}} - \lambda \mathbf{Q}_2^{\mathrm{T}} & \mathbf{A}_2 - \lambda \mathbf{S}_2 & \mathbf{B}_3 - \lambda \mathbf{Q}_3 & & \\ 0 & \mathbf{B}_3^{\mathrm{T}} - \lambda \mathbf{Q}_3^{\mathrm{T}} & \mathbf{A}_3 - \lambda \mathbf{S}_2 & \mathbf{B}_4 - \lambda \mathbf{Q}_4 & \\ & & & \ddots & \mathbf{A}_N - \lambda \mathbf{S}_N \end{vmatrix} = 0 \tag{4.80}$$

where $\mathbf{A}_i$ and $\mathbf{B}_{i-1}$ are, respectively, the diagonal and off-diagonal blocks of the Fock matrix, and $\mathbf{S}_i$ and $\mathbf{Q}_{i+1}$ are the corresponding blocks of the overlap matrix. Since the chain is disordered generally $\mathbf{A}_i \neq \mathbf{A}_j$, $\mathbf{B}_{i+1} \neq \mathbf{B}_{j+1}$, $\mathbf{S}_i \neq \mathbf{S}_j$, and finally $\mathbf{Q}_{i+1} \neq \mathbf{Q}_{j+1}$ $(i \neq j)$.

If we rewrite $\mathbf{M}(\lambda)$ in the form

$$\begin{aligned} \mathbf{M}(\lambda) &= \mathbf{F} - \lambda \mathbf{S} \\ &= \mathbf{S}^{1/2}[\mathbf{S}^{-1/2}\mathbf{F}\mathbf{S}^{-1/2} - \lambda \mathbf{1}]\mathbf{S}^{1/2} \\ &= \mathbf{S}^{1/2}[\tilde{\mathbf{F}} - \lambda \mathbf{1}]\mathbf{S}^{1/2} \end{aligned} \tag{4.81}$$

then

$$\begin{aligned} \det \mathbf{M}(\lambda) &= \det \mathbf{S}^{1/2} \det(\tilde{\mathbf{F}} - \lambda \mathbf{1}) \det \mathbf{S}^{1/2} \\ &= \det \mathbf{S}^{1/2} \det \mathbf{S}^{1/2} \det(\tilde{\mathbf{F}} - \lambda \mathbf{1}) \\ &= \det \mathbf{S} \det(\tilde{\mathbf{F}} - \lambda \mathbf{1}) \end{aligned} \tag{4.82}$$

The original determinant $\det(\mathbf{F} - \lambda\mathbf{1})$ can be easily brought to a diblock-diagonal form again with the help of successive Gaussian elimination. In this way one obtains for its diagonal blocks, in completely analogy to equation (4.79a),

$$\mathbf{U}_i(\lambda) = \mathbf{A}_i - \lambda\mathbf{S}_i - (\mathbf{B}_i^{\mathrm{T}} - \lambda\mathbf{Q}_i^{\mathrm{T}})\mathbf{U}_{i-1}^{-1}(\lambda)(\mathbf{B}_i - \lambda\mathbf{Q}_i)$$
$$\mathbf{U}_1(\lambda) = \mathbf{A}_1 - \lambda\mathbf{S}_1 \tag{4.83}$$

This means that the original expression for the value of $\det \mathbf{M}(\lambda)$, if expressed in the form given by the last expression in equation (4.82),[35] can be written instead as

$$\det(\mathbf{F} - \lambda\mathbf{S}) = \left(\prod_{i=1}^{n} s_i\right)\left[\prod_{j=1}^{n} (\lambda_j - \lambda)\right] = \prod_{i=1}^{N}\left[\prod_{k=1}^{l_i} u_{ik}(\lambda)\right]$$
$$\left(n = \sum_{i=1}^{N} l_i\right) \tag{4.84}$$

Here the quantities s_i are the eigenvalues of $\mathbf{S}$, and λ_j are the roots of the generalized eigenvalue equation

$$\mathbf{F}\mathbf{c}_j = \lambda_j \mathbf{S}\mathbf{c}_j \tag{4.85}$$

Further, $u_{ik}(\lambda)$ denotes the kth eigenvalue of the matrix block $\mathbf{U}_i(\lambda)$ defined by expression (4.83), and l_i is the dimension of the ith diagonal block.

The matrices $\mathbf{U}_i(\lambda)$ can be easily diagonalized for a given value of λ. The number of negative $u_{ik}(\lambda)$ must again equal the number of eigenvalues λ_j that are less than the chosen λ-value. On varying λ, the whole spectrum can again be scanned and the density-of-state curve of the disordered system (in this case, taking into account all the bands) obtained in this way to any desired accuracy.

The matrix blocks appearing in equation (4.80) can be built up by performing *ab initio* calculations both for the different units (to construct the diagonal blocks) and for their different clusters (to construct the off-diagonal blocks). For instance, for a first-neighbor-interaction approximation in a binary disordered chain one calculates the AA, AB, BA, and BB clusters, where A and B are the two different units.

In connection with NFC method in its matrix-block form, it is noteworthy that the inverse iteration technique[37] enables one to compute the wave function pertaining to any particular energy level of the disordered system (which can be determined to any desired accuracy with the help of the NFC method).

To solve the eigenvalue equation

$$(\mathbf{H} - \lambda_j \mathbf{S})\mathbf{c}_j = 0 \tag{4.86}$$

where now the eigenvalue λ_j is known but the eigenvector $\mathbf{c}_j$ is unknown, one can start with the trial equation

$$(\mathbf{H} - \lambda \mathbf{S})\mathbf{b}_2 = (\mathbf{H} - \lambda \mathbf{S}) \sum_j \frac{d_{1j}\mathbf{c}_j}{\lambda_j - \lambda} = \mathbf{S} \sum_j d_{1j}\mathbf{c}_j = \mathbf{S}\mathbf{b}_1 \tag{4.87}$$

Taking into account the equation $\mathbf{H}\mathbf{c}_j = \lambda_j \mathbf{S}\mathbf{c}_j$, one can rewrite equation (4.87) in the form

$$\sum_j \frac{d_{1j}(\lambda_j \mathbf{S} - \lambda \mathbf{S})\mathbf{c}_j}{\lambda_j - \lambda} = \mathbf{S} \sum_j d_{1j} \frac{\lambda_j - \lambda}{\lambda_j - \lambda} \mathbf{c}_j \tag{4.88}$$

which is identical with the right-hand side of equation (4.87). One can repeat this procedure by substitution

$$\mathbf{b}_3 = \sum_j (d_{1j}\mathbf{c}_j)/(\lambda_j - \lambda)^2$$

in the left-hand side of equation (4.87) and

$$\mathbf{b}_2 = \sum_j (d_{1j}\mathbf{c}_j)/(\lambda_j - \lambda)$$

in the right-hand side. One can show again that the equation

$$\begin{aligned}(\mathbf{H} - \lambda \mathbf{S})\mathbf{b}_3 &= (\mathbf{H} - \lambda \mathbf{S}) \sum_j \frac{d_{1j}\mathbf{c}_j}{(\lambda_j - \lambda)^2} \\ &= \mathbf{S} \sum_j \frac{d_{1j}\mathbf{c}_j}{\lambda_j - \lambda} \\ &= \mathbf{S}\mathbf{b}_2\end{aligned} \tag{4.89}$$

is valid.

Repetition of this procedure n times yields

$$(\mathbf{H}-\lambda\mathbf{S})\mathbf{b}_n = (\mathbf{H}-\lambda\mathbf{S})\cdot\sum_j \frac{d_{1j}\mathbf{c}_j}{(\lambda_j-\lambda)^{n-1}}$$

$$= \mathbf{S}\sum_j \frac{d_{1j}\mathbf{c}_j}{(\lambda_j-\lambda)^{n-2}}$$

$$= \mathbf{S}\mathbf{b}_{n-1} \tag{4.90}$$

If the trial value of λ is close enough to an eigenvalue $\lambda_{j'}$ (closer than to any other eigenvalue λ_j) and n is large enough (the procedure has converged), then $(\lambda_{j'}-\lambda)^{n-1}$ $(j' \neq j)$ will be so small that after normalization ($\langle \mathbf{b}_n \mid \mathbf{b}_n \rangle = 1$), of the expressions $d_{1j}/(\lambda_j-\lambda)^{n-1}$, the j'th expression $d_{1j'}/(\lambda_{j'}-\lambda)^{n-1}$ will equal unity to any desired accuracy and all the other terms vanish. In other words, the vector $\mathbf{b}_n = \mathbf{c}_{j'}$ will be the eigenvector corresponding to the eigenvalue $\lambda_{j'}$.

By determining the eigenvector corresponding to any desired energy level of the disordered system, one can form the charge-bond-order matrix $\mathbf{P}^{(v)} = \Sigma_j^{\text{val}}\, \mathbf{c}_j\mathbf{c}_j^+$ for those energy levels of the disordered system that fall within the region of the valence bands. If we maintain constant that electronic density (charge-bond-order matrix) arising from the core levels and lower-lying valence levels [known from the cluster calcultions necessary to construct the matrix (4.80)], then the valence electrons can be treated separately. That is, we can write $\mathbf{P} = \mathbf{P}^{(c)} + \mathbf{P}^{(v)}$ and correspondingly $\mathbf{F} = \mathbf{F}^{(c)} + \mathbf{F}^{(v)}$, where the superscript c stands for the constant and the v for the variable part of $\mathbf{P}$ and $\mathbf{F}$ (in $\mathbf{F}^{(v)}$, of course, only the two-electron part depends on $\rho^{(v)}$). By reconstructing the Fock matrix blocks, for which one needs a cluster calculation of a trimer in the case of a binary system and the first-neighbor-interaction approximation (see the paper by Gazdy *et al.*[38]), one can solve the problem by a combination of the NFC and inverse iteraction methods again, and can repeat this procedure until self-consistency is reached for the valence electrons of the whole disordered system.[38] Some results obtained in this way will be mentioned in the next section.

4.4.3. Application of the NFC Method to Aperiodic Polypeptides and Polynucleotides

It is well known that both DNA and proteins are aperiodic, so in order to treat their electronic structure one must apply the theory of disordered systems.

We consider first disordered single DNA helices. Simple (tight-binding) NFC calculations have been performed for the valence-band regions of the four nucleotide base stacks with their compositions given by $C_A = C_T = 0.50$ and $C_G = C_C = 0$, $C_A = C_T = 0.45$ and $C_G = C_C = 0.05$, and $C_A = C_T = 0.25$ $C_G = C_C = 0.25$.[39] The parameters α_i and β_{ii} used were obtained from a fit of the equation

$$\varepsilon_i(k) = \alpha_i + 2\beta_{ii} \cos ka \tag{4.91}$$

to the *ab initio* STO-3G valence bands of the four nucleotide base stacks.[40] The chemical formulas of the four nucleotide bases (two nucleotide base pairs) are given in Figure 4.17.

The value $\beta_{ij} = (|\beta_{ii} + \beta_{jj}|)/2$ has been applied to the case for which, in the random sequence of 5000 units (generated by a Monte Carlo program), two different bases follow each other. The histogram interval in Figures 4.18, 4.19, and 4.20 is 0.01 eV. It is clear from these figures

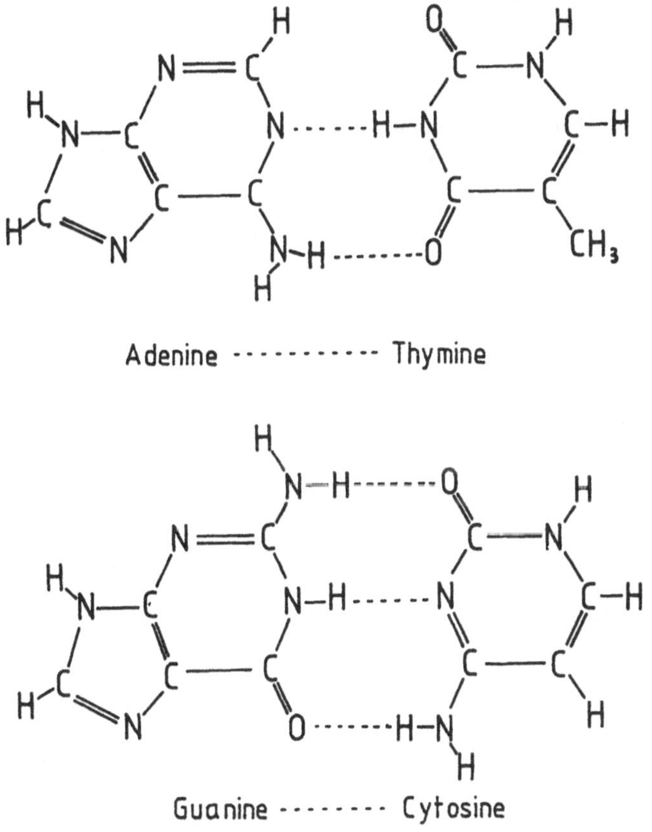

FIGURE 4.17. Chemical formulas of the adenine-thymine (A-T) and guanine-cytosine (G-C) base pairs.

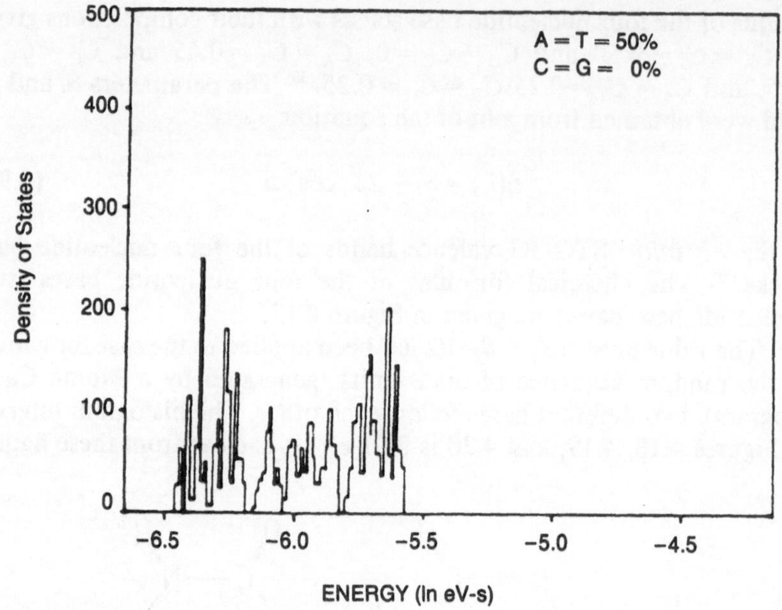

FIGURE 4.18. Density of states of the valence-band region of a single stranded aperiodic DNA chain containing only the bases A and T.

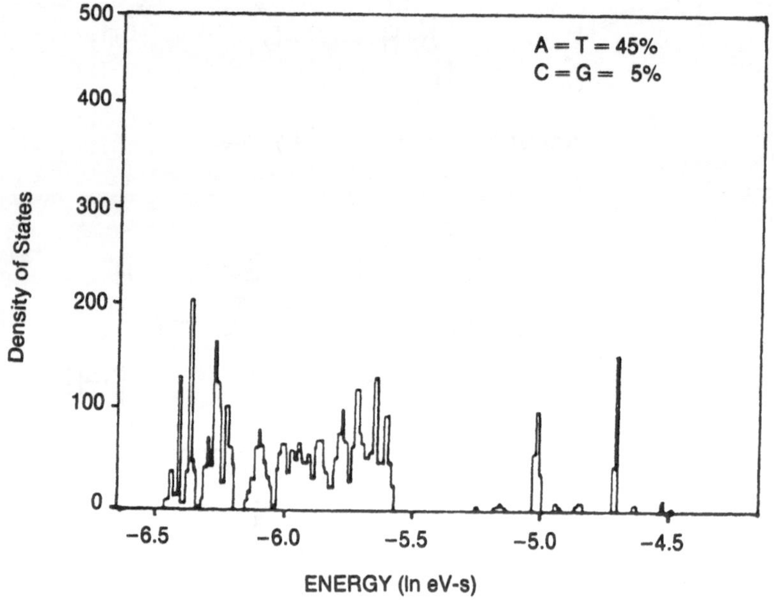

FIGURE 4.19. Density of states of the valence-band region of a single stranded aperiodic DNA chain containing all the four nucleotide bases.

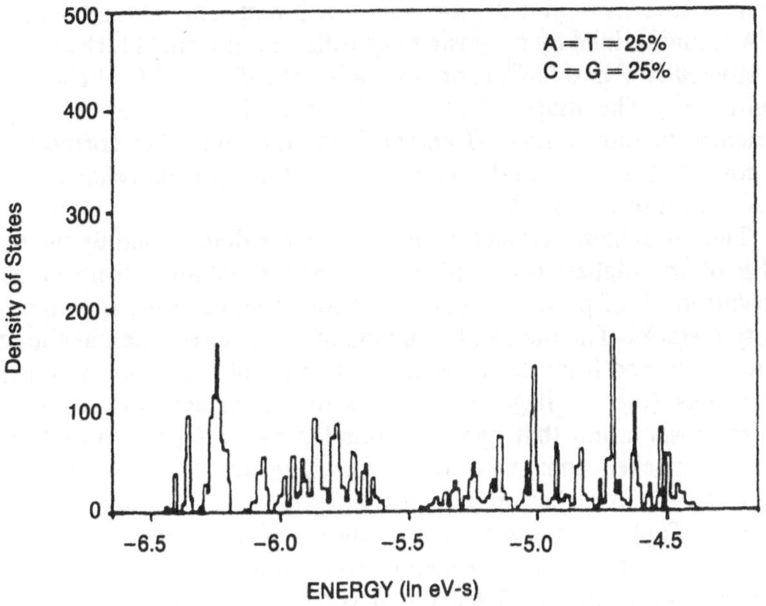

FIGURE 4.20. Density of states of the valence-band region of a single stranded aperiodic DNA chain containing all the four nucleotide bases. (The composition is indicated in the figures and density-of-state histograms were obtained with the help of the simple NFC technique.)

that disorder strongly influences the electronic structure and causes many gaps. In the case of the largest disorder (all four bases with a relative concentration of 0.25; see Figure 4.20) the valence bands of the nucleotide base stacks almost completely disappear and one has only a large number of isolated high peaks.

Next, two interacting nucleotide base stacks coupled with nearest-neighbor interactions were considered.[41] Generalization to larger numbers of coupled chains is straighforward. The coupled double chain is particularly interesting because it is suggestive of DNA, which is composed of a "random" sequence of the nucleotide base pairs adenine-thymine (A-T) and guanine-cytosine (G-C). The model Hamiltonian matrix for the coupled two chains is set up as a block-tridiagonal matrix

$$\mathbf{H} = \begin{bmatrix} \cdots & & & & \\ \mathbf{W}_\alpha & \boldsymbol{\varepsilon}_\alpha & \mathbf{W}_{\alpha+1} & & \\ & \mathbf{W}^t_{\alpha+1} & \boldsymbol{\varepsilon}_{\alpha+1} & \mathbf{W}_{\alpha+2} & \\ & & \mathbf{W}^t_{\alpha+2} & \boldsymbol{\varepsilon}_{\alpha+2} & \mathbf{W}_{\alpha+3} \\ & & & \cdots & \end{bmatrix} \tag{4.92}$$

where subblocks ε_α and $\mathbf{W}_\alpha$ are now 2×2 matrices. The analogy with DNA is maintained by proposing two different diagonal blocks ε_α, which are labeled A-T and G-C. (For simplicity, the T-A and C-G choices are suppressed.) The diagonal blocks also contain, in their off-diagonal elements, the interchain (A-T and G-C) interactions. They correspond to the possible sequences of the base pair, with three possible choices for the intrachain interactions $\mathbf{W}_\alpha$.

The intrachain parameters in the $\mathbf{W}_\alpha$ were determined by the bandwidth of the highest occupied valence bands obtained from *ab initio* calculations[40] of periodic single-chain adenine, guanine, thymine, and cytosine stacks. The diagonal elements of the ε_α were taken as the band centers obtained from the same calculations. The interchain interaction parameters (the off-diagonal elements of ε_α), in accordance with the general observation that hydrogen bonded systems perturb each other less than stacked ones, were taken for these model calculations to be about half the intrachain parameters ($-$ 006 eV for A-T and $-$ 0.09 eV for G-C). Table 4.4 summarizes the choice of parameters that we have used to construct the Hamiltonian matrix. Admittedly, this Hamiltonian matrix is simplistic for DNA, but it is sufficient to demonstrate some features of the matrix block NFC method. It is fairly straightforward to generalize to many coupled chains, or to vibrational problems in a single chain with random springs and masses.

TABLE 4.4. Parameters Used to Represent the Aperiodic DNA Double Helix (in eV)

Cluster–cluster interaction parameters			
	AT AT	AT/GC GC/AT	GC GC
W_{11}	-0.1185	-0.1618	-0.2052
W_{22}	-0.1485	-0.1820	-0.2155
W_{12}	0	0	0
W_{21}	0	0	0

Intracluster parameters		
	AT	GC
ε_{11}	-5.8037	-4.7541
ε_{22}	-6.1815	-5.0816
ε_{12}	-0.0600	-0.0900

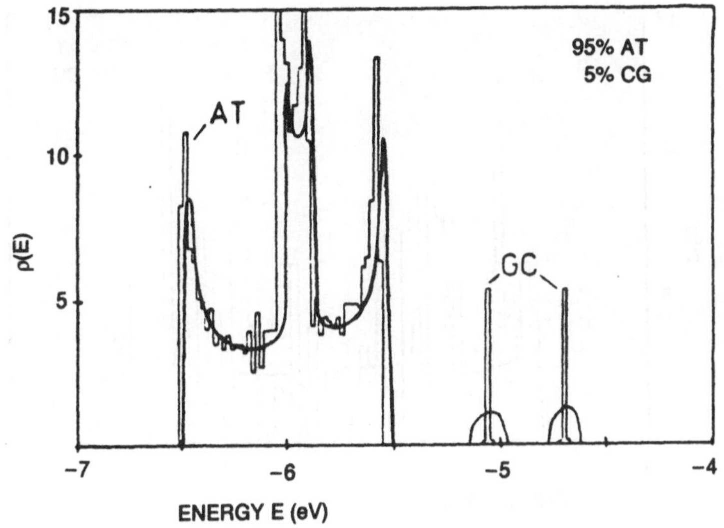

FIGURE 4.21. Density of states for two random coupled chains by 95% AT and 5% GC composition. The histogram is the result of the NFC calculation for a finite chain composed of 1000 nucleotide base pairs. The smooth line is the prediction of a two-site cluster CPA calculation.

The coupled-chains model for DNA was investigated for two limiting cases. The low impurity concentration limit was taken as a double chain that consists of 95% A-T and 5% G-C units. The high concentration limit was taken as a 50% A-T–50% G-C chain.

First, the density of states in the coupled chain of 2000 units in the high and low concentration limits were investigated. Figures 4.21 and 4.22 show comparisons between the NFC density of states and the two-site cluster CPA in the "diagonal disorder" approximation. There is excellent agreement for the low concentration limit in Figure 4.21, but poor agreement in the high concentration limit. The very spikey structure predicted by the NFC calculation for the high concentration limit is a consequence of the strong disorder (very different diagonal elements in the matrices ε_α). When the different interchain and intrachain interactions are increased or the site energies (diagonal elements of ε_α) reduced, the curve contains fewer spikes and a smoother structure of the density of states appears. To determine this, Figure 4.23 shows for comparison the high concentration limit between the NFC and two-site cluster CPA when the diagonal elements of ε_α in Table 4.4 are divided by 10. Agreement is again excellent. Thus, if the cluster CPA is to provide

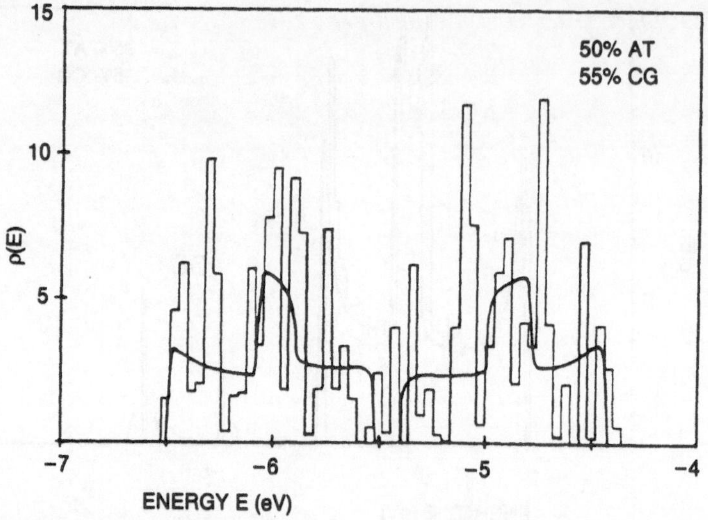

FIGURE 4.22. Density of states for two random coupled chains by 50% AT and 50% GC composition. The histogram is the result of the NFC calculation for a finite chain composed of 2000 nucleotide base pairs. The smooth line is the prediction of a two-site cluster CPA calculation.

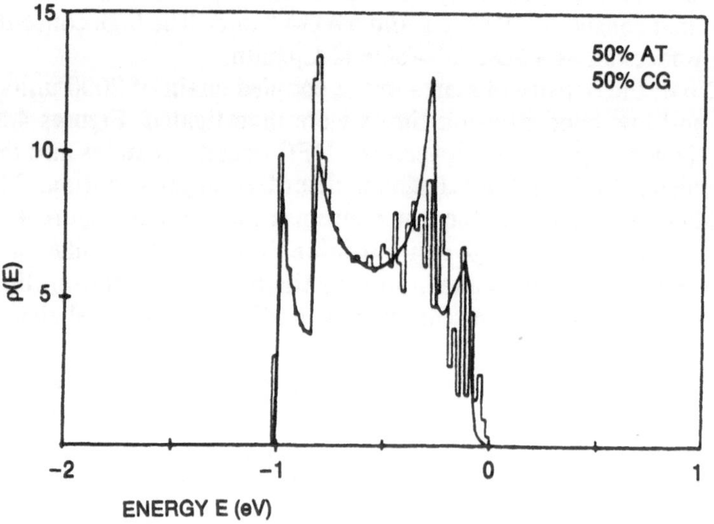

FIGURE 4.23. Density of states for two random coupled chains in the high concentration limit with weakened site diagonal energies. The legend is the same as in Figure 4.22.

reasonable results in the high concentration limit for a small cluster size, the differences between the elements of matrices $\boldsymbol{\varepsilon}_\alpha$ and $\mathbf{W}_\alpha$ must not be too great.

In connection with calculations on nonperiodic DNA it should be noted that in recent years a larger number of RNA and DNA fragment sequences have been determined experimentally. The sequences are stored in computers and are available.[(42)] With the aid of a detailed mathematical statistical analysis of these sequences one could find some regularities in them.[(43)] Therefore, one hopes that with the aid of these regularities it will be possible to construct so-called representative sequences for different classes of biological systems. These effective sequences could then be used instead of the presently employed random sequences to treat aperiodic DNA by the NFC method.

We now turn to nonperiodic polypeptides. Seel[(31)] has used the simple tight-binding NFC method to determine the DOS of the valence bands of binary glycine-alanine chains, the total number of monomer units being 1000. To demonstrate that chains of length 1000 units give as much information as is required for this model study, a calculation of the spectrum of a chain 10,000 units in length is added for the 50% alanine concentration case and shows little difference in the major characteristics of the spectrum of the 1000-unit chain. The parameters α and β were chosen to match the locations and widths of the valence bands of the corresponding periodic glycine and alanine chains of Suhai *et al.*[(3)] obtained in the MINDO/3 approximation. The values $\alpha^A = -9.6719$ eV, $\beta^{AA} = -0.1239$ eV, $\alpha^B = -9.3983$ eV, and $\beta^{BB} = -0.1215$ eV were used. When in the aperiodic chains a B unit followed an A unit (or vice versa), $\beta^{AB} = \frac{1}{2}(\beta^{AA} + \beta^{BB})$ was used. The random two-component chains are generated by employing a random-number generator. The sequence of numbers $\{\varepsilon_i(\lambda)\}$ is then computed for $n = 200$ values of λ in the energy region between -9.920 eV and -8.925 eV (step length 0.005 eV) and the eigenvalue spectra are plotted by noting the numbers of negative values of ε_i.

The computer time required to calculate a spectrum of a chain with 1000 units was about 6 s on a CYBER 172, including the random chain generation (without the Monte Carlo routine, about 5 s); the time for the chain of 10,000 units was 42 s.

It was seen in Section 4.2.1 that, though alanine differs from glycine by only a CH_3 group, the valence band of polyalanine is shifted by about 0.27 eV as compared to the corresponding band of polyglycine. This causes a splitting, the gap width of which is about 0.25 eV for the valence band of the ordered poly(glyala) mixed chain.

Figure 4.24 and 4.25 show the eigenvalue spectra for randomly generated chains, again 1000 units in length; Figure 4.25c is for a chain of

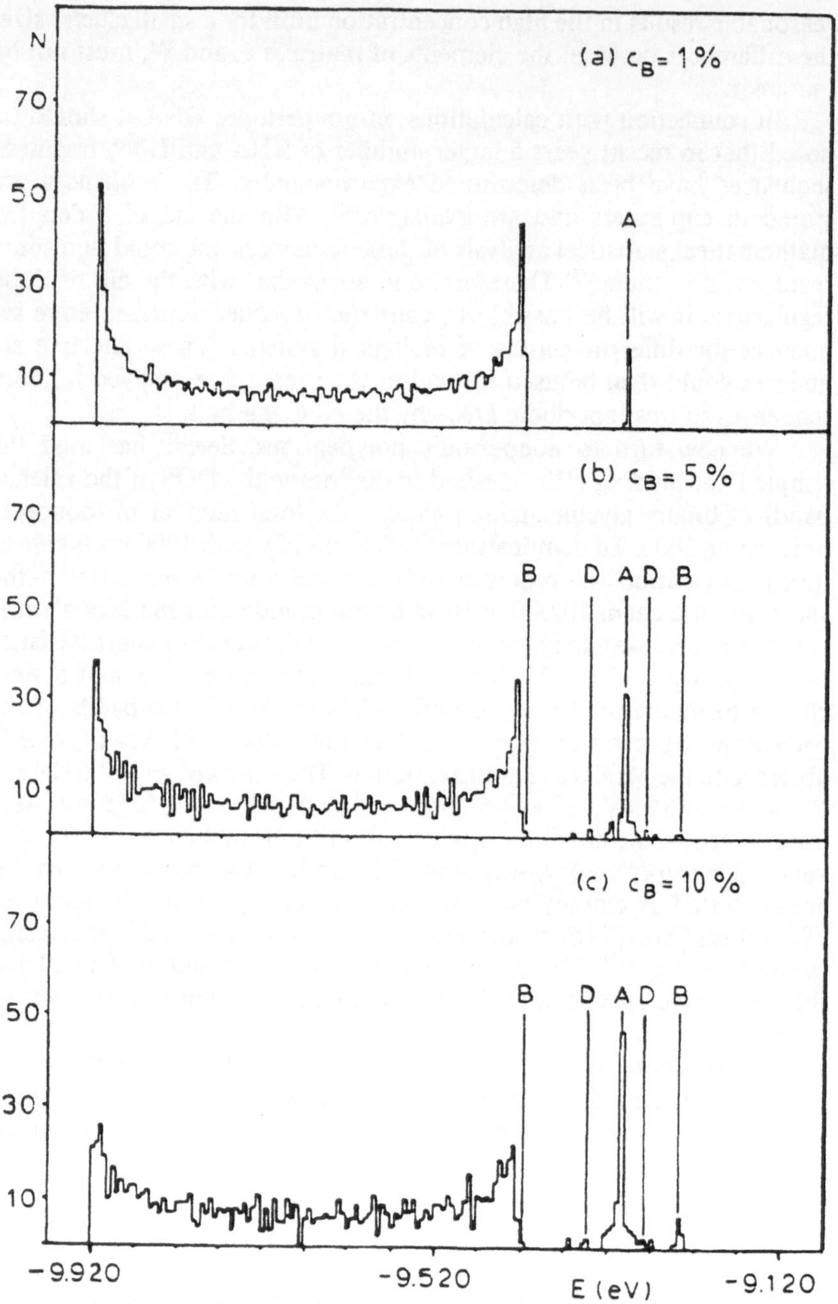

FIGURE 4.24. Eigenvalue spectra for two-component (glycine-alanine) disordered chains containing 1000 units. The parameter c_B refers to the percentage of alanine residues in the chain. The spectral lines labeled by letters A to D are associated with particular local chain sequences and are explained in the text. N is the number of eigenvalues in a histogram interval of 0.005 eV.

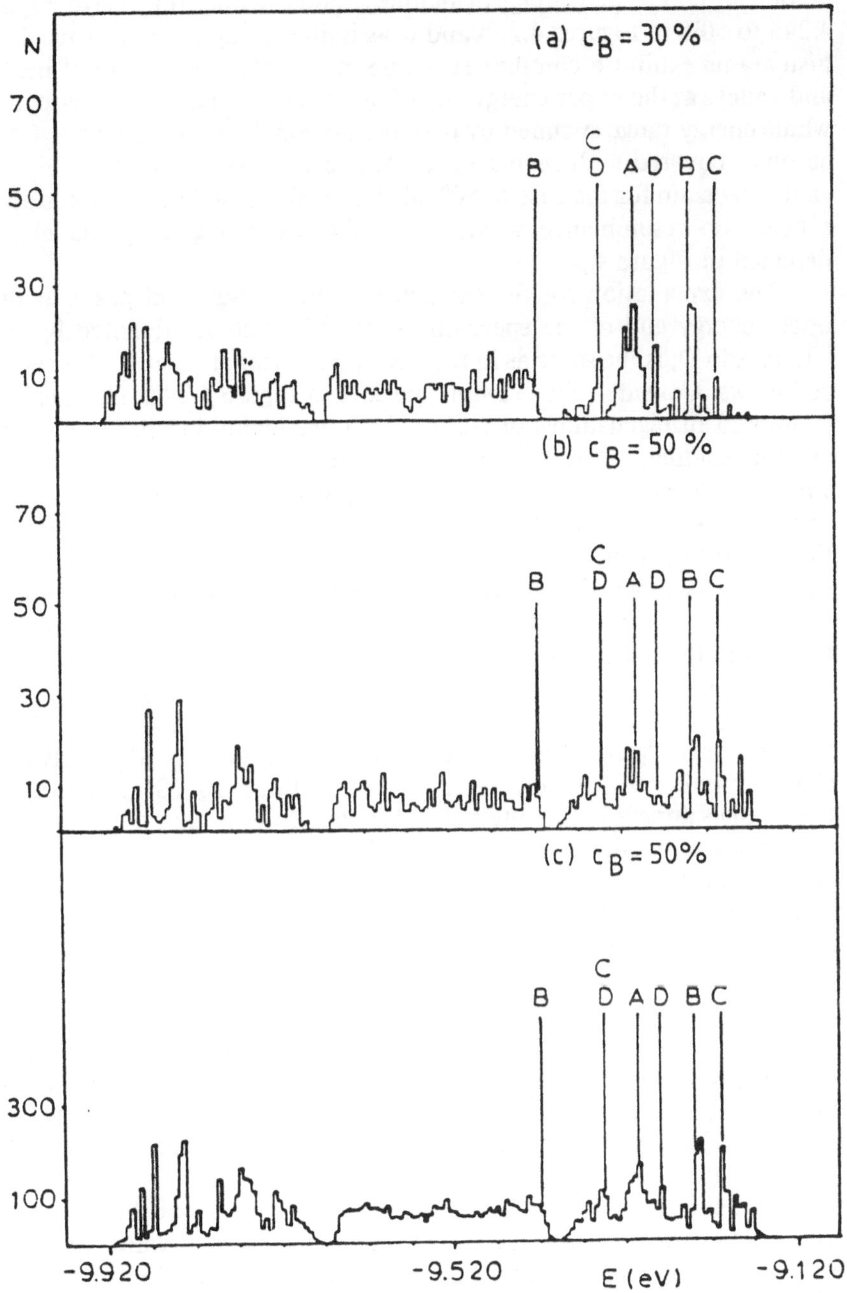

FIGURE 4.25. Eigenvalue spectra for two-component (glycine-alanine) disordered chains of length 1000 units in (a) and (b) and 10000 units in (c). c_B gives the fraction of alanine residues in the chain. The spectral lines A to D are explained in the text. N is the number of eigenvalues in a histogram interval of 0.005 eV.

10,000 units. The percentage of alanine residues varies from 1% in Figure 4.24a to 50% in Figures 4.25b and c, as indicated on the diagrams. The histograms exhibit a complex structure, a complicated system of peaks and valleys at the upper energy end of the spectrum starts to develop, the whole energy range spanned by periodic poly(gly) and periodic poly(ala) becomes covered with increasing alanine concentration, and the eigenvalue spectrum for the case of 50% alanine residues in Figures 4.25b and c bears no resemblance whatever to the ordered glyalaglyala chain depicted in Figure 4.3.

The explanation for the existence of the well-defined peaks at the upper energy end of the spectrum, some of which are denoted by the letters A to D, is the same as in the case of vibrational spectra,[29,44] and is as follows. Consider, for example, peak A in Figure 4.24a. This peak is composed of eigenvalues of states which are highly localized at single alanine residues surrounded by glycine units. At each point in the chain where the local structure — glyglyalaglygly — occurs, an eigenvalue is contributed to peak A. This statement can be made without calculating the eigenvectors on the basis of the spectra depicted in Figure 4.26. Figure 4.26a shows the spectrum of a glycine chain of 1000 monomers that contains a single alanine unit at position 500. One eigenvalue appears in the histogram interval between -9.305 eV and -9.300 eV (interval, peak A). In Figure 4.26b, the spectrum of a glycine chain containing 9 alanine residues at positions 100, 200, . . . , and 900 is given. In this case 9 eigenvalues lie in the histogram interval A. If the glycine chain contains 99 alanine residues at positions 10, 20, . . . , 990, 99 eigenvalues appear in the interval A. We therefore see that the eigenstate of an alanine residue surrounded locally by nine glycine units on each side is hardly affected by the sequence of the chain elsewhere: the eigenstates still lie in the energy interval of 0.005 eV, as in the case of a single alanine residue in an otherwise monomolecular chain of glycine units. A similar situation holds for the other clusters containing alanine units surrounded locally by glycine units. The other peaks labeled in the spectra in Figure 4.24 and 4.25 are identified by calculations similar to those depicted in Figure 4.26 and are associated with the following types of local sequences: peak B corresponds to the local sequence glyglyalaalaglygly–, C to the local sequence glyglyalaalaalaglygly, and D to the local sequence glyglyglyalaglyalaglyglygly.

This simple picture of associations between peaks of the spectrum and local chain sequence accounts fully for the changes which occur at the upper energy end of the spectrum as the concentration of alanine residues is increased. Thus, at low concentration of alanine units (Figure 4.24) the spectrum consists mainly of the well-known spectrum of the glycine valence band with a little structure in the alanine region domi-

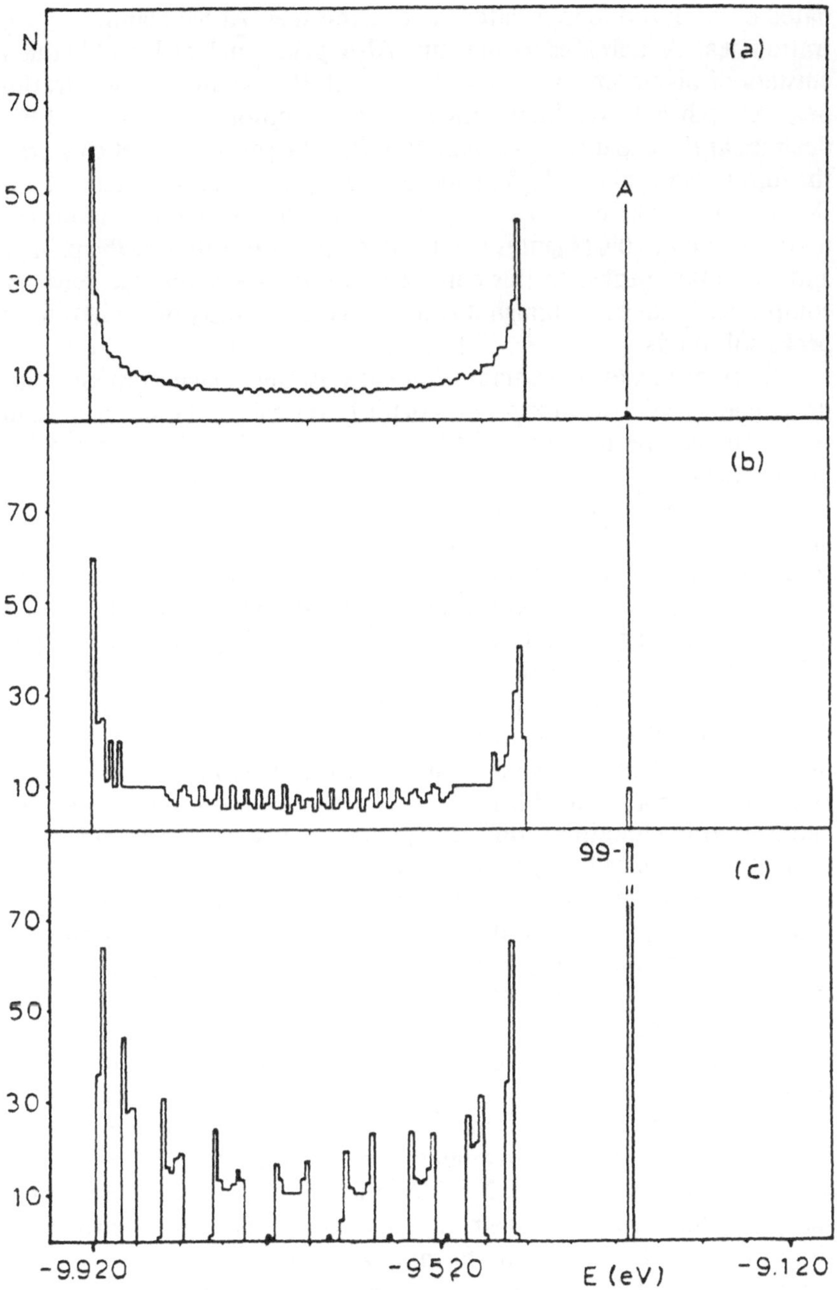

FIGURE 4.26. Eigenvalue spectra for a glycine chain of 1000 units which contains a single alanine residue at position 500 (a), 9 alanine residues at positions 100, 200, . . . , 900 (b), and 99 alanine units at positions 10, 20, . . . , 990 (c). The spectral line A is associated with the local chain sequence –glyglyalaglygly–.

nated by peak A due to isolated alanine residues. At 5% alanine concentration peak A increases in intensity. Also, peaks such as B and D due to clusters of alaala and alaglyala appear. At 10% alanine concentration, peak A reaches its maximum intensity in the computed spectra and then declines at the expense of secondary peaks: the probability of clusters of the forms alaala, alaglyala, and alaalaala increases. At 30% (Figure 4.25a) the tertiary peaks due to three-alanine clusters are already quite pronounced and at 50% (Figures 4.25b and c) they are as high as the primary and secondary peaks. At this concentration the spectrum becomes very complicated indeed, although it is clear that the identity of the individual peaks still holds.

Calculations for high alanine and low glycine concentrations would presumably lead to localized states in the lower part of the spectrum. This would change the picture qualitatively in contrast to the case of vibrational spectra.

One last aspect of the computed spectra should be noted. Since real proteins have a well-defined sequence of peptide units, it is interesting to compare, for example, Figure 4.26c with Figure 4.24c (both represent the case of 10% alanine concentration). When the alanine residues are at fixed positions and locally surrounded by nine glycine units (Figure 4.26c), the effect in the region of the valence band of glycine is very pronounced. The valence band is split into islands that contain eigenvalues, some of them as narrow as 0.015 eV, and forbidden gaps. In the random case, sharp peaks and valleys occur mostly in the upper energy region of the spectrum. Further calculations on more realistic protein model chains containing different peptide units in an experimentally defined sequence will very probably show that the spectra of these chains exhibit a structure lying between the random case and the ordered structures of Figure 4.26c, that is, a complicated system of peaks and valleys, the peaks being associated with particular types of local chain sequences. Thus, these well-defined peak structures can be used as "fingerprints" to identify different proteins. The effect of replacing a small sequence by other peptide units can be studied very easily.[31]

The glycine-serine (in serine we have a $-CH_2-OH$ side chain instead of the $-CH_3$ side group of alanine) binary chain has been treated[45] with the help of the matrix block NFC technique.[35] The binary polypeptide chain has the structure shown in Figure 4.27. The homopolypeptide chains are constructed by performing a simultaneous translation and screw rotation of 180° upon the basic glycine and serine subunits. The structure in Figure 4.27 corresponds to a chain in the antiparallel-β-pleated sheet structure of Pauling and Corey.[5]

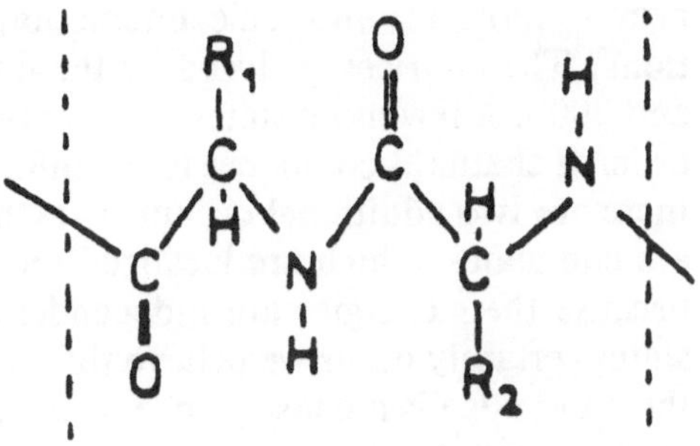

FIGURE 4.27. Structure of the polypeptide chain with side groups R_1 and R_2.

As an initial study the SCF all-valence electron MINDO/3 crystal-orbital calculations of Suhai *et al.*[3] for the periodic polypeptides were used as input for the calculations. The overlap matrix is equal to the identity matrix in these calculations, but in the NFC method it is no more difficult to employ *ab initio* input data when available. In this case the full overlap matrix **S** must be included.

Figure 4.28 shows the density-of-state histogram for the highest valence and lowest conduction bands of a finite poly(gly) chain. The arrows indicate the exact band edges from the periodic calculations of Suhai *et al.*[3] The matrix block NFC technique quite clearly reproduces the periodic results, within the error limits of the histogram interval (only

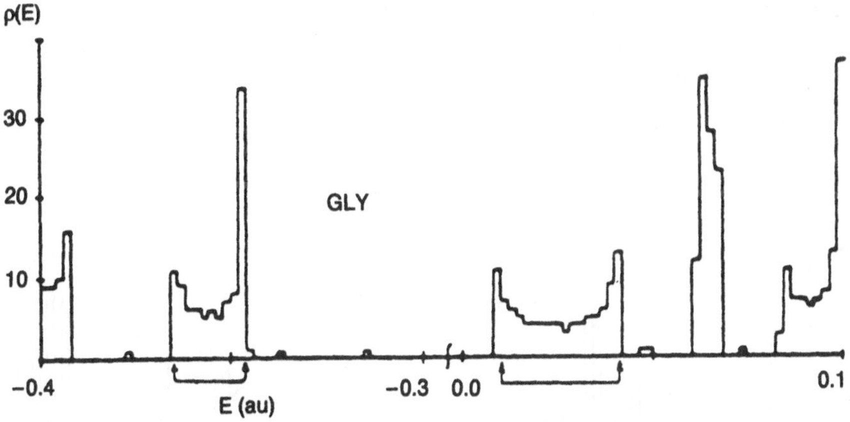

FIGURE 4.28. Approximate matrix block NFC results for the densities of states in the valence- and conduction-band region of a regular glycine chain of 100 units. The arrows denote the theoretical band limits from the crystal-orbital calculations of Suhai *et al.*[3] and the histogram interval is 0.002 au.

nearest molecule–molecule interactions were included in the calculations). The matrices analyzed for these results varied between rank 1000 and 2000. A few more states occur between the main bands in the finite ordered chains as compared to the infinite periodic chains. In particular, there are two additional occupied states above the valence band. These are end states, which are localized mostly around the end of the chain because their energies are independent of chain length. Though these states certainly occur we believe that their large energy separation from the band edge is a consequence of the approximate construction of the Fock matrices.

A further test of the reliability of the matrix block NFC method is shown in Figure 4.29. A calculation was performed for a sequence of 20 gly–gly–ser units and, to within about 0.002 au, we could reproduce the band limits of the exact periodic calculation of Suhai *et al.*[3] This is quite significant since only poly(gly), poly(ser), and poly(gly-ser) data were used as input for our calculations. This is an indication that these model investigations may have some quantitative predictive power.

Finally, a random sequence of gly and ser units is considered. Figure 4.30 shows the density of states in the highest valence- and lowest conduction-band regions. The overall location of the bands is quite similar to the positions in the periodic gly-ser chain. The similarity of the peptide units as well as their size is unquestionably responsible for this behavior. The singular behavior at the band edges in the periodic calculations is now missing and the bands appear to be slightly broadened. This broadening is enough that the split valence band loses the gap that existed in the periodic infinite chain. The overall location of the energy bands is, however, not dramatically different from the periodic chains.

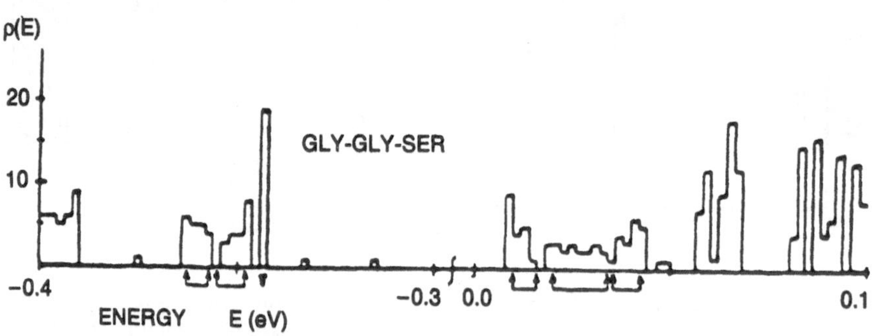

FIGURE 4.29. Approximate matrix block NFC results for the valence and conduction bands in a chain of 20 gly-gly-ser units. The legend is the same as in Figure 4.28.

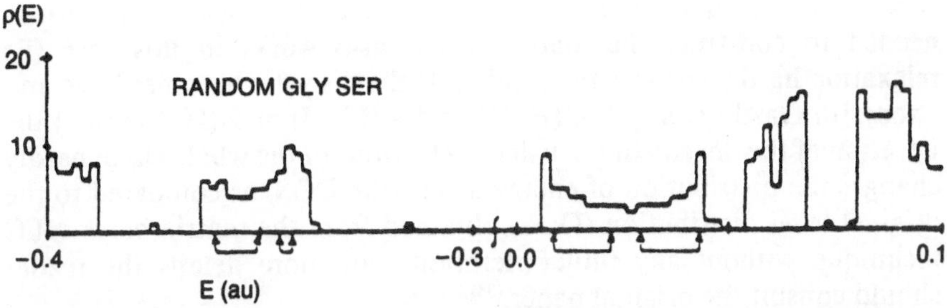

FIGURE 4.30. Approximate matrix block NFC results for the valence and conduction bands in a chain composed of a random sequence of 38 glycine and 42 serine units. The arrows indicate the theoretical band limits in a periodic gly-ser chain (after Suhai *et al.*[3]).

Owing to the large unit cell sizes in the polypeptide chains, one must work with quite large matrices in order to obtain reliable results. The calculations took from 20 min to over 1 hour of CPU time on a CYBER 173 computer, but it is possible that the programs can be further optimized to somewhat reduce these times. If one tried to perform the same calculations on a direct SCF level, however, one can conservatively estimate that they would require at least one order of magnitude more CPU time, even within the framework of the MINDO/3 scheme.

Finally, one should mention the results obtained by iterating until self-consistency at least the valence electron densities in the case of the whole finite binary chain in the framework of the matrix block NFC technique[38] (see the end of Section 4.4.2). The units were

$$X = \text{H–H} \quad \text{and} \quad Y = \begin{array}{c}\text{Li}\\ | \\ \text{H}\end{array} \quad \text{or H–F}$$

According to results published elsewhere,[38] one can conclude that for systems with small disorder (such as

$$13\ \text{H–H} \quad \text{and} \quad 3\begin{pmatrix}\text{Li}\\ \text{H}\end{pmatrix}$$

units) at least a few iteration steps to obtain an overall self-consistency of the valence electrons are needed to arrive at reasonable DOSs. At the same time, the frozen-core approximation (keeping the densities of the core electrons equal to those obtained from the cluster calculations needed to construct the matrix block) also works in this case (its relaxation hardly changes the results). If the disorder becomes large and strong [in Gazdy *et al.*,[38] 8 (H–H) and 9 (Li–H) or 9 (H–F) units] the iteration of the density of the valence electrons of the whole chain hardly changes the distribution of energy levels (the DOS) as compared to the original level distribution (DOS) obtained from the matrix block NFC technique without any futher iteration. For more details the reader should consult the original paper.[38]

4.5. INVESTIGATION OF THE LOCALIZATION OF THE ORBITALS IN DISORDERED CHAINS

The investigation of the localization properties (Anderson localization[8]) of the eigenstates of disordered chains is important primarily from the standpoint of their transport properties. If the Fermi energy (ε_F) falls into a more or less continuous region of allowed energy levels, one has to know whether the states around ε_F are localized or delocalized. If the wave functions are delocalized, a coherent, Bloch-type conduction is still possible. If, however, they are localized, one can expect only incoherent, hopping-type charge transport.

It was mentioned in Section 4.2.2 that the localization of the states starts at the band edges, and the boundary between the localized and delocalized states (the so-called mobility edge) moves with increasing disorder from both sides toward the middle of the band. Thus the final position of ε_F (whether it falls inside a region of allowed states) and its location relative to the mobility edge determine the conduction properties of a disordered chain.

In Section 4.2.2, dealing with the simple example of disordered finite hydrogen rings, the wave functions were calculated directly at the Hartree–Fock level and so their localization properties could be determined by simple inspection. The same is true for states determined by much larger chains with the aid of the inverse iteration technique (see Section 4.4.2).

However, to do this for every state is practically impossible in the case of really long chains ($N > 1000$, where N is the number of sites). Therefore, a technique is needed that gives a rough idea as to whether or not the states are localized in a certain energy region without explicitly calculating them. Such a method based on the calculation of Green matrix elements of the chain will be examined in the next section.

4.5.1. Green Matrix Method for the Study of the Localization Properties of the States Belonging to a Disordered Chain*

Suppose we describe a disordered polymer with the help of an LCAO wave function

$$|\psi_K\rangle = \sum_{i=1}^{N} C_i^K \,|\, i\rangle \tag{4.93}$$

where, as a simplication, it is assumed that there is only one basis function $|i\rangle$ per site. If these basis functions are not orthogonal, then the secular equation for the eigenstates $|\psi_K\rangle$ can be written in the form

$$(\mathbf{H} - \lambda_K \mathbf{S}) \,|\, C^K\rangle = 0 \tag{4.94}$$

where the overlap matrix elements are

$$S_{ij} = \langle i \,|\, j\rangle \tag{4.95}$$

and the vector $|C^K\rangle$ is composed of the expansion coefficients C_i^K of the eigenstate.

To determine the qualitative nature of the eigenstates in a given energy region it is convenient to introduce the folowing equation, which is closely related to the eigenvalue problem (4.94),[45]

$$(\mathbf{H} - \lambda \mathbf{S}) \,|\, \varphi^N\rangle = |\bar{N}\rangle \tag{4.96}$$

The vector $|\bar{N}\rangle$ is a state vector whose components are all zero except for the last, or Nth component, which is taken as 1. It is evident that equation (4.96) is identical to the eigenvector equation (4.94) except at one of the boundaries, and that $|\varphi^N\rangle$ can therefore be thought of as a "particular solution." The vector $|\varphi^N\rangle$ may be expressed formally as

$$|\varphi^N\rangle = (\mathbf{H} - \lambda \mathbf{S})^{-1} \,|\, \bar{N}\rangle \tag{4.97}$$

The components of $|\varphi^N\rangle$ can be obtained by multiplying on the left of equation (4.97) by vectors $\langle \bar{i} |$ that are similar to $|\bar{N}\rangle$ (they have a component 1 at the ith position and all the other components are zero). Thus formally the ith component of $|\varphi^N\rangle$ is given by

$$\langle \bar{i} \,|\, \varphi^N\rangle = G_{iN}(\lambda) \tag{4.98a}$$

where

$$G_{\bar{i}N}(\lambda) = \langle \bar{i} \,|\, (\mathbf{H} - \lambda \mathbf{S})^{-1} \,|\, \bar{N} \rangle \tag{4.98b}$$

We shall now establish a relationship between $|\varphi^N\rangle$ and the eigenvectors $|C^K\rangle$ by expanding $|\bar{N}\rangle$ in terms of $|C^K\rangle$. The orthogonality condition

$$\langle C^K \,|\, \mathbf{S} \,|\, C^J \rangle = \delta_{JK} \tag{4.99}$$

enables one to write

$$\langle C^J \,|\, \bar{N} \rangle = C^J_{\bar{N}} = \sum_K \langle C^J \,|\, \mathbf{S} \,|\, C^K \rangle \langle C^K \,|\, \bar{N} \rangle = \langle C^J \,|\, \mathbf{S} \,|\, C^J \rangle \langle C^J \,|\, \bar{N} \rangle$$
$$= \langle C^J \,|\, \bar{N} \rangle \tag{4.100}$$

from which it follows that

$$|\bar{N}\rangle = \sum_K \mathbf{S} \,|\, C^K \rangle \langle C^K \,|\, \bar{N} \rangle \tag{4.101}$$

Hence $|\varphi^N\rangle$ is given by

$$|\varphi^N_{(\lambda)}\rangle = \sum_K \frac{|C^K\rangle \langle C^K \,|\, \bar{N}\rangle}{\lambda_K - \lambda} \tag{4.102}$$

in terms of the eigenvectors and eigenvalues of equation (4.94). With the aid of the above equations equation (4.98a) can be expressed in the alternative form

$$\langle \bar{i} \,|\, \varphi^N \rangle = \sum_K \frac{C^{*K}_{\bar{i}} C^K_{\bar{N}}}{\lambda_K - \lambda} \tag{4.103}$$

Since our main interest is in quasi-one-dimensional systems defined in a set of localized basis orbitals, the coefficients $C^K_{\bar{i}}$ and $C^K_{\bar{N}}$ in equation (4.103) correspond to different physical locations within this linear system. Therefore, this equation gives a rough measure of the relative magnitude of the eigenvector components of two different locations in a quasi-one-dimensional chain.

There is a well-known review article by Ishii[48] that contains a proof that the "particular solutions" in a disordered one-dimensional system grow exponentially with probability approaching 1 as a function of the size of the system. The proof should also hold in a polymer or quasi-one-dimensional system. With this result in mind, we compare the first and last components of $|\varphi^N\rangle$, that is, the ends of our finite polymer chain.

The ratio of the first and last components of $|\varphi^N\rangle$, in an arbitrary local basis, is given by equations (4.96) and (4.101)[45]:

$$\frac{\langle\bar{1}\,|\,\varphi^N\rangle}{\langle\bar{N}\,|\,\varphi^N\rangle}=\sum_K\frac{\langle\bar{1}\,|\,C^K\rangle\langle C^K\,|\,\bar{N}\rangle}{\lambda_K-\lambda}\Big/\sum_K\frac{\langle\bar{N}\,|\,C^K\rangle\langle C^K\,|\,\bar{N}\rangle}{\lambda_K-\lambda} \tag{4.104}$$

This ratio should be a good measure of the eigenstate localization, and because of the energy denominators this measure is sensitive to the eigenstates in a particular energy region. We consider the case where the "particular solution" to equation (4.96) or (4.104) has finite components from the end N of the chain. By Ishii's theorem the exponential behavior of the particular solutions in a one- or quasi-one-dimensional system requires that the first component of $|\varphi^N\rangle$ should go to zero exponentially as a function of chain length in a disordered system for almost every energy. This means that the above ratio should approach zero as N becomes large for a disordered system. The ratio (4.104), expressed in terms of the eigenstates, will become small if the eigenstate is such that the product $\langle\bar{N}\,|\,C^K\rangle\langle C^K\,|\,\bar{1}\rangle$ is small. From this we infer that the eigenstate does not possess components of the basis functions from both ends of the chain simultaneously, and it is thus spatially localized.

We shall now consider a polymer chain. The Hamiltonian matrix of a polymer will be banded if a local basis is used. We can then apply the matrix inversion formula given by Butler[49]:

$$\mathbf{M}=\begin{bmatrix}\mathbf{X} & \mathbf{Y}\\ \mathbf{Y}^t & \mathbf{Z}\end{bmatrix}\quad(\mathbf{Y}^t\text{ is the transpose of }\mathbf{Y}) \tag{4.105}$$

$$\mathbf{M}^{-1}=\begin{bmatrix}(\mathbf{X}-\mathbf{Y}\mathbf{Z}^{-1}\mathbf{Y}^t)^{-1} & -\mathbf{X}^{-1}\mathbf{Y}(\mathbf{Z}-\mathbf{Y}^t\mathbf{X}^{-1}\mathbf{Y})^{-1}\\ -\mathbf{Z}^{-1}\mathbf{Y}^t(\mathbf{X}-\mathbf{Y}\mathbf{Z}^{-1}\mathbf{Y}^t)^{-1} & (\mathbf{Z}-\mathbf{Y}^t\mathbf{X}^{-1}\mathbf{Y})^{-1}\end{bmatrix} \tag{4.106}$$

Matrix $\mathbf{M}$ is selected as $z\mathbf{I}-\mathbf{H}$, where the Hamiltonian matrix $\mathbf{H}$ is expressed in terms of square subblocks as follows (this formulation allows more than one atom or/and orbitals per site):

$$\mathbf{H}=\begin{bmatrix}\mathbf{A}_1 & \mathbf{B}_2 & & \mathbf{0}\\ \mathbf{B}_2^t & \mathbf{A}_2 & \mathbf{B}_3 & \\ \cdots & \cdots & \cdots & \cdots\\ & & & \mathbf{B}_N\\ \mathbf{0} & & & \mathbf{A}_N\end{bmatrix} \tag{4.107}$$

It is rather tedious, but straightforward, to show that

$$\mathbf{G}_{N,N}(z) = (z\,\mathbf{I} - \mathbf{A}_N - \mathbf{B}_N^\dagger \mathbf{P}_N \mathbf{B}_N)^{-1} \tag{4.107a}$$

$$\mathbf{P}_i = (z\,\mathbf{I} - \mathbf{A}_{i-1} - \mathbf{B}_{i-1}^\dagger \mathbf{P}_{i-1} \mathbf{B}_{i-1})^{-1} \tag{4.107b}$$

$$\mathbf{P}_2 = (z\,\mathbf{I} - \mathbf{A}_1)^{-1}, \qquad z = \lambda + i\varepsilon \tag{4.107c}$$

and

$$\mathbf{G}_{1,N}(z) = \prod_{i=1}^{N} (\mathbf{P}_{i+1}\mathbf{B}_{i+1})\mathbf{G}_{N,N}(z) \tag{4.108}$$

These equations are ideally suited for evaluation on the computer. The N, N block of $\mathbf{G}$ can be obtained as a byproduct of evaluating the corner block element. Since all the quantities in the above equations are matrices, one must be careful to take the product in the correct order.

4.5.2. Application to a Model Hamiltonian

One can, of course, generalize $\mathbf{G}_{1,N}(z)$ to the case of $\mathbf{G}_{i,N}(z)$ and in this way investigate the localization properties of a particular solution $\varphi_N(z)$ at a certain energy $z = \lim_{\varepsilon \to 0} (\lambda + i\varepsilon)$ at an arbitrary site i, by inserting $\mathbf{G}_{i,N}(z)$ instead of $G_{1,N}(z)$ in the ratio (4.104). The expression for $G_{i,N}(z)$ can be found elsewhere.[47] In most practical cases, however, when the chains are not very long, equation (4.104) suffices for the analysis of the localization properties of a wave function.

As an example we consider a model Hamiltonian matrix in the case of orthogonal basis functions ($\mathbf{S} = \mathbf{I}$). The Hamiltonian for these systems is assumed to possess zero matrix elements H_{ij} when $|i - j| > 4$. It is trivial to modify the computer programs to consider a greater or lesser number of diagonals. The matrix half-bandwidth of 5 was chosen arbitrarily to demonstrate the technique for something other than the usual trivial case of the tridiagonal Hamiltonian of a one-dimensional system.

The Hamiltonian of the ordered system has matrix elements given by[45]

$$H_{ij} = \begin{cases} 0 & i = j \\ -2/(|i-j|+1) & i \neq j \end{cases} \tag{4.109}$$

The matrix elements of the disordered system are[45]

$$H_{ij} = \begin{cases} (0.5 - R_j)\Gamma & i = j \\ -2/(|i-j|+1) & i \neq j \end{cases} \tag{4.110}$$

where R_j is a random number distributed uniformly on the interval (0–1) and Γ is a constant that can be changed to yield various degrees of disorder. Figure 4.31 compares the ratio of the N, N and $1, N$ Green function elements in the ordered ($\Gamma = 0$) and disordered (in the calculations $\Gamma = 2$ was chosen) cases. We note that the ratio in the ordered system is, on the average, orders of magnitude larger than in the disordered system. One expects the ratio to be the order of unity in the ordered system when we are within the "allowed" band of states. In the disordered system, there is considerable variation in the ratio corresponding to different degrees of localization of the eigenstates. Indeed, if one calculates the eigenstates with the help of inverse iteration one finds a good correlation between the ratios in the figure and the localization of the wave functions.[45] Though not all eigenstates in any particular system were computed owing to the excessive amount of computer time required to sample a variety of systems, no contradictions were seen between the explicit eigenstates and the localization measure involving the Green matrix elements.

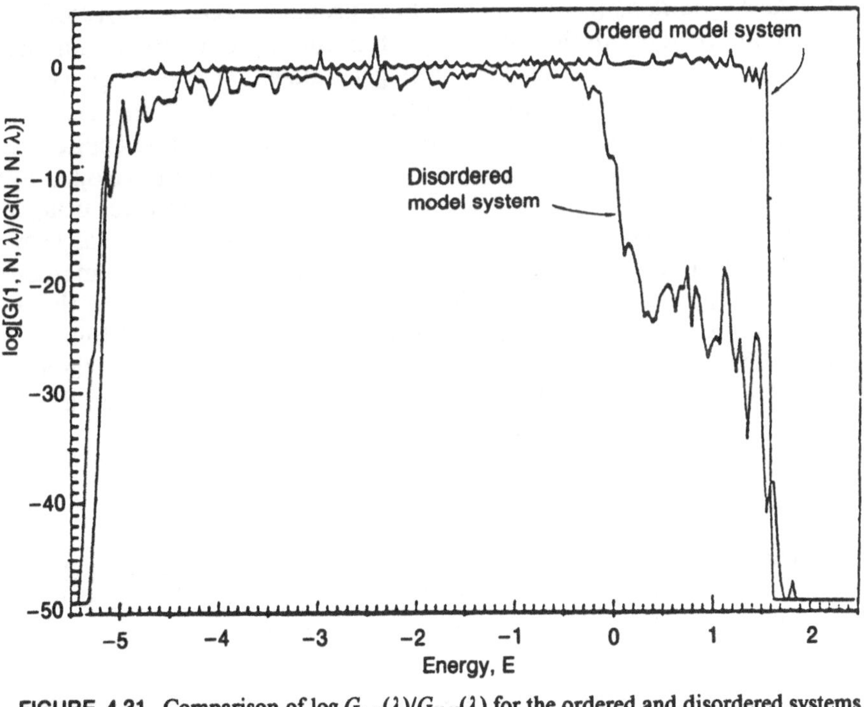

FIGURE 4.31. Comparison of log $G_{1,N}(\lambda)/G_{N,N}(\lambda)$ for the ordered and disordered systems defined by equations (4.109) and (4.110). Strong localization at the band edges of the disordered system and complete delocalized behavior in the ordered system is predicted.

It would be interesting to use this technique to study the localization in the case of the combination of diagonal and off-diagonal disorder, since this is the situation most akin to the actual situation in a large complicated polymer chain. Though the expressions for special elements of the Green matrix were written down, it is straightforward to generalize equations (4.107) and (4.108) for any arbitrary elements of the Green matrix.[47] These generalizations should be very useful in hopping transport calculations for finite disordered quasi-one-dimensional systems, where matrix elements of the Green function need to be calculated rapidly and efficiently.

4.6. TREATMENT OF A CLUSTER OF IMPURITIES IMBEDDED IN A PERIODIC CHAIN

In many cases an impurity (or a cluster of them) may be imbedded in a periodic chain. Special cases of this are a chemisorbed or covalently bound extra molecule attached to the chain [carcinogen binding to DNA is a frequent example of the latter, or a periodic chain *without periodic boundary conditions* (chain ends)]. These problems were first treated by Koster and Slater[50] in the simple tight-binding approximation with one orbital per unit cell. This has been generalized in the language of scattering theory by Callaway.[51] An LCAO matrix form (with arbitrary number of orbitals per unit cell) of the Koster–Slater method applicable to local perturbation phenomena and end states has been developed by Kertész and Biczó.[52] The SCF treatment of a periodic polymer using a nonorthogonal basis ($\mathbf{S} \neq \mathbf{I}$) containing a cluster of impurities has been formulated by Ladik and Seel.[53] The resolvent (Green matrix) method has been repeatedly applied to the problems of surface states of three-dimensional crystals and to chemisorption studies by Grimley,[54] Koutecký,[55] and Ladik[56] (in the latter case in an SCF form). The first numerical SCF applications of this method were reported by Baraff and Schlüter[57] and by Bernholc *et al.*[58]

4.6.1. Green Matrix Formulation of the Problem*

The best way to take into account also propagation during the iterations leading to self-consistency of a perturbation localized originally to one or a few cells[53] is a general Green matrix formulation of the problem.[59–61]

The elements of the Green matrix of the unperturbed periodic system $\mathbf{G}^0 = (z\mathbf{S}^0 - \mathbf{F}^0)^{-1}$ can be written as[59,60]

$$\mathbf{G}^0_{ai,bj}(z) = \frac{1}{2\pi} \sum_l \int_{-\pi}^{+\pi} \frac{c_{al}(k) c_{bl}(k)}{z - \varepsilon_l(k)} \exp[ik(i-j)]dk \qquad (4.111)$$

Here $\varepsilon_l(k)$ and $\mathbf{c}_l(k)$ are the eigenvalues (bands) and eigenvectors (LCAO expansion coefficients of the Bloch function) of the one-electron equation for the infinite periodic system (see Chapter 1)

$$\mathbf{F}^0(k)\mathbf{c}_l(k) = \varepsilon_l(k)\mathbf{S}(k)\mathbf{c}_l(k) \qquad (4.112)$$

i, j being site indices in the range $-N$ to $+N$, $N \to \infty$, while a and b are orbital indices. If $z = E$ and E falls within an energy band, $\mathbf{G}^0$ becomes singular so we must define a matrix $\mathbf{G}^0(E^+)$; E^+ denotes the limiting procedure $E^+ = E + i\varepsilon$, where ε is positive and infinitesimal and is allowed to vanish after the integration is performed.

Dyson's equation defines the Green matrix for the system with a perturbation in the form

$$\mathbf{G} = \mathbf{G}^0 + \mathbf{G}^0\mathbf{V}\mathbf{G} \qquad (4.113)$$

The impurity potential $\mathbf{V}$ describes the deviation from the perfect infinite system and, in the case of nonunit overlap, is given by

$$\mathbf{V} = (\mathbf{F}^p - \mathbf{F}^0) - z(\mathbf{S}^p - \mathbf{S}^0) = \mathbf{W} - z\tilde{\mathbf{S}} \qquad (4.114)$$

In the case of finite-range deviation potentials only a subblock of the infinite matrix V is nonzero. The linear combination of atomic orbital (LCAO) expressions for the Hartree–Fock case are given below.

Equation (4.113) can be solved formally to give

$$\mathbf{G} = (\mathbf{1} - \mathbf{G}^0\mathbf{V})^{-1}\mathbf{G}^0 \qquad (4.115)$$

For energies in the band gaps, $\mathbf{G}^0$ is real. The condition for the existence of bound states (additional poles in $\mathbf{G}$) is

$$D(E) = \det[\mathbf{1} - \mathbf{G}^0\mathbf{V}] = 0 \qquad (4.116)$$

because calculating the inverse matrix in (4.115) involves the determinant of this matrix in the denominator. Therefore the roots of this determinant provide additional poles that define the energies of such states. The size of this determinant reduces to the nonzero part of the deviation potential matrix

$$D(E) = \det[\mathbf{1} - \mathbf{G}^0_{00}\mathbf{V}_{00}] \tag{4.117}$$

if 0 denotes the cell containing the impurity (or a superblock when several cells are involved in the change of the potential).

The total density of states, in the presence of a nonunit overlap, is given by[59]

$$\rho(E) = -\pi^{-1}\,\mathrm{Tr}[\mathrm{Im}\,\tfrac{1}{2}(\mathbf{GS} + \mathbf{SG})] \equiv \mathrm{Tr}\,\mathbf{Q} \tag{4.118}$$

The diagonal elements of matrix $\mathbf{Q}$ are the local densities of states — those elements which, when integrated over E, give Mulliken's gross populations for the various basis orbitals.[59] The local density of states of unit i is given by

$$\rho_i(E) = \mathrm{Tr}\,\mathbf{Q}_{ii} \tag{4.119}$$

For the total change in the density of states,

$$\Delta\rho(E) = \rho(E) - \rho^0(E) = -\pi^{-1}\,\mathrm{Tr}\,\mathrm{Im}[\mathbf{S}^p\mathbf{G} - \mathbf{S}^0\mathbf{G}^0] \tag{4.120}$$

A very useful relation, first derived by Callaway,[51] can be applied, namely

$$\Delta\rho(E) = \frac{1}{\pi}\frac{d\delta(E)}{dE} = \frac{1}{\pi}\frac{d}{dE}\left[-\arctan\frac{\mathrm{Im}\,D(E)}{\mathrm{Re}\,D(E)}\right] \tag{4.121}$$

where $\delta(E)$ is the phase shift and $D(E)$ the determinant defined in equation (4.117). If $\mathrm{Re}\,D(E) = 0$, the phase shift passes through an odd multiple of $\pi/2$ and one has a resonance or antiresonance in $\Delta\rho(E)$.

The charge-bond-order matrix P, of central importance for self-consistent calculations, is given by[59]

$$\mathbf{P} = -\frac{1}{\pi}\int_{-\infty}^{E_F} \mathrm{Im}\,\mathbf{G}(E)dE \tag{4.122}$$

Outside the bands, Im $\mathbf{G}$ exhibits a δ-function singularity at the energies E_h of the bound states. The weight factor of this singularity gives the contribution to the charge-bond-order matrix and to Mulliken's gross population for the various basis orbitals from the localized states at E_h.

This contribution can be calculated as follows[60]:

$$\operatorname{Im} G_{aj,bj}(E_b) = -\pi\delta(E - E_b)\left[\frac{d}{dE}D(E)\Big|_{E=E_b}\right]^{-1}[\mathbf{G}^0_{j0}\mathbf{V}_{00}\mathcal{M}_{00}\mathbf{G}^0_{0j}]_{ab} \quad (4.123)$$

where $\mathcal{M}_{00}$ is the cofactor matrix of the inverse $\mathbf{M}_{00} = (\mathbf{1} - \mathbf{G}^0_{00}\mathbf{V}_{00})^{-1} = \mathcal{M}_{00}/D(E)$.

An alternative procedure that can be used to calculate the contribution from a bound state to the matrix $\mathbf{P}$ can be derived in the following way,[61] by generalizing for nonunit overlap a method proposed in the second paper of Bernholc *et al.*[58]

The Fock matrix equation for the imperfect infinite system is

$$\mathbf{F}^p\mathbf{c} = (\mathbf{F}^0 + \mathbf{W})\mathbf{c} = E(\mathbf{S}^0 + \tilde{\mathbf{S}})\mathbf{c} = E\mathbf{S}^p\mathbf{c} \quad (4.124)$$

The definitions for $\mathbf{G}^0 = (z\mathbf{S}^0 - \mathbf{F}^0)^{-1}$ and $\mathbf{V} = \mathbf{W} - z\tilde{\mathbf{S}}$ enable one to immediately obtain $\mathbf{c}$ for energies in the band gaps (where $\mathbf{G}^0$ has no poles) as nontrivial solutions to the infinite homogeneous matrix equation (see also the paper by Ladik and Seel[53]),

$$(\mathbf{1} - \mathbf{G}^0\mathbf{V})\mathbf{c} = \mathbf{0} \quad (4.125)$$

with normalization condition

$$\mathbf{c}^{tr}\mathbf{S}^p\mathbf{c} = 1 \quad (4.126)$$

If $\mathbf{0}$ denotes the submatrix, which differs from the corresponding block of the unperturbed system, and $\mathbf{B}$ the submatrix of the remaining (infinite) subspace, the relevant matrices can be written as follows:

$$\mathbf{V} = \begin{pmatrix} \mathbf{V}_{00} & \mathbf{0} \\ \mathbf{0} & \mathbf{0} \end{pmatrix}, \qquad \mathbf{G}^0 = \begin{pmatrix} \mathbf{G}^0_{00} & \mathbf{G}^0_{0B} \\ \mathbf{G}^0_{B0} & \mathbf{G}^0_{BB} \end{pmatrix}$$
$$\mathbf{S}^p = \begin{pmatrix} \mathbf{S}^0_{00} & \mathbf{S}^0_{0B} \\ \mathbf{S}^0_{B0} & \mathbf{S}^0_{BB} \end{pmatrix} + \begin{pmatrix} \tilde{\mathbf{S}}_{00} & \mathbf{0} \\ \mathbf{0} & \mathbf{0} \end{pmatrix} \quad (4.127)$$

If the eigenvector $\mathbf{c}$ is expressed as $(\mathbf{c}^0, \mathbf{c}^B)$, where $\mathbf{c}^0$ and $\mathbf{c}^B$ are the components of $\mathbf{c}$ in subspace 0 and subspace B, respectively, then the decomposition of equation (4.125) yields $\mathbf{c}^0$ as the nontrivial solution to the (finite) matrix equation

$$(\mathbf{1} - \mathbf{G}^0_{00}\mathbf{V}_{00})\mathbf{c}^0 = \mathbf{0} \quad (4.128a)$$

$\mathbf{c}_B$ can then be calculated from the equation

$$\mathbf{c}_B = \mathbf{G}^0_{B0}\mathbf{V}_{00}\mathbf{c}^0 \tag{4.128b}$$

The correct normalization of $\mathbf{c}^0$ can be obtained without calculating $\mathbf{c}_B$ but by using equation (4.125) and the fact that $\mathbf{V}$ and $\mathbf{G}^0$ (for E in the band gaps) are real symmetric matrices:

$$1 = \mathbf{c}^{tr}\mathbf{S}^p\mathbf{c} = \mathbf{c}^{tr}\mathbf{V}\mathbf{G}^0\mathbf{S}^p\mathbf{G}^0\mathbf{V}\mathbf{c} \tag{4.129}$$

If the relationship

$$\frac{d}{dE}\mathbf{G}^0 = \mathbf{G}^{0'} = -(E\mathbf{S}^0 - \mathbf{F}^0)^{-1}\mathbf{S}^0(E\mathbf{S}^0 - \mathbf{F}^0)^{-1} = -\mathbf{G}^0\mathbf{S}^0\mathbf{G}^0$$

is used, one obtains, with the aid of decomposition (4.127),

$$\mathbf{G}^0\mathbf{S}^p\mathbf{G}^0 = -\mathbf{G}^0 + \begin{pmatrix} \mathbf{G}^0_{00}\tilde{\mathbf{S}}_{00}\mathbf{G}^0_{00} & \mathbf{G}^0_{00}\tilde{\mathbf{S}}_{00}\mathbf{G}^0_{0B} \\ \mathbf{G}^0_{B0}\tilde{\mathbf{S}}_{00}\mathbf{G}^0_{00} & \mathbf{G}^0_{B0}\tilde{\mathbf{S}}_{00}\mathbf{G}^0_{0B} \end{pmatrix} \tag{4.130}$$

Substitution in equation (4.129) finally yields the normalization condition for $\mathbf{c}^0$ in the form

$$1 = -\mathbf{c}^{0\,tr}\mathbf{V}_{00}\mathbf{G}^{0'}_{00}\mathbf{V}_{00}\mathbf{c}^0 + \mathbf{c}^{0\,tr}\mathbf{V}_{00}\mathbf{G}^0_{00}\tilde{\mathbf{S}}_{00}\mathbf{G}^0_{00}\mathbf{V}_{00}\mathbf{c}^0 \tag{4.131}$$

which is the generalization of the expression for unit overlap[58]

$$1 = -\mathbf{c}^{0\,tr}\mathbf{V}_{00}\mathbf{G}^{0'}_{00}\mathbf{V}_{00}\mathbf{c}^0$$

The contribution of an occupied bound state to the (i, j) element of the corresponding $\mathbf{P}$-matrix block $\mathbf{P}_{00}$ is therefore $c^0_{il}c^0_{jl}$, where $\mathbf{c}^0$ is calculated from equation (4.128a) with normalization (4.131).

The perturbation matrix $\mathbf{V}$ and, finally, the self-consistent procedure in the Hartree–Fock case are defined by using the expression of the Fock matrix elements of a periodic chain (see Section 1.1).

The procedure for determining the self-consistent solution of the Dyson equation can now be summarized as follows:

1. Calculate $\mathbf{G}^0(E)$, the Green matrix of the unperturbed system, as defined in equation (4.111) for the elements of $\mathbf{F}^0$, using as input the results of a Hartree–Fock band-structure calculation of the

infinite periodic system. An accurate numerical procedure for the computation of the real and imaginary parts of $\mathbf{G}^0(E)$ is described elsewhere.[60]

2. Calculate a first guess for $\mathbf{V}$, which is defined in equation (4.114), together with the LCAO expressions for the elements of $\mathbf{F}^0$. A reasonable guess for the **P**-matrix blocks of the perturbed system can be obtained from a cluster calculation: all the one- and two-electron integrals can be kept on file; they do not change during the iterations.
3. Solve the Dyson equation (4.115) for a number of E points in the band gaps [in order to find possible bound states with the use of equation (4.116)] and for a dense mesh of E points within the bands.
4. Calculate a new charge-bond-order matrix **P** from equation (4.122). The bound-state contributions can be computed either using equation (4.123), or from equation (4.128a) together with equation (4.131) (the second procedure turned out to be numerically more advantageous). These two contributions to the matrix **P** — from the band region and from the bound states — can add up such that the difference $\Delta\mathbf{P} = \mathbf{P}^p - \mathbf{P}^0$ becomes zero from some near neighbor on, even if the bound state is quite extended.
5. The steps are repeated beginning from (2) until the difference in the **P** matrices from two consecutive iterations is below a certain threshold.

A computer program was developed for this procedure.

4.6.2. Application to a Hydrogen Impurity in a Lithium Chain[61]

As a first numerical test of the method described, a chain of lithium atoms with alternating bond distances (the model for an insulating system) and a single hydrogen atom representing the local perturbation was studied. The Li–Li distance in the Li_2 molecule (2.67 Å) was chosen as the shorter distance and 3.39 Å [twice the nearest-neighbor distance in the *bbc* Li metal (3.03 Å) minus 2.67 Å] as the larger one. In the perturbed system, one Li atom is replaced by an H atom in a Li_2 molecule at a distance of 2.04 Å from the nearest Li atom, the distance in an LiH crystal (the bond distance in the LiH molecule is 1.59 Å). The calculations were performed in the first-neighbor-interaction aproximation.

The basis set consisted of two s functions for Li and two s functions for the H atom contracted from three Gaussians, as tabulated in Pople's program.[62] The unit cell of the unperturbed system and the cell with the

H impurity therefore contained four contracted Gaussian-type functions. At this point it should be emphasized that the purpose of this study is not to obtain quantitative numbers for the Li–H interaction where, of course, correlation becomes important and a much better basis set had to be used, but to investigate the effect of an impurity in an infinite system going beyond the simple, non-self-consistent tight-binding approximation with one orbital per unit cell. However, it should also be pointed out that the difference in binding energies between LiH and Li_2 in the aforementioned basis and geometry has been found to be approximately 1.1 eV [1.12 eV–(– 0.005) eV] while the Hartree–Fock limit is approximately 1.3 eV (1.5 eV–0.2 eV) and the correlated value is also about 1.3 eV (2.3 eV–1.0 eV).[63]

4.6.2.1. Energy Spectra and Charge Distributions

It became evident from self-consistent cluster calculations that the core contributions to the matrix **P** remain constant during the iterations. Therefore, by the self-consistent Green matrix calculation, the core contribution was kept constant (frozen-core approximation).

The density of states (DOS) of the valence band of the pure Li chain has the U-shape typical of one-dimensional systems with van Hove (square-root) singularities at the band edges. If one introduces a perturbation matrix **V** consisting of three blocks (impurity site plus two nearest neighbors), then due to the impurity there are three bound states outside the valence band, the one at – 0.2700 au being the eigenvalue corresponding to the Li–H bond. After six iterations the maximum difference in the **P**-matrix elements of two consecutive iterations was 4×10^{-2}. As result of the iterations only one bound state remains outside the valence-band region, at – 0.2695 au. The inclusion of the next-nearest neighbors in **V** (allowing the perturbation to extend further) gave, after seven iterations, the same **P**-matrix difference of 4×10^{-2} and did not change the location of the only bound state at – 0.2695 au, though the total change in DOS inside the valence band differs from the one obtained before and still did not converge. The iterations were stopped at a difference of 4×10^{-2} in the elements of the matrix **P** because of numerical inaccuracies in the integration necessary to obtain **P** from the imaginary part of **G** [see equation (4.122)]. This important point will be discussed in detail in the next section.

The changes in charge distribution in the infinite chain introduced by the impurity and calculated by using Mulliken's population analysis are compared in Table 4.5. In the pure Li chain each Li atom has a charge of 3.0000. If one Li atom is replaced by a H atom, a strong ionic charge

TABLE 4.5. Charge Distribution (Mulliken Population) in the Infinite Pure Li Chain and in the Infinite Perturbed Chain for Different Iterations[61]

		Perturbed system		
Unit cell	Pure	Iteration: 0th	6th	7th
−2				3.02 3.11
−1	3.0000 3.0000	2.6830 2.9632	2.86 3.21	2.81 3.13
0	3.0000(Li) 3.0000(Li)	2.3671(Li) 1.6493(H)	2.47 1.57	2.42 1.57
−1	3.0000 3.0000	3.0587 3.2511	2.84 3.02	2.78 2.95
−2				3.07 3.10

distribution is induced: 1.6493 on H and 2.3671 on Li after the zeroth iteration. The next neighbors try to screen this charge in an alternative fashion. The results of the iterations to self-consistency show the following qualitative features: the next neighbors screen the perturbation enough so neutrality has almost been reached on the next-nearest neighbors (3.1 ± 0.1, last column in Table 4.4). It was noted earlier that the integration procedure in the computation of matrix **P** must be improved in order to obtain more accurate charge distributions.

4.6.2.2. Accurate Determination of the Charge–Bond–Order Matrix and Error Propagation[61]

The bound-state contribution to matrix **P** was calculated by the two alternative methods presented in Section 4.6.1. Without too much numerical effort five significant digits (relative error 0.1%) are easily obtained. It turns out that the second method is more advantageous, because it enables the derivative of the cofactor matrix to be calculated at a point where the determinant passes through zero (via calculation of the inverse near this singular point and multiplication by the determinant), while in the first method the derivative of $\mathbf{G}^0$, which is usually a smooth function near this point, must be calculated.

The critical point in the calculation of **P** is integration over the imaginary part of **G** inside the band region. Until now, 150 points in the valence band were used and these correspond to an average spacing of 0.0005 au (0.013 eV). Of course, the points were not chosen equidistant but the mesh was denser near the band edges. The regions around possible resonances and antiresonances were treated separately and integrated analytically in the following way: the total change in the density of states $\Delta\rho(E)$ around such a point E_0 can be written [using a linear approximation for Re $D(E)$ in equation (4.121); see also the second paper by Bernholc *et al.*[58]] in the form

$$\Delta\rho(E) = \frac{\Gamma}{2\pi} \frac{1}{(E - E_0)^2 + (1/4)\Gamma^2}, \quad \Gamma = \frac{2\,\mathrm{Im}\,D(E_0)}{\mathrm{Re}\,D'(E_0)} \tag{4.132}$$

which yields

$$\int_{E_l}^{E_u} \Delta\rho(E)dE = 1/\pi \left[\mathrm{arctg}\left(\frac{2(E_u - E_0)}{\Gamma}\right) - \mathrm{arctg}\left(\frac{2(E_l - E_0)}{\Gamma}\right)\right] \tag{4.133}$$

A four-point finite-difference subroutine[64] was used to estimate the error remaining after the numerical integration. This subroutine gives the numerical value of an integral I_{num} and an error estimate $\tilde{E}$ by using a quartic interpolation. Hence

$$I = \int_a^b = f(x)dx = I_{\mathrm{num}} + \tilde{E} \tag{4.134}$$

A comparison of the values for matrix **P** obtained with I and I_{num} indicated that the absolute error in a **P**-matrix element after the zeroth iteration was 2×10^{-4}. After monitoring five iterations using both I and I_{num}, it turned out that this initial difference of 2×10^{-4} is sufficient to lead to a final difference of 3×10^{-2} in the **P**-matrix elements, and therefore to a maximal difference in the population of 1×10^{-1}. This error propagation is due to the accumulation of relative errors of **P** (and therefore of **V**) in the calculation of the inverse $(\mathbf{1} - \mathbf{G}^0\mathbf{V})^{-1}$ at each subsequent iteration step. The analysis shows that one needs at least six significant digits in the **P**-matrix elements which should represent the average subroutine and machine accuracy of the computation of the inverse of a complex matrix. Work to improve this critical step in the self-consistent Green matrix approach — the accurate numerical integration over a rather complicated function — is in progress.

4.6.2.3. Calculation of the Change in Total Energy Due to the Impurity[61]

The total electronic energy in the Hartree–Fock formalism is

$$E_{\mathrm{HF}} = \sum_i^{\mathrm{occ}} \varepsilon_i + \mathrm{Tr}\,\mathbf{Ph} \tag{4.135}$$

where ε_i are the one-electron energies and $\mathbf{h}$ is the one-electron part of the Fock matrix. The sum over the one-electron energies can be rewritten as

$$\sum^{\mathrm{occ}} \varepsilon_i = \begin{cases} \int^{E_{\mathrm{F}}} E \sum \delta(E-\varepsilon_i)\,dE & \text{discrete spectrum} \\ \int^{E_{\mathrm{F}}} E\rho(E)\,dE & \text{continuous spectrum} \end{cases} \tag{4.136}$$

where the lower integration limit is the lowest state in the spectrum and $\rho(E)$ is the total density of states per spin. The following derivation is analogous to the discussion of the metallic case with a local exchange approximation.[65]

We suppose that the total density of states of the perfect crystal is denoted by $\rho_0(E) = N\bar{\rho}(E)$, $\bar{\rho}(E)$ being the DOS per cell and N the number of unit cells. The total density of states of the perturbed system is then

$$\rho(E) = N\bar{\rho}(E) + \Delta\rho(E) + \sum_i \delta(E - \varepsilon_i^{\mathrm{bound}}) \tag{4.137}$$

where $\Delta\rho(E)$ is defined in equation (4.121). The Fermi energy E_{F} is determined by

$$2\int^{E_{\mathrm{F}}} \rho_0(E)\,dE = 2N\int^{E_{\mathrm{F}}} \bar{\rho}(E)\,dE = NZ \tag{4.138}$$

where Z is the number of electrons per unit cell. If the number of electrons in the perturbed cell is Z', then, for the perturbed system,

$$\begin{aligned} 2\int^{E'_{\mathrm{F}}} \rho(E)\,dE &= 2\int^{E'_{\mathrm{F}}} \left[N\bar{\rho}(E) + \Delta\rho(E) + \sum_i \rho(E - \varepsilon_i^{\mathrm{bound}})\right] dE \\ &= NZ - Z + Z' \end{aligned} \tag{4.139}$$

The Fermi energy of the perturbed system has been denoted by E'_{F}. If, for the insulating case under consideration, E_F is situated at the top of the valence band and we allow for possible bound states in the perturbed system to be situated above the valence band, then E'_F denotes the energy of the highest occupied bound state. Since $\bar{\rho}(E)$ and $\Delta\rho(E)$ are zero in the gap, equation (4.139) yields

$$2\int^{E_F}\Delta\rho(E)dE + 2\int^{E'_F}\sum_i \delta(E - \varepsilon_i^{\text{bound}})dE = Z' - Z \quad (4.140)$$

This equation defines whether and how many states above the valence band of the unperturbed system have to be occupied. In our example of H substitution in the Li chain, $Z' - Z$ is -2. The integration of $\Delta\rho(E)$ over the valence band yields -3 (after the zeroth iteration) and one Li core state is missing. That means the first term in equation (4.140) gives -8 and therefore all three bound states, including the one on top of the valence band, have to be occupied.

After the occupation of the bound states is defined, the change in total electronic energy is easily calculated:

$$\Delta E_{\text{HF}} = E_{\text{HF}}^{\text{pert}} - E_{\text{HF}}^{0} = \int^{E_F} E\Delta\rho(E)dE + \sum_i^{\text{occ}} \varepsilon_i^{\text{bound}} + \text{Tr}(\mathbf{P}_{00}\mathbf{h}_{00}^{p} - \mathbf{P}_{00}^{0}\mathbf{h}_{00}^{0}) \quad (4.141)$$

where the block over which the perturbation extends is again denoted by 00. The total energy is obtained by adding the difference in nuclear repulsion energy ΔE_{nucl}. In the Li/H case, equation (4.141) for ΔE_{HF} (after the zeroth iteration when matrix $\mathbf{P}$ is still reasonably accurate) yielded 10.2090 au.

It should be observed that the change in HF energy ΔE_{HF} discussed so far cannot be compared directly with experiment. The physically interesting situation is when one has a Li chain plus one free H atom, which becomes built into the chain and leaves a Li chain containing a H impurity and a free Li atom. The corresponding change in total energy is given by

$$\Delta E_{\text{HF}}^{\text{tot}} = \Delta E_{\text{HF}} + \Delta E_{\text{nucl}} + E^{\text{Li}} - E^{\text{H}} \quad (4.142)$$

with the same basis sets as above, an unrestricted HF calculation yields $E^{\text{Li}} = -7.3123$ au and $E^{\text{H}} = -0.4958$ au; ΔE_{nucl} is -3.4078 au. The value of $\Delta E_{\text{HF}}^{\text{tot}}$ is then -0.0153 au (or -0.42 eV).

This section demonstrates how changes in the total energies of infinite systems due to perturbation can be calculated at the Hartree–Fock level. Moreover, by using equation (4.141) at every iteration step, one can monitor possible divergences in the self-consistent solution of the Dyson equation.

4.6.2.4. Concluding Remarks

The study of the substitution of a hydrogen impurity into an infinite lithium chain was the first application of the newly developed computer program for the self-consistent solution of the Dyson equation at the Hartree–Fock level, almost 30 years after Koster and Slater formulated this approach. The changes in the energy spectrum and in the charge distribution caused by the perturbation as well as the change in total energy were calculated. Although approximate self-consistency is achieved, the accurate solution still requires improvement in the calculation of the charge-bond-order matrix, i.e., a more accurate evaluation of the integral over the imaginary part of **G**.

The results indicate that the perturbation is screened out within two neighbors from the site containing the impurity, because of a detailed balance of the contribution to matrix **P** from bound states and the band continuum. It corroborates the conclusion of an earlier investigation[60] that, in a unit cell with more than one oribtal, on-site screening through redistribution of the population of the local orbitals becomes quite effective, thus damping oscillations in the charge distribution and leading to a faster decay than in strict one-dimensional models with one orbital per unit cell.[66] There are two physical mechanisms present in the screening process: the number of neighbors (which favors screening in two and three dimensions) and the complexity of the unit cell (which promotes screening in quasi-one-dimensional organic polymers and molecular crystals, for example).

REFERENCES

1. S. Raines, *Many-Electron Theory*, North-Holland Publ. Co., Amsterdam–London (1972).
2. See, for instance, A. L. Fetter and J. D. Walecka, *Quantum Theory of Many-Particle Systems*, McGraw-Hill Book Co., New York (1971); C. Mavroyannis, in: *Physical Chemistry, An Advanced Treatise* (D. Henderson, ed.), Vol. XIA (*Mathematical Methods*) p. 488, Academic Press, New York–San Francisco–London (1975).
3. S. Suhai, J. Kasper, and J. Ladik, *Int. J. Quantum Chem.* **17**, 995 (1980).
4. R. C. Dingham, M. J. S. Dewar, and D. H. Lo, *J. Am. Chem. Soc.* **97**, 1285 (1975).
5. L. Pauling and R. B. Corey, *Proc. Natl. Acad. Sci. USA* **39**, 253 (1973).

6. G. Del Re, J. Ladik, and G. Biczó, *Phys. Rev.* **155**, 967 (1967).
7. S. Suhai (unpublished result).
8. P. W. Anderson, *Phys. Rev.* **109**, 1492 (1958); P. W. Anderson, *Rev. Mod. Phys.* **50**, 195 (1978).
9. N. Mott, *Rev. Mod. Phys.* **50**, 191 (1979).
10. J. T. Edwards and D. J. Thouless, *J. Phys. C* **5**, 807 (1972); D. C. Licciardello and E. N. Economou, *Phys. Rev. B* **11**, 3697 (1975); D. C. Licciardello and D. J. Thouless, *Phys. Rev. Lett.* **35**, 1475 (1975); *J. Phys. C* **8**, 4175 (1975); K. Schonhammer and W. Brenig, *Phys. Lett.* **42A**, 447 (1973); D. Weaire and A. R. Williams, *J. Phys. C* **9**, L461 (1976); S. Yoshino and M. Okazaki, *J. Phys. Soc. Jpn.* **43**, 415 (1977).
11. R. day, J. Ladik, and F. Martino, *Chem. Phys. Lett.* **81**, 494 (1981).
12. K.-F. Barggren and F. Martino, *Phys. Rev.* **84**, 484 (1969).
13. W. J. Hehre, R. F. Stuart, and J. A. Pople, *J. Chem. Phys.* **51**, 2567 (1969); W. J. Hehre, R. Ditchfield, R. F. Stuart, and J. A. Pople, *J. Chem. Phys.* **52**, 2769 (1970).
14. J. Čížek and J. Paldus, *J. Chem. Phys.* **47**, 3976 (1967); *Phys. Rev. A* **2**, 2268 (1970); M. Bernard and J. Paldus, *J. Chem. Phys.* **72**, 6546 (1980).
15. For an *ab initio* formulation and application of this method see M. Seel and J. Ladik, *Chem. Phys.* **45**, 349 (1980).
16. See, for instance; F. Martino, in: *Qunatum Theory of Polymers*, (J.-M. André, J. Delhalle, and J. Ladik, eds.), p. 169, D. Reidel Publ. Co., Dordrecht–Boston (1978).
17. A. J. Elliott, J. A. Krumhansl, and P. Leath, *Rev. Mod. Phys.* **46**, 465 (1974).
18. J. Delhalle, *Bull. Soc. Chim. Belg.* **84**, 135 (1975).
19. S. Kirkpatrick, B. Velicky, and H. Ehrenreich, *Phys. Rev. B* **11**, 3250 (1970).
20. M. Seel, T. C. Collins, F. Martino, D. K. Rai, and J. Ladik, *Phys. Rev. B* **18**, 6460 (1978).
21. B. Street (personal communication).
22. See, for instance, J. Callaway, *Quantum Theory of the Solid State*, p. 517, Academic Press, New York–London (1974).
23. S. Suhai (unpublished results).
24. For a review see P. Dean, *Rev. Mod. Phys.* **44**, 127 (1972).
25. W. H. Butler, *Phys. Rev. B* **8**, 4499 (1973).
26. See, for instance, G. Rickaysen, *Theory of Superconductivity*, Wiley–Interscience, New York–London (1965).
27. P. Mengel, P. M. Grant, W. E. Rudge, B. H. Schechtmann, and D. W. Rice, *Phys. Rev. Lett.* **35**, 1803 (1977).
28. R. A. Hein, J. W. Gibson, and R. D. Blaugher, *Rev. Mod. Phys.* **36**, 149 (1964).
29. P. Dean, *Proc. R. Soc. London, Ser. A* **254**, 507 (1960); P. Dean, *Proc. R. Soc. London, Ser. A* **260**, 263 (1961); P. Dean, *Rev. Mod. Phys.* **44**, 127 (1972).
30. D. Pumpernik, B. Borštnik, M. Kertész, and A. Ažman, *Z. Naturforsch., A* **32**, 295 (1977).
31. M. Seel, *Chem. Phys.* **43**, 103 (1979).
32. J. Ladik and J. Čížek (unpublished).
33. J. Čížek, J. Ladik, and W. Förner (unpublished).
34. See Elliott *et al.*[17] and F. Cyrot-Lackmann, M. C. Desjonqueres, and J. P. Gaspard, *J. Phys. C* **7**, 925 (1974); L. M. Falicov and F. Yndurain, *Phys. Rev. B* **12**, 5664 (1975).
35. R. S. Day and F. Martino, *Chem. Phys. Lett.* **84**, 86 (1981).
36. M. Kertész and Gy. Göndör, *J. Phys. C* **14**, 851 (1981).

37. J. H. Wilkinson, *The Algebraic Eigenvalue Problem*, p. 633, Clarendon Press, Oxford (1965).
38. B. Gazdy, M. Seel, and J. Ladik, *Chem. Phys.* **86**, 41 (1984).
39. M. Seel and J. Ladik (unpublished).
40. J. Ladik and S. Suhai, *Int. J. Quantum Chem., Quantum Biol. Symp.* **7**, 181 (1980).
41. R. S. Day and J. Ladik, *Int. J. Quantum Chem.* **21**, 917 (1982); S. Suhai, R. S. Day, M. Seel, and J. Ladik (unpublished).
42. M. O. Dayhoff, R. M. Schwartz, H. R. Chen, W. C. Barker, L. T. Hunt, and B. C. Orcutt, *DNA*, Vol. 1, p. 51, Mary Ann Liebert Inc. (1982).
43. J. F. Gentleman, M. S. Shadboldt-Forbes, J. W. Hawkins, J. Ladik, and W. Forbes, *Math. Scientist* **9**, 125 (1984).
44. J. Hari, *Spectral Properties of Disordered Chains and Lattices*, p. 34, Pergamon Press, Oxford (1968).
45. R. S. Day, S. Suhai, and J. Ladik, *Chem. Phys.* **62**, 165 (1981).
46. R. S. Day and F. Martino, *J. Phys. C* **14**, 4247 (1980).
47. F. Martino, in: *Quantum Chemistry of Polymers; Solid State Aspects* (J. Ladik and J.-M. André, eds.), p. 279, D. Reidel Publ. Co., Dordrecht–Boston–Lancester (1984); B. Gazdy, R. S. Day, M. Seel, F. Martino, and J. Ladik, *Chem. Phys. Lett.* **88**, 220 (1982).
48. K. Ishii, *Prog. Theor. Phys., Suppl.* **53**, 77 (1973).
49. W. H. Butler, *Phys. Rev. B* **8**, 4489 (1973).
50. G. F. Koster and J. C. Slater, *Phys. Rev.* **96**, 1208 (1954).
51. J. Callaway, *J. Math. Phys.* **5**, 783 (1964); *Phys. Rev.* **154**, 515 (1967); J. Callaway and A. J. Hughes, *Phys. Rev.* **156**, 860 (1967); J. Callaway, *Phys. Rev. B* **3**, 2556 (1971).
52. M. Kertész and G. Biczó, in: *Proceedings on Computers in Chemical Research and Education* (D. Hadži, ed.), p. 4159, Ljubljana (1973).
53. J. Ladik and M. Seel, *Phys. Rev. B* **13**, 5338 (1976).
54. B. Grimley, *CRC Solid State Sci.* **6**, 238 (1976).
55. See, for instance, J. Koutecký, *Prog. Surf. Membr. Sci.* **11**, 1 (1976).
56. J. Ladik, *Phys. Rev. B* **17**, 1663 (1978).
57. G. A. Baraff and M. Schlüter, *Phys. Rev. Lett.* **41**, 892 (1978); *Phys. Rev. B* **19**, 4965 (1979).
58. J. Bernholc, N. Lipari, and S. T. Pantelides, *Phys. Rev. Lett.* **41**, 895 (1978); *Phys. Rev. B* **21**, 3545 (1980).
59. G. Del Re and J. Ladik, *Chem. Phys.* **49**, 32 (1980).
60. M. Seel, G. Del Re, and J. Ladik, *J. Comput. Chem.* **3**, 451 (1982).
61. M. Seel, *Int. J. Quantum Chem.* **26**, 753 (1984).
62. W. J. Hehre, R. F. Stewart, and J. A. Pople, *J. Chem. Phys.* **51**, 2657 (1969).
63. H. F. Schaefer III, *The Electronic Structure of Atoms and Molecules*, Addison-Wesley Publ. Co., Reading, Mass (1972).
64. P. E. Gill and G. F. Miller, *Comput. J.* **15**, 80 (1972).
65. S. P. Singhal and J. Callaway, *Phys. Rev. B* **19**, 5049 (1979).
66. S. Canuto, J.-L. Calais, and O. Goscinski, *J. Phys. B* **14**, 1409 (1981).

Chapter 5

Electronic Correlation in Polymers

It is well known that the Hartree–Fock method gives only an approximation to the correct total energy and wave function of a system. Unless heavy atoms are present in the system, which would require a relativistic treatment, the exact energy and wave function can be obtained only by solving the Schrödinger equation of the problem

$$\hat{H}\Psi = E\Psi \tag{5.1}$$

This is possible only in the case of one electron, and therefore both shell physics (atomic, molecular, and solid-state physics) as well as nuclear physics must employ approximation methods in any problem involving more than one body. Most of these methods use as a starting wave function a Slater determinant constructed from Hartree–Fock (HF) orbitals, while various approximation methods are used for the difference of the exact total energy (Schrödinger energy) and the Hartree–Fock energy for a given state of a given system. (By definition, this is the correlation energy.) These methods fall into two categories. The most frequently used configuration interaction (CI) method is not size-consistent, that is, a given level of approximation (for instance, CI with only double excitations) gives less and less correlation energy per electron if the number of electrons increases, and in the limit, $E_{corr}/n \to 0$ if $n \to \infty$, where n is the number of electrons.[1] However, other methods like many-body perturbation theory[2] and coupled cluster theory[3] are size-consistent. Therefore, for the calculation of correlation in infinite, or finite but very long chains, only the latter types of method are applicable.

One should emphasize that in the case of polymers, as in the case of molecules, the HF method with a good basis set works reasonably well until one is interested only in average ground-state properties (like charge distribution) in the equilibrium nuclear configuration of a closed-shell system. If, however, one excites or ionizes the system or changes the position of the nuclei, one cannot obtain even tolerable agreement with experiment unless one takes into account the major part (at least between 70 and 80%) of the correlation energy. This means that if one wishes to calculate the electronic or vibrational spectra, the ESCA spectra, the fundamental gap, the transport or mechanical properties of polymers, one must be in a position to treat the correlation problem of polymers.

There are basically two possible approaches to computing. In the case of insulators or semiconductors with a noncrossing completely filled valence band and empty conduction band, one can always use not the delocalized Bloch functions but localized Wannier functions (for their construction see Section 5.1). On the other hand, in the case of metallic polymers like $(SN)_x$ with a partially filled valence band, one cannot construct Wannier functions and so one must resort to other methods (see Section 5.3).

It is also noteworthy that one can obtain 70–80% of the correlation energy comparatively easily even for a metallic polymer, if the unit cell is small. On the other hand, if the polymer consists of large molecules as unit cell, the problem is still not well solved (as is the case also in molecular physics). Suitable methods for approaching this problem will be discussed in Section 5.6.

5.1. CONSTRUCTION OF WANNIER FUNCTIONS

As early as 1934 Wannier[4] suggested using not Bloch functions, which are delocalized and periodic in the reciprocal space with a continuous **k**, but rather their Fourier transforms to obtain localized functions. Therefore a Bloch function of the form

$$\phi_n^{\mathbf{k}}(\mathbf{r}) = \exp(i\mathbf{k}\cdot\mathbf{r})u_n^{\mathbf{k}}(\mathbf{r}) \tag{5.2}$$

where $u_n^{\mathbf{k}}(\mathbf{r})$ shows the periodicity of the lattice,

$$u_n^{\mathbf{k}}(\mathbf{r}+\mathbf{R}_h) = u_n^{\mathbf{k}}(\mathbf{r}) \tag{5.3}$$

can be expanded as a Fourier series in the form

$$\phi_n^{\mathbf{k}}(\mathbf{r}) = N^{-1/2}\sum_l w_n(\mathbf{r}-\mathbf{R}_l)\exp(i\mathbf{k}\cdot\mathbf{R}_l) \tag{5.4}$$

The coefficients of this series expansion are the Wannier functions. For instance, the Wannier function corresponding to the nth band and localized at site $\mathbf{R}_l$ is given by

$$w_n(\mathbf{r}-\mathbf{R}_l) = N^{-1/2}\sum_{\mathbf{k}} \phi_n^{\mathbf{k}}(\mathbf{r})\exp(-i\mathbf{k}\cdot\mathbf{R}_l) = N^{-1/2}\sum_{\mathbf{k}} \phi_n^{\mathbf{k}}(\mathbf{r}-\mathbf{R}_l) \quad (5.5)$$

For the second equality in equation (5.5) one has to take into account equations (5.2) and (5.3), and the summation over $\mathbf{k}$ must be extended over the whole first Brillouin zone (BZ).

The Wannier functions (5.5) are not defined uniquely because each Bloch function $\phi_n^{\mathbf{k}}(\mathbf{r})$ can be multiplied by an arbitrarily chosen phase factor of absolute value unity.

The Bloch functions (5.2) form a complete orthonormal function system

$$\langle \phi_n^{\mathbf{k}} \mid \phi_{n'}^{\mathbf{k}'} \rangle = \delta_{n,n'}\delta(\mathbf{k}-\mathbf{k}') \quad (5.6)$$

From this property of the Bloch functions one can derive for the Wannier functions

$$\begin{aligned}\langle w_n(\mathbf{r}-\mathbf{R}_l) \mid w_{n'}(\mathbf{r}-\mathbf{R}_{l'})\rangle &= N^{-1}\sum_{\mathbf{k}}\sum_{\mathbf{k}'} \exp[i(\mathbf{k}\cdot\mathbf{R}_l - \mathbf{k}'\mathbf{R}_{l'})]\langle \phi_n^{\mathbf{k}}(\mathbf{r}) \mid \phi_{n'}^{\mathbf{k}'}(\mathbf{r})\rangle \\ &= N^{-1}\sum_{\mathbf{k}} \exp[i\mathbf{k}\cdot(\mathbf{R}_l - \mathbf{R}_{l'})\delta_{n,n'} \\ &= \delta_{l,l'}\delta_{n,n'} \end{aligned} \quad (5.7)$$

Therefore the system of Wannier functions possesses the completeness property in the same space in which the system of Bloch functions is complete,

$$\sum_l \sum_n w_n^*(\mathbf{r}'-\mathbf{R}_l)w_n(\mathbf{r}-\mathbf{R}_l) = \delta(\mathbf{r}-\mathbf{r}') \quad (5.8)$$

We now assume that the Boch functions are eigenfunctions of an effective one-electron Hamiltonian $\hat{H}_{\text{eff}}$,

$$\hat{H}_{\text{eff}}\phi_n^{\mathbf{k}} = \varepsilon_n^{\mathbf{k}}\phi_n^{\mathbf{k}} \quad (5.9)$$

On substituting for $\phi_n^{\mathbf{k}}$ from equation (5.4) one can write

$$\sum_l \hat{H}_{\text{eff}} w_n(\mathbf{r}-\mathbf{R}_l)\exp(i\mathbf{k}\cdot\mathbf{R}_l) = \sum_l w_n(\mathbf{r}-\mathbf{R}_l)\varepsilon_n^{\mathbf{k}}\exp(i\mathbf{k}\cdot\mathbf{R}_l) \quad (5.10)$$

The quantity $\varepsilon_n^{\mathbf{k}}$ can also be expanded in a Fourier series:

$$\varepsilon_n^{\mathbf{k}} = \sum_{l'} \varepsilon_n(\mathbf{R}_{l'})\exp(i\mathbf{k}\cdot\mathbf{R}_{l'}) \tag{5.11}$$

If this expansion is substituted into equation (5.10), the $l = 0$ term ($\mathbf{R}_0 = \mathbf{0}$) selected from the sums over l on both sides, and use made of definition (5.5) of the Wannier functions, one obtains[5]

$$\hat{H}_{\text{eff}} w_n(\mathbf{r}) = \sum_{l'} \varepsilon_n(\mathbf{R}_{l'}) w_n(\mathbf{r} - \mathbf{R}_{l'}) \tag{5.12}$$

This result also demonstrates clearly that the Wannier functions are not eigenfunctions of the Hamiltonian of a periodic crystal. Therefore in many applications, when one applies the Wannier functions to describe local perturbations, it is more advantageous to start from a mixed representation in which some matrix elements are expressed with the help of Wannier functions and others with the aid of Bloch functions.

We shall now consider a nondegenerate band. In this case it can be seen from equation (5.9) that besides $\phi_n^{\mathbf{k}}$, all functions

$$\tilde{\phi}_n^{\mathbf{k}} = \phi_n^k \exp(i\lambda_n^{\mathbf{k}}) \tag{5.13}$$

are also eigenfunctions of $\hat{H}_{\text{eff}}$ with the same eigenvalue $\varepsilon_n^{\mathbf{k}}$. The fundamental question to be answered is, assuming $\hat{H}_{\text{eff}} = -\frac{1}{2}\nabla_r^2 + V(\mathbf{r})$, how are the real and complex parts of $w_n(\mathbf{r})$ and its spatial extension (localization) dependent on the choice of phase factor $\lambda_n^{\mathbf{k}}$? If we express $w_n(\mathbf{r})$ in the form

$$w_n(\mathbf{r}) = w_n^{(\text{Re})}(\mathbf{r}) + i w_n^{(\text{Im})}(\mathbf{r}) \tag{5.14}$$

and substitute it into equation (5.4), we obtain

$$\begin{aligned}\phi_n^{-\mathbf{k}}(\mathbf{r}) &= N^{-1/2} \sum_l w_n^{(\text{Re})}(\mathbf{r} - \mathbf{R}_l)\exp(-i\mathbf{k}\cdot\mathbf{R}_l)\\ &\quad + iN^{-1/2} \sum_l w_n^{(\text{Im})}(\mathbf{r} - \mathbf{R}_l)\exp(-i\mathbf{k}\cdot\mathbf{R}_l)\end{aligned} \tag{5.15}$$

$$\begin{aligned}\phi_n^{\mathbf{k}}(\mathbf{r})^* &= N^{-1/2} \sum_l w_n^{(\text{Re})}(\mathbf{r} - \mathbf{R}_l)\exp(-i\mathbf{k}\cdot\mathbf{R}_l)\\ &\quad - iN^{-1/2} \sum_l w_n^{(\text{Im})}(\mathbf{r} - \mathbf{R}_l)\exp(-i\mathbf{k}\cdot\mathbf{R}_l)\end{aligned} \tag{5.16}$$

On the other hand, it follows from equations (5.2) and (5.9), and from the above definition of $\hat{H}_{\text{eff}}$, that

$$\tfrac{1}{2}(k^2 - \nabla_r^2)u_n^{\mathbf{k}} - i\mathbf{k}\nabla_r(u_n^{\mathbf{k}}) + V(\mathbf{r})u_n^{\mathbf{k}} = \varepsilon_n^{\mathbf{k}} u_n^{\mathbf{k}} \tag{5.17a}$$

and for its complex conjugate

$$\tfrac{1}{2}(k^2 - \nabla_r^2)(u_n^{\mathbf{k}})^* + i\mathbf{k}\nabla_r(u_n^{\mathbf{k}})^* + V(\mathbf{r})(u_n^{\mathbf{k}})^* = \varepsilon_n^{\mathbf{k}}(u_n^{\mathbf{k}})^*[(\varepsilon_n^{\mathbf{k}})^* = \varepsilon_n^{\mathbf{k}}] \tag{5.17b}$$

Since equation (5.17b) is identical with the equation we would obtain for $u_n^{-\mathbf{k}}$, it follows that if the condition

$$\lambda_n^{-\mathbf{k}} = -\lambda_n^{\mathbf{k}} \tag{5.18}$$

is fulfilled, the Bloch function will possess the property

$$\tilde{\phi}_n^{-\mathbf{k}}(\mathbf{r}) = (\tilde{\phi}_n^{\mathbf{k}})^* \tag{5.19}$$

It follows from equations (5.15) and (5.16) that this choice of phase $\lambda_n^{\mathbf{k}}$ yields $w_n^{(\mathrm{Im})}(\mathbf{r} - \mathbf{R}_{l'}) = 0$.

The form of phase $\lambda_n^{\mathbf{k}}$, even if it has to satisfy equation (5.18), is still undefined in the half of the BZ. It can be selected such that the Wannier functions should possibly be optimally localized. This requirement is not unique, but a physically meaningful condition is that $|w_n|^2$ should be maximal in a certain region of the crystal (for instance, in the reference cell and in a few of its neighbors). When the Bloch function is expressed in the form

$$\phi_n^{\mathbf{k}}(\mathbf{r}) = N^{-1/2} \sum_{l=1}^{N} \exp(i\mathbf{k}\cdot\mathbf{R}_l)\Lambda_n^{\mathbf{k}}(\mathbf{r} - \mathbf{R}_l) \tag{5.20}$$

and this expression substituted into equation (5.5), one obtains

$$\begin{aligned} |w_n|^2 &= N^{-1} \sum_{\mathbf{k}} \sum_{\mathbf{k}'} (\phi_n^{\mathbf{k}})^* \phi_n^{\mathbf{k}'} \\ &= N^{-2} \sum_{\mathbf{k}} \sum_{\mathbf{k}'} \sum_{l=1}^{N} \sum_{l'=1}^{N} \exp[i(\mathbf{k}'\cdot\mathbf{R}_{l'} - \mathbf{k}\cdot\mathbf{R}_l)]\Lambda_n^{\mathbf{k}}(\mathbf{r} - \mathbf{R}_l)\Lambda_n^{\mathbf{k}'}(\mathbf{r} - \mathbf{R}_{l'}) \\ &\equiv \sum_{l=1}^{N} \Gamma_n^l \end{aligned} \tag{5.21}$$

We now require that the integral

$$I_n = \int d\mathbf{r}\left[\sum_l{}' \Gamma_n^l\right] \tag{5.22}$$

should be maximal if $\phi_n^{\mathbf{k}}$ in equation (5.21) is replaced by $\tilde{\phi}_n^{\mathbf{k}}$ [see equation (5.13)], where the prime at the summation sign indicates summation only over the neighboring cells of the reference cell. In this case I_n will become a function of $\lambda_n^{\mathbf{k}}$ and the variational problem

$$\delta I_n[\lambda_n^{\mathbf{k}}] = 0 \tag{5.23}$$

leads to the integral equation[6,7]

$$\lambda_n^{\mathbf{k}} = \arg\left[\int d\mathbf{r} \sum_l{}' \exp(-i\mathbf{k}\cdot\mathbf{R}_l)\Lambda_n^{\mathbf{k}}(\mathbf{r}-\mathbf{R}_l)^* \times \frac{V_c}{8\pi^3}\int dk \sum_{l'}{}' \exp(i\mathbf{k}'\cdot\mathbf{R}_{l'})\Lambda_n^{\mathbf{k}'}(\mathbf{r}-\mathbf{R}_{l'})\exp(i\lambda_n^{\mathbf{k}'})\right] \pm n\pi \tag{5.24}$$

where V_c denotes the volume of the unit cell. Equation (5.24) can be solved in an interative manner for the unknown function $\lambda_n^{\mathbf{k}'}$. According to experience gained by Suhai[5] on polyacetylene, polydiacetylene, polyethylene, and polyglycine, this procedure converges after only a few steps if one chooses as the starting phase $\lambda_n^{\mathbf{k}'} = -\lambda_n^{\mathbf{k}'}$ (orig) (the original phase of the Bloch function). Blount[6] has also suggested that instead of $|w_n|^2$ one can search for the maximum of the matrix elements $\langle w_n | \hat{O} | w_n \rangle$, where $\hat{O}$ is a one-electron operator. Suhai has found that the choice $\hat{O} = z^2$ is very advantageous (z is the coordinate in the direction of the polymer axis), because the matrix elements $\langle \Lambda_n^{\mathbf{k}}(\mathbf{r}-\mathbf{R}_l)| z^2 |\Lambda_n^{\mathbf{k}'}(\mathbf{r}-\mathbf{R}_{l'})\rangle$ also occur in other problems. Substitution of such matrix elements into equation (5.24) has led to convergence in 4 or 5 steps with excellent localization properties of the Wannier functions.

As an example we examine some results obtained for *trans*-polyacetylene.[5] The Wannier function of this polymer can be expressed in the LCAO form

$$w_n(\mathbf{r}) = N^{-1/2}\sum_l \sum_a d_{a,n}\chi_a(\mathbf{r}-\mathbf{R}_l-\mathbf{R}_a) \tag{5.25}$$

where the coefficient $d_{a,n}$ belongs to the AO $\rightarrow \chi_a$ centered in the cell characterized by vector $\mathbf{R}_l$. Function $w_n(\mathbf{r})$ itself belongs to the reference cell ($l = 0$). Further, it was assumed that the polymer chain lies in the Y–Z plane, i.e., the $2p_x$ orbitals of carbon form the Wannier functions corresponding to the highest filled (π) and lowest unfilled (π') bands. The linear combination of $2p_y$ orbitals gives rise to the highest filled σ band and to the lowest virtual one (σ').

Figure 5.1[5] shows on a logarithmic scale the absolute values of the LCAO coefficients for the valence (π) band as a function of l (in C_2H_2 units). The curve △ was obtained with the original phase of the Bloch functions (without any phase transformation). The fulfillment of the requirement $\lambda_n^{\mathbf{k}} = -\lambda_n^{\mathbf{k}}$ (orig) in itself leads to quite a good localization (curve □), which can be further improved to a small extent with the aid of equation (5.24) (curve ○). One should point out that these results are not general, because in the case of the π' conduction band (see Figure 5.2) only the curve ○ obtained by solving equation (5.24) provides good

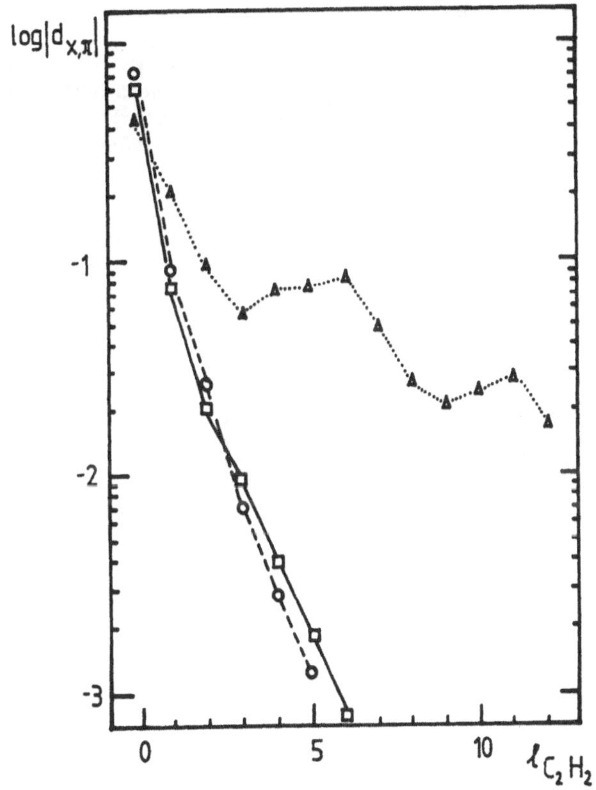

FIGURE 5.1. Coefficients of the Wannier function of the valence (π) band of alternating *trans*-polyacetylene (after Suhai[5]).

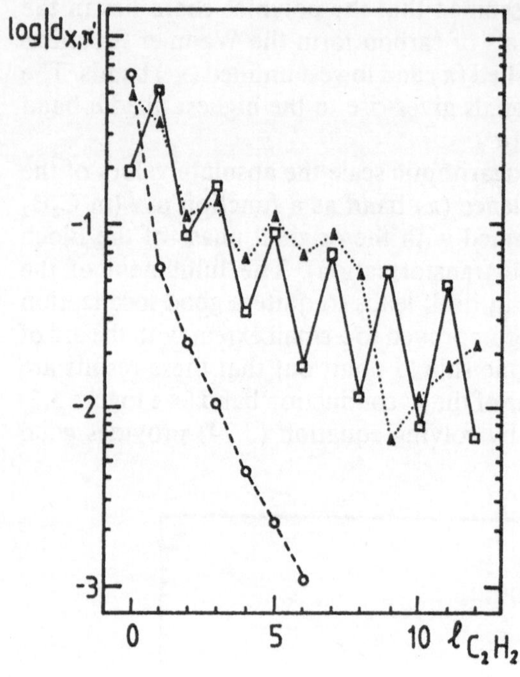

FIGURE 5.2. Coefficients of the Wannier function of the conduction (π^*) band of *trans*-polyacetylene (after Suhai[5]).

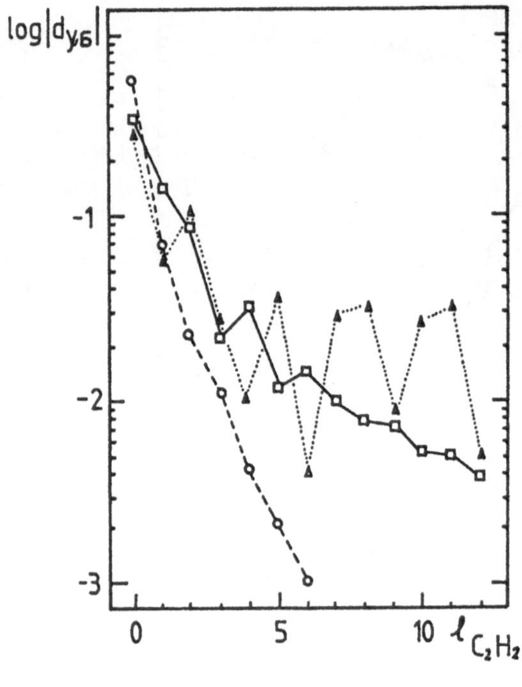

FIGURE 5.3. Wannier coefficients in the highest filled σ-band of *trans*-polyacetylene (after Suhai[5]).

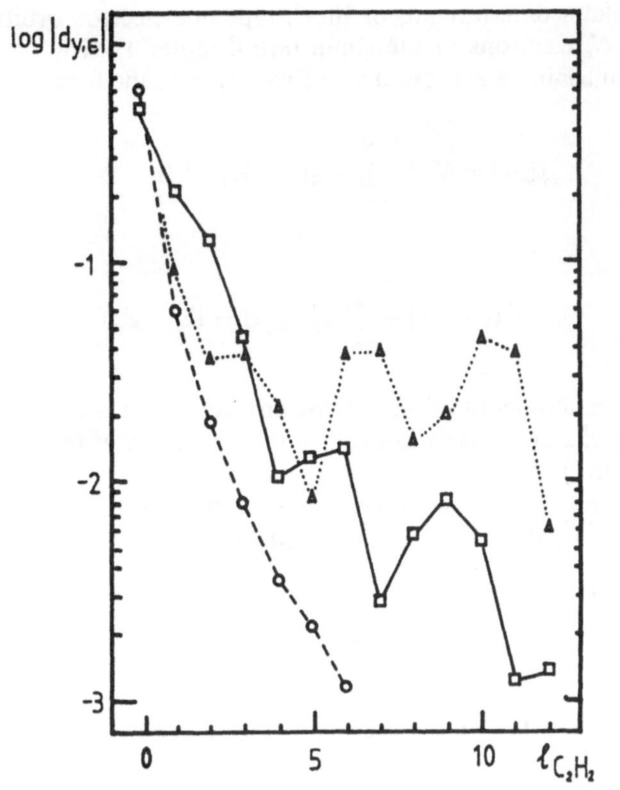

FIGURE 5.4. Wannier coefficients in the lowest unfilled σ^*-band of *trans*-polyacetylene (after Suhai[5]).

localization, while in the other two cases the coefficients oscillate as a function of l.

Figures 5.3 and 5.4 show that the σ electrons can be localized with the help of equation (5.24) as well as the π electrons. Finally, it should be noted that the results shown in Figures 5.1 to 5.4 were obtained with the aid of an STO-3G (minimal-basis) *ab initio* band-structure calculation of polyacetylene, but Suhai's unpublished results have shown that one obtains similar results with more refined basis sets as well.

5.2. SECOND-ORDER MØLLER–PLESSET MANY-BODY PERTURBATION THEORY FOR INFINITE SYSTEMS

We start with the HF ground-state wave function as an unperturbed wave function that can be written in the form of an antisymmetrized product (Slater determinant) of Bloch-type one-electron orbitals of the system of N_e electrons of the chain (see Chapter 1). The one-electron orbitals can again be expressed as a Bloch sum in the form

$$\phi_m^{\mathbf{k}}(\mathbf{r}) = N^{-1/2} \sum_{h=1}^{N} \exp(i\mathbf{k}\cdot\mathbf{R}_h) \Lambda_m^{\mathbf{k}}(\mathbf{r}-\mathbf{R}_h) \tag{5.20}$$

where

$$\Lambda_n^{\mathbf{k}}(\mathbf{r}-\mathbf{R}_h) = \sum_{p=1}^{\nu} c_{p,m}^{\mathbf{k}} \chi_p(\mathbf{r}-\mathbf{R}_h-\mathbf{R}_p) \tag{5.26}$$

Here, ν is the number of orbitals in the unit cell and it is assumed that the coefficients $c_{p,m}^{\mathbf{k}}$ are already known from the solution of the HF problem (see Chapter 1).

Following Møller and Plesset[2] we can take as the unperturbed Hamiltonian $\hat{H}_0$ the sum of the Fock operators,

$$\hat{H}_0 = \sum_{i}^{\text{occ}} \hat{F}_i \tag{5.27}$$

It is easy to show that in the case $\Phi_0 = \Phi_{\text{HF}}$ the eigenvalue equation

$$\hat{H}_0 \Phi_0 = E_0 \Phi_0$$

is valid.

From Φ_0 we can construct further eigenfunctions of $\hat{H}_0$, if one of the (ground-state) occupied orbitals is replaced by an empty one. To express this we introduce the composite index $I = (i, \mathbf{k})$, where i indicates the band and k the quasi-momentum, and denote occupied orbitals by $I, J, K, \ldots$ and unfilled ones by $A, B, C, \ldots$. In this way the eigenfunctions and eigenvalues of $\hat{H}_0$ can be classified as

$$\Phi_0 = \Phi_{\text{HF}}, \quad E_0 = \sum_{K}^{\text{occ}} \varepsilon_K \tag{5.28a}$$

$$\Phi_I^A = \Phi_{\text{HF}}(I \to A), \quad E_I^A = \sum_{K \neq I}^{\text{occ}} \varepsilon_K + \varepsilon_A \tag{5.28b}$$

where the notation $I \to A$ means that the Ith filled orbital is substituted by the Ath unfilled one, and

$$\Phi_{IJ}^{AB} = \Phi_{\mathrm{HF}}\begin{pmatrix} I \to A \\ J \to B \end{pmatrix}, \quad E_{IJ}^{AB} \sum_{K \neq I,J}^{\mathrm{occ}} \varepsilon_K + \varepsilon_A + \varepsilon_B \tag{5.28c}$$

Application of the Raleigh–Schrödinger perturbation theory yields

$$\hat{H} = \hat{H}_0 + \lambda \hat{V}$$

where

$$\hat{V} = \hat{H} - \hat{H}_0 = \hat{H} - \sum_{i}^{\mathrm{occ}} \hat{F}_i \tag{5.29}$$

The complete set of eigenfunctions of $\hat{H}_0$ is given by Φ_0, Φ_I^A, Φ_{IJ}^{AB}, ... If these eigenfunctions and the above given choice of perturbation operator $\hat{V}$ are employed, then for the second-order perturbation theory we only require matrix elements between the ground state and doubly excited states. From Brillouin's theorem, the matrix elements

$$\langle \bar{\Phi}_0 \,|\, \hat{V} \,|\, \Phi_I^A \rangle = 0 \tag{5.30}$$

and also the matrix elements between the ground states and triply or greater excited states disappear. If Φ is the eigenfunction of the exact operator $\hat{H}$, namely

$$\hat{H}\Phi = E\Phi \tag{5.31}$$

we can develop it as

$$\Phi^{\lambda} = \Phi_0 + \lambda\Phi_1 + \lambda^2\Phi_2 + \cdots \tag{5.32}$$

with the corresponding eigenvalue

$$E^{\lambda} = E_0 + \lambda E_1 + \lambda^2 E_2 + \cdots \tag{5.33}$$

For $\lambda = 1$, one obtains in the second order

$$\begin{aligned} E &= E_0 + E_1 + E_2 \\ &= E_0 + \langle \Phi_0 \,|\, \hat{V} \,|\, \Phi_0 \rangle + \sum_{I,J} \sum_{A,B}{}' \frac{\langle \Phi_0 \,|\, \hat{V} \,|\, \Phi_{IJ}^{AB} \rangle \langle \Phi_{IJ}^{AB} \,|\, \hat{V} \,|\, \Phi_0 \rangle}{E_0 - E_{IJ}^{AB}} \end{aligned} \tag{5.34}$$

where the prime after the summation sign means that only nonequivalent matrix elements should be considered.

The expression

$$\langle \Phi_0 | \hat{V} | \Phi_0 = \left\langle \Phi_0 \middle| \hat{H} - \sum_i^{\text{occ}} \hat{F}_i \middle| \Phi_0 \right\rangle = E_{\text{HF}} - E_0 \tag{5.35}$$

shows that first-order perturbation theory yields the HF energy. Therefore only second- and higher-order terms contribute to the correlation energy.

Application of the Slater–Condor rules to the matrix elements occurring in equation (5.34) gives

$$E_2 = \sum_{I,J} \sum_{A,B}{}' \frac{[\langle \phi_I(1)\phi_J(2) | (I - \hat{P}_{12}) r_{12}^{-1} | \phi_A(1)\phi_B(2) \rangle]^2}{\varepsilon_I + \varepsilon_J - \varepsilon_A - \varepsilon_B} \tag{5.36}$$

The central problem in calculating E_2 is to express the two-electron integrals, which were calculated when applying HF theory in an AO basis, in terms of the Bloch functions (5.20) (the so-called four-index transformation). Before attacking this problem one must employ the translational symmetry of the crystal (or chain) to simplify equation (5.36). For this purpose we shall consider the Coulomb integral

$$I = \langle \phi_I(1)\phi_J(2) | r_{12}^{-1} | \phi_A(1)\phi_B(2) \rangle \tag{5.37}$$

If, for the quasi-one-dimensional case, the Bloch functions ϕ_A and ϕ_B are expressed in the form

$$\phi_A(1) = \phi_a^{\mathbf{k}_a}(\mathbf{r}_1) = N^{-1/2} \sum_{\mathbf{R}_a} \exp(i\mathbf{k}_a\mathbf{R}_a) \Lambda_a^{\mathbf{k}_a}(\mathbf{r}_1 - \mathbf{R}_a)$$

$$\phi_B(2) = \phi_b^{\mathbf{k}}(\mathbf{r}_2) = N^{-1/2} \sum_{\mathbf{R}_b} \exp(i\mathbf{k}_a\mathbf{R}_b) \Lambda_b^{\mathbf{k}_b}(\mathbf{r}_2 - \mathbf{R}_b)$$

and substituted into equation (5.36), one obtains

$$I = N^{-1} \sum_{\mathbf{R}_a} \sum_{\mathbf{R}_b} e^{i\mathbf{k}_a\mathbf{R}_a + i\mathbf{k}_b\mathbf{R}_b} \langle \phi_i^{\mathbf{k}_i}(\mathbf{r}_1)\phi_j^{\mathbf{k}_j}(\mathbf{r}_1) | r_{12}^{-1} \Lambda_a^{\mathbf{k}_a}(\mathbf{r}_1 - \mathbf{R}_a) \Lambda_b^{\mathbf{k}_b}(\mathbf{r}_2 - \mathbf{R}_b) \rangle \tag{5.38}$$

This integral can be simplified by first introducing the transformations

$$\mathbf{r}_1' = \mathbf{r}_1 - \mathbf{R}_a, \quad \mathbf{r}_2' = \mathbf{r}_2 - \mathbf{R}_a, \quad r_{12}' = r_{12}$$

Bloch's theorem is then used to write

$$\phi_i^{\mathbf{k}_i}(\mathbf{r}_1') = \exp(-i\mathbf{k}_i\mathbf{R}_a)\phi_i^{\mathbf{k}_i}(\mathbf{r}_1' + \mathbf{R}_a) \tag{5.39a}$$

$$\phi_j^{\mathbf{k}_j}(\mathbf{r}_2') = \exp(-i\mathbf{k}_j\mathbf{R}_a)\phi_j^{\mathbf{k}_j}(\mathbf{r}_2' + \mathbf{R}_a) \tag{5.39b}$$

Next, these expressions are substituted into equation (5.38) and summation carried out over $\mathbf{R}_a - \mathbf{R}_b$ and $\mathbf{R}_b$ instead of $\mathbf{R}_a$ and $\mathbf{R}_b$. By dropping the primes on $\mathbf{r}_1'$ and $\mathbf{r}_2'$ and applying some algebraic manipulations, the quantity I can be expressed in the form

$$I = N^{-1} \sum_{\mathbf{R}_a - R_b} \sum_{\mathbf{R}_b} \exp[i(\mathbf{k}_a - \mathbf{k}_i + \mathbf{k}_b - \mathbf{k}_j)\mathbf{R}_b]\exp[i(\mathbf{k}_a - \mathbf{k}_i)(\mathbf{R}_a - \mathbf{R}_b)]$$
$$\times \langle \phi_i^{\mathbf{k}_i}(\mathbf{r}_1)\phi_j^{\mathbf{k}_j}(\mathbf{r}_2 + \mathbf{R}_a - \mathbf{R}_b) \,|\, \mathbf{r}_{12}^{-1} \,|\, \Lambda_a^{\mathbf{k}_a}(\mathbf{r}_1)\Lambda_b^{\mathbf{k}_b}(\mathbf{r}_2 + \mathbf{R}_a - \mathbf{R}_b)\rangle \tag{5.40}$$

If the variable $\mathbf{R} = \mathbf{R}_a - \mathbf{R}_b$ is introduced and the definition of the delta function taken into account, one can write

$$I = \sum_{\mathbf{R}} \delta_{\mathbf{k}_a - \mathbf{k}_i, \mathbf{k}_j - \mathbf{k}_b} \exp[i(\mathbf{k}_a - \mathbf{k}_i)\mathbf{R}]$$
$$\times \langle \phi_i^{\mathbf{k}_i}(\mathbf{r}_1)\phi_j^{\mathbf{k}_j}(\mathbf{r}_2 + \mathbf{R}) \,|\, r_{12}^{-1} \,|\, \Lambda_a^{\mathbf{k}_a}(\mathbf{r}_1)\Lambda_b^{\mathbf{k}_b}(\mathbf{r}_2 + \mathbf{R})\rangle \quad (\mathbf{R} = \mathbf{R}_a - \mathbf{R}_b) \tag{5.41}$$

This expression must now be substituted into the Coulomb part of E_2 in equation (5.36). If one neglects the band indices for the moment, then the above procedure yields

$$I = \sum_{\mathbf{k}_i}\sum_{\mathbf{k}_j}\sum_{\mathbf{k}_a}\sum_{\mathbf{k}_b}\sum_{\mathbf{R}}\sum_{\mathbf{R}_i}\sum_{\mathbf{R}_j} \exp[i(\mathbf{k}_a - \mathbf{k}_i)\mathbf{R}]\exp(-i\mathbf{k}_i\mathbf{R}_i)\exp(-i\mathbf{k}_j\mathbf{R}_j)$$
$$\times \langle \Lambda_i^{\mathbf{k}_i}(\mathbf{r}_1 - \mathbf{R}_i)\Lambda_j^{\mathbf{k}_j}(\mathbf{r}_2 - \mathbf{R}_j + \mathbf{R}) \,|\, r_{12}^{-1} \,|\, \Lambda_a^{\mathbf{k}_a}(\mathbf{r}_1)\Lambda_b^{\mathbf{k}_b}(\mathbf{r}_2 + \mathbf{R})\rangle$$
$$\times \delta_{\mathbf{k}_a - \mathbf{k}_i, \mathbf{k}_j - \mathbf{k}_b} \tag{5.42}$$

In a similar way the contribution of the exchange term can be attained.

One can express E_2 in terms of the pair correlation energies ε_{IJ} of the electrons described by the Bloch orbitals $\phi_i^{\mathbf{k}_i}$ and $\phi_j^{\mathbf{k}_j}$, namely

$$E_2 = \sum_I \sum_J{}' \varepsilon_{IJ} \tag{5.43}$$

We now substitute expression (5.42) and the corresponding expression for the exchange term into equation (5.37). If the transformation $\mathbf{q} = \mathbf{k}_a - \mathbf{k}_i$ is introduced, and one sums over $\mathbf{q}$ and $\mathbf{k}_b$ with allowance for the delta function in expression (5.42), then ε_{IJ} is obtained in the form

$$\varepsilon_{IJ} = \sum_a \sum_b \sum_{\mathbf{q}}{}' N^{-2} \frac{|A(i, j, a, b, \mathbf{k}_i, \mathbf{k}_j, \mathbf{q})|}{\varepsilon_i^{\mathbf{k}_i} + \varepsilon_j^{\mathbf{k}_j} - \varepsilon_a^{\mathbf{k}_i+\mathbf{q}} - \varepsilon_b^{\mathbf{k}_j-\mathbf{q}}} \tag{5.44}$$

with

$$\begin{aligned} &A(i, j, a, b, \mathbf{k}_i, \mathbf{k}_j, \mathbf{q}) \\ &\quad = \sum_{\mathbf{R}} \sum_{\mathbf{R}_i} \sum_{\mathbf{R}_j} \{\exp[i(\mathbf{qR} - \mathbf{k}_i\mathbf{R}_j - \mathbf{k}_j\mathbf{R}_j)] \\ &\qquad \times \langle \Lambda_i^{\mathbf{k}_i}(\mathbf{r}_1) \Lambda_j^{\mathbf{k}_j}[\mathbf{r}_2 - (\mathbf{R}_j - \mathbf{R}_i - \mathbf{R})] \,|\, r_{12}^{-1} \,|\, \Lambda_a^{\mathbf{k}_i+\mathbf{q}}[\mathbf{r} - (-\mathbf{R}_i)] \\ &\qquad \times \Lambda_b^{\mathbf{k}_j-\mathbf{q}}[\mathbf{r}_2 - (-\mathbf{R}_i - \mathbf{R})]\rangle \\ &\qquad - \exp\{i[(\mathbf{k}_j - \mathbf{k}_i - \mathbf{q})\mathbf{R} - \mathbf{k}_i\mathbf{R}_i - \mathbf{k}_j\mathbf{R}_j] \\ &\qquad \times \langle \Lambda_i^{\mathbf{k}_i}(\mathbf{r}_1) \Lambda_j^{\mathbf{k}_j}[\mathbf{r}_2 - (\mathbf{R}_j - \mathbf{R}_i - \mathbf{R})] \,|\, r_{12}^{-1} \,|\, \Lambda_b^{\mathbf{k}_j-\mathbf{q}}[\mathbf{r}_1 - (-\mathbf{R}_1)] \\ &\qquad \times \Lambda_a^{\mathbf{k}_i+\mathbf{q}}[\mathbf{r}_2 - (-\mathbf{R}_i - \mathbf{R})]\rangle\}\} \end{aligned} \tag{5.45}$$

In this derivation one should observe that the periodic symmetry of the crystal or chain enables one to obtain the conservation of the sum of momenta by double excitations, $\mathbf{k}_i + \mathbf{k}_j = \mathbf{k}_a + \mathbf{k}_b$, which reduces the originally fourfold summation in **k**-space to a threefold summation. Further, the correlation energy per unit cell E_2/N is independent of N if $N \to \infty$ because the threefold summation in **k**-space [see (5.43) and (5.44)] leads to a factor N^3, while equation (5.44) contains N^{-2} explicitly as a factor. Thus $E_2 \sim N$ and E_2/N is independent of N.

The Møller–Plesset second-order perturbation theory (MP/2) is comparatively simple because only matrix elements of the form $\langle \Phi_0 \,|\, \hat{V} \,|\, \Phi_{IJ}^{AB} \rangle$ need to be calculated, while in third order there are already matrix elements of type $\langle \Phi_{IJ}^{AB} \,|\, \hat{V} \,|\, \Phi_{KL}^{CD} \rangle$ (as in a CI calculation). There are many more of these matrix elements (even for an atomic basis of middle quality there are many more virtual than occupied bands) than those corresponding to excitations between the ground state and the different doubly excited states.

Another advantage of MP/2 is the ease with which E_2 can be partitioned into the pair correlation energies ε_{IJ} [see equations (5.43)–(5.45)]. This provides the possibility of a clear physical interpretation of the different contributions to the correlation energy, for not only can it be calculated in the ground state, but the HF band structure can also be redefined by taking correlation effects into account (electronic polaron theory; see the subsequent Section 5.3).

If one passes to higher-order perturbation theory this partitioning is unfortunately no longer possible (although it is possible in the framework of coupled cluster theory[8]) if one uses only single and double excitations and takes into account only the second order; see Section 5.5. For this reason, instead of pushing the calculations to higher order it seems to be more convenient to retain MP/2 but to use good basis sets (at least a double ζ set with polarization function on both the heavy and the hydrogen atoms). In our experience this gives 70 to 75% of the total correlation energy of a polymer (see Section 5.4). Moreover, one can use for the virtual orbitals not those obtained from the HF calculation, but other basis orbitals that are orthogonal to the filled ones. One such possibility is to follow the suggestion of Boys[9] and use the so-called oscillatory orbitals. These orbitals can be generated by mutiplying, for instances, an $1s$ function by x or y or z, or a $2p$ function again by the cartesian coordinates. Since this procedure does not change the orbital exponents, one generates in this way "$1p$," "$2d$," etc., type orbitals, which means that the virtual orbitals have their maxima in the same regions as the filled ones. One hopes in this way to obtain a larger part of the correlation energy. Preliminary studies on small and medium-size molecules using the coupled-cluster method have shown that this is so in most cases[10] (details can be found in Section 5.5), but the computational requirements become prohibitively large when such a basis is applied to larger molecules. On the other hand, if a polymer consists of small units, the oscillatory orbitals can be used to obtain a larger percent of the correlation even in the framework of MP/2.

On the one hand, the Møller–Plesset partitioning of $\hat{H}$ into $\hat{H}_0$ and $\hat{V}$ is not unique and therefore the different orders of perturbation theory are also not uniquely defined. Various other choices of $\hat{V}$ were proposed[11] but they all led to different variants of the Epstein–Nesbet[12] perturbation theory with a shifted denominator. This procedure also seems to be feasible for infinite systems, so there is hope that in the future more than 70 to 75% of the correlation energy will be obtained even in the second order.

On the other hand, the Møller–Plesset partitioning has the great advantage that the energy obtained in any order is invariant under unitary transformation of the filled spin orbitals ($\hat{H}_0$ is also invariant). The individual pair correlation energies ε_{IJ}, however, are not invariant and can be altogether different for canonical HF Bloch orbitals and for localized Wannier functions. This means that one can select the undefined phase factors of the Bloch orbitals in such a way that the intramolecular contributions to the correlation energy should be maximal, if one works with Wannier functions as a basis. In this way one can obtain pair correlation energies, that are transferable between structurally similar polymers or crystals and also facilitate the investigation of correlation in disordered systems (see Section 5.6).

Returning to the four-index transformation of two-electron integrals discussed above, in second order one has $\nu^4 n$ operations (ν is the number of orbitals per unit cell and n is the number of occupied bands), if one disregards the dispersion of the bands. However, neglect of **k**-dependence leads to serious errors. If one wishes to take into account many excitations in each region of the Brillouin zone of each band, this would lead to an astronomical number of matrix elements, even in the case of a simple $\varepsilon(\mathbf{k})$ curve. A possible solution to this problem could be found if one were to divide the BZ into different regions, choose only one Bloch function (with an appropriate **k** vector) for each region, and assign a weight factor to the matrix elements calculated with the aid of these selected Bloch functions. A plausible weight factor is the square root of the integral of the density of states, the integration being performed between the limits (in **k**-space) of the region under consideration.[13]

In the cases of completely filled or empty bands (semiconductors or insulators), the numerical work can be reduced if one does not substitute the Bloch functions in integral (5.37) in their LCAO form [see equations (5.20) and (5.21)], but replaces the AOs by Wannier functions (see Section 5.1). In this way matrix elements (5.37) retain their **k**-dependence (which is necessary for the correct treatment of the translational symmetry), but this procedure does not involve any special difficulties. By writing the Bloch functions in the form[14]

$$\phi_i^{\mathbf{k}_i}(\mathbf{r}) = N^{-1/2} \sum_h \exp[i\mathbf{k}\cdot\mathbf{R}_h] w_i(\mathbf{r}-\mathbf{R}_h) \tag{5.46}$$

one calculates only integrals of the type

$$\langle w_i(\mathbf{r}_1) w_j(\mathbf{r}_2 - \mathbf{R}_h) \,|\, r_{12}^{-1} \,|\, w_a(\mathbf{r}_1 - \mathbf{R}_{h'}) w_b(\mathbf{r}_2 - \mathbf{R}_{h''}) \rangle \tag{5.47}$$

which are **k**-independent. After computing expression (5.47) only once, one divides the BZ with the help of a fine mesh to take the **k**-dependence of equation (5.37) resulting from the exponential factors into account as well. The advantage of this procedure is that the numerical integration is already independent of the AO indices and therefore it can be performed very easily and rapidly. Only should note, however, that this method can be used only in the case of nonmetallic polymers.

In the case of metallic systems, it should be taken into account that E_2/N is divergent for two- and three-dimensional infinite systems,[15] but it can be shown to converge for one-dimensional chains.[16] To be able to use a larger number of **k** points for each band in a metallic system in which no Wannier functions can be defined, one can divide the BZ again into several regions and expand the coefficients of the LCAO functions A in each region with the help of a simple trigonometric series

$$c_{p,m}^{\mathbf{k}(i)} = \sum_{\alpha=0}^{n_i} A_{p,m}^{\alpha}(i)\cos[\alpha\mathbf{k}(i)a] \tag{5.48}$$

Here n_i denotes the number of terms in the series expansion in the ith region, for which the equation

$$|\mathbf{k}(i)|_{\min} \leqslant |\mathbf{k}(i)| \leqslant |\mathbf{k}(i)_{\max}| \tag{5.49}$$

is valid. One can compute coefficients $A_{p,m}^{\alpha}(i)$ on the basis of constants $c_{p,m}^{\mathbf{k}(i)}$ (obtained from the HF calculation for different **k**-vectors in each region) with the help of equation (5.48).

The four-index transformation can then be performed with the aid of the **k**-independent coefficients $A_{p,m}^{\alpha}(i)$, $A_{q,n}^{\beta}(j)$, $A_{r,a}^{\gamma}(k)$, and $A_{s,b}^{\vartheta}(l)$. Only after this transformation (which has to be performed only once, as in an atom or molecule) will the **k**-dependence be taken into account through multiplication by the trigonometric and exponential **k**-dependent factors. Separation of the **k**-dependent part of the integral enables one to use many **k** points in each band, because multiplication by the **k**-dependent factors is very rapid.[14] In this way a rather high degree of accuracy is achieved in the numerical integrations.

Finally one should note that the number of different regions defined in the BZ and the length of the various series expansions (the values of n_i) are, of course, inversely proportional to each other and both are rather system-dependent.[14]

5.3. ELECTRONIC POLARON MODEL AND THE QUASI-PARTICLE BAND STRUCTURE OF POLYMERS

The physical content of the HF energy bands is given by Koopmans' theorem,[17] which states that the orbital energies calculated for the conduction and valence bands, respectively, are equal to the corresponding electron affinities and ionization potentials:

$$\varepsilon_c^{\mathbf{k}_c}(\mathrm{HF}) = E_{\mathrm{HF}}^{(N+1)} - E_{\mathrm{HF}}^{(N)} \tag{5.50}$$

$$\varepsilon_v^{\mathbf{k}_v}(\mathrm{HF}) = E_{\mathrm{HF}}^{(N)} - E_{\mathrm{HF}}^{(N-1)} \tag{5.51}$$

If one assumes that during excitation of an electron from state $\phi_v^{k_v}$ to $\phi_c^{k_c}$ the distribution of the other electrons remains unchanged, i.e., no relaxation takes place (which is quite a reasonable assumption for an infinite system), and, furthermore, if the excited electron and the remaining hole are infinitely separated, then the excitation is a simple one-electron transition over the single-particle energy-band gap:

$$\Delta\varepsilon_g(\text{HF}) = \varepsilon_c^{k_c}(\text{HF}) - \varepsilon_v^{k_v}(\text{HF}) \tag{5.52}$$

Following the suggestion of Toyozawa,[18] who introduced the concept of the electronic polaron (an excess electron surrounded by a cloud of virtual longitudinal excitons) and whose idea was further developed by Kunz, Collins, and co-workers,[19-22] one can also retain the above picture if one goes beyond the HF model. By analogy to equations (5.50) and (5.51) one can define quasi-particle (QP) states using, instead of the HF total energies, the correlated energies from the preceding Section 5.2,[24]

$$\varepsilon_c^{k_c}(\text{QP}) = E^{(N+1)} - E^{(N)} \tag{5.53}$$

$$\varepsilon_v^{k_v}(\text{QP}) = E^{(N)} - E^{(N-1)} \tag{5.54}$$

By writing $E = E_{\text{HF}} + E_2$ for the total energy one obtains

$$\varepsilon_c^{k_c}(\text{QP}) = \varepsilon_c^{k_c}(\text{HF}) + E_2^{(N+1)} - E_2^{(N)} \tag{5.55}$$

$$\varepsilon_v^{k_v}(\text{QP}) = \varepsilon_v^{k_v}(\text{HF}) + E_2^{(N)} - E_2^{(N-1)} \tag{5.56}$$

These expressions can be interpreted by decomposing E_2 into the sum of independent pair correlations. We note that the application of the results obtained for the N-particle system in Section 5.2 involves a further approximation here, since the $(N + 1)$- and $(N - 1)$-particle states are not closed-shell configurations. Thus Brillouin's theorem does not exactly apply and therefore single-particle excitations should also be included in the perturbation theory. It can be shown,[19,23] however, that the contribution of these singly excited configurations is negligible compared to the doubly excited ones, which are included here. A further problem should be also mentioned, though it is a general problem in band-structure theory: the functions ϕ used to construct the $(N + 1)$- and $(N - 1)$-particle states are eigenfunctions of the N-particle Fock operator, respectively. In principle, one should again use an open-shell formalism for their determination,[24] but it is hoped that in infinite systems the error introduced by this approach is not serious. The overall success of band theory supports this view.

We employ the convention that, e.g., $E^{(N-1)V}$ denotes the correlation energy of the $(N-1)$-particle state, which is obtained by removing an electron from the valence-band state V. Hence equation (5.43) yields

$$E_2^{(N)} = \sum_I \sum_J{}' \varepsilon_{IJ}^{(N)} = \sum_{I \neq V} \sum_{J \neq V}{}' \varepsilon_{IJ}^{(N)} + \sum_{I \neq V} \varepsilon_{IV}^{(N)} \tag{5.57}$$

$$E_2^{(N-1)V} = \sum_{I \neq V} \sum_{J \neq V}{}' \varepsilon_{IJ}^{(N-1)V} \tag{5.58}$$

$$E_2^{(N+1)C} = \sum_{\substack{I \\ (C)}} \sum_{\substack{J \\ (C)}}{}' \varepsilon_{IJ}^{(N+1)C}$$

$$= \sum_I \sum_J{}' \varepsilon_{IJ}^{(N+1)C} + \sum_I \varepsilon_{IC}^{(N+1)C} \tag{5.59}$$

[The first two summations in equation (5.59) involve also the extra occupied conduction-band state C.] On substituting these expressions into equations (5.55) and (5.56) we obtain for the quasi-particle energies[24]

$$\varepsilon_C(\text{QP}) = \varepsilon_C(\text{HF}) + \sum_I \varepsilon_{IC}^{(N+1)C} + \sum_I \sum_J{}' (\varepsilon_{IJ}^{(N+1)C} - \varepsilon_{IJ}^{(N)}) \tag{5.60}$$

$$\varepsilon_V(\text{QP}) = \varepsilon_V(\text{HF}) + \sum_{I \neq V} \varepsilon_{IV}^{(N)} + \sum_{I \neq V} \sum_{J \neq V}{}' (\varepsilon_{IJ}^{(N)} - \varepsilon_{IJ}^{(N-1)V}) \tag{5.61}$$

Following the suggestion of Pantelides *et al.*,[23] one can interpret the correction terms appearing in addition to the HF band energies in these equations as electron and hole self-energies [Σ(e) and Σ(h), respectively]. For this purpose one can introduce the notation

$$\varepsilon_C(\text{QP}) = \varepsilon_C(\text{HF}) + \Sigma_C^{(N+1)}(\text{e}) + \Sigma_C^{(N+1)}(\text{h}) \tag{5.62}$$

$$\varepsilon_V(\text{QP}) = \varepsilon_V(\text{HF}) + \Sigma_V^{(N)}(\text{e}) + \Sigma_V^{(N)}(\text{h}) \tag{5.63}$$

Recalling Toyozawa's electronic-polaron model, one can identify the origin of these self-energy corrections (as noted above) as a cloud of virtual excitons dressing the "bare" HF particles (in complete analogy to the lattice polaron problem, where virtual optical phonons accompany the polarizing particle). Equation (5.44) indicates that each pair correlation between two HF particles consists of a sum of momentum-conserving virtual scattering pairs ($\phi_i^{\mathbf{k}_i} \to \phi_a^{\mathbf{k}_i+\mathbf{q}}$ and $\phi_j^{\mathbf{k}_j} \to \phi_b^{\mathbf{k}_j-\mathbf{q}}$, respectively).

With this physical model in mind one can visualize the formation of an electronic polaron according to equations (5.60) and (5.61) in two steps. Initially, the extra particle put into the HF conduction-band state $\phi_c^{\mathbf{k}_c}$ must establish its new "correlation bonds" with the other N particles present in the $(N+1)$-particle system $\Phi^{(N+1)C}$, giving rise to the electronic self-energy correction $\Sigma_C^{(N+1)}(\mathrm{e})$. At the same time, owing to the occupation of the previously empty state $\phi_c^{\mathbf{k}_c}$, the pair correlations $\varepsilon_{IJ}(I \neq C, J \neq C)$ are reduced in $\Phi_c^{\mathbf{k}_c}$, compared to $\Phi^{(N)}$, since scatterings to $\phi_c^{\mathbf{k}_c}$ are excluded. From the point of view of $\Phi^{(N+1)}$, the terms $\varepsilon_{IJ}^{(N+1)C} - \varepsilon_{IJ}^{(N)}$ in equation (5.60) destabilize the hole in $\phi_c^{\mathbf{k}_c}$ (before its occupation by the extra electron); their sum therefore results in a hole self-energy $\Sigma_C^{(N+1)}(\mathrm{h})$. The formal analysis of these self-energies can be made clearer if one denotes the contribution of each double excitation to the correlation energy by τ_{IJ}^{AB}, i.e., equation (5.36) is expressed in the form[24]

$$E_2 = \sum_I \sum_J \sum_A \sum_B{}' \tau_{IJ}^{AB} \tag{5.64}$$

In terms of these quantities τ the self-energies are obtained as

$$\Sigma_C^{(N+1)}(\mathrm{e}) = \sum_I \varepsilon_{IC}^{(N+1)C} = \sum_I \sum_{A \neq C} \sum_{B \neq C}{}' \tau_{IC}^{AB} \tag{5.65}$$

$$\Sigma_C^{(N+1)}(\mathrm{h}) = \sum_I \sum_J{}' \left[\sum_{A \neq C} \sum_{B \neq C}{}' \tau_{IJ}^{AB} - \sum_A \sum_B{}' \tau_{IJ}^{AB} \right] = -\sum_I \sum_J \sum_A{}' \tau_{IJ}^{AC} \tag{5.66}$$

Before explicitly calculating these quantities some of their qualitative trends can be observed.[27] For the bottom of the conduction band the denominators of τ_{IC}^{AB} in equation (5.65) are always negative (see Figure 5.5), therefore the electron self-energy shifts this state downward. Also, since the values of τ_{IJ}^{AB} in equation (5.66) are negative, the hole self-energy produces a positive (upward) shift. [We note that in the calculation of $\Sigma_C^{(N+1)}(\mathrm{h})$, C is regarded as an empty state that will be occupied only as a second step after the correlation stabilization of the corresponding "hole state."] Furthermore, one can expect that $|\Sigma_C^{(N+1)}(\mathrm{e})|$ will usually be larger than $|\Sigma_C^{(N+1)}(\mathrm{h})|$, since in the latter case both scattering events occur across the gap and therefore, as a net result, the bottom of the conduction band will move downward as a result of polaron formation.[24] It must be noted, however, that this consideration is very approximate, since in most cases the symmetry of the wave functions

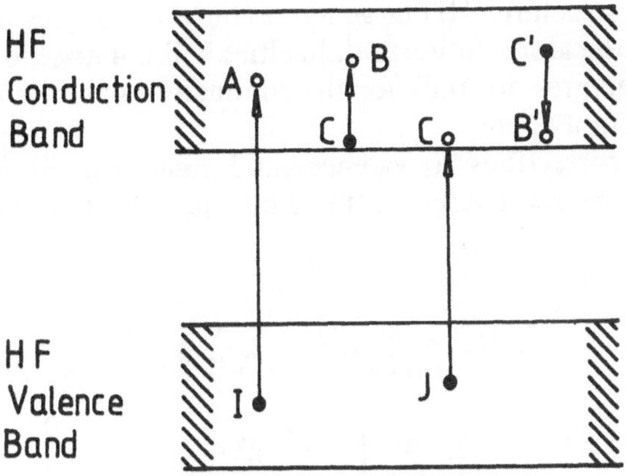

FIGURE 5.5. Virtual excitations contributing to the electron and hole self-energies [$\Sigma_C^{(N+1)}$(e) and $\Sigma_C^{(N+1)}$(h), respectively] in the $(N + 1)$-particle state.

entering the calculation of quantities τ (which can be different for different conduction bands) may strongly modify this picture or even reverse it (higher-lying conduction bands may have larger correlation contributions than the lowest conduction band, despite the larger denominators).

A further difficulty may arise if one calculates polaron states belonging to the upper part of the conduction band. It can be seen from Figure 5.5 that the self-energy correction that for state C' is larger than for state C, since the energy differences for the scatterings $I \rightarrow A$ and $C' \rightarrow B'$ have opposite signs [and at the same time the denominators of the $J \rightarrow C'$ scatterings increase in equation (5.66)]. This situation results in a stronger shift for the upper conduction-band states than for the lower states, i.e., in an effective band narrowing (in analogy to the Franck–Condon factor appearing in phonon-polaron theory). One must bear in mind, however, that if the width of the HF conduction band is equal to or larger than the gap, MP/2 breaks down for the top of the band and one must either sum to higher orders or choose another partitioning scheme. For the same reason, higher-lying empty bands cannot be corrected by this scheme either. It should be pointed out, however, as was noted in Section 5.2, that if one stops after the term $\frac{1}{2}\hat{T}_2\hat{T}_2$, where $\hat{T}_2$ is the excitation operator for all the double excitations, coupled-cluster theory[3] also allows partitioning of the total correlation energy into pair correlation contributions and thus in this case also one can define a quasi-particle band structure.[8] If one generates the self-energies with the aid of this method, the aformentioned difficulties will not arise; one can define QP band structures not only for the conduction band, but also for the higher-lying empty bands.

Polaron corrections to valence-band states can be computed by decomposing the self-energies defined by equations (5.61) and (5.63) in the form[24]

$$\Sigma_C^{(N)}(\mathrm{e}) = \sum_{I \neq V} \varepsilon_{IV}^{(N)} = \sum_{I \neq V} \sum_{A} \sum_{B}{}' \tau_{IV}^{AB} \tag{5.67}$$

$$\Sigma_V^{(N)}(\mathrm{h}) = \sum_{I \neq V} \sum_{J \neq V}{}' \left[\sum_{A} \sum_{B}{}' \tau_{IJ}^{AB} - \sum_{\substack{A \\ (V)}} \sum_{\substack{B \\ (V)}} \tau_{IJ}^{(AB)V} \right]$$

$$= -\sum_{I \neq V} \sum_{J \neq V} \sum_{A}{}' \tau_{IJ}^{AV} \tag{5.68}$$

By employing the same reasoning used above, one can conclude from Figure 5.6 that at the top of the valence band $\Sigma_V^{(N)}(\mathrm{e})$ produces a negative shift and $\Sigma_V^{(N)}(\mathrm{h})$ a positive shift; in this case the latter is generally larger in magnitude. The same caution as discussed above is again applicable when using Møller–Plesset partitioning.

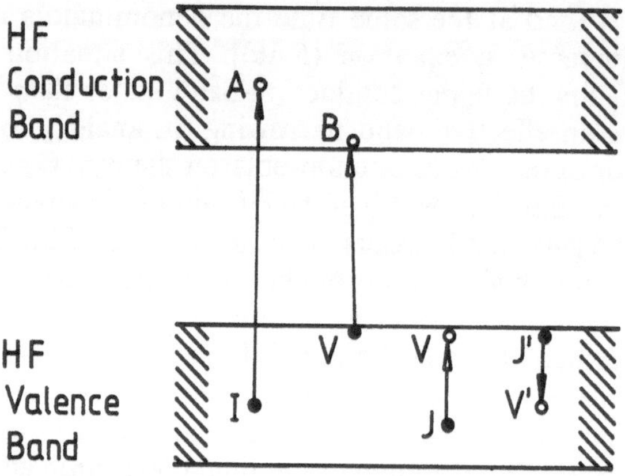

FIGURE 5.6. Virtual excitations contributing to the electron and hole self-energies in the $(N-1)$-particle state.

5.4. SELECTED CORRELATION ENERGY CALCULATIONS ON POLYMERS

5.4.1. Ground-State Energy of an Infinite Metallic Hydrogen Chain

In the case of an equidistant (metallic) H-atom chain, the BZ was divided into two parts in each of which the coefficients were expressed as a three-term trigonometric expansion [see equation (5.48) and the paper by Suhai und Ladik[26]].

Five different basis sets were used for the calculation at both the RHF and correlated levels, as shown in Tables 5.1 and 5.2, where the notation *ms*/*ns* means that on each hydrogen atom *m s*-type Gaussians are centered and are divided into *n* groups (contractions). The exponents and contraction coefficients of the first two sets were taken from the literature,[27] and the exponents of the six *s*-type Gaussians from another work.[28] In the fourth basis set these functions were supplemented by two *s*-type bond functions ($2\bar{s}$) centered in the middle between the hydrogen atoms, while in the fifth basis a set of three *p*-type atomic polarization functions (p_x, p_y, p_z) was used. The contraction coefficients and the exponents of the bond and polarization functions were optimized previously.

The quality of these basis sets can be demonstrated by the example of the H_2 molecule. The HF energy extrapolated to the complete basis set is $E_{HF} = -30.847$ eV at the optimal interatomic distance of $R = 1.386$ au.[29] Suppose, for instance, that the fifth basis set in Table 5.1 is used, then we obtain $E_{HF} = -30.814$ eV and $R = 1.398$ au. Application of second-order Møller–Plesset theory yields for hydrogen in this basis set a

TABLE 5.1. Some Physical Properties of the Infinite Metallic Hydrogen Chain Obtained by the RHF Method Using Different Atomic Basis Sets[a]

Basis set	E_{RHF}	E_{coh}	ε_F[b]	δE[c]	R[d]
$4s/1s$	−14.232	0.669	−1.384	55.153	1.782
$4s/2s$	−14.525	0.939	−3.690	42.513	1.839
$6s/3s$	−14.543	0.940	−3.646	45.135	1.843
$6s, 2\bar{s}/3s, 1\bar{s}$	−14.580	0.977	−3.574	44.863	1.826
$6s, 1p/3s, 1p$	−14.599	0.993	−3.536	50.587	1.834

[a] The total energy E_{RHF} and the cohesion energy E_{coh} are calculated per hydrogen atom in the infinite system; all energy values are given in eV.
[b] Position of the Fermi level.
[c] Width of the half-filled valence band.
[d] Interatomic distance in au optimized at the RHF level.

correlation energy of -0.756 eV as compared with the exact value of -1.110 eV.[30]

Table 5.1 presents the HF energy per H atom in the infinite chain at the optimized lattice constant R, as well as values of the Fermi level (ε_F), the width of the half-filled valence band ($\delta\varepsilon$), and the cohesion energy E_{coh} (defined as the difference between the free atomic energy and the lattice energy per cell in the infinite chain) at the HF level. The computations involved 64 atomic neighbors in the crystal, so that all infinite lattice sums converged properly for each basis set. We can also see that the cohesion energy is saturated with respect to the size of the atomic basis and lies somewhat above 1 eV/H atom. The equilibrium lattice constant is larger than the interatomic distance in the hydrogen molecule, as predicted earlier by cluster calculations.[31]

One of the most interesting ground-state properties of a crystal is the cohesion energy. Therefore, the central goal of this work was to calculate the correlation contribution to this quantity. The total energy per hydrogen atom in the infinite chain obtained by second-order perturbation theory using different atomic basis functions is given in the first column of Table 5.2. The second column shows the correlation energy E_{corr} obtained by this method. It is noteworthy that E_{corr} is not simply the difference between the corresponding values in the first columns of Tables 5.1 and 5.2, since the optimized H–H distance is slightly different in the HF and correlated cases. Its value in Table 5.2 is calculated by minimizing $E_{RHF} + E^{(2)}_{RHF,MP}$.

Several interesting conclusions can be drawn from these results:

1. The lattice expands to a small extent owing to correlation. In the case of a greater electron density, as in polymers constructed from heavy atoms, this effect is more pronounced.[32]

TABLE 5.2. Results of Correlation Energy Calculations for the Infinite Metallic Hydrogen Chain Using Second-Order Møller–Plesset Perturbation Theory and RHF Reference Functions[a]

Basis set	$E_{RHF} + E^{(2)}_{RHF,MP}$ [b]	E_{corr} [b]	E_{coh} [b]	R_{corr} [b,c]
$4s/1s$	-14.583	-0.351	1.020	1.810
$4s/2s$	-14.961	-0.418	1.375	1.866
$6s/3s$	-14.962	-0.428	1.358	1.852
$6s, \tilde{2}s/3s, \overline{1s}$	-15.090	-0.510	1.487	1.840
$6s, 1p/3s, 1p$	-15.223	-0.626	1.618	1.844

[a] All energy values are in eV.
[b] Calculated at the minimum of $E_{RHF} + E^{(2)}_{RHF,MP}$ with respect to R.
[c] Interatomic distance in au optimized at the correlated level.

2. The calculated value of the correlation energy is very sensitive to the size of the basis set. We can see that it has approximately converged at the level of the $6s/3s$ basis with respect to s-type atomic functions. The presence of polarization functions with nodes (p-type atomic functions) increases its value, however, by approximatley 50% ($= 0.2$ eV) although these functions improved the HF energy by only 0.05 eV. The performance of this method in the case of the H_2 molecule allows us to predict the correlation energy of the one-dimensional metallic hydrogen chain at about -0.9 eV/electron. The value of -0.626 eV/electron obtained by our method is larger than the CI results for a H_{14} ring obtained by Liskov *et al.*[33] ($E_{corr} = -0.32$ eV/electron) and Seel *et al.*[34] ($E_{corr} = -0.42$ eV/electron).
3. Since our HF energy is very close to the estimated HF limit, we expect that the overwhelming part of the energy missing from the computed cohesion is due to missing correlation. Thus adding about 0.3 eV to the calculated value of 1.618 eV, we can say that the HF theory predicts only about half of the cohesion in metallic hydrogen, which can be estimated to be around 1.9 eV/atom. It must be stressed, however, that the contribution of higher-order terms may be different for an infinite system; therefore, these will have to be investigated carefully in the future.

5.4.2. The Quasi-Particle Energy Gap of Alternating *trans*-Polyacetylene

It is known experimentally that polyacetylene (PA) is more stable if it has alternating *trans*-chains (see Figure 5.7) than chains in the *cis*-conformation. On the other hand, owing to the lack of single crystals and to the fibrous morphology of PA, the detailed geometry of the chain (the magnitude of bond alternation, individual bond lengths and bond angles) is not known precisely.[35]

The absence of precise experimental structural parameters of PA gives added interest to theoretical calculations optimizing single-chain bond lengths and bond angles both at the *ab initio* HF level (using different levels of sophistication for the basis set) and also with inclusion of the major part of correlation. The results show that the alternating *trans*-form of *trans*-PA is always the most stable[36–38] independent of the basis set applied, and inclusion of interchain interactions and correlation effects does not change this conclusion.[39]

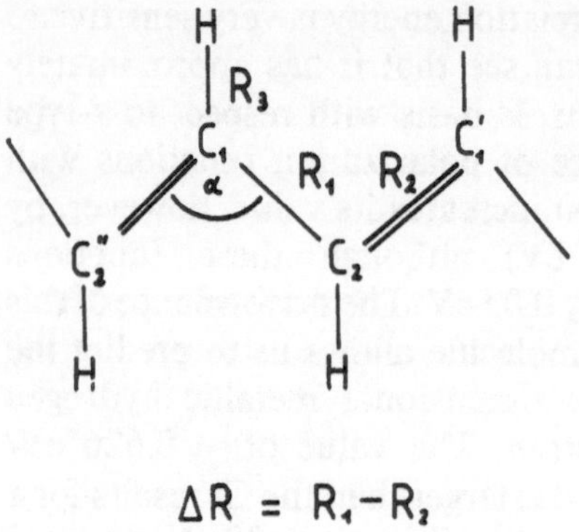

FIGURE 5.7. Geometrical structure of alternating *trans*-PA showing the different geometrical variables.

MP/2 correlation calculations with five different basis sets were used for the ground state; the specification of the basis sets and references are given in Table 5.3. For each basis, not only the bond alternation but also the values of all four geometrical variables have been optimized both at the HF and at the correlated level.[43]

One can see that the value of ΔR converges well at both levels of approximation with respect to the basis-set extension, and its correlated value (if projected on the polymer axis) is in good agreement with experiment.[35] We observe at the same time that, though E_{HF} has converged reasonably, a considerable part of E_2 is still missing. This is understandable physically since our *spd* sets still do not possess enough oscillations for the electrons to be able to avoid each other efficiently. For

TABLE 5.3. Optimized Single–Double Bond-Length Difference ΔR in *trans*-PA as Obtained with the HF Approximation and Using MP/2 ($E = E_{HF} + E_2$)[24][a]

Basis set	ΔR_{HF} (Å)	E_{HF} (au)	ΔR_E (Å)	E (au)
a: STO-3G[40]	0.156	− 75.947121	0.122	− 76.065389
b: 4-31G[41][b]	0.112	− 76.776843	0.091	− 76.955733
c: 6-31G[41][c]	0.109	− 76.859247	0.085	− 77.036179
d: 6-31G*[42][d]	0.103	− 76.886912	0.084	− 77.137486
e: 6-31G**[42][e]	0.103	− 76.892675	0.084	− 77.168100

[a] The total energies per elementary C_2H_2 units, E_{HF} and E belong to the optimum bond alternation for a given basis set.
[b] Double-ζ for valence shell.
[c] Complete double-ζ.
[d] Polarization (d) functions only on the carbon.
[e] Polarization functions on both carbon and hydrogen atoms.

this purpose further basis functions with higher angular momenta would be required which could increase (through their nodes) the kinetic energy, and thus result in a larger correlation contribution by virtue of the virial theorem.

One would like to know, of course, the percent of the full valence-shell correlation energy included in E_2 with the best *spd* basis used. One can obtain an approximate answer to this question if we recall that the valence-shell correlation energy of an acetylene unit was estimated to be about -10 eV[44]; therefore our best energy (~ 7.5 eV) should cover 70 to 75% of the total value. Nearly the same result was also obtained recently for an infinite atomic-hydrogen model chain (see Table 5.2).

Since various π-electron Hamiltonians have already been applied very extensively to the polyene problem,[45] it is also of interest to ask what part of the correlation originates from purely π-electron interactions. (Both the highest filled and lowest unfilled bands have π symmetry. This question is therefore decisive for doping and conductive properties.) Owing to symmetry, the σ- and π-type Bloch functions can be completely separated, so we can evaluate the $\pi - \pi$, $\sigma - \sigma$, and $\pi - \sigma$ contributions to E_2 individually. Almost independently of the atomic basis set, we found that $E_2^{\pi-\pi}$ contributes only 15 to 20%, showing that for this kind of polymer the whole valence shell must be treated as an entity; a simple model separating bands with π-type symmetry would not work.

The most important part of Suhai's calculations on the correlation energy of *trans*-PA[24] is the evaluation of the self-energy corrections to the HF band energies using equations (5.60) and (5.61). These quantities are collected in Table 5.4 for basis sets a, c, and e. (The results obtained for b and d are very similar to c and e, respectively, therefore they are not shown here.) The values of Σ_C are calculated at the bottom of the lowest conduction band, while the values of Σ_V refer to the top of the highest valence band. The shifts for both bands are seen to result from a positive and a negative term, but as a net effect the conduction band is shifted downward and the valence band upward.

We can follow the formation of the polyene energy-band gap in four consecutive steps, as shown in Figure 5.8. The uppermost curve is obtained at the HF level using the fixed geometry obtained for the minimal basis (a). In this case, the extension of the atomic basis does not influence the value of $\Delta\varepsilon_g$(HF) significantly. On the other hand, the value is reduced by about 4 eV if the bond alternation is optimized (at the HF level) with each basis set (second curve from the top). An approximately 0.7 eV further reduction is obtained if the structural optimization is performed with the correlated total energy (third curve from the top). The lowest curve shows the energy-band gap values after polaron forma-

TABLE 5.4. Different Physical Quantities Contributing to the Formation of the Quasi-Particle Energy-Band Gap in the Alternating *trans* PA: One-Particle Energies ε(HF), Electron and Hole Self-Energies, Σ(e) and Σ(h), Respectively, and Quasi-Particle Energies ε(QP)[24][a]

Quantity	STO-2G	6-31G	6-31G**
$\varepsilon_{cond-min}$(HF)	3.719	−0.806	−1.322
$\Sigma^{(N+1)}_{cond\ min}$(h)	0.112	0.193	0.274
$\Sigma^{(N+1)}_{cond\ min}$(e)	−0.452	−0.623	−0.938
$\varepsilon_{cond-min}$(QP)	3.379	−1.236	−1.996
$\varepsilon_{val-max}$(HF)	−4.563	−5.732	−5.749
$\Sigma^{(N)}_{val-max}$(h)	0.583	0.896	1.497
$\Sigma^{(N)}_{val-max}$(e)	−0.312	−0.436	−0.724
$\varepsilon_{val\ max}$(QP)	−4.293	−5.361	−4.976
Gap(HF)	8.282	4.926	4.427
Gap(QP)	7.672	4.125	2.980

[a] All quantities are given in eV.

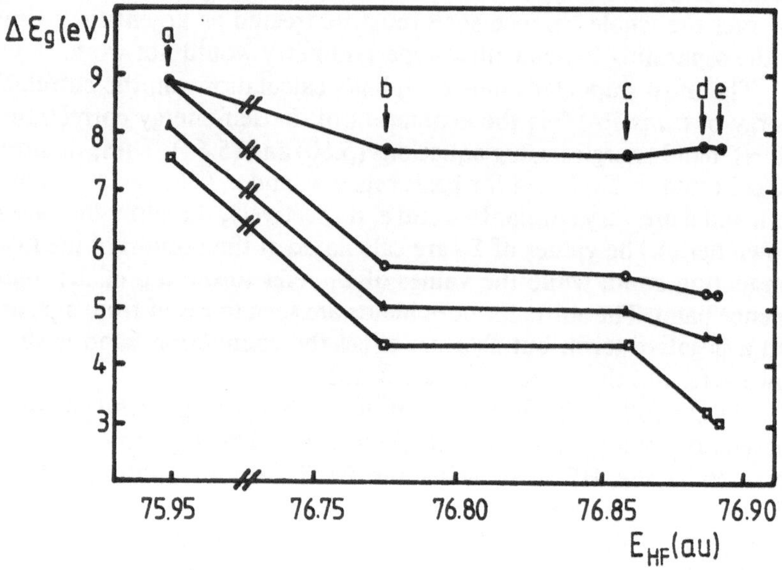

FIGURE 5.8. Energy-band gap of *trans*-PA vs. the HF energy per C_2H_2 unit obtained with the five different basis sets (a–e) defined in Table 5.3. ● denotes HF calculation with fixed bond alternation, ○ HF calculation with optimized bond alternation, △ HF + MP/2 calculation with optimized bond alternation but using $\Delta\varepsilon_g = \varepsilon_{cond\ min}(HF) - \varepsilon_{val\ max}(HF)$, and □ the electronic polaron model $\Delta\varepsilon_g = \varepsilon_{cond\ min}(QP) - \varepsilon_{val\ max}(QP)$.[24]

tion [$\Delta\varepsilon_g$(QP)]. It can be seen here that the amount of correlation included at this stage plays a rather important role: the polaron effect is more than twofold greater in the case of the extended *spd* basis than for the minimal one. The best value obtained with this method is $\Delta\varepsilon_g = 2.98$ eV. For the ionization potential we obtain at the same time 4.976 eV (if one applies Koopmans' theorem[(17)]), which is in reasonable agreement with the experimental value of 4.6 eV.[(46)]

Since even the applied best wave function contains only 75% of the total valence-shell correlation, it is interesting to extrapolate the obtained value of $\Delta\varepsilon_g$(QP) to the case of full correlation. Suhai did this in Figure 6 of his recent paper[(24)] and obtained in this way a gap value of about 2.5 eV for 100% correlation energy, showing that the estimated theoretical value of $\Delta\varepsilon_g$(QP) would lie at about 2.5 eV, i.e., about 0.5 eV higher than the position of the first peak in the absorption spectrum of pure *trans*-PA.[(46)] The tail of the experimental spectrum reaches 1.4–1.5 eV, probably because of the structural disorder present in PA. This tail is, of course, not represented in the above, fully periodic calculation, for which the one-dimensional density-of-state curve has a sharp maximum at the band edges. The theoretical result must therefore be compared with the position of the corresponding peak in experiment, which lies at about 2 eV.[(46)]

The *spd* basis employed is still not altogether satisfactory when approaching 100% correlation. Further polarization functions as well as higher orders of the perturbation theory are needed to cover the missing 25–30% of correlation energy in this system. Since the short- and long-range parts of the correlation are to be equally important,[(24)] high-lying unfilled bands must, without doubt, also be included in such calculations on solids.

One can think of two reasons to account for the remaining approximately 0.5 eV discrepancy between theory and experiment. On the one hand, other polarization processes, such as high-frequency lattice vibrations (phonon polaron), should be also included in the calculations; on the other hand, in the work described the ($N \pm 1$) states were constructed from the HF N-particle ground state, which of course is an approximation. One should actually use SCF open-shell crystal orbitals (COs for periodic chains or crystals in which the unit cell is an open-shell system) for these ionized states. The formalism to generate open-shell wave functions for crystals or periodic chains was developed more than 10 years ago[(47)] but has not been applied until now.

Nevertheless, the calculation described above for the quasi-particle gap in alternating *trans*-polyacetylene proves that one can start from a high-quality HF wave function for infinite systems as well (in contrast to the opinion shared by some solid-state physicists) and by taking into account the major part of the correlation can achieve reasonably good agreement with experiment. The same is also true for the excitonic spectra and mechanical properties of polymers (see Part II).

5.4.3. Correlation Energy and Quasi-Particle Gap in a Cytosine Stack

In another calculation a cytosine stack was investigated.[48] The superimposed nucleotide bases were kept in the same relative position (3.36 Å stacking distance, 36° rotation) as in DNA-B.[49] A Gaussian 4-31G basis set was used for the calculation.[41] In other words a double-ζ basis was employed for the valence electron by contracting all the four Gaussians for the $1s$ function of carbon, but contracting them into a $3+1$ group in the case of the $2s$ and $2p$ functions of a carbon and into a $3+1$ s-type function for hydrogen. Subsequently, a MP/2 correlation calculation was again performed with this basis.[48]

In this way $E_{HF/N} = -392.036284$ H was obtained for the infinite polymer. The valence-shell correlation energy per monomer is -0.663012 H (~ 18.5 mH per valence electron), which from previous experience indicates that about half of the estimated full correlation energy can be achieved with the given method and basis set.[50]

Turning now to the quasi-particle energies in Table 5.5, the values of the correction terms obtained at the extrema of the conduction and valence bands are presented. The self-energies are seen to follow the previously discussed general trends and, according to equations (5.60) and (5.61), together shift the conduction band to lower energies and the valence band to higher energies. We also observe that the shift is

TABLE 5.5. Single-Particle Energies (ε[HF]) and Self-Energy Correlations (Σ) Contributing to the Quasi-Particle Energies (el pol) in the Valence and Conduction Bands of PolyC[41][a]

	Valence band	Conduction band	Gap
ε_{max} (HF)	-8.203	3.084	10.499
ε_{min} (HF)	-9.039	2.296	
Σ(e) at band max	-0.574	-1.470	
Σ(e) at band min	-0.631	-1.461	
Σ(h) at band max	1.846	0.203	
Σ(h) at band min	2.007	0.254	
ε_{max} (el pol)	-6.931	1.817	
ε_{min} (el pol)	-7.663	1.089	8.020

[a] Results obtained using a 4-31G atomic basis set and second-order Møller–Plesset/2 perturbation theory; energies are in eV.

somewhat larger at the top than at the bottom of the conduction band, leading to an effective band narrowing. At the same time, the Σ terms produce a stronger shift at the bottom of the valence band than at its top. (This situation is again analogous to the Franck–Condon effect in phonon theory.) The HF conduction band is shifted in this way by about 1.2 eV downward and the valence band by about 1.3 eV upward, reducing the fundamental gap from approximately 10.5 to 8 eV.

One can conclude from this calculation that though only a medium-level basis was used the correlation correction to the gap is rather large (~ 2.5 eV), essentially larger than in PA where, as was seen in the case of the best applied basis (6-31 G**), it was only about 1.45 eV; see Table 5.4. This is most probably due to the fact that a cytosine molecule, with eight non-H atoms, is much larger than a C_2H_2 unit, with only two non-H atoms. Therefore one would expect that by using the same refined basis, as a result of both basis set effects and 70–75% of the correlation, one would obtain a gap between 6.0 and 7.0 eV. If this gap were to be extrapolated again to 100% correlation, one would most probably end up with a gap of about 6.0 eV. This value is not far from the values 5.0–5.5 eV that one can expect on the basis of the exciton spectra of the nucleotide base stacks. In other words the lower edge of the first singlet exciton band lies at approximately 4.3 eV and has a width of about 0.6 eV.[51,52] When in the future still larger computers will be available with which to undertake these very-large-scale calculations with allowance for phonon polaron and relaxation effects (see the previous section), one should expect rather good agreement between the theoretical and (until that time, hopefully, directly measured) experimental gap of the different nucleotide base stacks (at least in DNA B).

5.5. CORRELATION IN POLYMERS WITH LARGE UNIT CELLS

If the unit cell in a periodic polymer is large (like the nucleotide bases in their stacks), any treatment of their correlation becomes rather difficult because there are no good methods to treat electronic correlation in a larger single molecule (with more than 50 electrons). One can approach the problem only by working out methods to handle larger single molecules, which most probably can be achieved only by a combination of different methods, and then repeat periodically the single molecule under consideration.

As a first step in this direction, an investigation of the localization properties of the filled and virtual (both σ and π) orbitals of the four nucleotide bases has been performed.[53] It has been found that the filled *ab initio* canonical HF orbitals of cytosine, uracil, adenine, and guanine can be very well localized with the aid of both the Boys (B)[54] and Edminston–Ruedenberg (ER)[55] procedures. Since the two localization procedures have given almost identical results and the B method is much faster, the latter has been applied in most cases. This means that the integral

$$I_1 = \sum_{i=1}^{n^*} \langle \varphi_i \varphi_i \,|\, r_{12} \,|\, \varphi_i \varphi_i \rangle \tag{5.69}$$

was minimized[54] (n^* denotes the highest filled, doubly occupied canonical HF orbital of a closed-shell system) instead of maximizing the expression

$$I_2 = \sum_{i=1}^{n^*} \langle \varphi_i \varphi_i \,|\, 1/r_{12} \,|\, \varphi_i \varphi_i \rangle \tag{5.70}$$

as carried out by the ER method.[55]

The localization was very good (leading to one-center lone pairs and two-center localized orbitals even for π electrons) in both cases if one applies a σ–π separation or treats all electrons together. A rather good localization for the virtuals has also been obtained (again, even for the π electrons) with essentrally two-center orbitals with a small component at a third and fourth center in some cases.[53]

These results open the way to possibly reducing drastically the number of excitations to be taken into account in any size-consistent method (such as MBPT[2] or the coupled-cluster method[3]). In other words, one can classify the excitations as excitations within the same localization region, excitations between neighboring regions, and so on, and take into account only excitations within the same region and between neighboring (eventually, between second-neighbor) regions. For the definition of a localization region not on a purely intuitive basis, Pipek[56] has given a more precise criterion based essentially on the overlap of the localized orbitals.

As the next step in handling the correlation problem in larger molecules, the coupled-cluster doubles (CCD), also called coupled-pair many-electron theory (CPMET), introduced into the electronic correlation problem by Čížek,[3] was first applied to smaller molecules[10] and then to the nucleotide bases[57] using localized orbitals.

In the coupled-cluster theory the exponential ansatz

$$|\psi\rangle = e^{\hat{T}} \,|\phi\rangle \tag{5.71}$$

is used, where $|\psi\rangle$ is the exact wave function, which is the solution of the Schrödinger equation

$$\hat{H}|\psi\rangle = E|\psi\rangle \tag{5.72}$$

Further, $|\phi\rangle$ is the HF ground-state Slater determinant, and $\hat{T}$ is an excitation operator given by

$$\hat{T} = \sum_{j=1}^{n} \hat{T}_j \tag{5.73}$$

where n is the number of electrons in the system and $\hat{T}_j$ denotes the excitation operator of all j-fold excitations. Form (5.71) of the wave function was introduced in nuclear physics by Coester and Kümmel,[58] and its use in molecular physics (without working out the details) was first suggested by Sinanoğlu.[59] By the Brillouin theorem, the operator $\hat{T}_1$ in equation (5.73) is neglected, as are higher excitation operators $\hat{T}_i$ $(i \geqslant 3)$ in CCD.[3] However, it should be noted that through the use of $\exp(\hat{T}_2)$ in equation (5.71) the major part of the fourfold excitations is included in the CCD approximation, because of the term $\frac{1}{2}\hat{T}_2\hat{T}_2$ in the Taylor expansion of $\exp(\hat{T}_2)$. So the CCD wave function is given by

$$|\psi\rangle = \sum_{\nu=0}^{\infty} \frac{1}{\nu!} \hat{T}_2^{\nu} |\phi\rangle \tag{5.74}$$

Substitution into equation (5.72) yields

$$(\hat{H} - E)(1 + \hat{T}_2 + \tfrac{1}{2}\hat{T}_2\hat{T}_2 + \cdots)|\phi\rangle = 0 \tag{5.75}$$

which gives the correlation energy E_c projected on the HF ground state as $E_{HF} - E + \langle\phi|\hat{H}\hat{T}_2|\phi\rangle = 0$, or

$$E_c = \langle\phi|\hat{H}\hat{T}_2|\phi\rangle \tag{5.76}$$

Projection of equation (5.75) on the space of doubly excited Slater determinants $|\phi_{ij}^{rs}\rangle$, where indices i, j denote occupied MOs and r, s denote virtual MOs, yields

$$\langle\phi_{ij}^{rs}|(\hat{H} - E)(1 + \hat{T}_2 + \tfrac{1}{2}\hat{T}_2\hat{T}_2 + \cdots)|\phi\rangle = 0 \tag{5.77}$$

If the MOs contained in the Slater determinants occurring in system (5.77) are assumed to be orthogonal, then all terms $(1/\mu!)(\hat{T}_2)^{\mu}$ $(\mu \geqslant 3)$

vanish exactly and the final system of equations for the matrix elements of $\hat{T}_2$ is given by

$$\langle \phi_{ij}^{rs} | \hat{H}(1 + \hat{T}_2 + \tfrac{1}{2}\hat{T}_2\hat{T}_2) | \phi \rangle - \langle \phi_{ij}^{rs} | \hat{T}_2 | \phi \rangle E = 0 \qquad (5.78)$$

Evaluation of equations (5.76) and (5.78) via diagrammatic[3] or conventional quantum-mechanical methods[59] results in the explicit spin-independent CCD equations, given elsewhere.[60] The definitions

$$V_\alpha \equiv \langle rs | ij \rangle \quad \text{and} \quad T_\alpha \equiv T_{ij}^{rs} \equiv \langle rs | \hat{T}_2 | ij \rangle \qquad (5.79)$$

where α is a combined index, enable the CCD equations to be expressed in a condensed form:

$$0 = V_\alpha + \sum_\beta A_{\alpha\beta} T_\beta + \sum_{\beta,\beta'} B_{\alpha\beta\beta'} T_\beta T_{\beta'} \qquad (5.80)$$

where ϕ, β, β' run over all double excitations, and **A** and **B** are matrices composed from two-electron integrals and Fock matrix elements according to equation (60) in Förner *et al.*[60]

Starting from a guess vector for **T**, which corresponds to many-body second-order perturbation theory [MBPT(2)], one can solve equation (5.80) in an iterative way. At an iteration step m, we obtain a better approximation $T_\alpha^{(m)}$ for each α,

$$T_\alpha^{(m)} = -\frac{1}{A_{\alpha\alpha}} \left(V_\alpha + \sum_{\beta,\beta'} B_{\alpha\beta\beta'} T_\beta^{(m-1)} T_{\beta'}^{(m-1)} + {\sum_\beta}' A_{\alpha\beta} T^{(k)} \right) \qquad (5.81)$$

where the prime on the second sum indicates that β runs over all excitations except $\beta = \alpha$, while $k = m$ for $\beta < \alpha$ and $k = m - 1$ for $\beta > \alpha$. Matrices **A** and **B** are not stored explicitly, but are computed directly from the two-electron integrals and the Fock matrix elements.

The four-index transformation of the two-electron integrals from the AO to the MO basis was carried out by a formalism given in equation (5.61), which divides the procedure, which is of $O(v^8)$ when v is the number of basis functions, into four subsequent $O(v^5)$ summations. Similarly, the nonlinear part in equation (5.80), which is of $O(N_o^4 N_v^4)$ when N_o is the number of occupied and N_v the number of virtual MOs, can be computed by two summations of $O(N_o^3 N_v^3)$ and $O(N_o^4 N_v^2)$, and two sums of $O(N_o^3 N_v^2)$ and $O(N_o^2 N_v^3)$ (further details are given elsewhere[10]). In the computations one obtains from the first guess the MBPT(2) result, then one performs the iteration on the linear part of equation (5.80)

(CCL), and finally one solves the full nonlinear CCD system. The CCD problem can be solved routinely on a CYBER-845 computer up to about 30,000 excitations with the program in Erlangen.

The SCF problem was solved with the GAUSS 76 program[42] and some of the basis sets used are from the basis-set library of that program. In general, the HF virtual MO space was used for the calculations, but in the cases of minimal STO-3G and valence split 6-31G bases[42] also oscillatory orbitals, proposed by Boys,[9] were constructed as an alternative to the HF virtual space. These orbitals are computed by multiplying each occupied orbital φ_i by coordinates x_i, y_i, and z_i located at the center of charge of φ_i and pointing in the direction of the principal axis of the quadrupole moment tensor of the charge distribution represented by φ_i. To expand these orbitals in an LCAO form, the basis space used for the expansion of φ_i must be enlarged by a set of p-functions for each s-type AO and by a set of six d-functions for each p-type AO. The resulting orbitals are orthogonalized to the occupied ones by using Schmidt's procedure, and among themselves by Löwdin's symmetric orthogonalization procedure. Then the sum of diagonal overlap between the original nonorthogonal set and the orthogonal set is maximized to obtain the final set of oscillatory orbitals.

The CCD method was implemented on an *ab initio* basis, making possible the use of localized orbitals as well. Calculations of the potential curves of H_2, LiH, and NH_3 in different basis sets show that the method can compete with other schemes for the calculation of the correlation energy. We found that in valence split-type basis sets generally about 50%, and if polarization functions are included, about 75% of the total valence correlation energy can be obtained, somewhat more than with MP/2 as shown by the detailed results.[10] The core correlation is described only to a negligible degree with these types of basis sets, and for a better description, more s-type and p-type functions with exponents corresponding to the core region must be included. However, bond lengths, the stretching force constant of LiH, and the inversion potential of NH_3 are described in fair agreement with experimental data with the help of valence split-basis sets, including polarization functions.

The introduction of oscillatory orbitals as an alternative to the virtual HF MO space leads to a considerable increase in the computed correlation energies and also takes into account parts of the core correlation energy. However, if oscillatory orbitals are used, starting from a minimal basis, the improvement in the total energy is negligible compared even to HF energies obtained with extended basis sets, because of the poor HF energies in the minimal basis. In addition, the LCAO expansion of the oscillatory orbitals involves a large atomic basis set, which leads to computational difficulties. So there is some doubt that oscillatory orbitals are really useful for correlation calculations on large molecular systems.

Calculations on formaldehyde in the extended 6-31G basis using localized orbitals (LO) show that even in this small molecule the CH bonds are well separated from the O lone pair orbitals, giving only small off-diagonal pair correlation energies. Also, a set of virtual LOs that is considerably more extended than the rest of the virtual space contributes only a small amount to the correlation energy. This calculation provides about 90% of the correlation energy, if one takes into account only excitations within the subregions or between neighboring regions as compared to the full calculation with all the excitations.[10]

The next step was to use the method for a metallic and molecular hydrogen ring and for a medium-size molecule (1-oxy-3-aza-butadiene), which was divided into three subunits as shown in Figure 5.9. Both for metallic and molecular hydrogen rings $4n + 2$ ($n = 0, 1, 2, 3, 4$, and 5) H atoms were taken into account.[57] For the molecular hydrogen rings and for the medium-size molecule the application of LOs with only first-neighbor excitations again provided about 90% correlation energy in the given basis set as compared to the full calculation. On the other hand, in the case of metallic hydrogen rings, as expected, one cannot use LOs to reduce the number of excitations but must take into account all the excitations.[57]

Finally, the CCD with the LO procedure was also applied to the four nucleotide bases: cytosine, thymine, adenine, and guanine. Since these are already rather large molecules a STO-3G basis was applied with the exception of cytosine, for which the valence split 4-31G basis was also used.

Table 5.6 presents the results obtained from cytosine with both basis sets using LOs. In the intersystem contribution part, AB, AC, etc. indicate excitations between neighboring regions (see Figure 5.10). Table 5.6 lists not only data on CCD and its linear version (CCL), but also MP/2

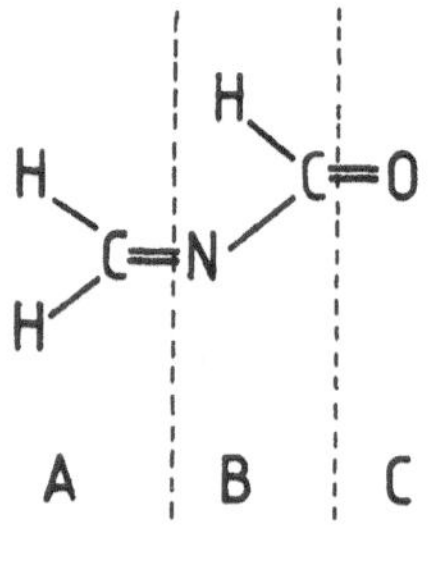

FIGURE 5.9. The chemical formula of 1-oxy-3-aza-butadiene.

perturbation theoretical results when the last nonlinear term in equation (5.80) is neglected. One can see that the excitations within the regions provide the major part of the correlation, but that excitations to neighboring regions also make nonnegligible contributions. A calculation on

TABLE 5.6. The Intra- and Intersystem Contributions to the Correlation Energy of Cytosine, Calculated by the MP/2, CCL, and CCD Methods Using Localized Orbitals in STO-3G and Valence Split 4-31G Basis[57][a]

	Intrasystem contributions[b]				Intersystem contributions		
	MP/2	CCL	CCD		MP/2	CCL	CCD
STO-3G Basis							
A	37.90	50.78	50.32	AB	23.61	25.88	25.10
B	44.05	59.29	58.49	AC	0.93	1.20	1.16
C	59.64	84.28	82.41	AD	0.06	0.09	0.08
D	33.44	44.91	44.59	AE	1.17	1.87	1.78
E	62.25	90.36	88.62	BC	19.64	26.26	24.51
				BD	2.63	3.09	3.01
Sum	237.28	329.62	324.43	BE	18.75	26.68	24.66
				CD	24.68	27.17	26.51
				CE	1.03	2.03	1.75
				DE	23.07	27.10	25.96
				Sum	115.57	141.37	134.52
				Total	352.85	470.99	458.95
4-31G Basis							
A	27.95	37.24	36.98	AB	45.93	50.24	50.04
B	29.11	38.04	37.80	AC	1.69	2.03	2.00
C	64.35	86.57	85.60	AD	54.44	58.65	58.19
D	40.28	50.82	50.42	AE	1.53	1.88	1.85
E	86.18	103.31	102.86	AF	35.05	41.78	41.12
F	106.34	115.55	115.29	BC	32.03	39.10	37.67
				BD	1.61	2.15	2.10
Sum	344.21	431.53	428.95	BE	0.13	0.18	0.16
				BF	4.32	5.10	4.99
				CD	40.93	48.38	46.77
				CE	1.76	2.40	2.31
				CF	0.86	1.28	1.21
				DE	33.98	38.57	37.63
				DF	5.44	7.95	7.63
				EF	0.36	0.39	0.38
				Sum	260.06	300.08	294.05
				Total	614.27	731.61	723.00

[a] All energies are in −mH.
[b] For the division of C into subregions using the STO-3G basis see Figure 5.10. For a different (more refined) division of C into subregions in the case of the 6-31G basis see Figure 5.11.

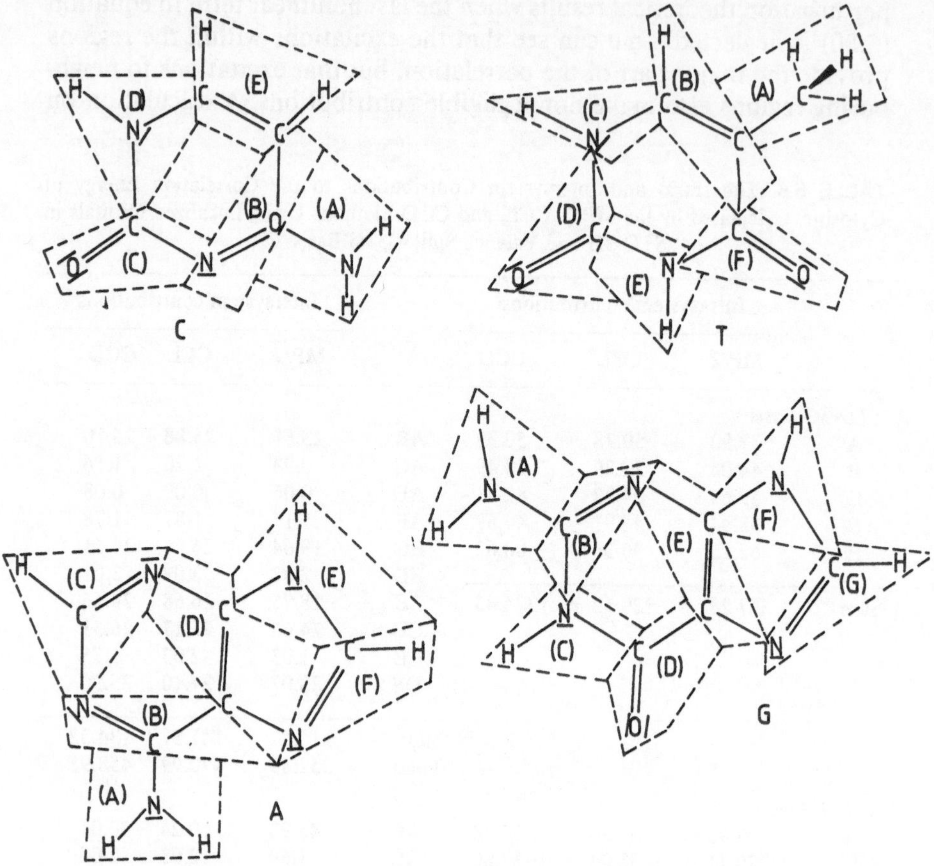

FIGURE 5.10. Division of the four nucleotide bases, cytosine (C), thymine (T), adenine (A), and guanine (G), into different subregions (after Förner *et al.*[57]).

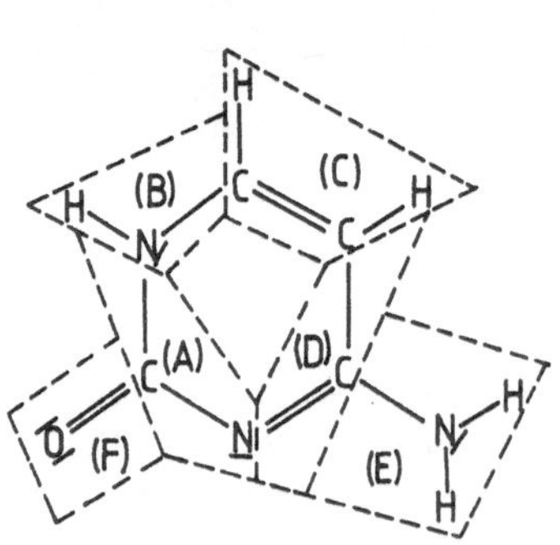

FIGURE 5.11. Subdivision of cytosine when using a valence split 6-31G basis set (after Förner *et al.*[57]).

the π system of cytosine has shown that excitations to second and third neighbors contribute less than 10% of the total π electron correlation obtained by taking into account all the excitations,[57] as was also found in the case of smaller molecules (see above). Therefore, one can be rather confident that by using LOs and restricting the excitations only to neighboring regions no more than 10% of the correlation energy in a given basis is lost by applying a given method to larger molecules, where no direct checks are possible.

It is also seen from Table 5.6 that CCD gives about 20–25% more correlation energy than MP/2. The fact that CCL generally gives more correlation energy than CCD is an artifact of the method that can occur because the coupled-cluster method is not based on variation theory (like CI) and therefore does not necessarily provide an upper bound for the energy.

Finally, Table 5.6 shows that the better valence split 4-31G basis provides about 60% more correlation energy than the STO-3G basis (actually, the factor $E_{corr}^{STO\text{-}3G}/E_{corr}^{6\text{-}31G}$ for CCDs in the case of cytosine is 0.635).

The contribution of excitations within the regions and between neighboring regions was also investigated in the cases of T, A, and G (with the STO-3G basis, the division of these molecules is shown in Figure 5.10 and again the MP/2, CCL, and CCD methods were applied). The results show the same general picture as described above in some detail for C.[57]

Finally, Table 5.7 presents STO-3G and valence split 4-31G basis correlation calculation results for the four nucleotide bases. The results to be expected in the case of the 4-31G basis were estimated by multiplying the corresponding numbers by factors $E_{corr}^{STO\text{-}3G}/E_{corr}^{6\text{-}31G}$ obtained for cytosine in the different methods. If one uses the valence split 4-31G basis or estimates the results, it is seen that the correlation energy per valence electron provides about 50% of the valence shell correlation using the CCD method with LOs. (One usually estimates the total correlation energy as 1 eV/electron, although this value may actually be too high in the case of valence electrons.)

We hope that in the near future, with the advent of yet larger computers, these types of calculations for the nucleotide bases can be repeated with a much better 6-31G** (double-ζ + polarization functions) basis, but again applying LOs and taking into account only excitations to neighboring regions. Hence one should be able to obtain about 70% of the valence electron correlation.

Next, one could use the LOs obtained from the canonical HF orbitals (approximated by a good basis) to construct Bloch orbitals of the periodic

TABLE 5.7. HF Total Energies (in − H), MP/2, CCL, and CCD Correlation Energies (in − mH; in − eV/valence electron in parentheses), and (HF + CCD) Total Energies (in − H) for the Nucleotide Bases in STO-3G and 4-31G Basis[57]

System	HF	MP/2	CCL	CCD	(HF + CCD)
STO-3G Basis					
Cytosine	387.51563	352.85 (0.23)	470.99 (0.31)	458.95 (0.30)	387.91458
Thymine	445.62393	394.09 (0.22)	544.23 (0.31)	528.43 (0.30)	446.15236
Adenine	458.59269	440.99 (0.24)	582.84 (0.32)	568.61 (0.31)	459.16130
Guanine	532.42247	473.66 (0.23)	628.47 (0.31)	612.49 (0.30)	533.03496
Valence Split 4-31G Basis					
Cytosine	392.42518	614.27 (0.40)	731.61 (0.47)	723.00 (0.47)	393.14818
Thymine[a]	451.27161	686.06 (0.39)	845.38 (0.48)	832.45 (0.47)	452.10406
Adenine[a]	464.28044	767.71 (0.42)	905.35 (0.49)	895.75 (0.49)	465.17619
Guanine[a]	539.11936	824.59 (0.40)	976.23 (0.47)	964.87 (0.47)	540.08423

[a] Correlation energies estimated with the help of the factors:

(STO-3G/4-31G) = 0.57442 (MP/2) derived from cytosine,
0.64377 (CCL) derived from cytosine,
0.63479 (CCD) derived from cytosine.

nucleotide base stacks and use them for MP/2 or CCD calculations in the case of an infinite stack. Here, however, the basis should most probably be supplemented by *p* functions with very small orbital exponents in order to better describe the space between the stacked bases. In this type of calculation with Bloch functions constructed from LOs, one could restrict the excitations within a molecule in the same way as was done for a single molecule,[57] and apply only excitations between LOs belonging to the nearest-lying regions of two stacked first-neighbor molecules. (These will not necessarily be the same regions because of the 36° rotation around the long axis of the neighboring nucleotide bases with respect to each other in their stacks in DNA B.)

When these calculations are carried out (work has already begun at Erlangen) we will be able to compute at least 70% of the correlation of nonmetallic polymers with larger units. Since this amount of correlation already suffices to describe most physical properties of a system, we will also be able to calculate phonon spectra, magnetic properties, mechanical properties, for example, for polymers with larger unit cells. However, there will still be the difficulty of calculating good exciton spectra of such polymers, because good quasi-particle energies will be required (see Part II). LOs, on the other hand, are very useful for obtaining the correlation part of the total energy, but cannot be used to generate band structures (neither on the HF nor at the quasi-particle level) because they are no longer canonical HF orbitals with corresponding one-electron energies. (After the unitary transformation leading to the LOs, although the total energy remains invariant, the nondiagonal Lagrangian multipliers reappear in the HF equations and thus the one-electron energies lose their meaning.)

A further problem is imposed by metallic polymers. First, we already noted that for the metallic hydrogen chain, if one neglects excitations between nonneighboring regions using LOs, one obtains unreasonable results.[57] Further, the CCD method itself, as shown recently for equidistant (metallic) polyacetylene in the PPP approximation,[63] does not converge at all. Only by an approximate inclusion of the connected part of the operator $\hat{T}_4^{(3)}$ could convergence be achieved. (The approximation is based on the fortuitous fact that some matrix elements of $\hat{T}_2\hat{T}_2$ and $\hat{T}_4$ cancel each other out.) One could either take this into account or return to many-body perturbation theory, but the problem of restricting the number of necessary excitations by using LOs still remains unsolved in the case of metallic polymers with large unit cells.

5.6. REMARKS ON CORRELATION IN DISORDERED CHAINS

If one has a disordered polymer, one expects that some or all of the states are localized, depending on the degree of disorder. This Anderson localization, which was shown with the help of a tight-binding Hamiltonian,[64] also takes place at the Hartree–Fock level (see Section 4.2 and the work of Day *et al.*[65]), starting at the band edges and extending with increasing disorder further and further into the band.

Let us assume that we have a disordered chain, but it is not so strongly disordered that all the states are localized. The localization starts from the band edges and further let us assume that there are no gaps in the band (see Figure 5.12). In such a system three types of excitations can occur between the two different band regions in a two-band model: (1) excitations between the localized states of both bands, indicated by arrows labeled 1 on the right-hand side of Figure 5.12; (2) excitations

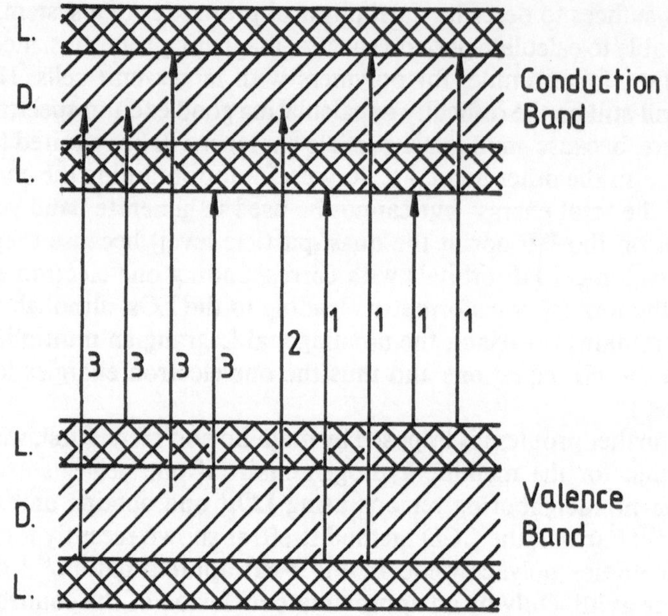

FIGURE 5.12. Localized and delocalized states in two bands in a not very strongly disordered chain. The mobility edges are indicated by arrows (L indicates localized states and D delocalized ones).

between the delocalized states of both bands, indicated by the arrow in the middle labeled 2, and, finally, (3) excitations between localized states in one band and delocalized ones in the other, or from delocalized to localized states, indicated by the four arrows with number 3 on the left-hand side.[66]

It is comparatively easy to calculate the excitations of type (1) (L→L), if one knows the localization properties of the corresponding wave functions (see Section 4.5): one takes into account only excitations between states in the two bands localized in the same or neighboring geometrical regions. The D→D type (2) of excitations can be handled in the same way as between Bloch states of a periodic polymer, taking in both bands a grid in **k** (see Section 5.2). The real difficulty arises in the treatment of the L→D or D→L type (3) excitations. For the delocalized states in both bands one can again take a grid in **k**. However, as energetically close-lying localized states are usually localized in completely different geometrical regions of the chain, in principle all of them must be taken into account in the L→D or D→L excitations for every value of **k** characterizing the delocalized states.[66] In practice, if one has determined the localization of all the states belonging to the regions L (which can be done most rapidly by using the ratio of the corresponding Green matrix elements; see Section 4.5 and Day, Gadzy *et al.*[67]), one can determine which are localized in the same or neighboring geometrical regions and try to group them together with an approximate weighting factor. One hopes that in this way also the L→D and D→L excitations can be handled within a reasonable amount of computer time and so open the way for correlation energy calculations of large, not very strongly disordered states. If, on the other hand, a chain is strongly disordered, all its energy levels in all band regions become localized, so one has to perform only L→L excitations, which can be undertaken in a way described above. (If the number of states is very large, that is, one has a long disordered chain, states localized in the same region can be regrouped applying again a weighting factor.)

In order to test the relative weight of the difficult L→D or D→L excitations as compared to the others, a model calculation was recently performed on a H_2 molecule ring introducing disorder by varying the intermolecular distances.[69] The number of H_2 molecules was 11, and a minimal STO-3G basis set was used for the calculation. In this way one obtains a simple two-band model with one completely filled and one unfilled band.

Standard geometries[67] were used for the ordered systems, and distance variations were obtained from a random number generator with a Gaussian probability distribution. The degree of disorder was then measured by the standard deviation σ of the Gaussian distribution.

The criterion for localization of the orbitals $\phi_{\mathbf{k}}$ to be used here is that proposed by Day *et al.*,[65] namely to calculate the quantity

$$R_{\mathbf{k}} = \left| \left(\sum_{\mu} r_{\mu} c_{\mu\mathbf{k}}^2 \right) \left(r \sum_{\mu} c_{\mu\mathbf{k}}^2 \right)^{-1} \right| \tag{5.82}$$

where r_{μ} are the position vectors of the atoms, r is the radius of the ring, and $c_{\mu\mathbf{k}}$ are the LCAO coefficients of $\phi_{\mathbf{k}}$. The quantity $R_{\mathbf{k}}$ assumes values in the interval between 0 and 1, the limiting cases being $R_{\mathbf{k}} = 1$ for perfect localization of $\phi_{\mathbf{k}}$ and $R_{\mathbf{k}} \approx 0$ ($R_{\mathbf{k}} = 0$ for an infinite system) for perfect delocalization. According to this there is a rather clear division of $\phi_{\mathbf{k}}$ into two sets of either localized or delocalized orbitals. The number of occupied (l_o) and unoccupied (l_u) localized orbitals is given in Table 5.8 for various values of σ. It should be noted, however, that not only the number of localized states but also the degree of localization changes as σ changes.[65] It can be seen that, by letting σ increase, it is possible to follow the Anderson transition the whole way from complete delocalization ($\sigma = 0$) to complete localization ($\sigma = 2.0$ au) for the alternating rings.

The second-order ground-state correlation energy has been calculated in Table 5.8 for each value of σ by the well-known expression

$$E^{(2)} = \sum_{\substack{i,j \in \text{occ} \\ a,b \notin \text{occ}}} \frac{V_{ijab}(2V_{ijab} - V_{ijba})}{\varepsilon_i + \varepsilon_j - \varepsilon_a - \varepsilon_b} \tag{5.83}$$

where i, j, a, b are spatial orbital indices (occ is the set of orbitals occupied in the Hartree–Fock ground state), V_{ijab} are the two-electron integrals, and ε_i are the orbital energies.

The contributions of the localized orbitals to $E^{(2)}$ can be displayed by breaking up the sum over the double replacements $(i,j) \rightarrow (a,b)$ in equation (5.83) into nine parts, according to whether ϕ_i and ϕ_j are both localized (L), one of them is localized and one delocalized (LD), or whether both are delocalized (DD), with a corresponding classification for (a,b).[66] The corresponding contributions are listed separately in Table 5.8. It is noteworthy that up to $\sigma = 1.5$ au the DD→DD excitations constitute the largest part of $E^{(2)}$. Since for $\sigma = 1.5$ au there are already more localized than delocalized states, one could have expected the LL→LL contribution to be dominant. That this is not the case can be understood from the fact that the localized states spread mainly from the lowest and highest levels toward the inside of the band, so that the energy denominators for the LL→LL excitations are mostly larger than for the DD→DD excitations.

For the alternating rings, $E^{(2)}$ first increases with increasing disorder. Obviously here the correlation interaction involving both localized and delocalized states is stronger than the correlation among only delocalized states. On the other hand, $E^{(2)}$ does not always increase monotonically but has a maximum at $\sigma = 1.25$ au and then decreases again. This means that the so-called localized–delocalized correlation is stronger than the correlation among only delocalized states. The importance of the localized–delocalized correlation is further supported by the observation that the sum of "mixed" contributions (LL→DD, LD→DD, etc.) in Table 5.8 has a local extremum at the same σ value as $E^{(2)}$.

We can conclude that the second-order correlation energy has an extremum at $\sigma = 1.25$ au, which corresponds to an intermediate disorder leading only to partial localization (12 out of 22 states). The "mixed" (localized and delocalized) contributions to the correlation energy dominate at this stage. The purely localized contributions to the correlation energy, however, remain relatively small almost until the critical disorder (complete localization).

Table 5.8. Second-Order Correlation Energy for Rings of H_{22} Atoms (11 H_2 Molecules) for Various Degrees of Disorder[a]

	Alternating (H_2 molecule) rings								
	$\sigma = 0.0$	$\sigma = 0.25$	$\sigma = 0.5$	$\sigma = 0.75$	$\sigma = 1.0$	$\sigma = 1.25$	$\sigma = 1.5$	$\sigma = 1.75$	$\sigma = 2.0$
l_o[b]	0	4	4	7	7	7	7	11	11
l_u	0	4	4	4	4	5	7	10	11
$E^{(2)}$	−0.1808	−0.1811	−0.1826	−0.1869	−0.1946	−0.2001	−0.1934	−0.1824	−0.1679
LL DD	—	−0.0160	−0.0166	−0.0161	−0.0463	−0.0353	−0.0168	−0.0067	—
LD DD	—	−0.0222	−0.0174	−0.0154	−0.0286	−0.0213	−0.0107	—	—
DD DD	−0.1808	−0.0618	−0.0734	−0.0870	−0.0555	−0.0615	−0.0466	—	—
LL LD	—	−0.0030	−0.0030	−0.0030	−0.0293	−0.0146	−0.0181	−0.0022	—
LD LD	—	−0.0386	−0.0365	−0.0317	−0.0121	−0.0330	−0.0246	—	—
DD LD	—	−0.0211	−0.0173	−0.0163	−0.0041	−0.0119	−0.0234	—	—
LL LL	—	−0.0040	−0.0050	−0.0058	−0.0153	−0.0114	−0.0306	−0.1534	−0.1679
LD LL	—	−0.0024	−0.0021	−0.0018	−0.0023	−0.0032	−0.0093	—	—
DD LL	—	−0.0118	−0.0113	−0.0097	−0.0011	−0.0078	−0.0131	—	—
E_{el}^{SCF}	−47.2228	−47.2884	−47.4982	−47.8826	−48.4866	−49.2952	−50.1013	−51.0499	−52.2167
E_{total}^{SCF}	−11.9364	−11.9179	−11.8433	−11.6358	−10.9670	−5.4280	−8.1816	−3.1669	−1.4064
$E_{toal}^{(2)}$[c]	−12.1172	−12.0990	−12.0259	−11.8227	−11.1616	−5.6281	−8.3750	−3.3493	−1.5743

[a] All energies in hartrees, σ in au.
[b] l_o and l_u denote the number of localized occupied and unoccupied orbitals, respectively.
[c] $E_{total}^{(2)} = E_{total}^{SCF} + E_{\sigma}^{(2)}$.

The importance of localized–delocalized correlation must be stressed. It implies that, in general, it should not be possible to completely omit either the localized or the delocalized states from a correlation energy calculation at intermediate disorder. (Still, it could be allowed, in particular for larger basis sets, to omit a few low-lying occupied and a few high-lying unoccupied localized levels from the calculation.) This may have consequences for the development of approximate approaches to electron correlation calculations in disordered systems.[69]

REFERENCES

1. K. A. Brueckner, *Phys, Rev.* **100**, 36 (1955); J. Goldstone, *Proc. R. Soc. London, Ser. A* **239**, 267 (1957); H. Primas, in: *Modern Quantum Chemistry* (O. Sinanoğlu, ed.), Vol. 2, Academic Press, New York (1965); R. Ahlrichs, *Theor. Chim. Acta* **35**, 59 (1974); J. A. Pople, R. Seeger, and R. Krishnan, *Int. J. Quantum Chem., Quantum Chem. Symp.* **11**, 149 (1977).
2. C. Møller and M. S. Plesset, *Phys. Rev.* **46**, 618 (1934); J. Paldus and J. Čížek, *Adv. Quant. Chem.* **9**, 105 (1975); E. Kapuy, Z. Csépes, and C. Kozmutza, *Int. J. Quantum Chem.* **23**, 981 (1983); J. A. Pople, M. J. Frisch, B. T. Luke, and J. S. Binkley, *Int. J. Quantum Chem., Quantum Chem. Symp.* **17**, 307 (1983).
3. J. Čížek, *J. Chem. Phys.* **45**, 4256 (1966); J. Čížek, *Adv. Quantum Chem.* **3**, 35 (1969); J. Čížek and J. Paldus, *Int. J. Quantum Chem.* **5**, 359 (1971); J. Paldus and J. Čížek, *Adv. Quantum Chem.* **9**, 105 (1975).
4. G. H. Wannier, *Phys. Rev.* **52**, 191 (1934).
5. S. Suhai, *Quantenmechanische Untersuchungen an quasi-eindimen-sionalen Festkörpern* (*Quantum Mechanical Investigations of Quasi-One-Dimensional Solids*), point 2, p. 39, Habilitation Thesis, University Erlangen–Nürnberg (1983).
6. E. I. Blount, *Solid State Phys.* **13**, 305 (1962).
7. J. Des Cloizeaux, *Phys. Rev.* **135A**, 685, 598 (1964); M. Kertész and G. Biczó, *Phys. Status Solidi B* **60**, 249 (1973).
8. W. Förner, *Beiträge zur Untersuchung der Elektronenkorrelation in großen Molekülen und Polymeren* (*Contributions to the Investigation of Electron Correlation in Large Molecules and Polymers*), Thesis, University Erlangen–Nürnberg (1985).
9. S. F. Boys, in: *Quantum Theory of Atoms, Molecules and the Solid State* (P.-O. Löwdin, ed.), p. 253, Academic Press, New York–London (1966).
10. W. Förner, J. Čížek, P. Otto, J. Ladik, and E. O. Steinborn, *Chem. Phys.* **97**, 235 (1985).
11. For a review see W. Kutzelnigg, in: *Methods of Electronic Structure Theory* (H. F. Schaefer III, ed.), p. 129, Plenum Press, New York–London (1977).
12. S. T. Epstein, in: *Perturbation Theory and its Application in Quantum Mechanics* (C. H. Wilcox, ed.), p. 49, Wiley–Interscience, New York–London (1974).
13. For further details see: J. Ladik, in: *Recent Advances in the Quantum Theory of Polymers* (J.-M. André, J.-L. Brédas, J. Delhalle, J. Ladik, G. Leroy, and C. Moser, eds.), p. 155, Springer-Verlag, Berlin–New York–Heidelberg (1980).
14. S. Suhai, Point 3, p. 49 in Reference 5; and Reference 26.

15. D. PINES, *Elementary Excitations in Solids*, p. 146, W. A. Benjamin Inc., New York (1964).
16. F. BELEZNAY, S. SUHAI, AND J. LADIK, *Int. J. Qunatum Chem.* **20**, 683 (1981).
17. T. KOOPMANS', *Physica* **1**, 104 (1933).
18. Y. TOYOZAWA, *Prog. Theor. Phys.* (*Kyoto*) **12**, 422 (1954).
19. A. B. KUNZ, *Phys. Rev. B* **6**, 2427 (1972).
20. J. T. DEVREESE, A. B. KUNZ, AND T. C. COLLINS, *Solid State Commun.* **11**, 670 (1972).
21. D. J. MICKISH, A. B. KUNZ, AND T. C. COLLINS, *Phys. Rev. B* **9**, 446 (1974).
22. S. T. PANTELIDES, D. J. MICKISH, AND A. B. KUNZ, *Phys. Rev. B* **10**, 2602 (1974).
23. M. INONE, C. K. MANUTTE, AND S. WANG, *Phys. Rev. B* **2**, 539 (1970).
24. S. SUHAI, *Phys. Rev. B* **27**, 3506 (1983).
25. J. LADIK, in: *Electronic Structure of Polymers and Molecular Crystals* (J.-M. André and J. Ladik, eds.), Plenum Press, New York–London (1975).
26. S. SUHAI AND J. LADIK, *J. Phys. C* **15**, 4327 (1982).
27. W. J. HEHRE, R. F. STEWART, AND J. A. POPLE, *J. Chem. Phys.* **51**, 2657 (1960); R. DITCHFIELD, W. HEHRE, AND J. A. POPLE, *J. Chem. Phys.* **54**, 1724 (1971).
28. S. HUZINAGA, *J. Chem. Phys.* **51**, 2657 (1969).
29. W. KOŁOS AND C. C. ROOTHAAN, *Rev. Mod. Phys.* **32**, 219 (1960).
30. W. KOŁOS AND L. WOLNIEWICZ, *J. Chem. Phys.* **41**, 3663 (1964).
31. L. F. MATTHEIS, *Phys. Rev.* **123**, 129 (1961); J. W. MOSKOWITZ, *J. Chem. Phys.* **38**, 677 (1963).
32. S. SUHAI (unpublished result).
33. D. H. LISKOV, J. M. MCKELVEY, C. F. BENDER, AND H. F. SCHAEFER III, *Phys. Rev. Lett.* **32**, 933 (1974).
34. M. SEEL, P. BAGUS, AND J. LADIK, *J. Chem. Phys.* **77**, 3123 (1982).
35. C. R. FINCHER, JR., C. E. CHEN, A. J. HEEGER, A. G. MACDIARMID, AND J. B. HASTINGS, *Phys. Rev. Lett.* **48**, 100 (1982).
36. S. SUHAI, *J. Chem. Phys.* **73**, 3843 (1980).
37. A. KARPFEN AND R. HÖLLER, *Solid State Commun.* **37**, 179 (1981).
38. J. L. BRÉDAS, R. R. CHANCE, R. SILBEY, AND G. P. DURAND, *J. Chem. Phys.* **75**, 255 (1981).
39. S. SUHAI, *Chem. Phys. Lett.* **96**, 619 (1983) (effect of correlation); unpublished results (interaction among PA chains).
40. W. J. HEHRE, R. F. STEWART, AND J. A. POPLE, *J. Chem. Phys.* **51**, 2657 (1969).
41. R. DITCHFIELD, J. W. HEHRE, AND J. A. POPLE, *J. Chem. Phys.* **54**, 726 (1971).
42. J. BRINKLEY, R. A. WHITESIDE, P. C. HARIHARAN, R. SEEGER, AND J. A. POPLE, Gaussian 76 program, *QCPE* **368**.
43. S. SUHAI (unpublished).
44. P. S. BAGUS, J. PACANSKY, AND W. WAHLGREN, *J. Chem. Phys.* **67**, 619 (1977).
45. See, for instance, B. HUDSON AND B. KOHLER, *Annu. Rev. Phys. Chem.* **25**, 437 (1974).
46. S. ETEMAT, A. J. HEEGER, L. LANCHLAN, T.-C. CHUNG, AND G. MACDIARMID, *Mol. Cryst. Liq. Cryst.* **77**, 431 (1981).
47. J. LADIK, in: *Electronic Structure of Polymers and Molecular Crystals* (J.-M. André and J. Ladik, eds.), p. 23, Plenum Press, New York–London (1975).
48. S. SUHAI, *Int. J. Quantum Chem., Quantum Biol. Symp.* **11**, 223 (1984).
49. For references on the geometry of DNA B see Chapter 2.
50. S. SUHAI, *Phys. Rev. B* **27**, 3506 (1982); *Int. J. Quantum Chem.* **23**, 1239 (1983).
51. See, for instance, I. TINOCO, JR., *J. Chem. Phys.* **33**, 1352 (1960); **34**, 1067 (1961).
52. H. DEVOE AND I. TINOCO, JR., *J. Mol. Biol.* **4**, 500 (1962).
53. J. ČÍŽEK, W. FÖRNER, AND J. LADIK, *Theor. Chim. Acta* **64**, 107 (1983).

54. J. M. Foster and S. F. Boys, *Rev. Mod. Phys.* **32**, 300 (1969).
55. C. Edminston and K. Ruedenberg, *Rev. Mod. Phys.* **34**, 457 (1963).
56. J. Pipek, *Int. J. Quant. Chem.* **27**, 527 (1983).
57. W. Förner, J. Ladik, P. Otto, and J. Čížek, *Chem. Phys.* **97**, 251 (1985).
58. F. Coester and H. Kümmel, *Nucl. Phys.* **7**, 477 (1960).
59. O. Sinanoğlu, *J. Chem. Phys.* **36**, 706 (1962); *Adv. Chem. Phys.* **6**, 315 (1964).
60. See the third Reference of 10.
61. See the first Reference of 10.
62. C. F. Bender, *J. Comput. Phys.* **9**, 547 (1972).
63. J. Paldus, M. Takahashi, and R. W. H. Cho, *Phys. Rev. B* **30**, 4267 (1984); M. Takahashi, J. Čížek, and J. Paldus, *Phys. Rev. A* **30**, 2193 (1984).
64. P. W. Anderson, *Phys. Rev.* **109**, 1492 (1958).
65. R. S. Day, J. Ladik, and F. Martino, *Chem. Phys. Lett.* **81**, 494 (1981).
66. J. Ladik, *Int. J. Quantum Chem.* **23**, 1073 (1983).
67. R. S. Day and F. Martino, *Chem. Phys. Lett.* **84**, 86 (1981); B. Gazdy, M. Seel, and J. Ladik, *Chem. Phys.* **86**, 41 (1984).
68. K. F. Berggren and F. Martino, *Phys. Rev.* **184**, 484 (1969); M. Bénard and J. Paldus, *J. Chem. Phys.* **72**, 6546 (1980); M. Seel, P. S. Bagus, and J. Ladik, *J. Chem. Phys.* **77**, 3123 (1982).
69. C.-M. Liegener and J. Ladik, *Phys. Lett.* **107A**, 79 (1985).

Chapter 6

Interaction between Polymers

6.1. PERTURBATION THEORETICAL CONSIDERATIONS

Murell *et al.*[1] have shown that if one develops a perturbation theory with overlap to be applied if the overlap integral $S < 0.1$, namely the two interacting molecules are at least 2.8–3.0 Å apart, then one obtains the interaction energy between two molecules A and B as a power-series expansion in the intermolecular potential U and the overlap integral S. Up to the order of U^2S^2 the interaction energy terms are

$$\Delta E^{\mathrm{PT}} = E_{\mathrm{el\,st}} + E_{\mathrm{pol}} + E_{\mathrm{exch}} + E_{\mathrm{ch\,tr}} + E_{\mathrm{disp}} + E_{\mathrm{exch\,pol}} + E_{\mathrm{exch\,disp}} \quad (6.1)$$

Here the terms $E_{\mathrm{el\,st}}$, E_{pol}, and E_{disp} are of zeroth order in overlap and have the same form as in zero-overlap perturbation theory. The contribution of order US^2 leads to the exchange energy E_{exch} and the energy contributions of order U^2S^2 may be subdivided into the charge transfer energy $E_{\mathrm{ch\,tr}}$ and the exchange polarization and exchange dispersion terms due to the exchange. Since the latter two terms are much smaller than the others, one can neglect them in the calculations.

In the monopole approximation the electrostatic term $E_{\mathrm{el\,st}}$ is given by

$$E_{\mathrm{el\,st}} = \sum_{\alpha=1}^{M_{\mathrm{A}}} \sum_{\beta=1}^{M_{\mathrm{B}}} (Z_\alpha^{\mathrm{A}} - q_\alpha^{\mathrm{A}})(Z_\beta^{\mathrm{B}} - q_\beta^{\mathrm{B}})/r_{\alpha\beta} \quad (6.2)$$

where Z_α^{A} is the nuclear charge of the αth nucleus in molecule A, M_{A} and M_{B} denote the number of nuclei in molecules A and B, respectively, q_α^{A} is the total electronic charge of the αth atom in molecule A and can be taken from Mulliken's population analysis[2] of the *ab initio* SCF results, and finally $r_{\alpha\beta}$ is the distance $|\mathbf{r}_\alpha - \mathbf{r}_\beta|$ between the αth atom in molecule A and the βth atom in molecule B. Of course, in more sophisticated calculations one could use instead of the monopole approximation,[2] a more refined analysis of the charge distributions of the two molecules applying either multipole expansions, or employing a three-dimensional mesh to describe the diagonal elements of the first-order density matrices.

The polarization term E_{pol} can be expressed in the form[3]

$$E_{\text{pol}} = 2 \sum_{i=1}^{n_A^*} \sum_{j=n_A^*+1}^{\infty} \frac{|\langle \varphi_i^A(1) | V_B | \varphi_j^A(1) \rangle|^2}{\varepsilon_i^A - \varepsilon_j^A + J_{i,j}^A - 2K_{i,j}^A}$$

$$+ 2 \sum_{k=1}^{n_B^*} \sum_{l=n_B^*+1}^{\infty} \frac{|\langle \varphi_k^B(1) | V_A | \varphi_l^B(1) \rangle|^2}{\varepsilon_k^B - \varepsilon_l^B + J_{k,l}^B - 2K_{k,l}^B} \tag{6.3}$$

where

$$V_B = -\sum_{\beta=1}^{M_B} \frac{Z_\beta^B}{|\mathbf{r}_1 - \mathbf{r}_\beta|} + 2 \sum_{k=1}^{n_B^*} \left\langle \varphi_k^B(2) \left| \frac{1}{r_{12}} \right| \varphi_k^B(2) \right\rangle$$

and (6.4)

$$V_A = -\sum_{\alpha=1}^{M_A} \frac{Z^A}{|\mathbf{r}_1 - \mathbf{r}_\alpha|} + 2 \sum_{i=1}^{n_A^*} \left\langle \varphi_i^A(2) \left| \frac{1}{r_{12}} \right| \varphi_i^A(2) \right\rangle$$

n_A^* and n_B^* are the number of filled MOs in molecules A and B, respectively, ε_i^A, ε_j^A, ε_k^B, and ε_l^B are one-electron energies of molecules A and B, respectively, φ_i^A, φ_j^A, φ_k^B, and φ_l^B are the corresponding MOs, and $J_{i,j}^A$, $K_{i,j}^A$, etc. denote the Coulomb and exchange integrals.

The exchange term E_{exch} has the form[1]

$$E_{\text{exch}} = -2 \sum_{i=1}^{n_A^*} \sum_{j=1}^{n_B^*} \left[\left\langle \varphi_i^A \varphi_j^B \left| \frac{1}{r_{12}} \right| \varphi_j^B \varphi_i^A \right\rangle \right.$$

$$+ \langle \varphi_i^A | \varphi_j^B \rangle \left(-\left\langle \varphi_i^A \varphi_j^B \left| \frac{1}{r_{12}} \right| \varphi_j^B \varphi_j^B \right\rangle - \left\langle \varphi_i^A \varphi_i^A \left| \frac{1}{r_{12}} \right| \varphi_i^A \varphi_j^B \right\rangle \right.$$

$$+ \left\langle \varphi_i^A \left| -\sum_{\beta=1}^{M_B} \frac{Z_\beta^B}{|\mathbf{r} - \mathbf{r}_B|} \right| \varphi_j^B \right\rangle + 2 \sum_{k=1}^{n_B^*} \left\langle \varphi_i^A \varphi_k^B \left| \frac{1}{r_{12}} \right| \varphi_j^B \varphi_k^B \right\rangle$$

$$\left. \left. + \left\langle \varphi_i^A \left| -\sum_{\alpha=1}^{M_A} \frac{Z_\alpha^A}{|\mathbf{r} - \mathbf{r}_\alpha|} \right| \varphi_j^B \right\rangle + 2 \sum_{l=1}^{n_A^*} \left\langle \varphi_i^A \varphi_l^A \left| \frac{1}{r_{12}} \right| \varphi_j^B \varphi_l^A \right\rangle \right) \right.$$

$$+(\langle\varphi_i^{\rm A}\mid\varphi_j^{\rm B}\rangle)^2\Bigg(\left\langle\varphi_i^{\rm A}\varphi_j^{\rm B}\left|\frac{1}{r_{12}}\right|\varphi_i^{\rm A}\varphi_j^{\rm B}\right\rangle$$

$$-\left\langle\varphi_i^{\rm A}\left|-\sum_{\beta=1}^{M_{\rm B}}\frac{Z_\beta^{\rm B}}{|\mathbf{r}-\mathbf{r}_\beta|}\right|\varphi_i^{\rm A}\right\rangle$$

$$-2\sum_{k=1}^{n_{\rm B}^*}\left\langle\varphi_i^{\rm A}\varphi_k^{\rm B}\left|\frac{1}{r_{12}}\right|\varphi_i^{\rm A}\varphi_k^{\rm B}\right\rangle-\left\langle\varphi_j^{\rm B}\left|-\sum_{\alpha=1}^{M_{\rm A}}\frac{Z_\alpha^{\rm A}}{|\mathbf{r}-\mathbf{r}_\alpha|}\right|\varphi_j^{\rm B}\right\rangle$$

$$-2\sum_{l=1}^{n_{\rm A}^*}\left\langle\varphi_j^{\rm B}\varphi_l^{\rm A}\left|\frac{1}{r_{12}}\right|\varphi_j^{\rm B}\varphi_l^{\rm A}\right\rangle\Bigg)\Bigg] \tag{6.5}$$

The $E_{\rm ch\,tr}=E_{\rm ch\,tr}^{\rm A^+B^-}+E_{\rm ch\,tr}^{\rm A^-B^+}$ terms can be derived from the expressions given by Murrell *et al.*,[1]

$$E_{\rm ch\,tr}^{\rm A^+B^-}=2\sum_{i=1}^{n_{\rm A}^*}\sum_{j=n_{\rm B}^*+1}^{\infty}\Bigg\{\left\langle\varphi_i^{\rm A}\left|-\sum_{\beta=1}^{M_{\rm B}}\frac{Z_\beta^{\rm B}}{|\mathbf{r}-\mathbf{r}_\beta|}\right|\varphi_j^{\rm B}\right\rangle+2\sum_{j_m=1}^{n_{\rm B}^*}\left\langle\varphi_i^{\rm A}\varphi_{j_m}^{\rm B}\left|\frac{1}{r_{12}}\right|\varphi_j^{\rm B}\varphi_{j_m}^{\rm B}\right\rangle$$

$$-\langle\varphi_i^{\rm A}\mid\varphi_j^{\rm B}\rangle\Bigg(\left\langle\varphi_i^{\rm A}\left|-\sum_{\beta=1}^{M_{\rm B}}\frac{Z_\beta^{\rm B}}{|\mathbf{r}-\mathbf{r}_\beta|}\right|\varphi_i^{\rm A}\right\rangle$$

$$+2\sum_{j_m=1}^{n_{\rm B}^*}\left\langle\varphi_i^{\rm A}\varphi_{j_m}^{\rm B}\left|\frac{1}{r_{12}}\right|\varphi_i^{\rm A}\varphi_{j_m}^{\rm B}\right\rangle\Bigg)$$

$$-\sum_{i_n\neq i}^{n_{\rm A}^*}\langle\varphi_{i_n}^{\rm A}\mid\varphi_j^{\rm B}\rangle\Bigg(\left\langle\varphi_i^{\rm A}\left|-\sum_{\beta=1}^{M_{\rm B}}\frac{Z_\beta^{\rm B}}{|\mathbf{r}-\mathbf{r}_\beta|}\right|\varphi_{i_n}^{\rm A}\right\rangle$$

$$+2\sum_{j_m=1}^{n_{\rm B}^*}\left\langle\varphi_i^{\rm A}\varphi_{j_m}^{\rm B}\left|\frac{1}{r_{12}}\right|\varphi_{i_n}^{\rm A}\varphi_{j_m}^{\rm B}\right\rangle\Bigg)$$

$$-\sum_{j_m=1}^{n_{\rm B}^*}\langle\varphi_i^{\rm A}\mid\varphi_{j_m}^{\rm B}\rangle\Bigg(\left\langle\varphi_j^{\rm B}\left|-\sum_{\alpha=1}^{M_{\rm A}}\frac{Z_\alpha^{\rm A}}{|\mathbf{r}-\mathbf{r}_\alpha|}\right|\varphi_{j_m}^{\rm B}\right\rangle$$

$$+2\sum_{i_n\neq i}^{n_{\rm A}^*}\left\langle\varphi_j^{\rm B}\varphi_{i_n}^{\rm A}\left|\frac{1}{r_{12}}\right|\varphi_{i_m}^{\rm B}\varphi_{i_n}^{\rm A}\right\rangle-\left\langle\varphi_j^{\rm B}\varphi_i^{\rm A}\left|\frac{1}{r_{12}}\right|\varphi_{j_m}^{\rm B}\varphi_{j_m}^{\rm B}\right\rangle\Bigg)$$

$$+\sum_{i_n\neq i}^{n_{\rm A}^*}\sum_{j_m=1}^{n_{\rm B}^*}\langle\varphi_{i_n}^{\rm A}\mid\varphi_{j_m}^{\rm B}\rangle\left\langle\varphi_i^{\rm A}\varphi_j^{\rm B}\left|\frac{1}{r_{12}}\right|\varphi_{i_n}^{\rm A}\varphi_{j_m}^{\rm B}\right\rangle\Bigg\}^2(\varepsilon_i-\varepsilon_j)^{-1} \tag{6.6}$$

from which the term $E_{\rm ch\,tr}^{\rm A^-B^+}$ follows if we interchange A and B everywhere in equation (6.6),

$$E_{\text{ch tr}}^{\text{A}^-\text{B}^+} = E_{\text{ch tr}}^{\text{A}^+\text{B}^-} \qquad (\text{A} \leftrightarrow \text{B}) \tag{6.7}$$

The convention $\langle \varphi_i \varphi_j | r_{12}^{-1} | \varphi_k \varphi_l \rangle = \langle \varphi_i(1)\varphi_j(2) | r_{12}^{-1} | \varphi_k(1)\varphi_l(2) \rangle$ was used in equations (6.6) and (6.7).

Finally, the dispersion energy has the form[4]

$$E_{\text{disp}} = -\sum_{i=1}^{n_A^*} \sum_{j=1}^{n_B^*} \sum_{a=n_A^*+1}^{\infty} \sum_{b=n_B^*+1}^{\infty} \frac{|\langle \varphi_i^A \varphi_j^B | (1/r_{12}) | \varphi_a^A \varphi_b^B \rangle|}{{}^1\Delta E_{i\to a}^A + {}^1\Delta E_{j\to b}^B} \tag{6.8}$$

where i and j refer to an occupied orbital of molecules A and B, respectively, while a and b stand for virtual orbitals of A and B, respectively.

By expressing all MOs in equations (6.3)–(6.6) and (6.8) in LCAO form it is easy to derive the corresponding expressions in terms of the AOs and the LCAO coefficients. To save space these rather lengthy relationships will not be written here, but in actual calculations they must be used.

If one wishes to generalize these theoretical perturbation expressions for the interaction between two periodic linear chains instead of each MO, a Bloch function must be substituted. For instance, $\varphi_i^A \to \phi_i^A(k_i)$, and the molecular excitation energies

$${}^1\Delta E_{i\to a}^A = \varepsilon_a - \varepsilon_i - J_{ia} + 2K_{ia}, \quad {}^1\Delta E_{j\to b}^B = \varepsilon_b - \varepsilon_j - J_{jb} + 2K_{jb} \tag{6.9a}$$

become

$${}^1\Delta E_{i\to a}^A(k_i, k_a) = \varepsilon_a(k_a) + \varepsilon_i(k_i), \quad {}^1\Delta E_{j\to b}^B(k_j, k_b) = \varepsilon_b(k_b) - \varepsilon_j(k_j) \tag{6.9b}$$

because the individual Coulomb integrals J_{ia}, J_{jb} and exchange integrals K_{ia}, K_{jb} disappear in an infinite system and the eigenvalues ε_i, ε_j, etc., become k-dependent. In this way, for instance, the dispersion energy (6.8) for the interaction of one unit cell of one chain with the whole other chain will have the form

$$\frac{E_{\text{disp}}}{N} = -\sum_{i=1}^{n_A^*} \sum_{k_i} \sum_{j=1}^{n_B^*} \sum_{k_j} \sum_{a=n_A^*+1}^{\infty} \sum_{k_a} \sum_{b=n_B^*+1}^{\infty} \sum_{k_b} \times \frac{|\langle \phi_{i(k_i)}^A \phi_{j(k_j)}^B | (1/r_{12}) | \phi_{a(k_a)}^A \phi_{b(k_b)}^B \rangle|^2}{\varepsilon_a^A(k_a) + \varepsilon_b^B(k_b) - \varepsilon_i^A(k_i) - \varepsilon_j^B(k_j)} \tag{6.10}$$

where N is the number of unit cells.

One can introduce again (as in Section 5.2) the composite indices $I = i, k_i$, $J = j, k_j$, $A = a, k_a$, $B = b, k_b$, in which case equation (6.10) is obtained in the more condensed form

$$\frac{E_{\text{disp}}}{N} = -\sum_I \sum_J \sum_A \sum_B \frac{|\langle \phi_I^{\text{A}} \phi_J^{\text{B}} | (1/r_{12}) | \phi_A^{\text{A}} \phi_B^{\text{B}} \rangle|}{\varepsilon_A^{\text{A}} + \varepsilon_B^{\text{B}} + \varepsilon_I^{\text{A}} - \varepsilon_J^{\text{B}}} \tag{6.11}$$

The evaluation of expression (6.11) can be performed in a similar way to that described in detail in Section 5.2. The conservation of momenta $k_i + k_j = k_a + k_b$ also holds in this case, so reducing the number of summations over k from 4 to 3. Here again a grid in k can be taken for each band and those integrals dependent on the three different values of k can be separated again from the rest (see Section 5.2).

It is of interest to show that if both linear polymers have a partially filled valence band (they are not insulators, but conductors) the dispersion interaction increases by orders of magnitude[5] as compared to the interaction between two insulator polymers with completely filled valence and empty conduction bands. The same argument can also be applied to the polarization term [generalization of equations (6.3) and (6.4) to periodic polymers].

To prove this we first replace the summations over k in equation (6.10) by integrations,

$$\frac{E_{\text{disp}}}{N} = -\left(\frac{a}{2\pi}\right)^4 \sum_{i=1}^{n_{\text{A}}^*} \sum_{j=1}^{n_{\text{B}}^*} \sum_{a=n_{\text{A}}^*+1}^{\infty} \sum_{bn_{\text{B}}^*+1}^{\infty} \int\int\int\int_{-\pi/a}^{-\pi/a} dk_i dk_j dk_a dk_b$$

$$\times \frac{|\langle \varphi_i^{\text{A}}(k_i) \varphi_j^{\text{B}}(k_j) | (1/r_{12}) | \varphi_a^{\text{A}}(k_a) \varphi_b^{\text{B}}(k_b) \rangle|^2}{\varepsilon_a^{\text{A}}(k_a) + \varepsilon_b^{\text{B}}(k_b) - \varepsilon_i^{\text{A}}(k_i) - \varepsilon_j^{\text{B}}(k_j)} \delta_{k_a+k_b,\, k_i+k_j} \tag{6.12}$$

where a is the elementary translation.

Equation (6.12) can easily be simplified to the case when we take into account only excitations from the valence bands to the conduction bands (two-band system):

$$\frac{E_{\text{disp}_2}}{N} = -\left(\frac{a}{2\pi}\right)^4 \int_{-\pi/a}^{\pi/a} \frac{dk_{n_{\text{A}}^*} dk_{n_{\text{B}}^*} dk_{n_{\text{A}}^*+1} dk_{n_{\text{B}}^*+1}}{\varepsilon_{n_{\text{A}}^*+1}^{\text{A}}(k_{n_{\text{A}}^*+1}) + \varepsilon_{n_{\text{B}}^*+1}^{\text{B}}(k_{n_{\text{B}}^*+1}) - \varepsilon_{n_{\text{A}}^*}^{\text{A}}(k_{n_{\text{A}}^*}) - \varepsilon_{n_{\text{B}}^*}^{\text{B}}(k_{n_{\text{B}}^*})}$$

$$\times |\langle \varphi_{n_{\text{A}}^*}^{\text{A}}(k_{n_{\text{A}}^*}) \varphi_{n_{\text{B}}^*}^{\text{B}}(k_{n_{\text{B}}^*}) | (1/r_{12}) | \varphi_{n_{\text{A}}^*+1}^{\text{A}}(k_{n_{\text{A}}^*+1}) \varphi_{n_{\text{B}}^*+1}^{\text{B}}(k_{n_{\text{B}}^*+1}) \rangle|^2$$

$$\times \delta_{k_{n_{\text{A}}^*+1}+k_{n_{\text{B}}^*+1},\, k_{n_{\text{A}}^*}+k_{n_{\text{B}}^*}} \tag{6.13}$$

In the case when the valence bands of the two chains are only partially filled, besides equation (6.13)* excitations over the Fermi level within the same band will also make a contribution. Hence

$$\frac{E_{\text{disp}}}{N} = \frac{E_{\text{disp}_1}}{N} + \frac{E_{\text{disp}_2}}{N} \tag{6.14}$$

where the new term is given by

$$\begin{aligned} \frac{E_{\text{disp}_1}}{N} = -(a/2\pi)^4 & \left\{ \begin{matrix} \left\{ \begin{matrix} \displaystyle\int_{-k_{F1}}^{k_{F1}} \left(\int_{-\pi/a}^{-k'_{F1}} + \int_{k'_{F1}}^{\pi/a} \right) \\ \displaystyle\left(\int_{-\pi/a}^{-k_{F1}} + \int_{k_{F1}}^{\pi/a} \right) \int_{-k'_{F1}}^{k'_{F1}} \end{matrix} \right\}_1 \end{matrix} \right\}_2 \\ & \times dk_1 dk'_1 dk_2 dk'_2 \frac{|\langle \varphi^{A}_{n_A^*}(k_1) \varphi^{B}_{n_B^*}(k_2) \,|\, (1/r_{12}) \,|\, \varphi^{A}_{n_A^*}(k'_1) \varphi^{B}_{n_B^*}(k'_2) \rangle|}{\varepsilon^{A}_{n_A^*}(k'_1) + \varepsilon^{B}_{n_B^*}(k'_2) - \varepsilon^{A}_{n_A^*}(k_1) - \varepsilon^{B}_{n_B^*}(k_2)} \\ & \times \delta_{k'_1 + k'_2,\, k_1 + k_2} \end{aligned} \tag{6.15}$$

Here the brackets { }, $i = 1, 2$, indicate that the curves $\varepsilon^{j}_{n_j^*}(k_i)$ ($i = 1, 2$; $j = \text{A, B}$) in Figure 6.1 have their minimum either at 0 or at π/a. In the former case the upper line has to be read (the occupied part of the band between $-k_{F_1} < 0 < k_{F_1}$, the empty parts between $k_{F_1} < \pi/a$ and $-k_{F_1} > -\pi/a$), while in the latter case the lower line describes the opposite situation. Also, k_i ($i = 1, 2$) is always a reciprocal lattice vector in the occupied part of the band, while k'_i ($i = 1, 2$) is such a vector in the empty part of the band. [For the sake of simplicity, the notations of the crystal momenta have been changed in the following way: $k_{n_A^*} \rightarrow k_1$ [occupied part of the Brillouin zone (BZ)], $k_{n_A^*} \rightarrow k'_1$ (empty part of the BZ), $k_{n_B^*} \rightarrow k_2$ (occupied part of the BZ), and $k_{n_B^*} \rightarrow k'_2$ (empty part of the BZ).] Since the intraband excitation energies in the denominator of equation (6.15) are much smaller than the corresponding interband values in equation (6.13), it follows that

$$\frac{E_{\text{disp}_1}}{N} \gg \frac{E_{\text{disp}_2}}{N} \tag{6.16}$$

* In this case the only change in equation (6.13) is that we must integrate k_1 and k_2 not over the whole interval $-\pi/a$ to π/a, but only over the filled parts of the valence bands.

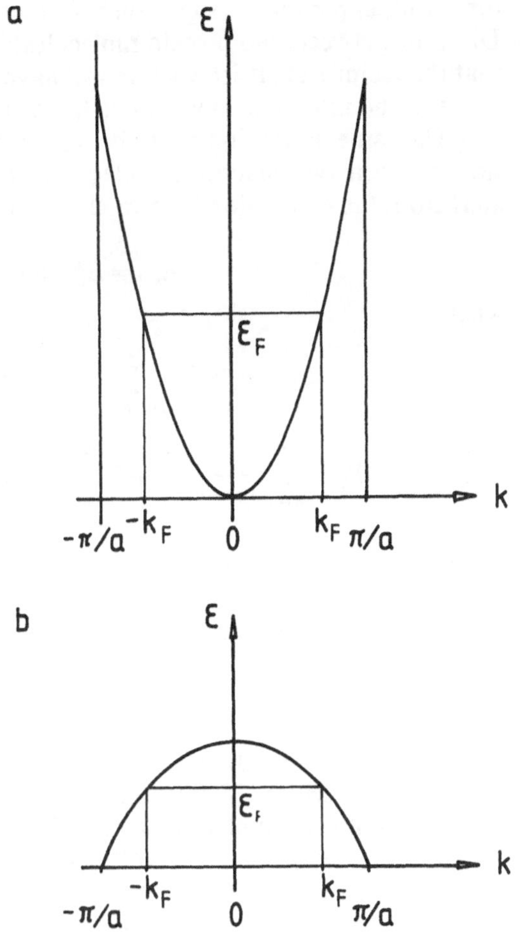

FIGURE 6.1. Dispersion curve $\varepsilon^*_{n_j}(k_i)$ ($i = 1, 2$; $j = \mathrm{A, B}$), and position of the Fermi level (ε_F) and of the Fermi momentum (k_F) (a) if the curve has its minimum at 0 and (b) if the curve has its minimum at π/a.

This means that the dispersion interactions between polymers with partially filled bands are much stronger than those between polymers with completely filled valence bands and empty conduction bands.

It is noteworthy that in a very similar way it can be shown that the dispersion interaction between a partially filled valence band in macromolecule A and a partially filled conduction band in macromolecule B (the case of intermacromolecular charge transfer) is again much stronger than in the case of two "closed-shell" macromolecules. Furthermore, the same derivation can also be applied for the case when one or two of the interacting macromolecules are not a one-dimensional, but a two-dimensional periodic system (such as interaction between a protein and DNA, or between two protein molecules). The only change then will be that the reciprocal lattice vectors will have two components (see Section 1.1) and the integrations will have to be changed accordingly.

The same derivation could be repeated for the polarization interaction between two macromolecules with partially filled bands. We can start from the expression for two closed-shell molecules:

$$E_{\text{pol}} = E_{\text{pol}}^{\text{A}} + E_{\text{pol}}^{\text{B}} \tag{6.17}$$

where

$$E_{\text{pol}}^{\text{A}} = -2 \sum_{i=1}^{m_{\text{A}}^*} \sum_{j=n_{\text{A}}^*+1}^{\infty} \frac{|\langle \varphi_i^{\text{A}}(1) | V_{\text{B}} | \varphi_j^{\text{A}}(1) \rangle|^2}{{}^1\Delta E_{i\to j}^{\text{A}}} \tag{6.18}$$

$$V_{\text{B}} = \sum_{\beta=1}^{M_{\text{B}}} \frac{Z_\beta}{|\mathbf{r}_1 - \mathbf{r}_\beta|} + 2 \sum_{l=1}^{n_{\text{B}}^*} \langle \varphi_l^{\text{B}}(2) | (1/r_{12}) | \varphi_l^{\text{B}}(2) \rangle \tag{6.19}$$

M_{B} is the number of nuclei in molecule B, Z_β is the charge, and $\mathbf{r}_\beta$ is the position vector of nucleus β. The expression $E_{\text{pol}}^{\text{B}}$ can be obtained by interchanging A and B in equations (6.18) and (6.19):

$$E_{\text{pol}}^{\text{B}} = E_{\text{pol}}^{\text{A}} \qquad (\text{A} \leftrightarrow \text{B}) \tag{6.20}$$

Equations (6.17)–(6.20) can be easily generalized again for the case of two macromolecules. If only their valence and conduction bands are again taken into account and the case of partially filled bands considered, one finds once more that due to the much smaller intraband excitation energies $\varepsilon_{n_{\text{A}}^*}^{\text{A}}(k_1 \to k_1')$ and $\varepsilon_{n_{\text{B}}^*}^{\text{B}}(k_2 \to k_2')$, the contribution of these excitations to the polarization term will be dominant.

It should be noted that the expression E_{disp_1}/N possesses singularity owing to the excitations through the Fermi level with energy tending to 0. A detailed mathematical analysis of the problem[6] using the so-called "**kp**" perturbation has shown, however, that the total expression for E_{disp_1}/N is finite, increases with increasing chain length, and is some orders of magnitude larger than E_{disp_2}/N. This fact probably plays rather an important role by DNA–protein interactions, where charge can easily be transferred from one chain to the other and this CT can be influenced by molecules (such as mutagens or carcinogens) binding to DNA or proteins. A detailed discussion of this problem will be given in Chapter 11 of this book.

The application of expressions (6.3)–(6.8) to interacting linear chains was made on the assumption that both chains are periodic and their unit cells have the same size (commensurable polymers). In many cases, however, the interacting chains are noncommensurable (like an α-helix of a polypeptide interacting with a polynucleotide in DNA B configuration). In such cases the problem is still manageable if m_A unit cells of one chain have the same length as m_B unit cells of the other, i.e.,

$$m_A \mathbf{a}_A = m_B \mathbf{a}_B = \mathbf{a} \tag{6.21}$$

where $\mathbf{a}$ is the elementary translation vector of the supercell (nearly commensurable polymers). For such cases Imamura *et al.*[7] have worked out a theoretical perturbation treatment for the interaction energy and charge distribution starting from the *ab initio* SCF LCAO COs (generalized Bloch functions of the individual chains). Since the derivation and resulting expressions are rather complicated, we do not reproduce them here but refer the reader to the original paper.[7]

When even equation (6.21) is not satisfied for the two chains, after the calculation of Bloch orbitals of the individual chains one has to apply the theory of disordered systems (see, in Section 6.3, the treatment of the interaction of a polyglycine chain with polynucleotides).

6.2. THE MUTUALLY CONSISTENT FIELD (MCF) METHOD

We saw in the previous section that the perturbation theoretical expressions governing two molecules (or linear chains) at medium distances (where a multipole expansion for the electrostatic term alone is insufficient) are rather complicated even in second order. On the other hand, perturbation theory in this form cannot describe the simultaneous interactions of more than two molecules (only with the aid of the still more complicated double perturbation theory), and it is also not very accurate. Therefore one must develop a new method, which is nearly as accurate as the supermolecule approach (which, for larger interacting molecules, is not feasible because of the prohibitively large amount of computer time), can treat an arbitrary number of interacting molecules (or linear chains) at medium intermolecular (interchain) distances, and is much faster than perturbation theory (PT). This problem was solved at Erlangen in a series of papers for both molecules and linear chains.

The basic idea[8] is that in the case of, say, three interacting molecules A, B, C (selected here only for the sake of simplicity) one solves not the HF equations of the three free molecules, namely

$$\hat{F}^A \varphi_i^A = \varepsilon_i^A \varphi_i^A, \quad \hat{F}^B \varphi_i^B = \varepsilon_i^B \varphi_i^B, \quad \hat{F}^C \varphi_i^C = \varepsilon_i^C \varphi_i^C \tag{6.22}$$

but rather the HF equation of each molecule in the presence of the two others, i.e.,

$$\tilde{\hat{F}}^{A}\tilde{\varphi}_i^{A} = (\hat{F}^{A} + V^{B} + V^{C})\tilde{\varphi}_i^{A} = \tilde{\varepsilon}_i^{A}\tilde{\varphi}_i^{A} \tag{6.22a}$$

$$\tilde{\hat{F}}^{B}\tilde{\varphi}_i^{B} = (\hat{F}^{B} + V^{A} + V^{C})\tilde{\varphi}_i^{B} = \tilde{\varepsilon}_i^{B}\tilde{\varphi}_i^{B} \tag{6.22b}$$

$$\tilde{\hat{F}}^{C}\tilde{\varphi}_i^{C} = (\hat{F}^{C} + V^{A} + V^{B})\tilde{\varphi}_i^{C} = \tilde{\varepsilon}_i^{C}\tilde{\varphi}_i^{C} \tag{6.22c}$$

Starting with unperturbed potentials V^A, V^B, and V^C one has to solve this system iteratively [for instance, recalculating V^A with the changed HF wave functions $\tilde{\varphi}_i^A$ obtained from equation (6.22a) thus yielding a new potential $\tilde{V}^A$, which is then substituted into equations (6.22b) and (6.22c), etc.] until a mutually consistent solution is obtained [mutually consistent field (MCF)]. This means that the charge distributions and potentials of *all* the molecules would no longer vary if one were to continue the iterations. The real problem lies in how the classical electrostatic potentials

$$V^X(\mathbf{r}) = \int \frac{\rho^X(\mathbf{r}')}{|\mathbf{r} - \mathbf{r}'|}\, d\mathbf{r}' \qquad (X = A, B, C) \tag{6.23}$$

are represented in order to allow an accurate and sufficiently rapid solution, where ρ^X denotes the charge density of molecule X. If equation (6.23) is used for the potential, the MCF solution automatically provides not only the electrostatic but also the polarization term. However, equation (6.23) can also be supplemented by an exchange term (using the local exchange $\sim \rho^{1/3}$, though in most cases this term is rather small at medium intermolecular distances).

Two different methods have been developed for the successful representation of the potentials V^X, one in 1978[9] and the other in 1984.[10] Both methods give rather accurate results as compared to the supermolecule (SM) calculations, but the latter version is much faster. In order to understand the new version of the MCF method it is necessary to be familiar with the older one, which is the only one that has been applied to two interacting chains; see Section 6.3. Therefore, both versions are described in the next two subsections.

6.2.1. The MCF Method in the Point-Charge Representation of the Potentials

The first step was to represent the potential of a molecule by point charges situated at the positions of the nuclei.[8] This monopole approximation was much more rapid than the SM or PT calculation, but has yielded only about 60% of the SM results. The next step was to subdivide the molecule into a large number of cubes, and a point charge was placed at the center of each cube (details are given elsewhere[11]). This brute-force method was accurate enough, but the necessary computer time was much longer than in the SM or PT cases.

Finally, Otto[9] solved the problem as follows. First, the canonical HF orbitals of the molecules were transformed into localized orbitals using Boys' method.[12] For instance, in the case of molecule B the potential for every localized orbital was computed at points situated on spheres with different radii relative to the center of charge of the considered orbital. The range of the radii was varied from about 1.5 to 15 au, so that the short-, intermediate-, as well as long-range regions were included. Thus one can write

$$V_l^{\mathrm{B}}(\mathbf{r}_{i_l}) = \int \frac{\rho_l^{\mathrm{B}}(\mathbf{r}_2)}{|\mathbf{r}_{i_l} - \mathbf{r}_2|} d\mathbf{r}_2 \quad (l = 1, 2, \ldots, n^*; i = 1, \ldots, m) \tag{6.24}$$

where $\mathbf{r}_{i_l}$ is the position vector of the ith point on the sphere belonging to the lth localized orbital, m is the number of points used for the calculation of the potential, n^* is the number of filled orbitals, and ρ_l^{B} is the charge density of the lth localized orbital.

Next, the solution of the nonlinear system of equations

$$\sum_{j=1}^{M_{\mathrm{B}}} \frac{q_{j_l}}{|\mathbf{r}_{i_l} - \mathbf{r}_{j_l}|} = V^{\mathrm{B}}(\mathbf{r}_{i_l}) \qquad \text{for all } i = 1, \ldots, m \tag{6.25}$$

provided the values of parameters q_{j_l} and $\mathbf{r}_{j_l}$, so that the sum of squares

$$F_l = \sum_{i=1}^{m} \left[\sum_{j=1}^{M_{\mathrm{B}}} \frac{q_{j_l}}{|\mathbf{r}_{i_l} - \mathbf{r}_{j_l}|} - V^{\mathrm{B}}(\mathbf{r}_{i_l}) \right]^2 \tag{6.26}$$

has a minimum. Here M_{B} is the number of point charges for each localized orbital of molecule B. In order to ensure that the q_{j_l} values are obtained with a minus sign, the condition

$$\sum_{j=1}^{M_{\mathrm{B}}} |q_{j_l}| - \int \rho_l^{\mathrm{B}}(\mathbf{r}) d\mathbf{r} = 0 \tag{6.27}$$

has been added to equation (6.26). An efficient procedure for solving the least-squares problem was applied.[13] Calculations have been carried out for different numbers of point charges and different numbers of potential points for each localized orbital l. Equation (6.26) was solved with an accuracy of 10^{-7}–10^{-10} for the value of F_l. This criterion ensures a deviation of less than $10^{-4}e_0$ for the total charge.

In order to consider also the exchange term between the molecules, the local X_α potential for each localized orbital was calculated at the same points as the Coulomb potential,

$$V_{X_\alpha}^{B}(\mathbf{r}_{i_l}) = -6\alpha[\tfrac{3}{4}\pi^{-1}\rho_l^B(\mathbf{r}_{i_l})]^{1/3} \quad (l = 1, 2, \ldots, n^*;\ i = 1, 2, \ldots, m) \quad (6.28)$$

and expanded as a sum of Gaussian functions,

$$\sum_{j=1}^{s} c_{j_l} \exp[-\alpha_{j_l}(\mathbf{r}_{j_l} - \mathbf{r}_{i_l})^2] = V_{X_\alpha}^{B}(\mathbf{r}_{i_l}) \qquad (i = 1, \ldots, m) \qquad (6.29)$$

where c_{j_l}, α_{j_l}, and $\mathbf{r}_{j_l}$ are the parameters to be determined and s is the number of Gaussian functions for each localized orbital. The same least-squares procedure as in the case of the Coulomb potential was used, namely

$$F_l^{X_\alpha} = \sum_{i=1}^{m} \left[\sum_{j=1}^{s} c_{j_l} \exp[-\alpha_{j_l}(\mathbf{r}_{j_l} - \mathbf{r}_{i_l})^2] - V_{X_\alpha}^{B}(\mathbf{r}_{i_l}) \right]^2$$

$$= \text{minimum} \qquad (6.30)$$

The Fock operators of the molecules were modified by the Coulomb potential and, in the case of two H_2O molecules, also by the exchange potential of the partner molecule, and iterated until mutually consistent solutions were obtained. The interaction energy can then be expressed in the form

$$\Delta E_{AB}^{MCF} = \tilde{E}_A + \tilde{E}_B - E_A - E_B + \tilde{E}_{Coul} + \tilde{E}_{exch} + \tilde{E}_{ch\,tr} \qquad (6.31)$$

Here

$$\tilde{E}_{Coul} = \sum_{j=1}^{M_B} \sum_{l=1}^{n_B^*} \sum_{k=1}^{M_A} \sum_{l=1}^{n_A^*} \frac{q_{k_{l'}}^{A}(\mathbf{r}_{k_{l'}}) q_{j_l}(\mathbf{r}_{j_l})}{|\mathbf{r}_{k_l} - \mathbf{r}_{j_{l'}}|} + \tilde{E}_{el\,n}^{AB} + E_{el\,n}^{BA} + E_{n\,n} \qquad (6.32)$$

and $\tilde{E}_{exch}$ is the perturbational expression (6.5) calculated not with the MOs φ_i^X of the free molecule, but with the MCF MOs $\tilde{\varphi}_i^X$ (X = A, B, C). When the local exchange potential has been included in the Fock operator,

$$\tilde{E}_{exch} = \frac{1}{2}\left(\sum_{kl} \tilde{P}_{kl}^{A} \langle \chi_k^A \mid \tilde{V}_{X_\alpha}^{B} \mid \chi_l^A \rangle + \sum_{mn} \tilde{P}_{mn}^{B} \langle \chi_m^B \mid \tilde{V}_{X_\alpha}^{A} \mid \chi_m^M \rangle \right) \qquad (6.33)$$

where the $\tilde{P}_{mn}$ are elements of the charge-bond-order matrices, and $\tilde{V}^{A}_{X_\alpha}$ and $\tilde{V}^{B}_{X_\beta}$ are the local exchange potentials of A and B, respectively, calculated with the MCF MOs; $\tilde{E}_X$ and E_X denote the HF total energy of molecule X (X = A, B) in the field of its partners, or of the free molecule; $\tilde{E}^{AB}_{el\,n}$ represents the classical electrostatic interaction between the electrons of one molecule (calculated again with the help of the MCF point charges) and the nuclei of the other; and $E^{AB}_{n\,n}$ is the electrostatic repulsion between the nuclei of the two molecules. Finally, the perturbation theoretical expressions (6.6) and (6.7) can again be used for $\tilde{E}_{ch\,tr}$, but with MCF orbitals.

If one also wishes to take the dispersion interaction into account, which of course is not included in an MCF calculation at the HF level, one can use either equation (6.8), again with MCF orbitals, or apply the empirical expression of London,[14] namely

$$E^{AB}_{disp} = -\frac{3}{2}\sum_{i=1}^{N_A}\sum_{j=1}^{N_B}\frac{\alpha_i^A\alpha_j^B I_i^A I_j^B}{(I_i^A + I_j^B)R_{ij}^6} \tag{6.34}$$

Here N_A and N_B are the numbers of atoms in molecules A and B, respectively, α_i^X and I_i^X are the respective valence-state polarizabilities and ionization potentials[15] of the ith atom in molecule X (X = A, B), and finally R_{ij} is the distance between atoms i and j. Hence the total interaction energy between molecules A and B can be expressed as

$$\Delta E^{int}_{AB} = \Delta E^{MCF}_{AB} + E^{AB}_{disp} \tag{6.35}$$

The method described here has been applied successfully first for the interaction of two $HCONH_2$ molecules, for the system of $2H_2O + H_3O^+$, and for three H_2O molecules with medium intermolecular distances (details are given by Otto[9]). The calculations were later extended to the stacking energies of two and three cytosine molecules[16] in the same relative positions as in DNA B. In the case of two cytosine molecules $\Delta E^{MCF} = 3.79$ kcal/mol, while $\Delta E^{SM} = 3.46$ kcal/mol showing that the MCF method works really well also in the case of larger interacting molecules. In a further study the solvation energy of glycine was computed applying two model systems, glycine · $6H_2O$ and glycine · $12H_2O$ (the geometries and other details are given elsewhere[17]). In the case of the smaller glycine · $6H_2O$ system a supermolecule calculation was also performed and yielded -79.9 kcal/mol, while ΔE^{MCF} yielded -85.7 kcal/mol.[17]

The first version of the MCF method described can be generalized to interacting chains in a straightforward manner. The one-electron part of the Fock operator of chain A in the presence of T other chains can be expressed as[18]

$$\hat{\tilde{H}}^{N,A} = \hat{H}^{N,A} + \sum_{\beta=1}^{T} \sum_{q_2=-N}^{N} V_{q_2}^{\beta}(\mathbf{r}) \tag{6.36}$$

where

$$\hat{H}^{N,A} = -\tfrac{1}{2}\Delta - \sum_{q_1=-N}^{N} \sum_{\alpha_A=1}^{M_A} \frac{Z_{\alpha_A}}{|\mathbf{r} - \mathbf{R}_{\alpha_A}^{q_1}|} \tag{6.37}$$

and

$$V_{q_2}^{\beta}(\mathbf{r}) = \int \frac{\rho_{q_2}^{\beta}(\mathbf{r}')d\mathbf{r}'}{|\mathbf{r} - \mathbf{r}'|} + \sum_{\alpha_\beta=1}^{M_\beta} \frac{Z_{\alpha_\beta}}{|\mathbf{r} - \mathbf{R}_{\alpha_\beta}^{q_2}|} \tag{6.38}$$

with

$$\begin{aligned} \rho_{q_2}^{\beta} &= \sum_{r,s=1}^{m_\beta} \sum_{l=1}^{n_\beta^*} 2\left[(a/2\pi) \int_{-\pi/a}^{\pi/a} c_{l,r}^{\beta*}(k) c_{l,s}^{\beta}(k)dk\right] \chi_r^{\beta,q_2} \chi_s^{\beta,q_2} \\ &= \sum_{r,s=1}^{m_\beta} P_{r,s}^{\beta} \chi_r^{\beta,q_2} \chi_s^{\beta,q_2} \end{aligned} \tag{6.39}$$

Here m_β is the number of AOs per unit cell in chain β and n_β^* is the number of filled bands in the same chain. All the other quantities in equations (6.36)–(6.38) were defined earlier in Section 1.1. In this way the *ab initio* SCF LCAO CO method (see Section 1.1) can be reformulated to obtain MCF charge densities of the interacting chains. By taking the electrostatic interactions between these chains [which can be easily obtained by the appropriate generalization of equation (6.32)], one obtains automatically the polarization contribution as well. For further details see Section 6.3, in which the interaction of a polyglycine chain with different polynucleotide chains is discussed.

6.2.2. The MCF Method in the Pseudopolarization Tensor Formulation

Despite all its successes, the MCF method with the point-charge representation of the potentials described above has one major disadvantage, namely if we have a molecule A, the values and positions of the point charges that represent its potential must be recalculated in every case that this molecule interacts with one or more other molecules, and this has to be done for every relative position of the interacting molecules. Otto[10] has shown that one can develop a much faster version of the MCF method if one draws up an inventory of point charges (q_i^X) and their positions ($\mathbf{r}_i^X$) representing their Coulomb potentials for different free molecules (X = A, B, C, ...) and takes into account the effect of the electric field caused by its interaction with other molecules through the shifts $\Delta\mathbf{r}_i^X$ of the original positions of its point charges. In order to calculate these shifts, a library of the polarizability tensors of these molecules must also be established.

The total induced dipole moment in molecule X (X = A, B, C, ...) at position $\mathbf{r}_i^X$ due to the electric field caused by the other interacting molecules can be expressed in the form

$$\sum_{i=1}^{M_X} q_i^X \Delta\mathbf{r}_i^X = \sum_{i}^{M_X} \boldsymbol{\alpha}_i^X \mathbf{E}(\mathbf{r}_i^X) \tag{6.40}$$

where M_X is the number of point charges and $\boldsymbol{\alpha}_i^X$ the polarization tensor corresponding to point charge q_i^X.

For a given free molecule X the Coulomb potential at the HF level can be written as

$$V_{HF}^{0,X}(\mathbf{R}_k^X) = \int \frac{\rho_{HF}^{0,X}(\mathbf{r})}{|\mathbf{R}_k^X - \mathbf{r}|} d\mathbf{r} \tag{6.41}$$

with

$$\rho_{HF}^{0,X}(\mathbf{r}) = \sum_{j=1}^{n^{*X}} 2\varphi_j^{0,X^*}(\mathbf{r})\rho_j^{0,X}(\mathbf{r})$$

where $\varphi_j^{0,X}$ denotes the jth occupied HF orbital of the free molecule X. The vectors $\mathbf{R}_k^X$ are chosen again to cover the whole range of potentials for molecule X. The point charges and their positions are calculated again (as in the first version of the MCF method) from the least-squares expression

$$\sum_{k=1}^{m_X} \left| \sum_{i=1}^{M_X} \frac{q_i^X}{|\mathbf{R}_k^X - \mathbf{r}_i^X|} - V_{HF}^{0,X}(\mathbf{R}_k^X) \right|^2 = \min \tag{6.42}$$

In practice, the least-squares optimization has converged if

$$\left| \sum_{i=1}^{M_X} \frac{q_i^X}{|\mathbf{R}_k^X - \mathbf{r}_i^X|} - V_{HF}^{0,X}(\mathbf{R}_k^X) \right| \leq 10^{-5}$$

for all points m_X.

The next step is to calculate the perturbed HF orbitals $\tilde{\varphi}_j^{\mathrm{X}}(\mathbf{E}_J)$ in the presence of N_{E} different external electric fields ($J = 1, 2, \ldots, N_{\mathrm{E}}$). These can be represented again by point charges placed at various positions in molecule X such that the components of the electric field at position $\mathbf{R}_k$, namely $E_x^{\mathrm{X}}(\mathbf{R}_k)$, $E_y^{\mathrm{X}}(\mathbf{R}_k)$, and $E_z^{\mathrm{X}}(\mathbf{R}_k)$, should possess the same order of magnitude as the components of the electric field generated by the interacting molecules.

The N_{E} different perturbed wave functions can be employed to compute the perturbed HF Coulomb potentials $V_{\mathrm{HF}}^{\mathrm{X}}(\mathbf{E}_J, \mathbf{R}_k^{\mathrm{X}})$ at the same points $\mathbf{R}_k^{\mathrm{X}}$ with the aid of equation (6.41).

The last step is to compute the pseudopolarization tensors $\boldsymbol{\alpha}_i^{\mathrm{X}}$. The "ansatz"(10)

$$\Delta \mathbf{r}_i^{\mathrm{X},J} = \mathbf{d}^{\mathrm{X}} \mathbf{E}_J(\mathbf{r}_i^{\mathrm{X}}) \qquad (J = 1, 2, \ldots, N_{\mathrm{E}}) \tag{6.43}$$

immediately implies

$$\boldsymbol{\alpha}_i^{\mathrm{X}} = q_i^{\mathrm{X}} \boldsymbol{d}^{\mathrm{X}} \qquad (\mathrm{X} = \mathrm{A}, \mathrm{B}, \mathrm{C}, \ldots) \tag{6.44}$$

that is, the pseudopolarization tensors depend on i only through the charges. [Various other relations like equation (6.43) were tried out, but equation (6.43) proved to be the most successful.(10)]

The six components d_{mn}^{X} of the symmetric 3×3 tensor (d_{xx}, d_{yy}, d_{zz}, d_{xy}, d_{xy}, d_{yz}) were again determined with the help of two least-squares procedures (it proved advantageous to optimize the diagonal and off-diagonal tensor elements separately(10)):

$$\sum_{k=1}^{m_{\mathrm{X}}} \sum_{J=1}^{N_{\mathrm{E}}} | V_{\mathrm{PC}}^{J,\mathrm{X}}(\mathbf{R}_k^{\mathrm{X}}) - V_{\mathrm{HF}}^{J,\mathrm{X}}(\mathbf{R}_k^{\mathrm{X}}) |^2 = \min \tag{6.45}$$

where

$$V_{\mathrm{PC}}^{J,\mathrm{X}}(\mathbf{R}_k^{\mathrm{X}}) = \sum_{i=1}^{M_{\mathrm{X}}} q_i^{\mathrm{X}} / [(x_i^{\mathrm{X}} + d_{xx}^{\mathrm{X}} E_{Jx}(\mathbf{r}_i^{\mathrm{X}}) - X_k^{\mathrm{X}})^2 + (y_i^{\mathrm{X}} + d_{yy}^{\mathrm{X}} E_{Jy}(\mathbf{r}_i^{\mathrm{X}}) - Y_k^{\mathrm{X}})^2$$
$$+ (z_i^{\mathrm{X}} + d_{zz}^{\mathrm{X}} E_{Jz}(\mathbf{r}_i^{\mathrm{X}}) - Z_k^{\mathrm{X}})^2]^{1/2} \qquad (\mathrm{X} = \mathrm{A}, \mathrm{B}, \mathrm{C}, \ldots) \tag{6.46}$$

If the shifted coordinates are given by

$$\tilde{x}_i^{J,\mathrm{X}} = x_i^{\mathrm{X}} + d_{xx}^{\mathrm{X}} E_{J,x}(\mathbf{r}_i^{\mathrm{X}}) \tag{6.47a}$$

$$\tilde{y}_i^{J,\mathrm{X}} = y_i^{\mathrm{X}} + d_{yy}^{\mathrm{X}} E_{J,y}(\mathbf{r}_i^{\mathrm{X}}) \tag{6.47b}$$

$$\tilde{z}_i^{J,\mathrm{X}} = z_i^{\mathrm{X}} + d_{zz}^{\mathrm{X}} E_{J,z}(\mathbf{r}_i^{\mathrm{X}}) \tag{6.47c}$$

then one can write the following least-squares expression, which includes also the off-diagonal tensor elements,

$$\sum_{k=1}^{m_X} \sum_{J=1}^{N_E} |\tilde{V}_{PC}^{J,X}(\mathbf{R}_k^X) - V_{HF}^{J,X}(\mathbf{R}_k^X)|^2 = \min \tag{6.48}$$

where

$$\begin{aligned}\tilde{V}_{PC}^{J}(\mathbf{R}_k) = \sum_{i=1}^{M} q_i/[(\tilde{x}_i^J + d_{xy}E_{Jy}(\mathbf{r}_i) + d_{xz}E_{Jz}(\mathbf{r}_i) - X_k)^2 \\ + (\tilde{y}_i^J + d_{yx}E_{Jx}(\mathbf{r}_i) + d_{yz}E_{Jz}(\mathbf{r}_i) - Y_k)^2 \\ + (\tilde{z}_i^J + d_{zx}E_{Jx}(\mathbf{r}_i) + d_{zy}E_{Jy}(\mathbf{r}_i) - Z_k)^2]^{1/2}\end{aligned} \tag{6.49}$$

The capital X superscript indicating the Xth molecule has been suppressed here for the sake of simplicity.

The minimization problem (6.45) must be solved first and then the resulting diagonal elements of $\boldsymbol{d}$ substituted into expression (6.48). The resulting equation (6.48) is then solved for the potentials $\tilde{V}_{PC}^{J,X}$, which are substituted back into equation (6.46). This equation (6.46) is again solved with the new potentials to provide new diagonal elements of $\boldsymbol{d}$. The procedure is repeated for each molecule and value of J until self-consistency is attained.[10] Experience gained from calculations on several molecules shows that the iteration procedure converges to the same tensor $\boldsymbol{d}$ if $N_E > 3$; in the actual calculations $N_E = 8$ was used.[10]

Once quantities q_i, $\mathbf{r}_i$, and $\boldsymbol{d}$ are known for a series of, say, N molecules it is easy to establish the interaction energy between them at any relative positions of the ensemble of these molecules. For this reason, their resultant electric field acting on a chosen molecule X must be calculated at the positions $\mathbf{r}_i^X$ and is given by

$$\mathbf{E}(\mathbf{r}_i^X) = \sum_{J \neq X}^{N} \sum_{j=1}^{M_J} \frac{q_j^J \mathbf{r}_j^J}{|\mathbf{r}_i^X - \mathbf{r}_j^J|^3} \tag{6.50}$$

Substitution of this expression into equation (6.43) yields the shifts $\Delta\mathbf{r}_i^X[\mathbf{E}(\mathbf{r}_i^X)]$ of the point charges q_i^X from their original positions $\mathbf{r}_i^X$. The new position vectors $\tilde{\mathbf{r}}_i^X = \mathbf{r}_i^X + \Delta\mathbf{r}_i^X$ enable the electric field of molecule X to be recalculated at any other molecule. This procedure must be repeated until one obtains an MCF solution for the positions $\tilde{\mathbf{r}}_i^{J\,\mathrm{MCF}}$ of the point charges q_i^J ($J = \mathrm{A, B, C}, \ldots, N$), usually after five or six iteration steps.[10] These MCF position vectors very easily yield the MCF electrostatic and polarization interaction energy between all N molecules. The exchange energy can be calculated again with the help of the local $\rho^{1/3}$

potential as[10]

$$E_{\text{exch}}^{I(J)} = \sum_{i=1}^{M_I} V_{\text{exch}}^{J}(\tilde{\mathbf{r}}_i^{I\,\text{MCF}})\rho^I(\tilde{\mathbf{r}}_i^{I\,\text{MCF}})w_i \tag{6.51}$$

where

$$V_{\text{exch}}^{J}(\tilde{\mathbf{r}}_i^{I\,\text{MCF}}) = -6\alpha[\tfrac{3}{4}\pi^{-1}\rho^J(\tilde{\mathbf{r}}_i^{I\,\text{MCF}})]^{1/3} \tag{6.52}$$

with the weight factor $w_i = q^I{}_i/N_e^I$, N_e^I being the number of electrons in molecule I.

It has been proved that a good approximation[10] for the charge-transfer (CT) energy is given by a situation in which the transferred partial charge Δq is distributed among the point charges q_i^I of molecule I in a simple manner, given by

$$q_i^{I'} = q_i^I + \Delta q_i^I = q_i^I + \Delta q^I/M^I \tag{6.53}$$

Finally, the dispersion energy can be calculated most advantageously with the aid of London's expression (6.34).

The scheme described here was first applied to a water dimer and to three H_2O moecules in different relative positions using a triplet ζ basis set. In both cases good agreement with SM results (corrected for basis-set superposition errors) were obtained. Details can be found elsewhere.[10]

In a subsequent paper[19] the aforementioned, much more rapid new version of the MCF method was used in order to determine the stacking energies of larger segments of DNA single and double helices in both the B and Z conformation, proceeding to a nonamer in the case of a single helix and a pentamer in the case of a double helix. The results when compared with SM calculations (two interacting stacked bases) showed good agreement and seemed to be reasonable in the case of larger DNA segments. Some interesting conclusions could also be drawn from these extended calculations concerning the sequence dependence of the stability of the DNA helices (see Otto's paper[19] for details).

6.3. APPLICATION TO DNA–PROTEIN INTERACTIONS

The first application of the MCF method (in its first version) to the interaction of two infinite chains, namely to the interaction of polyglycine (in different conformations) and different polynucleotide chains, was performed by Otto *et al.*[20] This calculation represents the first step in the theoretical treatment (at the *ab initio* level) for DNA–protein interaction. In other words, interactions between proteins and nucleic acids are of central importance in molecular biology. Acidic proteins and basic proteins, such as histones and protamines, both rich in arginine, form complexes with single- and double-stranded polynucleotides, called chromatins, which form the fundamental structure of the chromosomes. Acidic proteins are found mainly in metabolically active chromosomal structures. Their high concentration at the nucleus and their high metabolism rate indicate their regulative biological activity, which may be the counterpart to the inhibitive function of the histones in the transcription.[21] A structure conserving and stabilizing function has been attributed to the basic proteins, since histones are associated mostly with dense-packed and metabolically inactive DNA sequences. It appears, however, that they are responsible for the more general function of gene repression. They exist with a nonspecific sequence of amino acids, especially in the heterochromatin part of the chromosomes. This region with apparently no active genes is characterized by a high concentration of proteins, which may act as multiple control elements. The lysine-rich histone I, for instance, has a special regulating function suppressing the transcription of certain genes.[22]

Despite the vast number of chemical and physical experiments, the available information is still insufficient to explain the detailed structure of the nucleohistones, with the exception of several DNA–repressor protein complexes. Some general tendencies, however, have been found: the α-helix conformation of the polypeptides is predominant in histones, as has been shown by ORD experiments[23]; lysine-rich histones mainly occupy the major groove of B-DNA,[24] while protamines are bound to DNA in the minor groove[25] (for further details see Otto *et al.*[20]).

The level of detailed knowledge is much higher for certain repressor proteins and their interactions with DNA. The structures of four polypeptides that specifically bind to DNA have been determined recently, namely the lac repressor protein,[26,27] the cro repressor protein from bacteriophage,[28] the catabolite gene activator protein from *Escheria coli*,[29] and the amino-terminal fragment of the CI repressor protein from bacteriophage λ.[30] In addition to the primary structure of the polypeptide, the sequence of the base pairs in that DNA fragment to which the relatively short polypeptide is bound, is known. Furthermore, structural data about the conformation of the peptide backbone are available, e.g., strands of α-helices followed by antiparallel β-pleated sheets.

Already this short and by no means complete survey reveals the general problem inherent in the theoretical investigations of the interaction between DNA and proteins. One of the main difficulties in the treatment of complicated systems like DNA–peptide complexes consists of the choice of the model. One can consider two different approaches, according to the state of structural knowledge. On the one hand we have the "general" nucleohistone complexes, while on the other we have the well-defined repressor protein–operator systems. In the former case we must assume certain conformations on the basis of the available information about the polypeptide backbone configuration. Then the amino-acid side groups must be adjusted and finally this geometry optimized relative to the DNA structure by calculating the interaction energy.[31] In principle, this cycle should be repeated for different starting configurations in order to obtain the energetically most stable complex. For those cases where the basic structural properties are known, there still remains the formidable task of fitting both constituents together stereochemically to form the complex.[28,32] Recent progress in the construction of complementary structures between flexible strands of DNA and polypeptides has been achieved with the help of new computer programs for model building and graphical representation.[33] These procedures are combined with the theoretical computation of interaction energies between those macromolecules,[34,35] mostly with the use of empirical potential functions.

Fundamentally, one can partition the interaction between DNA and polypeptides into contributions of the subgroups of DNA (phosphate, sugar, and base residues) interacting with the amino-acid side chains and the backbone of polypeptides. The study began with an investigation of the interaction between the polypeptide backbone in different conformations and periodic single and double helices of B-DNA. From the results one expects the answers to several questions: (1) Does there exist a stable complex between the polypeptide backbone and DNA, and if so, (2) which conformation is the most favorable one for the polypeptide, and (3) is the interaction between the backbone of the polypeptide and DNA large enough to be a dominant factor in the relative geometrical arrangement of both macromolecules?

In further investigations one should include the effect of neutral and charged amino-acid side chains. Detailed studies have been done to determine the interactions between side groups of amino acids with specific atomic sites of the DNA subunits (e.g., hydrogen bond formation with nucleotide bases[36] and ion–ion interactions between the negatively charged phosphate group of DNA and the cation arginine[37]).

6.3.1. Models for B-DNA and Polyglycine and Methods of Calculation

Periodic single-stranded DNA helices have been represented by the four homopoly (nucleotide bases stacks), with either cytosine, thymine, adenine, or guanine as a repeating unit, and the polynucleotide chains, polycytidine (CSP), poly(adenylic acid) (ASP) and polythymidine (TSP), with the sugar-phosphate base unit as elementary cell. The Watson–Crick base pairs G–C and A–T and the dinucleotide ASP–TSP build up the elementary cell of those macromolecules that have been chosen as models for the DNA double helix. For the molecular geometry of the elementary cell and the helical parameters, the structural data of B-DNA[38] have been used. Recently, *ab initio* LCAO-SCF Hartree–Fock crystal-orbital computations were carried out for the electronic band structure of the periodic DNA helices, described above, with the exception of poly(ASP–TSP).[39] In the last case we generated the model by superposition of the charge distributions of the corresponding single helices.

Since the binding properties of the protein backbone with DNA was of interest, polyglycine was taken as the polypeptide component and the elementary cell described by a diglycyl unit. The following conformations, for which the structural parameters are given in Table 6.1, have been selected: the fully extended form of polyglycine (PGFE), the α-helix (PGαH), and four helices with radii somewhat larger than the radius of the B-DNA helix (10.5 Å). These configurations were constructed with the help of a computer program for the calculation of helical parameters[40] applying the following criteria: (1) the radius of the helix should be about 15 Å, and (2) their turn length should equal a multipole or a fraction of the length of one complete winding of B-DNA (33.6 Å). The turn lengths of the selected polyglycine helices are 100.8. 67.2, 33.6, and 8.4 Å corresponding respectively to PG turn length/B-DNA turn length ratios of 3, 2, 1, and 0.25. It should be noted that in the latter four cases the helix axes of both macromolecules, DNA and polyglycine, coincide. Subsequently and in Table 6.1, which contains the parameters of the polyglycine chains in different conformations, the abbreviations PG101, PG67, PG33, and PG8 are used for these systems.

Table 6.2 summarizes the energy parameters of the highest occupied and lowest virtual bands of polyglycine in the conformations described. The *ab initio* band-structure calculations using Clementi's minimal basis set ($7s/3p$) were carried out in the second-neighbor-interaction approximation. The correct electrostatic balance[41] was taken care of by computing Coulomb and exchange integrals including atomic functions up to the fourth-neighbor cell. From the total energy per unit cell (column 6 in Table 6.2) it follows that the extended conformation of polyglycine is the most stable. PGαH is less stable by about 6 kcal/mol per unit cell, followed by the helices in the order PG33, PG67, PG8, and finally PG101. At first glance one would expect that with increasing turn length of the PG helix the total energy would approach the value for PGFE.

TABLE 6.1. Structural Parameters of the Six Investigated Polyglycine Models and the Definitions of the Supercells

Polyglycine model	Helical parameters a_{trans} (Å)	α_{rot}	Dihedral angles ψ_1	φ_1	ψ_2	φ_2	Supercell,[a] number of subunits in Polyglycine	B-DNA model
PGFE[b]	3.380	180.0	0	0	0	0	10	10
PGαH	6.450	197.59	123	132	123	132	11	20
PG101	5.130	18.65	180	180	200	180	19	30
PG67	3.581	19.03	140	240	300	60	19	20
PG33	2.039	21.76	200	180	320	40	17	10
PG8	0.604	25.48	0	20	200	140	55	10

[a] The supercell of the complex is defined as the smallest translationally symmetric subunit of the complex.
[b] In PGFE the subunit is the glycyl residue.

TABLE 6.2. Energy Parameters of Two Bands for Helices of Polyglycine[a]

Polyglycine model	Band	Band minima	Band maxima (eV)	Bandwidth	Gap	Total energy per unit cell (au)	E_{stab} (kcal/mol)	E_{intra} (kcal/mol)
PGFE	n^*+1	3.624	4.169	0.545	16.353	−412.102 70	0.0	—
	n^*	−12.548	−12.184	0.364				
PGαH	n^*+1	3.479	3.949	0.470	15.996	−412.093 19	5.97	—
	n^*	−12.071	−12.047	0.024				
PG101	n^*+1	3.636	3.659	0.023	15.401	−412.004 64	61.55	—
	n^*	−11.757	−11.742	0.015				
PG67	n^*+1	3.737	4.108	0.371	15.331	−412.073 71	18.20	−0.004
	n^*	−11.418	−11.223	0.195				
PG33	n^*+1	3.719	4.064	0.345	15.586	−412.080 79	13.75	−0.023
	n^*	−11.693	−11.522	0.171				
PG8	n^*+1	3.893	4.189	0.296	13.778	−412.029 35	46.04	2.99
	n^*	−9.912	−9.589	0.323				

[a] For comparison the results for the fully extended and the α-helix conformations of polyglycine are also reported. In the last two columns the difference in the total energy per elementary cell with respect to the result for PGFE and the interaction energy between subunits of one helix separated by one turn are given (after Otto *et al.*[20]).

However, in the systems described above this regularity is not observed. First, several sets of dihedral angles (ψ_1, ϕ_1, ψ_2, ϕ_2) exist to generate each of the helices, but there is no correlation between the values of these angles of the helices and their turn lengths. The molecular structure of the diglycyl unit of PG8 is such (for details see Otto *et al.*[20]) that there is a possibility of intramolecular hydrogen bonding, which may be responsible for the stabilization energy of about 15 kcal/mol compared with PG101. Another effect might play a role for the observed "reversed" order of stability. The interaction between two subunits being separated by a distance of one turn length is not taken into account in the crystal-orbital calculation. This fact, which might be important especially in the case of PG8, has been estimated by computing the electrostatic part of this energy contribution with the aid of atomic point charges. The values given in Table 6.2 in the E_{intra} column exhibit only a very small attractive correction for PG33 and PG67. However, the repulsive interaction of about 3 kcal/mol per subunit for PG8 cannot balance the difference in total energies for PG101 and PG8. Again, for the same reasons, no "regularity" is observed for the bandwidths of the conduction and valence bands (denoted by n^* and $n^* + 1$ in Table 6.2) of the four helices. In the case of PG101 they are much narrower than the other helix models (about one-tenth).

In principle, all methods that can be used to compute intermolecular interaction energies may be generalized for the investigation of the interaction between macromolecules. The superchain method, however, where the new elementary cell of the combined system is formed from the unit cells of the interacting polymers, can only be applied if the translation vectors of both macromolecules are parallel and of equal length. This method cannot be used to compute the interaction energies between DNA and polyglycine, because there is no common operation to generate the repeating units in both chains.

Another possibility consists in the application of the perturbation theoretical expressions for the various energy terms, appropriately modified for interacting polymers. The corresponding formalism has been derived also for the case of systems with incommensurable unit cells.[42] However, as noted earlier, this method assumes that the number of unit cells in one chain is a multiple of the unit cells in the other chain. Therefore its application would cause computational problems because of the large number of atoms in the unit cells of the DNA models.

Hence the MCF method[18,20,43] was applied in the form appropriate to treat the interaction between polymer chains without the condition of commensurable cells. By computing the interaction between the MCF point charges (and nuclear charges) of both macromolecules, one obtains both the electrostatic energy and the contribution of the polarization energy (see above).

Both one-dimensional periodic polymers, DNA and polyglycine, become aperiodic as soon as they interact. The symmetry operation for the combined system is now a simple translation in the direction of the DNA helix axis with the length corresponding to the turn length of B-DNA or a multiple of it, depending on the PG model. This translationally symmetric molecular fragment is defined as the supercell of the complex. It can be seen from the schematic illustration in Figure 6.2a that each unit cell of one chain experiences a different effective potential of the partner chain, and the aperiodicity results in a different charge distribution within each unit of both polymers. The change in the electronic distribution was taken into account with the help of the MCF method. The procedure can be described in more detail for the DNA–PGFE complex represented in Figure 6.2a, as follows: The translationally symmetric unit cell of the complex (the supercell) contains ten nucleotide and ten glycyl units. First, a starting configuration for the DNA–polyglycine complex is chosen. Then the energy band-structure calculation (only the one-electron integral computation and the SCF iteration) is repeated for each of the ten glycyl units on the assumption that a periodic potential field arises from the neighboring cell of the partner chain. This external potential originates from the atoms of the DNA helix that are located in a cylinder with height equal to the length of the internal translation vector length of PGFE. A suitable representation of the electronic distribution of the molecular unit cell can be found by computing the electrostatic potential at a given set of spatial positions. The next step is to fit the resulting potentials with the aid of point-charge potentials, by minimizing the deviation of the two electrostatic fields in the least-squares sense. Up to three SCF iterations were necessary, depending on the strength of the perturbation. In the same way the different charge distributions are determined for the DNA component. Finally, the point-charge representation of one complete winding of DNA and polyglycine is built up from the point-charge distributions of the single subunits.

The exchange and charge transfer energy contributions were computed by a procedure described in detail by Förner *et al.*[17] This energy term is calculated for several relative positions of the subunits of the two macromolecules with the help of perturbation expressions [see Murrell *et al.*[1] and equations (6.5)–(6.7)]. The results are then fitted to an exponential function in order to obtain the exchange energy for additional relative geometries.

The dispersion energy has been computed using London's formula

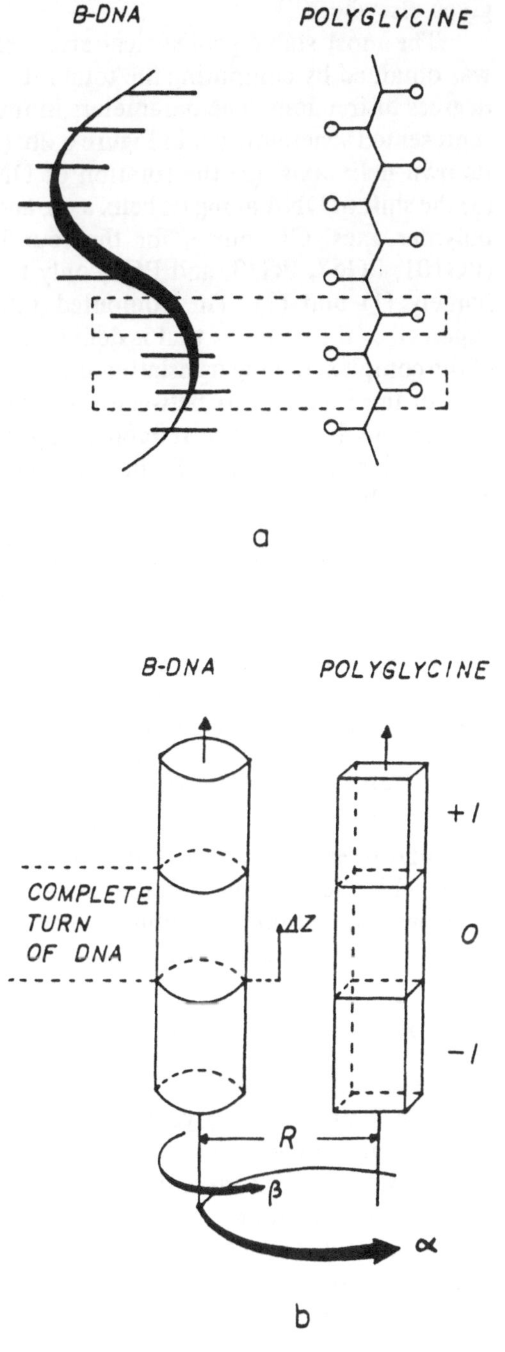

FIGURE 6.2. (a) Schematic graph of the complex PGFE and DNA single helix to demonstrate the creation of aperiodicity within each chain; (b) graphical representation of the degrees of freedom used to determine the most stable DNA-polypeptide structure.

with the atomic polarizabilities and valence-state ionization energies given elsewhere.[15]

The most stable geometrical arrangement of both macromolecules was obtained by computing the total interaction as a function of several degrees of freedom. The parameters in the case of PGFE and PGαH are represented schematically in Figure 6.2b: (1) the rotation of DNA around its own helix axis; (2) the rotation of DNA around the PG helix axis; (3) the shift of DNA along its helix axis; and (4) the distance between both polymer axes. Of course, for the four helical models of polyglycine (PG101, PG67, PG33, and PG8) only two degrees of freedom remain, namely (1) and (3). The computed interaction energy refers to the supercell of the complex that is defined as the smallest possible fragment of the complex, having translational symmetry. The energy is given as the sum of the interactions between the reference supercells ΔE^{00}, and the interactions between the reference supercell with the lower and upper neighboring supercells of the partner system, ΔE^{0-1} and ΔE^{0+1}, respectively:

$$\begin{aligned} \Delta E &= \Delta E^{00} + \Delta E^{0+1} + \Delta E^{0-1} + \Delta E^{10} + \Delta E^{-10} \\ &= \Delta E^{00} + (2\Delta E^{01} + \Delta E^{0-1}) \end{aligned} \tag{6.54}$$

The latter equality is valid because of the polymer symmetry. Interactions with further neighbors have been shown to be negligible. The general term ΔE^{01} in the expression for the energy can be written as

$$\Delta E^{0I} = E^{0I}_{\text{elst+pol}} + E^{0I}_{\text{exch}} + E^{0I}_{\text{ch tr}} + E^{0I}_{\text{disp}} \qquad (I = 0, 1, -1) \tag{6.55}$$

It should be noted that the mutually consistent band-structure calculations, which lead to the different electronic charge distributions within the originally periodic macromolecules, must be performed for different values of the degrees of freedom.

6.3.2. Results of B-DNA Polygly Calculations and their Discussion

Table 6.3 summarizes the calculated interaction energies given by equation (6.54) for polyglycine in the fully extended conformation with the periodic single- and double-stranded B-DNA models. The first column presents the sum of the energies $E_{\text{elst+pol}}$, E_{exch}, and $E_{\text{ch tr}}$, while the second coumn contains the dispersion E_{disp}; the sum of both is listed in the last column. Table 6.4 contains similar information for polyglycine in the investigated other helix conformations (PGαH, PG101, PG67, PG33, and PG8) with poly(ASP) and poy(ASP–TSP).

TABLE 6.3. Contributions to the Interaction Energies between Polyglycine in the Fully Extended Conformation and Periodic Single- and Double-Stranded B-DNA Helices (in kcal/mol)[20]

B-DNA model	Polyglycine fully extended		
	$(E_{elst+pol} + E_{exch} + E_{ch\,tr})$	E_{disp}[a]	ΔE
Poly(A)	−2.9	−2.4	−5.3
Poly(G)	−3.1	−2.0	−5.1
Poly(T)	−3.7	−4.4	−8.1
Poly(C)	−3.6	−3.9	−7.5
Poly(ASP)	−24.6	−1.7	−26.3
Poly(TSP)	−21.0	−1.7	−22.7
Poly(A–T)	−9.6	−6.9	−16.5
Poly(G–C)	−7.7	−5.8	−13.5
Poly(ASP–TSP)	−24.4	−2.8	−27.2

[a] $E_{disp} = E^{00}_{disp} + 2(E^{0+1}_{disp} + E^{0-1}_{disp})$, and an analogous expression for $(E_{elst+pol} + E_{exch} + E_{ch\,tr})$.

TABLE 6.4. Interaction Energies for the B-DNA Single and Double Helix, Poly(ASP) and Poly(ASP–TSP), with the Different Helical Conformations of Polyglycine (in kcal/mol)[20]

Poly-glycine helix	B-DNA model					
	Poly(ASP)			Poly(ASP–TSP)		
	$(E_{elst+pol} + E_{exch} + E_{ch\,tr})$	E_{disp}	ΔE	$(E_{elst+pol} + E_{exch} + E_{ch\,tr})$	E_{disp}	ΔE
PGαH	−14.4	−1.1	−15.5			
PG101	−4.8	−2.8	−7.6	−0.8	−5.1	−5.9
PG67	−10.7	−1.7	−12.4	−21.9	−3.3	−18.6
PG33	−55.3	−0.9	−56.2	−27.6	−4.0	−31.6
PG8	−5.6	−3.2	−8.8	−2.4	−6.1	−8.5

[a] See footnote of Table 6.3 for the definition of the interaction terms.

Despite several serious but well-defined simplifications and assumptions inherent in our DNA–protein models we are able to deduce general conclusions valid for the real system. Three main approximations have been made until now: (1) neglect of possible interactions between amino-acid side chains and subgroups of DNA, (2) assumptions of full periodicity in the polypeptide backbone, and finally, (3) neglect of the effects of the surrounding water and ions.

Computational studies on molecular subunits of the nucleohistone complex[31] indicate that the binding strength of a single hydrogen bond between nucleotide bases and suitable amino-acid residues is of the order of 5 to 9 kcal/mol. Unscreened ion–ion interactions between the phosphate group and charged amino acids yield about -60 kcal/mol.[37] Furthermore, experiments mainly on repressor proteins indicate that the polypeptide chain exists in the α-helix and extended β conformation,[28,32] both being more or less distorted, and the length of the respective sequences depends on the nature of the amino-acid sequence.

The fully extended polypeptide backbone conformation used in the investigations is optimal in some sense such that every second amino-acid residue is directed toward the DNA helix and could interact with the neighboring base or phosphate group, especially if a distortion as proposed by Feughelmann *et al.*[44] takes place. Since PGFE forms one strong local bond with DNA only in the case when the two chains fit together exactly, the probability of this will be rather small. When this happens, the peptide units belonging to this conformation and having neutral or hydrophobic side chains would have high flexibility with respect to rotations. A structure-supporting function of the protein cannot be attributed to these regions of loosely bound and mobile polypeptide sequences.

Similar conclusions may be drawn for the α-helix conformation investigated with the peptide helix being strictly parallel to the DNA helix axis. The single local interaction per DNA turn is still smaller than for PGFE. A longer sequence of the polypeptide, assuming this conformation relative to DNA, would make this region attackable for chemical reactions. In addition, only every fourth or fifth amino-acid side group may interfere with suitable binding sites of DNA subunits. Better binding properties can evidently be achieved by introducing a distortion into the linear polypeptide chain such that the α-helix to some extent follows the winding of the DNA helix.

Present work strongly supports the hypothesis that DNA is able to impose its helical symmetry on the conformation of the polypeptide chain. This hypothesis is suggested by the existence of a stoichiometry between peptide and nucleotide residues, corresponding to 1/1, as found in synthetic basic homopolypeptide–DNA complexes. However, more diglycyl than nucleotide units are needed to build up a PG33 helix because its radius is larger than the radius of the DNA helix. In cases where the turn length of the polypeptide differs from the corresponding B-DNA parameter, only weak binding or no binding at all is observed. Side-group interactions with DNA may stabilize these configurations but sequences containing mainly glycine and alanine would again cause regions of labile complexes. Contrary to this finding the polypeptide backbone of PG33, even in the absence of appropriate amino-acid residues, contributes essentially to the overall stabilization of the complex. Additional links between subgroups of both chains will, of course, enforce the strength of the bonding.

Only speculative predictions could be made until now about the effect of the surrounding solvent molecules on the stability and structure of the nucleohistones. Recently, Monte Carlo computer experiments of the water and counterion structure of periodic homopolynucleotide single helices were performed.[45] One of the main results was that the position of the sodium ions depends on the nature of the base. In the case of poly(GSP) all counterions were found to have moved near to the atoms of the bases, while for poly(TSP) they are located mainly in the vicinity of the phosphate–sugar residues. In these model systems the sodium counterions are bound to the free oxygen atoms of the phosphate group, determined to be the most stable geometry with the help of *ab initio* molecular calculations.[46] Keeping this in mind one can expect drastic changes in the stability and structure of those DNA–protein complexes where the binding is due to ion–dipole interactions between $PO_4^-Na^+$ and the glycyl unit (such as PGFE and PGαH). It can also be said that for the other helix models of polyglycine the effect will be less dramatic, because here the stability is based mainly on dipole–dipole interactions. One can even imagine that the binding energy will be increased as a result of more favorable distribution of the counterions relative to the polypeptide.

A more conclusive answer to this and the other problems noted above can only be given by further theoretical investigations of the nucleohistones. On the one hand, the model systems must be improved by constructing a polypeptide helix that fits into the major groove of B-DNA and, at the same time, takes into account the amino-acid side chains. This last improvement, of course, has to be paralleled by the introduction of chemical aperiodicity in the DNA helix. On the other hand, a more realistic system can be approached step by step by including the influence of water surrounding the complex (more details can be found in Otto *et al.*[20]).

REFERENCES

1. J. N. Murrell, M. Randić, and O. R. Williams, *Proc. R. Soc. London, Ser. A* **284**, 566 (1965).
2. R. S. Mulliken, *J. Chem. Phys.* **23**, 1833 (1955).
3. See, for instance, R. Rein, P. Claverie, and J. Pollak, *Int. J. Quantum Chem.* **2**, 129 (1968).
4. E. F. Haugh and J. O. Hirschfelder, *J. Chem. Phys.* **23**, 1778 (1955).

5. K. Laki and J. Ladik, *Int. J. Quantum Chem., Quantum Biol. Symp.* **3**, 51 (1976).
6. F. Beleznay, S. Suhai, and J. Ladik, *Int. J. Quantum Chem.* **20**, 683 (1981).
7. A. Imamura, S. Suhai, and J. Ladik, *J. Chem. Phys.* **76**, 6067 (1982).
8. P. Otto and J. Ladik, *Chem. Phys.* **8**, 192 (1975).
9. P. Otto, *Chem. Phys.* **33**, 407 (1978).
10. P. Otto, *Int. J. Quantum Chem.* **28**, 895 (1985).
11. P. Otto and J. Ladik, *Chem. Phys.* **19**, 209 (1977).
12. J. M. Foster and S. F. Boys, *Rev. Mod. Phys.* **32**, 300 (1960); S. F. Boys, *Rev. Mod. Phys.* **32**, 306 (1960); S. F. Boys, in: *Quantum Theory of Atoms, Molecules and the Solid State* (P.-O. Löwdin, ed.), p. 253, Academic Press, New York–London (1966).
13. B. Gaissmaier, W. Hohecker, R. Unbehauen, and W. Werhalm, *Frequenz*, Vol. 29, No. 5 (1975).
14. K. S. Pfitzer, *Adv. Chem. Phys.* **2**, 59 (1959).
15. Y. K. Kang and M. S. Jhon, *Theor. Chim. Acta* **61**, 41 (1982).
16. P. Otto, *Chem. Phys. Lett.* **62**, 538 (1979).
17. W. Förner, P. Otto, J. Bernhardt, and J. Ladik, *Theor. Chim. Acta* **60**, 269 (1981).
18. L. Ladik, *Int. J. Quantum Chem.* **S9**, 563 (1975).
19. P. Otto, *Int. J. Quantum Chem.* **30**, 275 (1986).
20. P. Otto, E. Clementi, J. Ladik, and F. Martino, *J. Chem. Phys.* **80**, 5294 (1984).
21. L. J. Kleinsmith, J. Heidema, and A. Carroll, *Nature* **22**, 1025 (1970); C. S. Teng, C. T. Teng, and V. G. Allfrey, *J. Biol. Chem.* **246**, 3597 (1971).
22. J. Hindley, *Biochem. Biophys. Res. Commun.* **12**, 175 (1963); G. C. Barr and J. A. V. Butler, *Nature* **199**, 1170 (1963).
23. G. Zubay and P. Doty, *J. Mol. Biol.* **1** (1959).
24. M. H. F. Wilkins, G. Zubay, and R. H. Wilson, *Trans. Faraday Soc.* **55**, 497 (1959).
25. V. Luzzati and N. Nicolaieff, *J. Mol. Biol.* **7**, 142 (1963).
26. K. Adler, K. Beyreuther, E. Fanning, N. Geisler, B. Gronenborn, A. Klemm, B. Müller-Hill, M. Pfahl, and A. Schmitz, *Nature* **237**, 322 (1972).
27. K. Beyreuther, K. Adler, N. Geisler, and A. Klemm, *Proc. Natl. Acad. Sci. U.S.A.* **70**, 3576 (1973).
28. W. F. Anderson, D. H. Ohlendorf, Y. Takeda, and B. W. Mattheus, *Nature* **290**, 754 (1981).
29. D. B. McKay and T. A. Steitz, *Nature* **290**, 744 (1981).
30. C. O. Pabo and M. Lewis, *Nature* **298**, 443 (1982).
31. G. M. Church, J. L. Sussman, and S. H. Kim, *Proc. Natl. Acad. Sci. U.S.A.* **74**, 1458 (1977).
32. D. H. Ohlendorf, W. F. Anderson, R. G. Geisler, Y. Takeda, and B. W. Matthews, *Nature* **298**, 718 (1982).
33. S. Nir, R. Garduno, R. Rein, Y. Coeckelenberg, and P. D. MacElroy, *Int. J. Quantum Chem., Symp.* **4**, 135 (1977); S. Nir, R. Garduno, and R. Rein, *Polymer* **18**, 431 (1976).
34. R. Rein, R. Garduno, J. T. Egan, and S. Columbano, *Biosystems* **9**, 131 (1977).
35. R. Rein, R. Garduno, J. T. Egan, S. Columbano, Y. Coeckelenberg, and R. D. MacElroy, *Orgins Life*, 265 (1978).
36. R. Garduno, K. Haydock, R. D. MacElroy, and R. Rein, *Ann. N.Y. Acad. Sci.* **281** (1981).
37. R. Day, F. Martino, and J. Ladik, *J. Theor. Biol.* **84**, 451 (1980).
38. R. Fieldman, *Document 13.2.1.1.1.*, National Institutes of Health, Bethesda, Washington (1976).

39. P. Otto, E. Clementi, and J. Ladik, *J. Chem. Phys.* **78**, 4547 (1983).
40. S. Columbano and R. Rein, *Comput. Programs* **11**, 3 (1980).
41. J. Ladik and S. Suhai, *Chem. Rev.* **80**, 263 (1980).
42. A. Imamura, S. Suhai, and J. Ladik, *J. Chem. Phys.* **76**, 6067 (1982).
43. J. Ladik, *Int. J. Quantum Chem., Symp.* **9**, 563 (1975); J. Ladik and S. Suhai, in: *Molecular Interactions* (H. Ratajczak and W. J. Orville-Thomas, eds.), p. 151, John Wiley and Sons, New York (1980); J. Ladik, in: *Recent Advances in the Quantum Theory of Polymers* (J.-M. André, J.-L. Brédas, J. Delhalle, J. Ladik, G. Leroy, and C. Moser, eds.), p. 155, Springer-Verlag, Berlin (1980).
44. R. Feughelmann, R. Langridge, W. E. Seeds, A. R. Stokes, H. R. Wilson, L. W. Hooper, M. H. F. Wilkins, R. K. Barcklay, and L. D. Hamilton, *Nature* **175**, 834 (1955).
45. P. Otto, E. Clementi, and G. Corongiu (unpublished).
46. E. Clementi, in: *Computational Aspects for Large Chemical Systems, Lecture Notes in Chemistry*, Vol. 19, Springer-Verlag, New York (1980).

Chapter 7

The Effect of Environment on the Band Structure of Polymers

Polymers, especially biopolymers, are not found in a vacuum but exist rather surrounded by a medium (DNA and proteins are in an aqueous solution, with K^+ counterions in the case of DNA, and a smaller number of other ions around both). Hence it is important to consider the effect of the environment on the electronic structure of a (bio)polymer.

The first step in such an approach was to recalculate the band structure of a cytosine (C) stack[1] in the field of the water molecules surrounding a cytidine unit (mean-field approximation).

7.1. GENERATION OF AN EFFECTIVE POTENTIAL FIELD OF THE ENVIRONMENT

The C stack calculation involved the positions (more precisely, the weighted averages of the probability distributions of the atomic nuclei) of five water clusters around a cytidine molecule (and five clusters in the planes above and below) computed by Clementi's group[2] with the aid of the Metropolis–Monte Carlo method.[3] The five water clusters together contain 37 water molecules (see Figure 7.1). If three of their planes are selected, one has to construct the resultant effective potential of 111 water molecules. This was carried out with the help of the first (point-charge) version of the MCF method (see Section 6.2.1).

The 37 water molecules surrounding one cytidine unit have been divided into five clusters on the basis of their geometrical positions. It should be noted that in this first model calculation of the effect of the water surrounding DNA on its band structure, we used the water

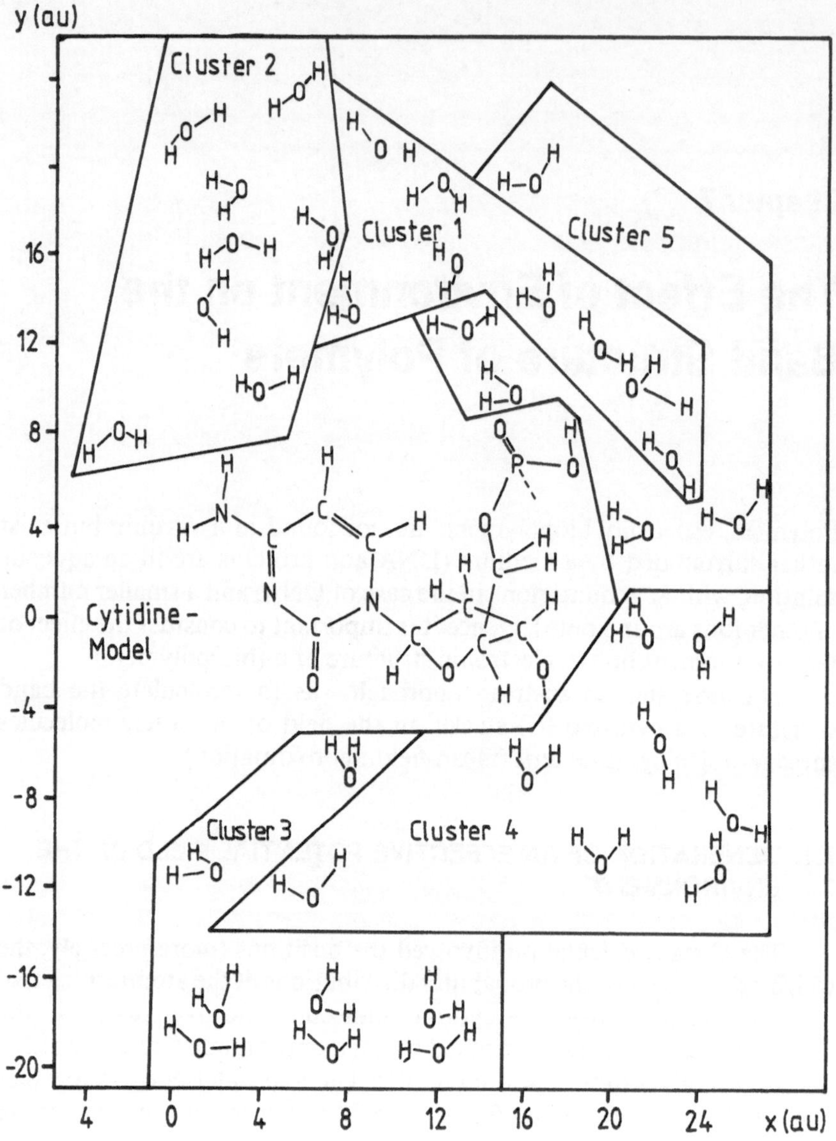

FIGURE 7.1. Five water clusters around a cytidine unit (after Corongiu and Clementi[2]).

structure around a cytidine (cytosine + sugar + phosphate) unit to generate an effective potential acting on a cytosine stack. This procedure, of course, does not give a realistic description of the effect of the water structure on the electronic structure of DNA, but gives in a relatively simple model calculation an order-of-magnitude estimate of the effects to be expected. In a more realistic calculation (which is planned as the next step) the water structure around a cytidine unit together with the counterions should be considered and the effect of its effective potential on the band structure of a cytidine polymer investigated.

In subsequent steps the aperiodicity of DNA as well as the role of the polypeptide chain in the deep groove of DNA has to be taken into account. In the light of these requirements this calculation (but also a more consistent one, which would use the water structure computed for a cytosine stack that was unavailable at the time of the calculation) can be regarded only as a rough order-of-magnitude estimate of the investigated effect. It does not describe completely the role of the first hydration shell around a cytosine stack and at the same time overestimates the effect of the water molecules situated further away by not introducing any screening to account for the sugar and phosphate molecules between the water and the cytosine stacks.

Inspection of the five water clusters (see Figure 7.1) around a cytidine unit shows, however, that most of the water molecules are situated in the immediate neighborhood of the cytosine molecules (forming the larger part of their first hydration shell) while only some of the water molecules are further apart than in the case of a cytosine stack. Therefore, the error is less than one would expect without looking at the details. When the motivation of this model calculation is understood in the sense discussed above, it can be regarded as the next step after the work of Clementi,[4] who calculated the band structure of polyethylene in the field of periodic point charges situated around it in different ways.

After determining the proper point-charge representations of the water clusters, the potentials of the other water molecules situated around neighboring DNA subunits could be simply calculated by transforming the vectors $\mathbf{r}_{ji}^{N}$, which represent the positions of the point charges with the symmetry operation of the helix. After the charge distribution of each water cluster was represented in this way, the clusters could be allowed to interact and the modification of their charge distribution as a result of the interactions had been taken into account via the first version of the MCF method (see Section 6.2.1).[5]

We have finally calculated the matrix elements of the electronic point-charge potentials together with those of the corresponding nuclei in terms of the AO basis functions also used to construct the Bloch orbitals of the polymers. These matrix elements were then added to the one-electron part of the Fock matrix.

This method clearly gives only a first approximation of the problem. In a more sophisticated future calculation (which appears possible by employing the faster second version of the MCF method; see Section 6.2.2) the mutual polarization of the environment and the periodic stack (or chain) should also be taken into account.

7.2. RESULTS FOR A CYTOSINE STACK

The Bloch functions of the periodic DNA models have again been selected as symmetry-adapted linear combinations of Gaussian atomic orbitals. An $8s4p$ (primitive Gaussian) minimal basis on heavy atoms and the corresponding $4s$ basis on hydrogens has been used,[6] while second-neighbor interactions have been taken into account when calculating the lattice sums.

Table 7.1 shows the upper and lower limits of the conduction and valence bands of a C stack without and with hydration. The results show that the potential of the hydration shell somewhat stabilized the base stack. All the bands are moving downward and the total energy is lowered in the presence of the water molecules by 8 eV per elementary cell. It is interesting to note that the band shifts are quite different and vary between 0.3 and 0.8 eV, depending on the structures of the corresponding wave function, i.e., on which atom(s) contributes most to the Bloch function in question and how many water molecules are situated in its neighborhood (further details are provided in Otto *et al.*[1]).

Though no dramatic changes can be observed in the band structure as a whole, it should be pointed out that the hydration together with positive ions (such as Na^+ and K^+) could play a very important role in determining the conduction properties of DNA. Preliminary calculations have shown that the presence of K^+ ions causes band shifts of many electron volts in DNA. Of course, these ions do not appear alone but are always in solvated form. It will therefore be very important to treat them together with the hydration shell in order to obtain a more realistic description of the electronic structure of DNA.

TABLE 7.1. Conduction and Valence Bands of a Cytosine Stack Calculated without and with Hydration[a(1)]

Without hydration			With hydration		
E_{min}	E_{max}	δE	E_{min}	E_{max}	δE
5.916	6.738	0.822	5.608	6.473	0.865
−5.739	−4.872	0.867	−6.592	−5.679	0.913

[a] All quantities are in eV.

REFERENCES

1. P. Otto, J. Ladik, G. Corongiu, S. Suhai, and W. Förner, *J. Chem. Phys.* **77**, 5026 (1982).
2. G. Corongiu and E. Clementi, *Biopolymers* **20**, 551 (1981).
3. N. Metropolis, A. W. Rosenbluth, A. H. Teller, and E. Teller, *J. Chem. Phys.* **21**, 1087 (1953).
4. E. Clementi, *J. Chem. Phys.* **54**, 2492 (1971).
5. P. Otto and J. Ladik, *Chem. Phys.* **8**, 192 (1975); P. Otto, *Chem. Phys.* **33**, 407 (1978).
6. W. Hehre, R. F. Stewart, and J. A. Pople, *J. Chem. Phys.* **51**, 265 (1969).

II

Theoretical Calculation of the Different Physical Properties of Polymers

II

Chapter 8

Excited and Ionized States of Polymers

8.1. INTERMEDIATE EXCITON THEORY WITH CORRELATION

It is well known that in the HF theory of a closed-shell atom or of a finite molecule, the singlet excitation energy from a filled level i to an unfilled level a is given by the expression

$$^{1}\Delta E_{i\to a} = \varepsilon_a^{\mathrm{HF}} - \varepsilon_i^{\mathrm{HF}} - J_{ia} + 2K_{ia} \tag{8.1}$$

where $\varepsilon_a^{\mathrm{HF}}$ and $\varepsilon_i^{\mathrm{HF}}$ are, respectively, the corresponding HF one-electron energies while J_{ia} and K_{ia} denote the Coulomb and exchange integrals between orbitals ψ_i and ψ_a, respectively. It can easily be shown that if we have a very long chain, J_{ia} and K_{ia} tend to zero (see, e.g., Ladik[1]) as the number of unit cells, N, approaches infinity:

$$\lim_{N\to\infty} J_{ia} = \lim_{N\to\infty} K_{ia} = \lim_{N\to\infty} \frac{\ln N}{N} \to 0 \tag{8.2}$$

In this way we are left with too large an excitation energy, because equation (8.1) for an infinite system reduces to the HF gap (which is too large anyway).

However, in insulator solids or long linear chains possessing a band structure with delocalized electrons, a new physical phenomenon must be taken into account, namely, interaction between the excited electron and the remaining, positive hole in the valence band. This interaction in molecular crystals in usually described with the help of the simplest form of exciton theory, so-called Frenkel exciton theory (see, e.g., Knox[2]) which assumes that the excited electron and the remaining positive hole can be found in the same unit cell.

Since in most polymers the units are strongly coupled (covalent chemical bonds or at least strong stacking interactions), this simple form of exciton theory cannot be used for them. Instead, one has to apply the intermediate (or charge-transfer) exciton theory first suggested by Takeuti.(3) [The other simpler form of exciton theory, Wannier exciton theory, cannot be applied for organic polymers either. Usually, due to the too weak dielectric screening (dielectric constant $\varepsilon \sim 3$), a weakly bound exciton with large radius between the electron and hole cannot be formed.] In the intermediate exciton theory it is assumed that the electron and hole are neither localized to the same unit, nor are they situated at a large distance from each other, but on average both can be found within a radius of a few unit cells. The method employs the resolvent formalism first suggested by Lax(4) for the treatment of vibrational problems and by Koster and Slater(5) for computing the effect of an impurity in a chain that contains one atom (and orbital) per unit cell (tight-binding approximation). We have already seen this method formulated in an *ab initio* form to treat a cluster of impurities (or chain ends) embedded in a periodic chain (see Section 4.6).

The intermediate exciton theory of Takeuti [which applied some empirical parameters and assumed parabolic curves $\varepsilon(k)$; effective mass approximation] was developed in an *ab initio* form without any empirical parameters, using not HF but quasi-particle (QP) one-electron energies (see Section 5.3) by Suhai.(4)

The mathematical formulation of the theory of interaction between an electron and a hole starts from the Hamiltonian operator $\hat{H}$ (in second quantization and atomic units) given by

$$\hat{H} = \int \hat{\psi}^+(x)[-\tfrac{1}{2}\Delta + V_p(x)]\hat{\psi}(x)d^3x + \tfrac{1}{2}\int \hat{\psi}^+(x)\hat{\psi}^+(x')\frac{1}{|x-x'|}\hat{\psi}(x')\hat{\psi}(x)d^3x d^3x' \quad (8.3)$$

where $x = (r, \sigma)$ and $V_p(x)$ is the periodic potential of the ions. The field operators fulfill the anticommutation relations of fermions (see e.g., Fetter and Valecka(5))

$$\hat{\psi}_i\hat{\psi}_j + \hat{\psi}_j\hat{\psi}_i \equiv [\hat{\psi}_i, \hat{\psi}_j]_+ = [\hat{\psi}_i^+, \hat{\psi}_j^+]_+ = 0 \quad (8.4a)$$

$$[\hat{\psi}_i^+, \hat{\psi}_j]_+ = \delta_{i,j} \quad (8.4b)$$

and could be expanded with the help of Wannier spin orbitals (see

Section 5.1) in the form

$$\hat{\psi}(x) = \sum_l \hat{a}_{l\mathrm{v}} w(x-l) + \sum_l \hat{a}_{l\mathrm{c}} w_{\mathrm{c}}(x-l) \tag{8.5a}$$

$$\hat{\psi}^+(x) = \sum_l \hat{a}_{l\mathrm{v}}^+ w_{\mathrm{v}}^*(x-l) + \sum_l \hat{a}_{l\mathrm{c}}^+ w_{\mathrm{c}}^*(x-l) \tag{8.5b}$$

Here, v and c refer to the valence and conduction bands, respectively, while $w_{\mathrm{v\,or\,c}}(x-l) \equiv w_{\mathrm{v\,or\,c}}(\mathbf{r}-\mathbf{R}_l)$, $\mathbf{R}_l$ is the lattice vector of the cell in which the Wannier function is localized. The creation and annihilation operators $\hat{a}_{lm}^+$ and $\hat{a}_{lm}$ ($m = \mathrm{c, v}$), respectively, satisfy again the relations

$$[\hat{a}_{lm}, \hat{a}_{l'm'}]_+ = [\hat{a}_{l,m}^+, \hat{a}_{l'm'}^+]_+ = 0 \tag{8.6a}$$

$$[\hat{a}_{l,m}^+, \hat{a}_{l'm'}]_+ = \delta_{l,l'}\delta_{m,m'} \tag{8.6b}$$

If expansions (8.5) are substituted into expression (8.3) and the one- and two-electron parts of $\hat{H}$ separated, then one obtains

$$\hat{H} = \hat{h} + \hat{g} \tag{8.7}$$

with

$$\hat{h} = \sum_{l_1,l_2} \hat{a}_{l_1\mathrm{v}}^+ \hat{a}_{l_2\mathrm{v}} h_{l_1,l_2}^{\mathrm{v}} + \sum_{l_1,l_2} \hat{a}_{l_1\mathrm{c}}^+ \hat{a}_{l_2\mathrm{c}}^+ h_{l_1l_2}^{\mathrm{c}} \tag{8.8a}$$

$$\hat{g} = \frac{1}{2} \sum_{l_1,l_2,l_3,l_4} \sum_{m_1,m_2,m_3,m_4} \hat{a}_{l_1m_1}^+ \hat{a}_{l_2m_2}^+ \hat{a}_{l_3m_3} \hat{a}_{l_4m_4} \times \langle w_{m_1}^{l_1} w_{m_2}^{l_2} \mid w_{m_3}^{l_3} w_{m_4}^{l_4} \rangle, \qquad m_i = \mathrm{v, c} \tag{8.8b}$$

The matrix elements in equations (8.8) are defined by

$$h_{l_1l_2}^m = \int w_m^*(x-l_1)[\tfrac{1}{2}\Delta + V_{\mathrm{p}}(x)] w_m(x-l_2) d^3x, \qquad m = \mathrm{v, c} \tag{8.9a}$$

$$\langle w_{m_1}^{l_1} w_{m_2}^{l_2} \mid w_{m_3}^{l_3} w_{m_4}^{l_4} \rangle = \int\int w_{m_1}^*(x-l_1) w_{m_2}^*(x'-l_2) \frac{1}{|x-x'|} \times w_{m_3}^{l_3}(x'-l_3) w_{m_4}(x-l_4) d^3x\, d^3x' \tag{8.9b}$$ †

† Owing to the second quantization approach in integral (8.9b) the bracket $\langle ab \mid cd \rangle$ has not the usual definition of a Coulomb integral $\langle a(1)b(2) | 1/r_{12} | c(1)d(2) \rangle$ but the definition $\langle a(1)b(2) | 1/r_{12} | c(2)d(1) \rangle$.

One obtains an easier overview of the situation if one introduces the particle–hole description.[4] We define, respectively, operators $\hat{a}_l^+ \equiv \hat{a}_{lc}^+$, $\hat{a}_l \equiv \hat{a}_{lc}$ for the creation or annihilation of an electron in the conduction band and $\hat{d}_l^+ \equiv \hat{a}_{lv}$, $\hat{d}_l \equiv \hat{a}_{lv}^+$ for the creation or annihilation of a hole in the valence band. These operators can be used to construct the wave function of an intermediate exciton by the following steps:

1. First, one creates an electron–hole pair from the completely filled valence band $|\Phi\rangle = 2\hat{d}_{l_1}\hat{d}_{l_2}\cdots\hat{d}_{l_N}|0\rangle$ (the valence band contains $n = 2N$ electrons, $|0\rangle$ is the physical vacuum, and the number of unit cells is N) with separation $\mathbf{R}_s$,

$$|\Psi_{l+s,l}\rangle = \hat{a}_{l+s}^+\hat{d}_l^+\,|\Phi\rangle \tag{8.10}$$

2. Next, one constructs stationary states with quasi-momentum K in Bloch-function form:

$$|\Psi_{s,\mathbf{K}}\rangle = N^{-1/2}\sum_l \exp[i\mathbf{K}\cdot\mathbf{R}_l]\hat{a}_{l+s}^+\hat{d}_l^+\,|\Phi\rangle \tag{8.11}$$

3. Finally, one writes a linear combination of the Bloch-type wave functions (8.11) for different values of $\mathbf{R}_s$,

$$|\Psi_\mathbf{K}\rangle = \sum_s \Omega_{s,\mathbf{K}}\,|\Psi_{s,\mathbf{K}}\rangle \tag{8.12}$$

The contribution of the various charge-transfer components (originating from different vectors $\mathbf{R}_s$) can be determined from the solution of the Schrödinger equation

$$\hat{H}\Psi_\mathbf{K} = E_\mathbf{K}\Psi_\mathbf{K} \tag{8.13}$$

which also provides the energy dispersion of the exciton band.

Substitution of the particle–hole operators also in equations (8.7) and (8.8) yields, after elementary calculation,[4]

$$\hat{H} = E_{\mathrm{HF}} + \hat{H}_\mathrm{e} + \hat{H}_\mathrm{h} + \hat{H}_\mathrm{e,h} \tag{8.14}$$

where $E_{\mathrm{HF}} = \langle\Phi\,|\,\hat{H}\,|\,\Phi\rangle$, and

$$\begin{aligned}\hat{H}_\mathrm{e} &= \sum_{l_1,l_2} \hat{a}_{l_1}^+\hat{a}_{l_2}\left\{\hat{h}_{l_1l_2}^\mathrm{c} + \sum_p [\langle w_\mathrm{c}^{l_1}w_\mathrm{v}^p\,|\,w_\mathrm{v}^p w_\mathrm{c}^{l_2}\rangle - \langle w_\mathrm{v}^p w_\mathrm{c}^{l_1}\,|\,w_\mathrm{v}^p w_\mathrm{c}^{l_2}\rangle]\right\}\\ &= \sum_{l_1,l_2} \hat{a}_{l_1}^+\hat{a}_{l_2}\langle w_\mathrm{c}(x-l_1)\,|\,\hat{F}\,|\,w_\mathrm{c}(x-l_2)\rangle\end{aligned} \tag{8.15a}$$

$$\hat{H}_{\rm h} = -\sum_{l_1,l_2} \hat{d}_{l_2}^{+} \hat{d}_{l_1} \left\{ \hat{h}_{l_1 l_2}^{\rm v} + \sum_{p} [w_{\rm v}^{l_1} w_{\rm v}^{p} \mid w_{\rm v}^{p} w_{\rm v}^{l_2}\rangle - \langle w_{\rm v}^{p} w_{\rm v}^{l_1} \mid w_{\rm v}^{p} w_{\rm v}^{l_2}\rangle] \right\}$$

$$= -\sum_{l_1 l_2} \hat{d}_{l_2}^{+} \hat{d}_{l_1} \langle w_{\rm v}(x - l_1) \mid \hat{F} \mid w_{\rm v}(x - l_2)\rangle \tag{8.15b}$$

$$\hat{H}_{\rm e,h} = -\sum_{l_1,l_2,l_3,l_4} \hat{a}_{l_1}^{+} \hat{a}_{l_4} \hat{d}_{l_3}^{+} \hat{d}_{l_2} [\langle w_{\rm c}^{l_1} w_{\rm v}^{l_2} \mid w_{\rm v}^{l_3} w_{\rm c}^{l_4}\rangle - \langle w_{\rm v}^{l_2} w_{\rm c}^{l_1} \mid w_{\rm v}^{l_3} w_{\rm c}^{l_4}\rangle] \tag{8.15c}$$

It should be mentioned that, in deriving equation (8.14), we neglected all terms in $\hat{H}$ that contain two particle or two hole operators (the wave function $\Psi_{\mathbf{K}}$ contains also only the operators of one electron–hole pair). If one wants to consider also double excitations (two electron–hole pairs) a two-particle Green function must be used and a diagrammatic technique. Further even at the one electron-hole case one has to introduce so-called vertex corrections. This has been done for the most important diagrams,[6] but a description of this rather complicated formalism lies outside the framework of this book.

One should point out that the Wannier functions are not eigenfunctions of the Fock operator $\hat{F}$. It is therefore advantageous in a calculation of the matrix elements of $\hat{H}_{\rm e}$ and $\hat{H}_{\rm h}$ not to employ Wannier functions but rather to transform them back to Bloch functions $\Psi_{s,\mathbf{K}}$. (For this transformation see Section 5.1.) In this way definitions (8.15a) and (8.15b) enable us to obtain the expressions

$$\langle \Psi_{r,\mathbf{K}} \mid \hat{H}_{\rm e} \mid \Psi_{s,\mathbf{K}}\rangle = N^{-1} \sum_{\mathbf{k}} \exp[i\mathbf{k} \cdot (\mathbf{R}_r - \mathbf{R}_s)] \varepsilon_{\rm c}(\mathbf{k}) \tag{8.16a}$$

$$\langle \Psi_{r,\mathbf{K}} \mid \hat{H}_{\rm h} \mid \Psi_{s,\mathbf{K}}\rangle = N^{-1} \sum_{\mathbf{k}} \exp[i\mathbf{k} \cdot (\mathbf{R}_r - \mathbf{R}_s)][-\varepsilon_{\rm v}(\mathbf{k} - \mathbf{K})] \tag{8.16b}$$

One needs further the matrix element of $E_{\mathbf{K}}$ to be able to solve (in an approximate way) equation (8.13). For this one obtains

$$\langle \Psi_{r,\mathbf{K}} \mid E_{\mathbf{K}} \mid \Psi_{s,\mathbf{K}}\rangle = N^{-1} E_{\mathbf{K}} \sum_{\mathbf{k}} \exp[i\mathbf{k} \cdot (\mathbf{R}_r - \mathbf{R}_s)] \tag{8.17}$$

If one puts the zero point of the energy scale at the HF energy, i.e., $E_{\rm HF} = 0$, and considers $\hat{H}_{\rm e,h}$ as a perturbation operator, then a resolvent approach can be applied to yield

$$\Psi_{\mathbf{K}} = [E_{\mathbf{K}} - (\hat{H}_{\rm e} + \hat{H}_{\rm h})]^{-1} \hat{H}_{\rm e,h} \Psi_{\mathbf{K}} \tag{8.18}$$

If we now substitute equation (8.12) for $\Psi_{\mathbf{K}}$, multiply from the left by $\Psi_{r,\mathbf{K}}$, and insert the identity operator $\hat{I} = \Sigma_s \mid \Psi_{s,\mathbf{K}}\rangle\langle \Psi_{s,\mathbf{K}} \mid$ on the right-hand side, one obtains the following equation for the determination of the unknown coefficients $\Omega_{t,\mathbf{K}}$:

$$\Omega_{r,\mathbf{K}} = \sum_s \sum_t \langle \Psi_{r,\mathbf{K}} | [E_\mathbf{K} - (\hat{H}_e + \hat{H}_h)]^{-1} | \Psi_{s,\mathbf{K}} \rangle \langle \Psi_{s,\mathbf{K}} | \hat{H}_{e,h} | \Psi_{t,\mathbf{K}} \rangle \Omega_{t,\mathbf{K}} \quad (8.19)$$

The matrix elements of $\hat{H}_{e,h}$ can be computed using equation (8.11):

$$\begin{aligned} &\langle \Psi_{s,\mathbf{K}} | \hat{H}_{e,h} | \Psi_{t,\mathbf{K}} \rangle \\ &\quad = -\sum_u \exp[-i\mathbf{K}\cdot\mathbf{R}_u][\langle w_c^{s+u} w_v^0 | w_v^u w_c^t \rangle - \langle w_v^0 w_c^{s+u} | w_v^u w_c^t \rangle] \\ &\quad \equiv V^{(v,c)}(\mathbf{R}_s, \mathbf{R}_t, \mathbf{K}) \end{aligned} \quad (8.20)$$

On substituting equations (8.16), (8.17), and (8.20) into expression (8.19), one obtains the final result in the form

$$\Omega_{r,\mathbf{K}}^{(v,c)} = \sum_s \sum_t G^{(v,c)}(\mathbf{R}_r, \mathbf{R}_s, E_\mathbf{K}) V^{(v,c)}(\mathbf{R}_s, \mathbf{R}_t, \mathbf{K}) \Omega_{t,\mathbf{K}}^{(v,c)} \quad (8.21)$$

where the Green matrix element

$$G^{(v,c)}(\mathbf{R}_r, \mathbf{R}_s, E_\mathbf{K}) \equiv N^{-1} \sum_\mathbf{k} \frac{\exp[i\mathbf{k}\cdot(\mathbf{R}_r - \mathbf{R}_s)]}{E_\mathbf{K} - (\varepsilon_{c,\mathbf{k}} - \varepsilon_{v,\mathbf{k}-\mathbf{K}})} \quad (8.22)$$

can be computed from the band structure of the periodic chain.

The above approach can be applied in order to determine the exciton spectra in the following steps:

1. With the help of the HF band structures one calculates elements (8.22) of the Green matrix $\mathbf{G}(E_\mathbf{K})$ in the region of the gap (where one expects exciton states).
2. The Bloch functions obtained from the HF CO calculation must be transformed to optimally localized Wannier functions (see Section 5.1) and the matrix elements of $\hat{H}_{e,h}$ (equation 8.20) computed at selected points.
3. Finally, one determines the values of $E_\mathbf{K}$ from the zero values of the determinant, $D = |\mathbf{GV} - \mathbf{I}| = 0$ where $\mathbf{I}$ is the unit matrix, and solves the system of equations (8.19) for the coefficients $\Omega_{r,\mathbf{K}}$.

From a numerical point of view step (1) is not difficult if one applies the Gauss–Legendre method to the oscillatory integrand (due to the exponential factor occurring in it). Most time-consuming is step (2), because to calculate $V^{(v,c)}$ one must compute the two-electron integrals obtained from the HF calculation in terms of AOs. A four-index transformation must be performed now to derive the integrals in terms of Bloch functions, and a further transformation to obtain them in terms of optimally localized Wannier functions (see Section 5.1).

The zeros of function $D(E_{\mathbf{K}})$ are found by first determining the regions where D changes its sign. Knowing these regions, a very fine grid in $E_{\mathbf{K}}$ gives the desired solutions in 8–10 interpolation steps [$D(E_{\mathbf{K}})$ is a slowly changing function of $\mathbf{E_K}$] with an accuracy of D approximately 10^{-7}; this provides values of $E_{\mathbf{K}}$ (in eV) to five significant figures.

Singlet and triplet exciton bands are distinguished by substituting in relation (8.20) not Wannier spin orbitals but only Wannier functions with spatial dependence. The only difference compared to equation (8.20) will be that, before its second term (the exchange term), one has to insert a factor $2\delta_{\mathrm{M}}$, where $\delta_{\mathrm{M}} = 1$ in the singlet case and $\delta_{\mathrm{M}} = 0$ in the triplet case.

Application of the above-described method to different polymers[4,7–9] has yielded reasonably good agreement with experiment only (see the next section) if one employs a good basis set and substitutes into the Green matrix elements (8.22) not the HF one-electron energies, but rather the quasi-particle energies, which contain also correlation contributions at least in the second order of many-body perturbation theory (see Section 5.3).

Finally, we note that this approach does not include screening of interaction between the excited electron and the remaining hole. To obtain this, one should apply perturbation theory (at least in first order) not only to the Hamiltonian, but also to the exciton wave function. Since this is not an easy task one can use alternatively a distance-dependent dielectric function $\varepsilon(r)$ (as Suhai[4] did). The interpolation scheme proposed by Hermanson and Phillips[10] enables this function to be expressed in the form

$$\varepsilon^{-1}(r) = \varepsilon_0^{-1} + [(\varepsilon_0 - 1)/\varepsilon_0]\exp(-Qr) \tag{8.23}$$

where Q^{-1} plays the role of a "breakdown" length for dielectric effects. Quantity Q is related to the Thomas–Fermi wave number[11,12] and Suhai, in his calculations on polydiacetylenes,[4] used the value $Q = a^{-1}$, a being the lattice constant. Since the precise value of ε_0 corresponding to different investigated systems was not known, calculations were performed with $\varepsilon_0 = 2$, 3, and 5,[4,7–9] which represent the usual values of ε_0 in the case of organic polymers.

8.2. APPLICATION OF INTERMEDIATE EXCITON THEORY TO UV SPECTRA OF DIFFERENT POLYMERS

8.2.1. Applications to Polydiacetylenes and to Polyethylenes

8.2.1.1. Polydiacetylenes

The approach described in Section 8.1 was first applied to the two experimentally most investigated forms PTS and PCDU of polydiacetylene.[4] In PTS the main chain has the form

$$(=\underset{\displaystyle R}{\underset{|}{C}}-C\equiv C-\underset{\displaystyle R}{\underset{|}{C}}=)_n$$

an acetylene structure with side group $R = -CH_2-SO_3-C_6H_4-CH_3$, while TCDU has in its main chain a butatriene structure

$$(-\underset{\displaystyle R}{\underset{|}{C}}=C=C=\underset{\displaystyle R}{\underset{|}{C}}-)_n$$

with side group $R = -(CH_2)_4OCONHC_6H_5$. The geometries of the unit cell of PTS and TCDU, respectively, are shown in Figure 8.1.

In the actual calculation the complicated side chains R were substituted by an H atom. For the first step (calculation of the HF band structures) a 6-31G** basis set (double ζ + polarization function on both the carbon and hydrogen atoms) was applied. The valence and conduction bands obtained in this way were then corrected using the generalized electronic polaron model [quasi-particle (QP) band structures; see Section 5.3]. The lowest singlet-exciton energies (at $K = 0$) were then calculated using the QP one-electron levels and performing the three steps described in the previous section after equation (8.22). Table 8.1 shows the results obtained in this way for both PTS and TCDU.[4]

One can see from Table 8.1 that the Frenkel exciton ($n = 0$) provides only about 30% of the electron–hole binding energy (the gap obtained in the QP framework is 3.7 eV for PTS and 3.2 eV for TCDU).[4] The exciton must be localized with a radius of 25–30 Å, $n = 5$ (also found by analyzing the excitonic wave function; see Figure 3 of Suhai[4]), to obtain acceptably convergent excitation energies. Further, one can see that the correlation corrections reduce the excitation energies for both polymers by approximately 2.1 eV. Finally, one should point out that the exci-

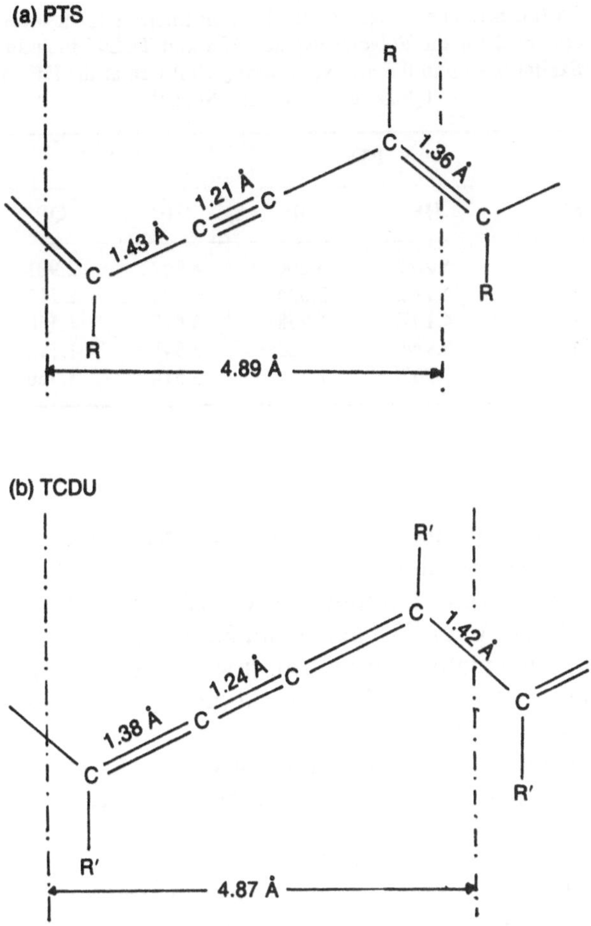

FIGURE 8.1. Geometries of the unit cell of (a) PTS and (b) TCDU.

tation energies are about 0.4 eV lower for TCDU than for PTS in both the HF and QP schemes.

If one also introduces screening for the e–h interaction by applying equation (8.23) with $\varepsilon_0 = 2$, 3, and 5, the ${}^1E_{\mathbf{K}=\mathbf{O}}$ singlet excitation energies increase monotonically with increasing ε_0 (see Figure 4 of Suhai[4]). At the most realistic looking value of $\varepsilon_0 \approx 3$, the binding energy of the e–h pair is substantially reduced and consequently the lowest singlet level ($\mathbf{K} = \mathbf{O}$) is shifted from 1.79 eV to 2.16 eV for PTS and from 1.37 eV to 1.74 eV for TCDU. The difference between the PTS and TCDU values, on the other

TABLE 8.1. The Lowest Singlet-Exciton Energies (at **K** = O) Obtained for the Polydiacetylenes PTS and TCDU Including Excitations up to the *n*th Neighboring Unit Cell at the HF and QP Level (in eV; after Suhai[4])

	PTS		TCDU	
n	HF	QP	HF	QP
1	5.282	3.206	4.762	2.901
2	4.718	2.624	4.275	2.317
3	4.107	1.958	3.672	1.550
4	3.994	1.823	3.548	1.392
5	3.965	1.791	3.514	1.366

hand, is almost unaffected by the screening. These results show that the first theoretical singlet absorption maximum of PTS should be around 2.1–2.2 eV and of TCDU around 1.7–1.8 eV. The width of the exciton band is about 2.9 eV for PTS and about 3.5 for TCDU.[4]

The experimentally determined first singlet absorption edge is approximately 2 eV in TPS.[13] The slight discrepancy from the theoretical results can be explained, if one takes into account that: (1) in the QP picture used only about 70% of the correlation has been considered, (2) no phonon polaron effects have been considered, and (3) in the formation of the QP band structure (calculating polaron corrections) both the $(N + 1)$ and $(N - 1)$ electron systems have been computed with the aid of the HF orbitals of the N electron system. In the case of polyacetylene these three effects still give a gap 1 eV too large (3 eV instead of 2 eV[14]). Therefore, one would expect that the real gap of PTS will be not 3.7 eV but around 2.5 eV. Moreover, one should take into account that by calculating the exciton energies, allowance should be made not only for the valence and conduction bands, but also for excitations between other bands, and should not neglect double excitations. It was noted in the previous section that this requires a more general two-particle Green function approach. Finally, we have seen that screening between the e–h pair has been taken into account in a rather approximate way, which is a further source of error.

In the case of TCDU the lowest experimental singlet excitation energies are about 1.9 eV at low temperatures (< 100 K).[15] Since the described calculations were performed in the fixed nuclei (Born–Oppenheimer) approximation, this is the experimental value (and not the value of about 2.3 found at higher temperatures[15]) with which the theoretical value of 1.7–1.8 eV should be compared. The slight discrepancy of 0.1–0.2 eV can be easily explained by the different approximations described above for PTS (actually, the different errors most probably somewhat compensate each other and provide good agreement with experiment).

8.2.1.2. Polyethylenes

As the next example we take the $(CH_2)_x$ polyethylene (PE) chain. In a subsequent calculation Suhai[8] also applied the intermediate exciton theory with QP one-particle energies to this chain. Using again a 6-31G** basis he calculated the HF and subsequently the QP band structures of this system. In the latter case he obtained for the fundamental gap a value of 10.3 eV, while the experimental value is 8.8 eV.[16] The reasons for this 1.5 eV discrepancy have already been discussed (see Sections 5.3 and 8.2.1.1).

Suhai subsequently employed the approach in Section 8.1 to determine the singlet excitation energies at $K = 0$ (at the lower edge of the excitation band). He found that one must proceed to $n = 6$ (in C_2H_4 units) to obtain reasonable convergence (with respect to the number of neighbors taken into account for the charge-transfer excitations) in the excitation energy, in this case 8.67 eV.[8] This value is again in reasonably good agreement with the experimentally observed strong absorption of PE in the region of 7.6–8.8 eV.[16] Finally, one should mention that in PE the Frenkel exciton ($n = 0$) provides only about 20% (0.32 eV) of the whole exciton binding energy (1.63 eV).[8]

8.2.2. Applications to a Cytosine Stack and to Polyglycine

8.2.2.1. Cytosine Stack

In a subsequent calculation the intermediate exciton theory with correlation was applied to a cytosine stack,[17] the superimposed cytosine molecules possessing the same relative geometry as in the *in vivo* stable DNA B.[18] In this calculation, only a double-ζ basis was applied (no polarization functions) owing to the rather large size of the unit cell (a cytosine molecule).[19]

In the case of polynucleotides, besides the position of the first exciton band, the substantially decreased intensity (hypochromicity) as compared to the intensity obtained by superimposing the intensities of the constituent nucleotide bases has for twenty years been of profound interest both from the theoretical[20] and experimental[21] points of view.

To calculate the intensity of the spectrum one must determine its oscillator strength which, in the case of singlet transition from the highest valence-band level to the bottom of the first exciton band, is given by

$$f^{v,c} = \tfrac{4}{3}E_{K=0}\left(\sum_{t} \Omega_{t,K=0}\langle \Lambda_{v,0} | \mathbf{r} | \Lambda_{c,t}\rangle\right)^2 \tag{8.24}$$

where

$$\Lambda_{j,t} = \sum_{q} f_q(\mathbf{R}_t)\chi_q(\mathbf{r} - \mathbf{R}_t - \mathbf{R}_q), \qquad j = \mathrm{v, c} \tag{8.25}$$

is the tth unit contribution to the Wannier function localized in the reference cell (0).[27]

Table 8.2 presents the calculated exciton energies and intensities of the first two exciton bands of a cytosine stack.[17]

The calculations strongly support the assignment of the observed emissions from states with longer lifetime in DNA[21] as originating from delocalized singlet exciton states. Both the significant bandwidth (about 0.6 eV) and the fact that a substantial part of the exciton binding energy is missing if only Frenkel excitons are calculated point to the importance of charge-transfer components in the excitation. In the case of the first singlet band, for instance, the binding energy of the Frenkel excitons is less than 80% of the total value. At the same time, owing to interbase interactions, the absorption of the polymer is somewhat red-shifted as compared with the separated monomers. For the first singlet band we can observe, for example, a shift of 0.3 eV toward lower energies.

The most interesting result in Table 8.2 is the significant reduction in the absorption intensity of the first singlet exciton band as compared again with the corresponding monomer transition. Though the quantum-

TABLE 8.2. Energies and Intensities of the First Two Singlet Charge-Transfer Exciton Bands of PolyC and the Corresponding Monomer Properties (energies in eV; after Suhai[17])

	1st	2nd
Exciton band minimum ($K = 0$)	4.732	5.912
Exciton band maximum ($K = \pi/a$)	5.384	6.184
Monomer excitation energy	5.041	6.072
Oscillator strength of polymer	0.257	0.327
Oscillator strength of monomer	0.346	0.291
Hypochromicity (in %)	26	−12

mechanical expression for the intensities is approximate and, therefore, the absolute values of the oscillator strengths are less important, the reduction by about 26% is certainly significant. As expected from the Kuhn–Thomas sum rule, it is partially compensated by the next transitions, which show a hyperchromic shift.

Though the classification of the electronic transitions of the DNA macromolecule according to π-type or lone-pair states of the monomers is only approximate, it is still possible to assign the exciton bands to monomer levels if the Wannier functions are contracted to molecular orbitals. As observed also in previous *ab initio* calculations on nucleotide bases[22] and polynucleotides,[23] the π and lone-pair levels are always mixed. In our case, for the three highest filled and three lowest unfilled bands, respectively, the ordering n, π, π, π^*, π^*, n has been found. The first two singlet monomer excitation energies obtained at 5.04 and 6.07 eV, respectively, must be compared with the experimentally observed values of 4.5 and 5.2 eV.

8.2.3. The Exciton Spectrum of Polyglycine and Polyalanine

In another paper,[24] the *ab initio* intermediate exciton theory with QP one-electron energies has been applied to the two simplest homopolypeptides: polyglycine and polyalanine. Both polypeptides were taken in the α-helical configuration[25] and the 6-31G* (double-ζ plus a set of d-functions on the nonhydrogen atoms) basis set applied.[26] The proper truncation of the infinite lattice sums of the polypeptide chain is a very sensitive point in the HF part of the calculations. Since the unit cell has a dipole moment, the electrostatic interactions also contain contributions of a long-range character besides the nonlocal exchange. Test calculations using first a minimal basis set have shown that, in order to obtain convergence of *ab initio* quality, interactions with five neighboring units must be taken into account in both directions along the helical backbone for each peptide unit. Using this truncation scheme with a careful balance between one- and two-electron contributions,[27] no numerical instabilities were observed.

The exciton spectrum of two α-helical polypeptide chains were calculated. The first system was a simple backbone without side groups (modeling polyglycine), the second was polyalanine. Physical intuition tells us that, apart from special side-chain groups containing a π-electron system, the absorption of an incoming photon basically takes place in the delocalized π-electron system of the covalently bound polypeptide backbone. This should, at the same time, be the channel for the transfer of excitation energy limited by different scattering processes. The role of side-chain groups attached to the backbone is that of impurity centers in this scheme. The difference between the exciton levels in the simple backbone and in polyalanine will give initial information about the influence of a side group on the spectrum. On the other hand, methods working with Green functions have the great advantage of easy extension also for aperiodic systems. Appropriate generalization of the present procedure will therefore also be capable of treating the visible spectrum of proteins.

In calculating the correlation contribution to the total energy per peptide unit, only virtual excitations starting from valence-band states were included. On the other hand, the virtual space was used to its full extent, since earlier calculations pointed to the important role of high-lying conduction bands in describing short-range correlation effects.[14] In this way, -0.641746 H for the correlation energy in MP/2 MBPT (-29.2 mH per valence electron) for the backbone model and -0.806458 H (-28.8 mH per valence electron) for polyalanine were obtained. Since corresponding experimental results are not known for peptides, it is difficult to tell what percent of the correlation energy is covered in this procedure. One should remark, however, that the 6-31G* basis set provided, in the same theoretical framework, about 70–75% of the correlation energy in other polymers.[4] One can expect a similar performance for polypeptides, too. With the results of these QP band-structure calculations, the first two singlet and first triplet exciton bands of polyglycine are shown in Table 8.3.[27] The same quantities are presented for polyalanine in Table 8.4.

In order to be able to judge the role of interpeptide interactions in building up the intensities of the individual transitions, the value of f for the separated residues, namely f (monomer), was also calculated. The midpoint energy in the corresponding exciton band was used as the approximate transition energy of the monomer. It is interesting to note in Tables 8.3 and 8.4 that the lowest singlet exciton for both polymers lost about 30% of its intensity as compared with the corresponding monomer transition. As expected from sum rules, the higher-lying excited states compensate for this hypochromicity. The order of magnitude of the observed effect can reasonably be compared with experimental hypochromicity values reported for polyglutamate (36%) and polylysine (30%).[28]

The importance of strong intermonomer interactions in building up the exciton is also reflected in the substantial delocalization of the excitonic wave function. Detailed results[24] show that charge-transfer excitations for a distance of 4–5 residues contribute significantly to the calculated binding energy of the exciton, E_0. The Frenkel exciton ($n = 0$)

TABLE 8.3. Singlet and Triplet Exciton Bands, $E(K)$, and the Corresponding Oscillator Strengths, f, Obtained for the α-Helical Polypeptide Backbone (energies in eV; after Suhai[24])[a]

	1st singlet	2nd singlet	1st triplet
$E(K=0)$	6.208	7.322	3.806
$E(K=\pi/a)$	6.602	7.830	3.918
δE	0.394	0.508	0.112
f(polymer)	0.246	0.162	—
f(monomer)	0.341	0.094	—

[a] For comparison, the approximately calculated value of f for the corresponding monomer transition is also shown.

TABLE 8.4. Singlet and Triplet Exciton Bands, $E(K)$, and the Corresponding Oscillator Strengths, f, Obtained for α-Polyalanine (energies in eV; after Suhai[24])[a]

	1st singlet	2nd singlet	1st triplet
$E(K=0)$	6.052	7.146	3.580
$E(K=\pi/a)$	6.398	7.628	3.688
δE	0.346	0.482	0.108
f(polymer)	0.228	0.157	—
f(monomer)	0.324	0.096	—

[a] For comparison, the approximately calculated value of f for the corresponding monomer transition is also shown.

possesses only about 70–80% of the total value. The obtained lowest singlet exciton energy regions of 6.0–6.2 and 7.1–7.3 eV can reasonably be related to the absorptions at wavelengths of 2060 and 1890 Å (6.0 and 6.5 eV), respectively, observed for polyglutamate,[28,29] and at 2050, 1900, and 1650 Å (6.0, 6.5, and 7.5 eV), respectively, observed for polyleucine.[30]

The lowest energy transitions show hypochromism for both polyglycine and poyalanine, and the higher ones hyperchromism as compared with the spectrum of Nylon 6, where the random conformation prevents the cooperative excitation of the peptide units.[30] The observed differences in these spectra are therefore related to the previously described delocalization of the excitation energy for the regularly spaced peptide groups. The absence of long-range order in globular protein sets, of course, a limit to this delocalization. On the other hand, at physiological temperatures, mostly due to the scattering on lattice vibrations, the traveling excitons have a finite free path anyway, comparable to or shorter than the length of helical segments in proteins. Therefore, the efficiency of energy transport via excitons in proteins seems to be primarily determined by the strength of the exciton–phonon interaction.

8.3. IONIZED STATES OF POLYMERS AND THEIR X-RAY PHOTOELECTRON SPECTRA

8.3.1. Theoretical Calculation of the Ionized States of Simple Periodic Polymer Chains

8.3.1.1. Elementary Treatment Based on Koopmans' Theorem

According to the simplest form of Koopmans' theorem[31] the first ionization potential of a molecule or polymer is the negative of its HOMO energy, or of the highest filled level of its valence band, respectively, obtained in the HF approximation

$$I_1 = -\varepsilon_{\mathrm{HOMO}}^{\mathrm{HF}}, \qquad I_1 = -\varepsilon_{\mathrm{v}}^{\mathrm{HF}}(k_{\max}) \tag{8.26}$$

Even if one uses a good-quality basis set, this oversimplified picture has several shortcomings. First, any ionization potential is the difference between the total energies of the neutral N electron and positively charged $N-1$ electron systems (if one restricts oneself to single ionizations),

$$I_i = E_i^{(N-1)} - E_G^{(N)} \tag{8.27}$$

where $E_i^{(N-1)}$ refers to the $N-1$ electron system from which an electron is ionized out from the ith state. Obviously, the total energies in equation (8.27) must contain correlation contributions in the case of both molecules and polymers [see equations (5.53) and (5.54) of Section 5.3].

Further, especially in the case of molecules, due to ionization all the other remaining electronic orbitals will change (relaxation effects). This effect is not so important in the case of a polymer chain, but it may still have a nonnegligible effect, especially if we consider the ionization even in a long polymer chain as a basically local event, as is the case of excitations; see Section 8.1.

Finally, one should not forget that even due to a valence-state ionization the geometry (again locally) of a polymer chain may undergo essential changes, which may again have a strong effect on the value of the ionization potential.

8.3.1.2. Calculation of Ionized States Including the Change of Correlation

One can define a generalized Koopmans' theorem if, by calculating quantity (8.27), one takes into account correlation effects and determines a quasi-particle (QP) band structure based on the generalized electronic polaron model (see Section 5.3). One can then write for a polymer

$$I_i = E_i^{(N-1)\,\mathrm{HF}} + E_i^{(N-1)\,\mathrm{corr}} - E_G^{(N)\,\mathrm{HF}} - E_G^{(N)\,\mathrm{corr}} = -\,\varepsilon_v^{\mathrm{QP}}(k_{\max}) \quad (8.28)$$

It was seen in Section 5.3 that to determine the QP band structures of a polymeric chain one must use a size-consistent method to determine the major part of the correlation [many-body perturbation theory (MBPT) in the Møller–Plesset partitioning, coupled-cluster theory, etc.]. Suhai,[32] in his QP band-structure calculation on polyacetylenes and polydiacetylenes,[4] used second-order (MP/2) Møller–Plesset MBPT. For polydiacetylenes he obtained 5.7 eV as first ionization potential (using the generalized Koopmans' theorem) for the PTS structure (see Figure 8.1), in reasonable agreement with experiment ($I_1 = 5.5 \pm 0.1$ eV[33]), while the HF value (the simple Koopmans' theorem) is 6.8 eV.[4] For the TCDU diacetylene structure the theoretical value is 5.0 eV (HF value, 6.2 eV). Unfortunately, there is no reliable experimental ionization-potential value available for the TCDU structure of polydiacetylene.

Of course, in generating a QP band structure of a polymer and applying in this way a generalized Koopmans' theorem one is not obliged to stop at MP/2 in the correlation calculations. In a subsequent paper Liegener[34] applied MP/3 for an alternating hydrogen chain and went beyond the generalized Koopmans' theorem by taking as ionization potential the negative solution of the inverse Dyson equation,[35]

$$I_1 = -\,\omega_1, \quad \omega_1 = \varepsilon_v^{\mathrm{HF}}(k_{\max}) + M_{11}(\omega_1) \quad (8.29)$$

If one includes in M_{11} only second order in the MP partitioning and substitutes $\omega_1 = \varepsilon_v(k_{\max})$, one obtains back Suhai's expressions, which he derived for the QP band structures of polyacetylene. On the other hand, if one proceeds to higher order and does not apply the $\omega_1 = \varepsilon_v^{\mathrm{HF}}(k_{\max})$ approximation, one obtains further corrections to the ionization potential. Since the resulting expressions are very complicated and an understanding of them would require some knowledge of diagrammatic techniques (coupled with the Green function approach) we do not write them down here but refer the reader to the original papers.[34,35]

A double-ζ basis set was used in the actual calculation, the shorter bond length $d_<$ was 1.326 au, and the longer one $d_> = 2.362$ au. The theoretical result for the first ionization potential is 0.4199 au (the HF value is 0.4324 au, the HF + MP/2 value is 0.4198 au, while for a free hydrogen atom it is 0.5000 au), which corresponds to 11.421 eV. One can see that the third-order correction *increases* the IP by 0.0001 au = 0.0027 eV, while the second order has decreased it by 0.0226 au = 0.6102 eV. From these results we can conclude that, also in the case of polymers with larger unit cells, the third-order corrections to the ionization potentials will most probably be very small.

8.3.2. Interpretation of the X-Ray Photoelectron Spectra of Polymers

The calculated ionization potentials enable one to interpret the measured X-ray photoelectron spectra (XPS) of polymers. The Namur group has performed a series of investigations in this field (a review has been published elsewhere[36]). To be able to compare an experimental valence electron XPS with a theoretical spectrum, the intensities of the different peaks are also needed for the construction of the theoretical spectrum besides the IPs (the energy difference between the $N-1$ and N electron states). (Actually, XPSs are more important in the identification and determination of different elements in a material. For an experimental check of the calculated band structures of an organic polymer, however, the valence electron XPSs have to be used.)

In comparing an experimental and theoretical XPS one should not forget that the experimental XPS contains three contributions, which originate from three different physical processes:

1. Excitation of an electron from an initial state $\varepsilon_i(k)$ to the final one $\varepsilon_f(x)$.
2. Transfer of the electron to the surface of the solid sample.
3. Escape of the electron from the surface.

However, since the structure (both bulk and surface) of most polymeric solids is rather ill-defined, there seems to be no hope of treating theoretically, in the near future, steps (2) and (3) and hence finding out the energies necessary for these steps.[36] Therefore, all authors restricted themselves to step (1) and nearly always applied all-valence semiempirical (extended Hückel CO, CNDO/2 CO) methods or, in the best case, Floating Spherical Gaussian Orbitals (FSGO).[37]

For the ionization potential, all these calculations have applied the Koopmans' theorem in its simple form

$$I_1 = E_i(k) - E_f(k) = -\varepsilon_v(k_{max}) \tag{8.30}$$

while the experimental ionization potential can be determined from the equation

$$E_i(k) + h\nu = E_f(k) + E_{kin} \tag{8.31}$$

or

$$E_{kin} = h\nu - [E_f(k) - E_i(k)] = h\nu + I_i$$

where $h\nu$ is the known energy of the incident photon, $I_i \approx E_b$ is the binding energy of the ionized electron in the sample, and E_{kin} is the measured kinetic energy of the ionized electron with which it leaves the surface of the polymeric solid.

In order to overcome the very drastic simplifications inherent in this procedure [using $I_i = E_b$, neglecting steps (2) and (3), application of semiempirical band structures, etc.] most authors determined theoretically only the relative positions of the absorption peaks in the XPS, superimposing the experimental and theoretical spectra by bringing into coincidence a well-identified peak. For this procedure, however, as noted above, the intensities of the theoretical spectrum as a function of energy E must also be computed.

The theoretical intensities can be calculated by starting from the expression[36]

$$I(E) \sim \sum_{i=1}^{n^*} \int_{BZ} \delta[\varepsilon_f(k) - \varepsilon_i(k) - h\nu]\delta[E - \varepsilon_i(k)]P_{fi}^{\nu}(k)dk \tag{8.32}$$

which is valid for a linear chain. Here BZ denotes the first Brillouin zone, n^* the highest occupied (valence) band, $\varepsilon_f(k) - \varepsilon_i(k) - h\nu = 0$ defines the different allowed excitations [we note that $\varepsilon_f(k) = E_f(k) + E_{kin}$], and finally $P_{fi}(k)$ is the probability of exciting an electron from state $|\psi_i(k)\rangle$ to state $|\psi_f(k)\rangle$ by a photon with energy $h\nu$. It is rather difficult to calculate $P_{fi}(k)$ directly and in most cases it has been estimated on the basis of the method suggested by Gelius.[38] According to this method, which applies Mulliken's populations and the relative photoionization cross sections of a particular atomic subshell (which can be obtained from fitting procedures on reference systems[38]),

$$P_{\mathrm{fi}}^{\mathrm{v}}(k) \sim \sum_{p=1}^{\infty} C_{ip}^{*}(k)\left[\sum_{q=1}^{m} S_{pq}(k)C_{ip}(k)\right]\sigma_p \tag{8.33}$$

where m is the number of basis functions in the unit cell of the polymer.

One can approximate the continuous function $I(E)$ by the histogram

$$\bar{I}(E_n) \sim \frac{1}{\Delta E}\sum_{i=1}^{n^*}\int_{E_n-\Delta E/2}^{E_n+\Delta E/2} \delta[E-\varepsilon_{\mathrm{i}}(k)]P_{\mathrm{fi}}^{\mathrm{v}}(k)dk \tag{8.34}$$

assuming that the equation $\varepsilon_{\mathrm{f}}(k)-\varepsilon_{\mathrm{i}}(k)-h\nu=0$ is fulfilled.[39] In most solid-state physical measurements the apparent width of the photoelectron peaks ranges from 0.6 to 0.9 eV. Since the experimental resolution function is reasonably well reproduced by a Gaussian with halfwidth $\Gamma = 2.345\ \sigma = 0.7$ eV, we can finally write[36]

$$I(E_n) \sim \int_{-\infty}^{\infty} \bar{I}(E)\exp[-(E-E_n)^2]/2\sigma^2 dE \tag{8.35}$$

Despite all these simplifications, one can simulate theoretically the experimental XPS quite well. As a first step, the XPS of the structurally rather well-defined all-*trans*-polyethylene[40] was investigated using a CNDO/2 band structure.[41] The theoretical spectrum was compared to the valence electron spectrum of the long paraffin chain $C_{36}H_{74}$, comparing the observed peaks in those energy regions where the first derivatives of the energy bands are small.[41] This investigation was made easier with the help of density-of-state calculations[42] [equation (8.34) provides the total density of states in the neighborhood of E_n if one sets factors $P_{\mathrm{fi}}^{\mathrm{v}}(k)$ equal to unity], but also the inclusion of convenient photoionization cross sections together with a correlation with a suitable experimental resolution function[43] was necessary in order to achieve reasonable agreement. In a later paper the same method was used, but the band structure of all-*trans*-polyethylene was computed with the help of FSGOs.[44] In this way the Namur group has achieved excellent agreement between the theoretical and experimental XPS of all-*trans*-polyethylene at large binding energies. The low-intensity region is more problematic, because it contains the C–H orbitals as dominant ones and these orbitals have low photoionization probabilities.

The dependence of the XPS of polymers on substitutions and conformational changes was examined by first performing an extended Hückel band-structure calculation on the series of linear fluoropolyethylenes $-CH_2-CH_2-$, $-CHF-CH_2-$, $-CF_2-CH_2-$, $-CHF-CHF-$, $-CHF-CFH-$, $-CF_2-CHF-$, and finally $-CF_2-CF_2-$,[45] taking them in the artificial planar zigzag conformation. The theoretical conclusions obtained indicate that the fluoro substituents should be observable in the XPS measurements. Subsequent experimental works have confirmed these predictions.[46]

Clark *et al*. performed XPS measurements on a series of polybutylcrystals, $-CH_2-CH(CO-O-R)-$ with $R = C_4H_9(t)$, $C_4H_9(i)$, and $C_4H_9(n)$. They found that XPS valence lines are very useful in distinguishing structurally isomeric polymer systems with identical core-level spectra. By again using an extended Hückel band structure, the Namur group was able to predict important differences in the distribution of valence levels of the three model chains $-CH_2-CH(OCH_3)-$, $-(CH_2)_3-O$, and $-CH_2-CH(CH_3)-O-$.[47] These predictions were again confirmed by experimental XPSs.[48]

Finally, it should be noted that even a not-too-large change in the geometry of a polymer can have a profound effect on its valence photoelectron spectrum,[49] as was demonstrated in the case of isotactic polypropylene. This example shows that the conformation of a polymer influences strongly its XPS (through both its band structure and its photoionization cross sections).

REFERENCES

1. J. Ladik, in: *Excited States in Quantum Chemistry* (C. A. Nicolaides and D. R. Beck, eds.), p. 495, D. Reidel Publ. Co., Dordrecht–Boston–New York (1979).
2. R. S. Knox, *Theory of Excitons*, Academic Press, New York (1963).
3. Y. Takeuti, *Prog. Theor. Phys. (Kyoto)* **18**, 421 (1957); *Prog. Theor. Phys. (Kyoto), Suppl.* **12**, 75 (1959).
4. S. Suhai, *Phys. Rev. B* **29**, 4570 (1984); *Quantum Mechanical Investigations on Quasi-One-Dimensional Solids*, Habilitation Thesis, Erlangen (1983) (in German).
5. A. L. Fetter and J. D. Valecka, *Quantum Theory of Many-Particle Systems*, McGraw-Hill Book Co., New York (1971).
6. C.-M. Liegener and J. Ladik, *Chem. Phys.* **106**, 339 (1986).
7. S. Suhai, *Int. J. Quantum Chem.* **11**, 223 (1984).
8. S. Suhai, in: *Quantum Chemistry of Polymers; Solid State Aspects* (J. Ladik and J.-M. André, eds.), p. 101, D. Reidel Publ. Co., Dordrecht–Boston–Lancaster (1984).
9. S. Suhai, *J. Mol. Struct.* **123**, 97 (1985).
10. J. Hermanson and J. C. Phillips, *Phys. Rev.* **150**, 652 (1960).
11. J. Hermanson, *Phys. Rev.* **150**, 660 (1966).
12. D. Pines, *Elementary Excitations in Solids*, p. 146, Benjamin, New York (1984).

13. D. Bloor, D. J. Ando, F. H. Preston, and G. C. Stevens, *Chem. Phys.* **24**, 1407 (1974); D. Bloor, F. H. Preston, and D. J. Ando, *Chem. Phys. Lett.* **38**, 33 (1976); D. Bloor and F. H. Preston, *Phys. Status Solidi A* **37**, 427, 607 (1976).
14. S. Suhai, *Phys. Rev. B* **27**, 3506 (1983).
15. Z. Igbal, R. R. Chance, and R. H. Baughman, *J. Chem. Phys.* **67**, 3616 (1977).
16. K. J. Less and E. G. Wilson, *J. Phys. C* **6**, 3110 (1973).
17. S. Suhai, *Int. J. Quantum Chem.* **11**, 223 (1984).
18. S. Arnott, S. Doven, and A. J. Wonacott, *Acta Crystallogr., Sect. B* **28**, 2198 (1969).
19. R. Ditchfield, J. W. Hehre, and A. J. Pople, *J. Chem. Phys.* **54**, 726 (1971).
20. H. Devoe, *Ann. N.Y. Acad. Sci.* **158**, 298 (1969); I. Tinoco, Jr., *J. Am. Chem. Soc.* **82**, 4785 (1960); **83**, 5047 (1969); W. Rhodes, *J. Am. Chem. Soc.* **83**, 3609 (1961); H. Devoe and I. Tinoco, Jr., *J. Mol. Biol.* **4**, 51881 (1968); J. Koutecký and J. Paldus, *Theor. Chim. Acta* **1**, 268 (1963); T. A. Hoffmann and J. Ladik, *J. Theor. Biol.* **6**, 26 (1964); J. Ladik and K. Sundaram, *J. Mol. Spectrosc.* **29**, 146 (1969); A. Pullman and B. Pullman, *Adv. Quantum Chem.* **4**, 287 (1970); V. I. Danilov, V. T. Pechenaga, and N. V. Zheltorsky, *Int. J. Quantum Chem.* **17**, 307 (1980).
21. J. P. Ballini, P. Vigny, and H. Daniels, *Biophys. Chem.* **18**, 61 (1983); the first four references of Reference 20.
22. E. Clementi, J.-M. André, M.-Cl. André, D. Klint, and D. Hahn, *Acta Phys. Hung. Acad. Sci.* **27**, 493 (1969); B. Mely and A. Pullman, *Theor. Chim. Acta* (Berlin) **13**, 278 (1969); E. Clementi, J. Mehl, and W. V. Niessen, *J. Chem. Phys.* **54**, 508 (1971).
23. S. Suhai, C. Merkel, and J. Ladik, *Phys. Lett.* **61A**, 487 (1977); J. Ladik and S. Suhai, *Int. J. Quantum Chem.* **QBS7**, 181 (1980).
24. S. Suhai, *J. Mol. Struct.* **123**, 97 (1985).
25. J. F. Yan, G. Vanderkooi, and H. Scheraga, *J. Chem. Phys.* **49**, 2713 (1968).
26. J. S. Binkley, R. A. Whiteside, P. C. Hariharan, R. Seeger, and J. A. Pople, QCPE, Bloomington, Indiana, Program No. 368.
27. S. Suhai, *J. Chem. Phys.* **73**, 3843 (1980).
28. I. Tinoco, Jr., A. Halpera, and W. T. Simpson, in: *Polyamino Acids, Polypeptides and Proteins* (M. A. Stahmann, ed.), p. 147, University of Wisconsin, Madison (1962).
29. W. C. Johnson, Jr. and I. Tinoco, Jr., *J. Am. Chem. Soc.* **94**, 4389 (1972).
30. S. Onari, *J. Phys. Soc. Jpn.* **29**, 528 (1970).
31. T. Koopmans', *Physica* **1**, 104 (1933).
32. S. Suhai, *Phys. Rev. B* **27**, 3506 (1983).
33. A. A. Murashovi, E. A. Silinsh, and H. Bässler, *Chem. Phys. Lett.* **93**, 1481 (1982).
34. C.-M. Liegener, *J. Phys. C* **18**, 6011 (1985).
35. D. J. Thouless, *The Quantum Mechanics of Many-Body Systems*, 2nd Edition, Academic Press, New York–London (1972); L. S. Cederbaum and W. Domcke, *Adv. Chem. Phys.* **36**, 205 (1977).
36. J. Delhalle and J.-M. André, in: *Quantum Chemistry of Polymers; Solid State Aspects* (J. Ladik and J.-M. André, eds.), D. Reidel Publ. Co., p. 23, Dordrecht–Boston–Lancaster (1984).
37. J. L. Brédas, J.-M. André, and J. Delhalle, *Chem. Phys.* **45**, 109 (1980).
38. U. Gelius, in: *Electron Spectroscopy* (D. A. Shirley, ed.), p. 311, North-Holland Publishing Company, Amsterdam (1972).
39. J. Delhalle and S. Delhalle, *Int. J. Quantum. Chem.* **11**, 349 (1977).

40. S. Kavash and J. M. Schultz, *J. Polym. Sci. A* **28**, 243 (1970); G. Avitebile, R. Napolitano, B. Pirozzi, K. D. Rouse, M. W. Thomas, and B. T. M. Willis, *J. Polym. Sci., Polym. Lett. Ed.* **13**, 351 (1975).
41. M. H. Wood, M. Barber, I. H. Hillier, and J. M. Thomas, *J. Chem. Phys.* **56**, 1788 (1972).
42. J. Delhalle, J.-M. André, S. Delhalle, J. J. Pireaux, R. Caudano, and J. J. Verbist, *J. Chem. Phys.* **60**, 595 (1974).
43. J. Delhalle, S. Delhalle, and J.-M. André, *Chem. Phys. Lett.* **34**, 430 (1975).
44. J. L. Brédas, J.-M. André, and J. Delhalle, *Chem. Phys.* **45**, 109 (1980).
45. J. Dehalle, *Chem. Phys.* **5**, 306 (1974).
46. J. J. Pireaux, J. Riga, R. Caudano, J. J. Verbist, J.-M. André, J. Delhalle, and S. Delhalle, *J. Electron Spectrosc.* **5**, 53 (1974); J. Delhalle, S. Delhalle, J.-M. André, J. J. Pireaux, J. Riga, R. Caudano, and J. J. Verbist, *J. Electron Spectrosc.* **12**, 293 (1977).
47. J. Delhalle, in: *Recent Advances in the Quantum Theory of Polymers* (J.-M. André, J. L. Brédas, J. Delhalle, J. Ladik, G. Leroy, and C. Moser, eds.), Lecture Notes in Physics **113**, p. 255, Springer-Verlag, Berlin–New York–Heidelberg (1980).
48. J. J. Pireaux, J. Riga, R. Caudano, and J. J. Verbist, in: *Photon, Electron and Ion Probes of Polymer Structure and Properties* (D. W. Dwight, T. C. Fabish, and H. R. Thomas, eds.), ACS Symp. Series, **162**, p. 169, Washington D. C. (1981).
49. J. Delhalle, R. Montigny, C. Demanet, and J.-M. André, *Theor. Chim. Acta* **50**, 343 (1979).

Chapter 9

Vibrational Spectra and Transport Properties of Polymers

9.1. METHODS FOR THE CALCULATION OF VIBRATIONAL SPECTRA OF POLYMERS

The standard way of calculating the normal coordınates of a molecule is the well-known GF method of Wilson.[1] In this method one starts from the harmonic force constants of the molecule in order to obtain the different normal coordinates of vibrations and their corresponding energies. These force constants are the second derivatives of the molecular total energies with respect to the different internal coordinates at equilibrium geometry. Therefore, the force constants are obtained by computing the energy hypersurface; in the case of a larger molecule this task is rather time-consuming. Differentiation of the total energy is carried out by calculating the first derivative (gradient), usually analytically, according to the method developed by Pulay,[2] while the second derivative is obtained by numerical methods.[3] It should be noted that, is some cases, both differentiations have been performed analytically after fitting the calculated potential hypersurfaces to quadratic potential-energy functions with the help of least-square methods.[4]

If one wishes to calculate the vibration of a crystal or of a periodic polymer [with both intermolecular (intercell) and intramolecular (intracell) couplings] it is more advantageous to apply Cartesian instead of internal coordinates.[5]

The theory of vibrations of a periodic chain can be described using Cartesian coordinates, following the fundamental paper of Piseri and Zerbi[6] as follows. We consider a periodic quasi-one-dimensional chain comprising an infinite number of units. Each unit cell contains N atoms and the cells themselves are labeled by n, where $-\infty \leqslant n \leqslant \infty$. The location of the nth cell is described by vector $\mathbf{t}(n)$ referred to a suitably chosen origin.

If the displacement vector of the nth cell in Cartesian and internal coordinates is denoted by $\mathbf{x}_n$ and $\mathbf{r}_n$, respectively, we can express the kinetic-energy term as a function of the vectors $\mathbf{x}_n$ in the form

$$2T = \sum_n \dot{\mathbf{x}}_n^+ \mathbf{M} \dot{\mathbf{x}}_n \tag{9.1}$$

and the potential energy in the harmonic approximation in terms of the internal coordinates as

$$2V = \sum_{n,n'} \mathbf{r}_n^+ \mathbf{F}_{n,n'} \mathbf{r}_{n'} \tag{9.2}$$

In these equations $\mathbf{M}$ is a diagonal matrix of order $3N \times 3N$ the elements of which are the masses of the atoms within one unit cell, suitably arranged (each mass appears consecutively three times in the main diagonal); $\mathbf{F}_{n,n'}$ is also a $3N \times 3N$ matrix containing harmonic force constants describing the interaction of the internal coordinates of cell n with those of unit n'. The potential energy in equation (9.2) has been expressed in terms of internal coordinates, because in the literature one finds harmonic force constants (especially for molecules and molecular crystals) calculated in this coordinate system. It should be noted that because the periodicity of the chain requires that $\mathbf{F}_{n,n'}$ depend only on the difference $|n - n'|$, one can set $\mathbf{F}_{n,n'} \equiv \mathbf{F}^{\tilde{n}}$; $\mathbf{F}_{n',n} = \mathbf{F}^{\tilde{n}+}$.

In order to express the potential energy in terms of Cartesian displacement coordinates, we first introduce vectors $\mathbf{X}(k_z)$ and $\mathbf{R}(k_z)$, respectively describing the phonons in terms of Cartesian and internal coordinates,[7] where

$$\mathbf{X}(k_z) = \frac{1}{\sqrt{2\pi}} \sum_{n=-\infty}^{\infty} \mathbf{X}_n e^{ik_z t_z(n)} \tag{9.3}$$

and

$$\mathbf{R}(k_z) = \frac{1}{\sqrt{2\pi}} \sum_{n=-\infty}^{\infty} \mathbf{r}_n e^{-ik_z t_z(n)} \tag{9.4}$$

with the Cartesian coordinate z in the direction of the chain. These definitions enable one to express equations (9.1) and (9.2) in the form

$$2T = \int_{-\pi/a}^{\pi/a} \dot{\mathbf{X}}(k_z)^+ \mathbf{M} \dot{\mathbf{X}}(k_z) dk_z \tag{9.5}$$

and

$$2V = \int_{-\pi/a}^{\pi/a} \mathbf{R}(k_z)^{+}\mathbf{F}_R(k_z)\mathbf{R}(k_z)dk_z \tag{9.6}$$

Owing to the periodic symmetry of a polymeric chain we have

$$\mathbf{F}_R(k_z) = \mathbf{F}^0 + \sum_{\tilde{n}} (\mathbf{F}^{\tilde{n}+} e^{-ik_z t_z(\tilde{n})} + \mathbf{F}^{\tilde{n}} e^{ik_z t_z(n)}) \tag{9.7}$$

The matrix $\mathbf{B}(k_z)$, which transforms Cartesian phonon coordinates into internal phonon coordinates [equations (9.3) and (9.4), respectively], can be determined by setting

$$\begin{aligned} \mathbf{r}_n &= \mathbf{B}_{-p}\mathbf{x}_{n-p} + \cdots + \mathbf{B}_{-1}\mathbf{x}_{n-1} + \mathbf{B}_0\mathbf{x}_n + \mathbf{B}_1\mathbf{x}_{n+1} + \cdots + \mathbf{B}_p\mathbf{x}_{n+p} \\ &= \sum_{l=1}^{p} (\mathbf{B}_{-l}\mathbf{x}_{n-l} + \mathbf{B}_l\mathbf{x}_{n+l}) \end{aligned} \tag{9.8}$$

Here $\mathbf{x}_{n+l}$ indicates the vector expressed in Cartesian coordinates of the l cell units apart from the reference cell n, and $\mathbf{B}_l$ is the corresponding transformation matrix between $\mathbf{r}_n$ and $\mathbf{x}_{n+l}$. Substitution of relation (9.8) into expression (9.4) with allowance for equation (9.3) yields

$$\mathbf{R}(k_z) = \sum_{l=-p}^{p} \mathbf{B}_l e^{-ik_z t_z(l)}\mathbf{X}(k_z) \tag{9.9}$$

which provides the transformation matrix $\mathbf{B}(k_z)$ in the form

$$\mathbf{B}(k_z) = \sum_{l=-p}^{p} \mathbf{B}_l e^{-ik_z t_z(l)} \tag{9.10}$$

This equation enables the potential-energy matrix elements for a linear periodic chain to be written in terms of Cartesian coordinates:

$$F_x(k_z) = \mathbf{B}(k_z)^{+}\mathbf{F}_R(k_z)\mathbf{B}(k_z) \tag{9.11}$$

One can introduce further mass-weighted Cartesian coordinates for the phonons via the equation

$$\mathbf{Q}(k_z) = \mathbf{M}^{1/2}\mathbf{X}(k_z) \tag{9.12a}$$

or

$$\mathbf{Q}(k_z) = \frac{1}{\sqrt{2\pi}} \sum_n \mathbf{q}_n e^{-k_z t_z(n)} \tag{9.12b}$$

[If expansion (9.3) is substituted into equation (9.12a) and new vectors $\mathbf{q}_n = \mathbf{M}^{1/2}\mathbf{x}_n$ introduced, one obtains equation (9.12b).] Equation (9.12a) can be used to write down finally the Hermitian dynamical matrix $\mathbf{D}(k_z)$ in the form

$$\mathbf{D}(k_z) = \mathbf{M}^{-1/2}\mathbf{F}_x(k_z)\mathbf{M}^{-1/2} \tag{9.13}$$

The phonon dispersion curves $\omega_i(k_z)$ of the linear chain (or, in the three-dimensional case, of a crystal) can then be obtained by the well-known eigenvalue equation of the GF method,[1]

$$[\mathbf{D}(k_z) - \omega_i^2(k_z)\mathbf{1}]\mathbf{L}_{x_i}(k_z) = \mathbf{0} \tag{9.14}$$

which can be solved numerically.

Starting from a set of harmonic force constants calculated in terms of internal coordinates, one can determine in this way the normal coordinates of the different vibrations (phonons) of an infinite chain as a function of k_z (the columns of matrices $\mathbf{L}_{x_i}$) as well as the corresponding phonon dispersion curves $\omega_i^2(k_z)$. One should note that, in order to determine the transport properties of polymers (see Section 9.3), one must calculate the electron–phonon interaction matrix elements. If simple Hermite polynomials are taken as phonon wave functions for these calculations, then one obtains a tolerable approximation, but knowledge of the dispersion relations $\omega_i^2(k_z)$ is very necessary to obtain reliable results.

After the different phonon dispersions have been determined, the density of vibrational states of a linear chain can be calculated as

$$N(\omega) = \sum_i \int_{-\pi/a}^{\pi/a} dk\delta[\omega - \omega(k_z)] \tag{9.15}$$

If the δ-function is replaced by a Lorentzian of half-width ε,

$$\delta[\omega - \omega(k_z)] \approx \frac{1}{\pi} \frac{\varepsilon}{\varepsilon^2 + [\omega - \omega(k_z)]^2} \tag{9.16}$$

and substituted into equation (9.15), we obtain

$$N(\omega) = \sum_i \int dk \frac{1}{\pi} \frac{\varepsilon}{\varepsilon^2 + [\omega - \omega(k_z)]^2} \tag{9.17}$$

which can be evaluated numerically. A trigonometric interpolation of the calculated frequencies for different wave vectors was used to compute the phonon spectra of alternating *trans*-polyacetylene[3] to obtain at least 50 points for each dispersion curve.

We note that the described method can also be applied to nonperiodic polymers consisting of a number of different units.[8] In order to obtain directly the density of states, the negative-factor counting (NFC) method in the matrix block form (9) (see Section 4.4) can be employed[10] applying Dean's negative-eigenvalue theorem[11] (which is actually based on Sturm's well-known mathematical theorem about the distribution of eigenvalues of a random matrix[12]). After determining in this way the density of vibrational states of the random chain, the inverse iteration technique[13] can be applied to determine any desired eigenvector.

The above technique is nowadays applied routinely for polymers with small unit cells. As we have seen, this method describes the determination of at least the important parts of the potential hypersurface (with an approximation involving a certain number of neighbors) when computing the harmonic force constants. In other words, one must therefore repeat the whole band-structure calculation of the polymer at very many different geometrical configurations. As long as their unit cells are small (and hence the number of geometrical variables is also small) this approach is possible, but prohibitively large computer times would be necessary in the case of polymers with large unit cells (such as poly-p-phenylene, periodic polynucleotides, and polypeptides containing a larger amino-acid residue).

This difficulty could be avoided by applying linear response theory, which is widely used in solid-state physics to determine directly the dynamic matrix, polarization, and frequency-dependent dielectric functions, as well as phonon dispersion curves in the harmonic approximation. This method has the great advantage that it requires only a band structure at the equilibrium geometry of the solid (chain), i.e., one does not have to determine a potential hypersurface. However, since this theory has not yet been applied to polymers and involves a rather complicated formalism, we cannot enter into details here but refer the reader to standard solid-state physical works[14] and an application to a simple solid (Si).[15]

9.2. PHONON CALCULATIONS FOR SELECTED ORDERED AND DISORDERED POLYMER CHAINS

The preceding methods were applied to a number of periodic polymers with a small unit cell. Many of these calculations used different all-valence electron semiempirical crystal-orbital methods (CNDO/2, – INDO, – MINDO, and MNDO CO[16] methods). At this point we shall discuss only some of the not very numerous *ab initio* vibrational calculations and also an MNDO CO work on *trans*-polyacetylene (PA), because in this investigation disorder in PA due to soliton and polaron formation was also taken into account.[3]

9.2.1. Polymethineimine

Teramae *et al.*[17] performed both minimal (STO-3G) and split valence (4-31G) *ab initio* CO calculations on *cis*-transoid-PA, *trans*-cisoid-PA, and all-*trans*-polymethineimine (see Figures 9.1 and 9.2). They optimized the geometry for all systems in both basis sets (fitting the points on the potential hypersurface by appropriate curves) and applied the gradient method[2] to obtain the first derivative of the total energy per unit cell, taking special care in the application of the method to the periodic boundary condition and also performing the differentiation $\partial E/\partial a$, where a is the translational period. The second derivative was obtained numerically with respect to the internal coordinates and the force constant matrix was $\mathbf{F}_R$ determined in terms of these coordinates. The GE method was then employed to derive the normal modes and phonon dispersion curves in the standard way. In comparing their theoretical results with experimental data, the authors always took the theoretical frequency lying at the bottom of the phonon band. HF calculation with a poor basis set and no correlation usually gives force constants, and consequently phonon frequencies, that are too high. To correct this failure they followed Blom[18] and scaled down the theoretical frequencies and hence obtained reasonably good agreement with experiment (in other words, relative frequencies within a given system are quite acceptable).

To illustrate this Table 9.1 presents theoretical and experimental vibrational spectra of *trans*-PA both without and with scaling.[17,19,20] It is seen that after appropriate scaling agreement between theory and experiment is quite good.

Another example is PMI (see Table 9.2). With a scaling factor of 0.89 agreement with experiment is quite satisfactory.

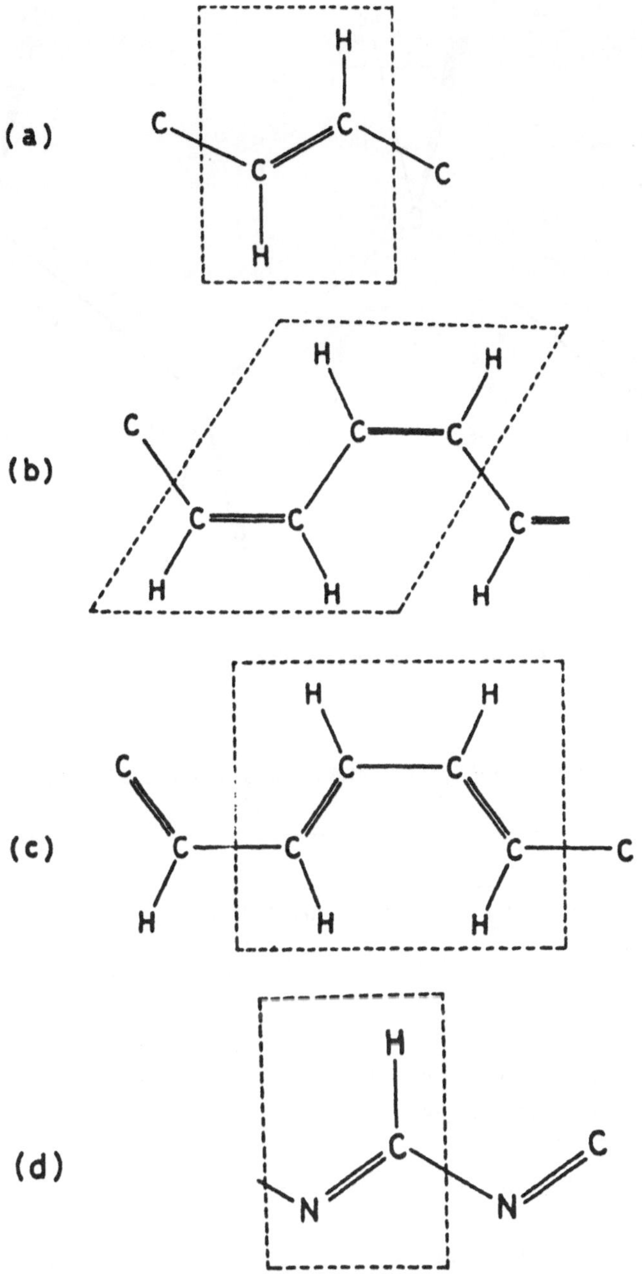

FIGURE 9.1. Structure of (a) *trans*-PA, (b) *cis*-transoid-PA, (c) *trans*-cisoid-PA, and (d) *trans*-polymethineimine (PMI).

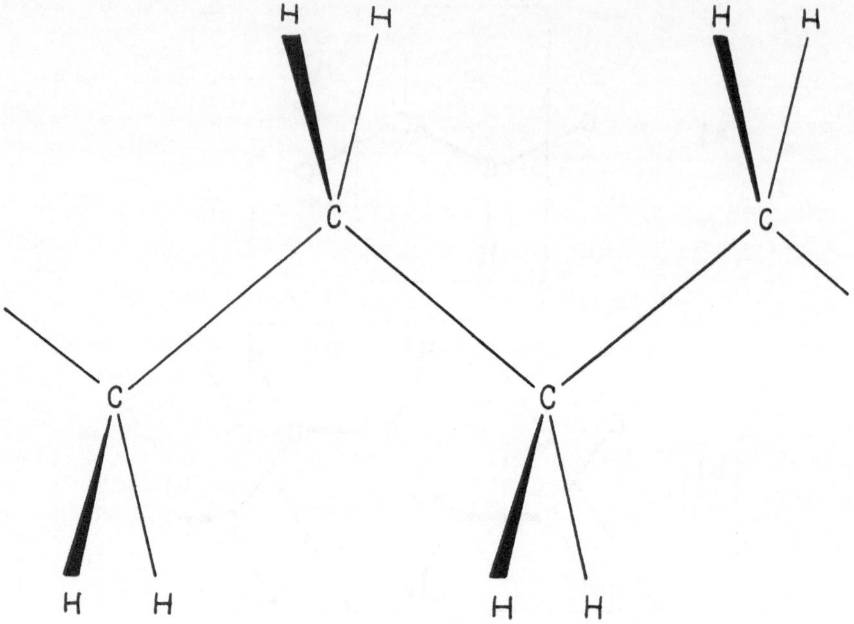

FIGURE 9.2. Structure of all-*trans*-polyethylene.

TABLE 9.1. Vibrational Frequencies of All-*trans*-Polyacetylene Obtained by Scaling (after Teramae *et al.*[17])[a]

	STO-3G			4-31G				
	(1.0)	(0.89)	(0.84)	(1.0)	(0.89)	(0.84)	Expt.[19]	Assignment for $(CH)_x$
A_g	1331	1190	1113	1310	1172	1096	1060	C–C stretching
	1528	1366	1278	1460	1306	1221	1285	C–C stretching with C–H deformation
	2006	1795	1679	1911	1709	1599	1450	C=C stretching
	3609	3228	3020	3264	2920	2731	2990	C–H stretching
A_u	1236	1106	1034	1149	1028	961	1015	C–H deformation out-of-plane
B_g	1088	973	910	1123	1005	940	1008	C–H deformation out-of-plane
B_u	1365	1220	1142	1336	1195	1118	1292	C–H deformation in-plane
	3568	3191	2985	3305	2956	2765	3013	C–H stretching

[a] All units are in cm^{-1}, and the values in parentheses are the scaling factor of the frequencies.

TABLE 9.2. Vibrational Frequencies of All-*trans*-Polymethineimine (after Teramae *et al.*[17])[a]

Symmetry	Calculation[b]	Experiment[20]	Assignment
A'	1152	—	C–N stretching
	1351	1410	C–H deformation in-plane
	1669	1620	C=N stretching
	3077	3170	C–H stretching
A''	1001	—	HCN deformation out-of-plane

[a] All units are in cm^{-1}.
[b] Scaling factor 0.89.

9.2.2. Polyethylene

As a further example we consider all-*trans*-polyethylene. In a comparatively recent paper Karpfen[21] undertook a rather extensive investigation of the band structure, DOS, vibrational spectra, and even Young's modulus (see Section 10.3) of this system. He investigated the effect of basis-set dependence, convergence with respect to lattice summations, and equilibrium conformations on these properties. The most stable conformation was found to be that using a roughly double-ζ, 7s3p/3s primitive basis contracted to a 4s2p/2s one.[22] It yielded for the C–C and C–H distances 2.95 and 2.08 au, respectively (the STO-3G minimal basis values are 2.92 and 2.05 au, respectively),[21] while for the valence angles 107.0° ($\sphericalangle$HCH) and 112.6° ($\sphericalangle$CCC) were obtained with the latter basis (see Figure 9.2). These values are in reasonable agreement with experiment.[21] Six-neighbor interactions were used in these calculations and gave rather good convergence for the total energy per unit cell and for the band structure.[21]

In order to find the equilibrium geometry and calculate the force constants, the four-dimensional (r_{C-C}, r_{C-H}, $\sphericalangle$CCC, and $\sphericalangle$HCH) energy hypersurface was calculated at 41 points. The 41 calculated energy values were fitted with third-order polynomials and the equilibrium geometry and force constants found. Some of the force constants determined with the better basis are summarized in Table 9.3; the data are systematically larger than their empirical values by 10–16%.[23] The reason for this is that the calculated force constants are purely harmonic while the empirical values contain also anharmonic contributions, which are especially important for the C–H stretching force constant.

Finally, it should be noted that the rather restricted part of the energy hypersurface, mainly used to determine the equilibrium structure

TABLE 9.3. Harmonic Force Constants of Polyethylene (after Karpfen[21])[a]

	r_{CC}	r_{CH}	∠HCH	∠CCC
r_{CC}	4.99			
r_{CH}	0.06	5.36		
∠HCH	−0.13	0.05	0.95	
∠CCC	0.28	−0.07	0.14	1.86

[a] Stretching–stretching force constants in mdyn/Å, stretching–bending force constants in mdyn/rad, and bending–bending force constants in mdyn Å/rad².

of all-*trans*-polyethylene, was insufficient to perform a complete vibrational analysis of this polymer.

9.2.3. Bent Chain of Hydrogen Fluoride Molecules

In another calculation[24] the electronic structure of a bent hydrogen fluoride chain was determined using three different Gaussian lobe basis sets. In basis I, 8s4p/5s3p was used for F and 4s/3s for H; basis II is characterized by 9s5p1d/5s3p1d for F and 5s1p/3s1p for H while the best basis applied (basis III) was 10s6p2d/6s4p2d for F and 6s1p/4s1p for H.[22] Truncation of the lattice sums was carried out in a quite sophisticated way following Piela and Dehalle[25] and Karpfen's own method (cut off according to the number of neighbor cells taken into account explicitly).[26]

HF forms a zigzag chain in the crystal and, in the chain direction, the unit cell contains 2HF molecules (see Figure 9.3). The internal coordinates at successive HF molecules obey a screw axis symmetry. Applying this symmetry, several authors[27] have suggested that a single HF molecule forms the unit cell and only four internal coordinates (r_1, r_3, α, and β) need be optimized. Application of the combined symmetry operation leads to a great decrease in the amount of computational work necessary to determine the equilibrium geometry.

On the other hand, when evaluating vibrational spectra of polymers, that part of the energy surface that determines the equilibrium geometry is often insufficient to provide the necessary information on various off-diagonal force constants. This is also the case for $(HF)_\infty$. Therefore, calculations with enlarged unit cells must be undertaken to increase the number of geometrical degrees of freedom. Doubling the unit cell, i.e., removing the screw axis symmetry, results in a twelve-dimensional energy surface. Within the harmonic approximation in-plane and out-of-

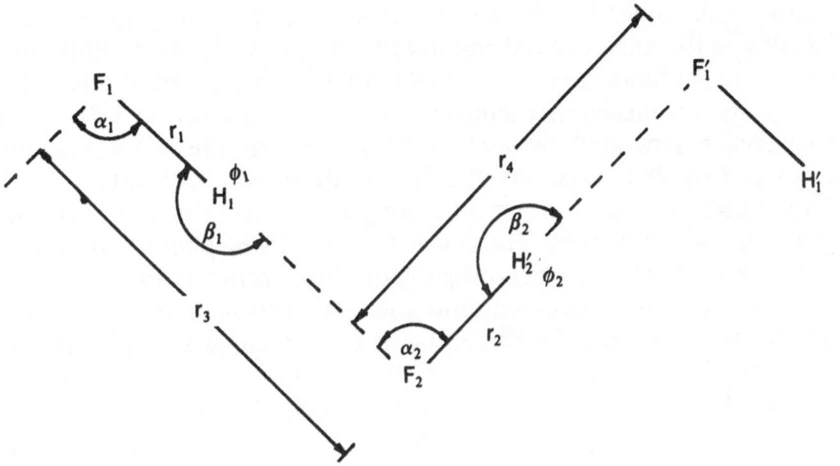

FIGURE 9.3. Structure and internal coordinates of $(HF)_\infty$; ϕ_1 and ϕ_2 and the FHF out-of-plane angles. Additional torsional coordinates $\tau_1(H_2''F_1F_2H_2)$ and $\tau_2(H_1F_2F_1'H_1')$ were used.

plane vibrations factorize, since the equilibrium structure of $(HF)_\infty$ is planar. Owing to technical reasons, the internal coordinates α_1 and α_2 could not be separated; tripling of the unit cell would have been necessary for that purpose. Hence only a seven-dimensional surface was considered. The results show that this approximation does not seem to cause serious errors. A total of 246 points were used for the fit. In the case of out-of-plane motions, an additional approximation was applied. The small values of the torsional force constants allow separation from the out-of-plane angle coordinates ϕ_1 and ϕ_2 and, equally, decoupling of the two torsional coordinates. All energy surfaces were approximated by several different cubic and quartic polynomials of appropriate dimension to obtain numerically reliable results for the force constants.

In contrast to the case of an isolated molecule, calculation of the force constants of a polymer is only possible if a further approximation is introduced. By definition, the crystal orbital method can only be applied to systems with perfect periodic symmetry. Energy surfaces were hence computed with all unit cells moving in phase.[24] The force constants $\bar{f}_{ij}$ obtained by the fitting procedure thus correspond formally to the infinite sums

$$\bar{f}_{ij} = \sum_{q-\infty}^{+\infty} f_{ij}^{0q} \tag{9.18}$$

where i and j refer to internal coordinates while 0 and q specify their location in the reference cell and in cell q, respectively. In the procedure applied the authors have omitted all but the leading terms of this series. Physically this means that interactions such as those between stretching coordinates separated by more than two bonds are not taken into account. Coupling constants of this type are usually exceedingly small. This approximation is irrelevant as long as only vibrational frequencies at $k = 0$ are considered. The shape of the derived phonon dispersion curves may, however, depend slightly on this approximation.

The harmonic force constants thus obtained have been used in a Wilson GF procedure[1] modified by periodic boundary conditions[6] to compute phonon dispersion curves. These curves enabled the phonon density of states to be evaluated in a manner similar to that employed in obtaining the electronic density of states from the electronic band structure.

The resulting harmonic force constants of $(HF)_\infty$ are shown in Table 9.4. The stretching force constant of the isolated HF molecule and a selected number of force constants of $(HF)_2$[28,29] are also included. In the case of the smaller basis set (8s4p/5s3p for F and 4s/3s for H) the harmonic force field is complete, with the exception of some separation of α_1 and α_2. Fortunately, however, most of the off-diagonal force constants are exceedingly small, particularly those involving coupling between α and other internal coordinates. If only the diagonal force constants and the coupling constants $f_{r_1r_2}$, $f_{r_3r_4}$, $f_{\beta_1\beta_2}$, and $f_{\phi_1\phi_2}$ are retained, the computed vibrational frequencies are shifted by less than 10 cm^{-1} compared to those obtained using the full force field. It therefore seems safe to work with this restricted force field, which contains the essential physical coupling terms. In view of the methodical limitations of the calculations (limited basis set, neglect of correlation, harmonic approximation, one-dimensional model) one cannot hope for quantitative agreement with experimental vibrational spectra. Therefore, the better basis set (9s5p1d/5s3p1d for F and 5s3p/3s1p for H) was employed to compute this restricted force field alone. With the largest basis set only certain sums of force constants could be computed which, however, allow further insight into the basis-set dependence.

Previous force fields of $(HF)_\infty$ used in lattice dynamical studies[30] have always been obtained by a fit to experimental frequencies, relying on a reasonable choice of fundamentals. These force fields are not directly comparable to the present ones, partly owing to a different choice of internal coordinates, partly because fewer force constants were considered or because interchain force constants were also included.[30] Nevertheless, some general features may be discussed.

TABLE 9.4. Internal Force Constants of Various $(HF)_x$ Species (after Beyer and Karpfen[24])[a]

Type of constant		Gaussian lobe basis sets I	II	III
HF	$f_{r_1r_1}$	9.31	11.32	11.19
$(HF)_2$	$f_{r_1r_1}$	9.17[b]	—	10.76[c]
	$f_{r_1r_2}$	−0.040	—	−0.068
	$f_{r_3r_3}$	0.25	—	0.13
	$f_{\beta_1\beta_1}$[d]	≈0.04	—	≈0.025
$(HF)_\infty$	$f_{r_1r_1}$	6.93	9.27	9.2 ($f_{r_1r_1} + 2f_{r_1r_2}$)
	$f_{r_3r_3}$	0.64	0.37	0.24 ($f_{r_3r_3} + 2f_{r_3r_4}$)
	$f_{r_1r_2}$	−0.31	−0.23	—
	$f_{r_3r_4}$	−0.077	−0.058	—
	$f_{\alpha_1\alpha_1}$	0.037	0.061	0.05
	$f_{\beta_1\beta_1}$	0.15	0.13	0.07 ($f_{\beta_1\beta_1} + 2f_{\beta_1\beta_2}$)
	$f_{\beta_1\beta_2}$	−0.026	−0.030	—
	$f_{\phi_1\phi_1}$	0.12	0.075	—
	$f_{\phi_1\phi_2}$	−0.0092	−0.0085	—
	$f_{\tau_1\tau_1}$	0.002	—	—

[a] Stretching–stretching force constants in mdyn/Å, stretching-bending force constants in mdyn/rad, and bending–bending force constants in mdyn Å/rad². For the definition of internal coordinates see Figure 9.3.
[b] 4-31G basis, free molecule value 9.55.[28]
[c] Free molecule value 11.17.[29]
[d] Originally in Curtiss and Pople[28] and Lischka[29] the angle between HF and FF was chosen as internal coordinate. The original values have therefore been reduced by a factor of four.

As expected, the large geometrical changes between the gas-phase dimer and the infinite polymer are reflected in the force constants in an even more pronounced way. The largest change occurs for the intramolecular force constant $f_{r_1r_1}$. Compared to the free-molecule value, $f_{r_1r_1}$ is reduced only slightly in the gas-phase dimer. At a near Hartree–Fock limit level the reduction is about 3.7%.[31] Using the 4-31G basis, the corresponding value is 4%.[28] A much more drastic effect is found for $(HF)_\infty$, where $f_{r_1r_1}$ is now reduced by 25.6 and 17% with basis set 4-31G and the best applied one, respectively, in good qualitative agreement with earlier attempts[31] to evaluate the stretching force constant in $(HF)_\infty$. Furthermore, coupling between neighboring HF molecules ($f_{r_1r_2}$) is enhanced in the chain by about a factor of 5 and 7, depending on the basis set applied. Equally, $f_{r_3r_3}$ becomes significantly larger in the polymer in accordance with the smaller intermolecular distance. The same effect is observed for $f_{\beta_1\beta_1}$.

Some important properties of the force field deserve further attention. Quantity $f_{\beta_1\beta_1}$, the force constant for in-plane bending, is larger than $f_{\phi_1\phi_1}$, representing out-of-plane bending. Both $f_{\beta_1\beta_2}$ and $f_{\phi_1\phi_2}$ are large in absolute value and cannot be neglected. The value of $f_{\beta_1\beta_2}$ is substantially larger than that of $f_{\phi_1\phi_2}$. All four coupling constants retained in the restricted force field have a negative sign.

The interpretation of the experimental vibrational spectra of crystalline hydrogen fluoride is difficult. In order to demonstrate the considerable disagreement between various research groups, vibrational frequencies stemming from infrared, Raman, and neutron spectroscopic studies are compared with the results of lattice dynamical calculations in Table 9.5. Vibrational frequencies obtained in this work are shown in Table 9.6 for $(HF)_\infty$.

Turning first to a discussion of experimental and lattice dynamical results one observes that, despite difficulties in the librational region, most authors agree about the assignment of the remaining modes. The two highest frequencies are unequivocally assigned to the antisymmetric and symmetric intramolecular HF stretching modes, with the antisymmetric B_1 mode having the larger frequency and smaller intensity. Actually, four lines can be observed experimentally in this region. The general interpretation is in favor of the coupling mechanism between intra- and intermolecular stretching vibrations,[30,32–34] interchain coupling being of no importance. Of the three out-of-phase translational modes, the out-of-plane A_2 mode is placed lowest. Most authors consider

TABLE 9.5. Survey of Experimental and Lattice Dynamical Studies on the Vibrational Frequencies of Crystalline HF (after Beyer and Karpfen[24])[a]

	Kittelberger and Hornig[32] infrared (88 K)	Turbino and Zerbi[30] calculation	Axmann *et al.*[30] neutron scattering and calculation	Anderson *et al.*[33] Raman (18 K)	Anderson *et al.*[30] calculation
$T(A_2)$	—	—	135.5	57	—
$T(A_1)$	202	214	221.8	187.5	202
$T(B_1)$	366	359	359.7	363.5	349
$L(A_1)$	962	970	233.9	569.5	625.5
$L(A_2)$	—	539	673.5	687	591
$L(B_2)$	552	552	552.5	943	605.5
$L(B_1)$	1025	1022	827.6	742	1006.5
$S(A_1)$	3065	3065	3053.1	3044.5	3063.5
$S(B_1)$	3404	3404	3404	3386	3368

[a] All values are in cm^{-1}.

TABLE 9.6. Vibrational Frequencies of $(HF)_\infty$ Obtained with Basis Sets I and II (after Beyer and Karpfen[24])[a]

	Basis set I	Basis set II
$T(A_1)$	160.9	149.5
$T(B_1)$	490.3	386.4
$L(A_1)$	701.6	604.3
$L(A_2)$	728.7	563.3
$L(B_2)$	866.7	715.6
$L(B_1)$	1024.6	1007.0
$S(A_1)$	3356.9	3966.6
$S(B_1)$	3674.6	4169.5

[a] All values are in cm^{-1}.

the following two lines as the in-plane translational modes A_1 and B_1, where the latter is considered to have the higher frequency and corresponds to the out-of-phase translation in the chain direction. Only the assignment by Axmann *et al.*[30] is different in this region.

The uncertainties concerning the interpretation of the librational region are considerable and are caused by at least two experimental difficulties. First, the lack of single crystals of HF prevents measurements of polarized infrared or Raman spectra; if they were available the puzzle would be resolved immediately. Second, more than four lines are observed in the region between 500 and 1300 cm^{-1} for HF. In the absence of polarized spectra or, alternatively, without the availability of a decent intermolecular potential, the choice of fundamentals is not at all obvious and has largely been based on intensity arguments. These have been derived assuming nonpolarizable dipoles. In view of the strong increase in the dipole moment of HF during polymer formation and considering the numerous possible mechanisms which may contribute to the infrared line shape and intensity[35] in hydrogen bonded systems, it is questionable whether the simple oriented gas mode is appropriate.

The negative sign of the four coupling constants $f_{r_1r_2}$, $f_{r_3r_4}$, $f_{\beta_1\beta_2}$, and $f_{\phi_1\phi_2}$ explains that all symmetric A modes lie lower than the corresponding antisymmetric B modes in the vibrational spectra computed. One observes reasonable qualitative agreement with experimental results on intramolecular stretching modes $S(A_1)$ and $S(B_1)$ and translational modes $T(A_1)$ and $T(B_1)$, consistent with most of the earlier assignments. For the case of stretching modes an interpretation of the sign of the coupling constant $f_{r_1r_2}$ has already been given earlier.[31] In the course of the $S(A_1)$ mode the hydrogen atoms move cooperatively in a double minimum potential for simple symmetry reasons. The potential barrier (about 10 kcal/mol with basis set I and certainly larger for the more extended ones) is, however, sufficiently high to prevent instability of the crystal structure with respect to this mode. The existence of a second minimum reduces the $S(A_1)$ force constant substantially. The $S(B_1)$ mode occurs in a symmetric single minimum potential and has therefore the higher frequency. Owing to the foregoing remarks it is impossible to obtain quantitative agreement with experimental frequencies within the framework of the harmonic approximation. The large splitting is, however, correctly predicted.

The relative ordering of $T(A_1)$ and $T(B_1)$ can be explained by similar arguments. Again the A mode occurs in a double minimum potential while the B mode has a symmetric single minimum potential. Therefore $T(A_1)$ lies lower than $T(B_1)$. Obviously, the frequency of $T(A_2)$ is zero in the one-dimensional model applied.

Unfortunately, difficulties are encountered in the case of the librational motions. A different relative order is obtained with basis sets I and II. Nevertheless, a number of qualitative trends can be extracted independent of the basis set applied. Both in-plane and out-of-plane librations exhibit considerable dispersion. Splitting of the in-plane modes is significantly larger. This behavior can again be understood qualitatively if the potential surfaces pertaining to the four librational modes are considered in detail and simple electrostatic approximations applied. Three of the four librational modes [$L(B_1)$, $L(A_2)$, and $L(B_2)$] are described by symmetric potential curves. In the course of the $L(B_1)$ mode close contact is established between hydrogen atoms of neighboring HF molecules, thus creating an energetically highly unfavorable situation from the standpoint of dipole–dipole interaction and for steric reasons. The $L(B_1)$ frequency is therefore by far the highest. The symmetric counterpart $L(A_1)$ has a weakly asymmetric potential. On the flatter side, when hydrogen atoms move away from the chain axis, the dipole–dipole interaction is still attractive although the configuration for hydrogen bonding becomes worse. The potential is slightly more repulsive if both hydrogens move into the interior of the chain. On inspecting the out-of-plane librations the relative strength of dipole–dipole interactions immediately explains the observed order.

Comparison of these results with earlier interpretations enables one to exclude the assignment by Kittelberger and Hornig[32] and by Turbino and Zerbi,[30] where the latter argued that the bending modes are too local to have any appreciable dispersion. On the other hand, Axmann *et al.*[30] have correctly predicted large splitting for the librational modes.

The uncertainty in the computed frequencies lies in the relative position of the center of the in-plane and out-of-plane librations, which may also be sensitive to interchain coupling. In the current stage it seems impossible to decide definitely on the basis of these results for one of the alternatives concerning the relative order of the three lower librational frequencies.

Finally, one should note that the phonon energy bandwidths (dispersion curves) obtained from the GE analysis of HF_∞ lie in the valence band region at 150–200 cm^{-1} if calculated by the poorest basis set; computed DOS curves show the typical behavior of the quasi-one-dimensional system (see Figure 5 of Huzinaga[22]).

9.2.4. Periodic and Nonperiodic Alternating *trans*-Polyacetylene

The literature contains such minimal basis (STO-3G) CO calculations on periodic all-*trans*-polyacetylene, in the course of which the most important force constants have also been determined.[36] However, no attempt was made to compute the phonon dispersion curves $\omega_i(k)$ from the force constants of this polymer, because a not large enough part of the energy hypersurface was determined. The calculated force constants in the case of second-neighbor interactions and a better (8s4p/5s3p for carbon and 4s/3s for hydrogen) basis agree tolerably well with the experimental force constants of butadiene[37] [$f_{r_1r_1} = 8.2$ dyn/A (theoretical), 8.15 (experimental); $f_{r_2r_2} = 4.6$ (theretical), 6.07 (experimental); $f_{r_1r_2} = 0.9$ (thoretical), 0.58 (experimental); $f_{r_3r_3} = 6.0$ (theoretical), 5.07 (experimental); the geometrical variables are defined in Figure 9.4]. The discrepancies between the theoretical and experimental force constants are due primarily (besides neglect of correlation effects in the theoretical calculation) to the fact that the experimental values refer to butadiene, while the theoretical values pertain to an infinite *trans* alternating PA chain.

In two subsequent papers[3,8] the semiempirical all-valence electron MNDO (modified MINDO method[38]) was used to calculate clusters of CH_2=CH–CH=CH–CH=CH_2 (modeling periodic alternating PA), CH_2=CH···CH···CH=CH_2 (soliton in PA), and CH_2=CH···CH···CH···CH=CH_2 (polaron in PA). The MNDO method, though approximate, takes into account electron–electron interaction in an SCF way and is known to give the correct trend for vibrational properties of organic molecules.[38] In the case of the soliton cluster the unrestricted version of the MNDO method with an odd number of electrons (DODS) has been applied.

In order to obtain a reasonable force field for *trans*-PA, the force

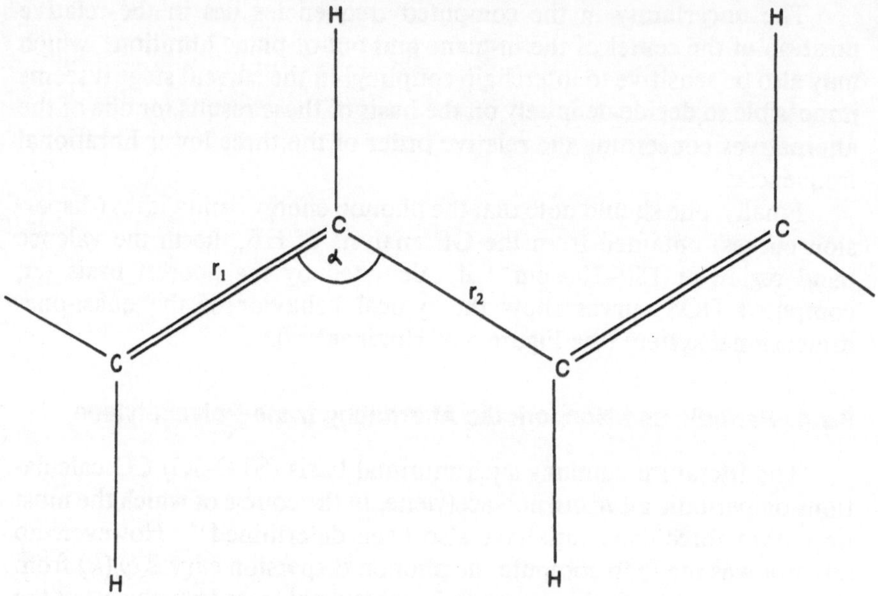

FIGURE 9.4. Geometry of all-*trans*-polyacetylene.

constants for the *trans*-1,3,5-hexatriene molecule have been calculated in a geometry shown in Figure 9.5. The geometrical parameters used are slightly different from the experimental geometry, but more appropriate to simulate a *trans*-PA chain. The carbon–carbon double-bond and single-bond lengths have been assumed to be 1.36 and 1.44 Å, respectively, with alternation parameter $r = 0.08$ Å, as suggested by recent calculations.(39) This molecular model therefore also differs slightly from the one used by Zannoni and Zerbi(40) and Inagaki.(41)

The geometries used for the soliton and polaron cluster are shown in Figure 9.6.

Owing to the semiempirical nature of the MNDO method, the theoretical force constants are between 10 and 20% larger than the experimental values. Therefore, following Török and Pulay(42) empirical scale factors have been calculated in the case of the MNDO force field of butadiene(3) for which molecule enough experimental information is available.(36) The suggestion followed in the computation of these scale factors was that the geometrical mean value of the diagonal force constants be used.(42)

Table 9.7 presents the most important force constants of butadiene and hexatriene. As noted before, the errors in the unscaled theoretical values with respect to the experimental values are between 10 and 20%; only the carbon–carbon double-bond force constant is somewhat more overestimated. On the other hand, if one introduces an appropriate scale factor in the case of butadiene, the trends of the different force constants

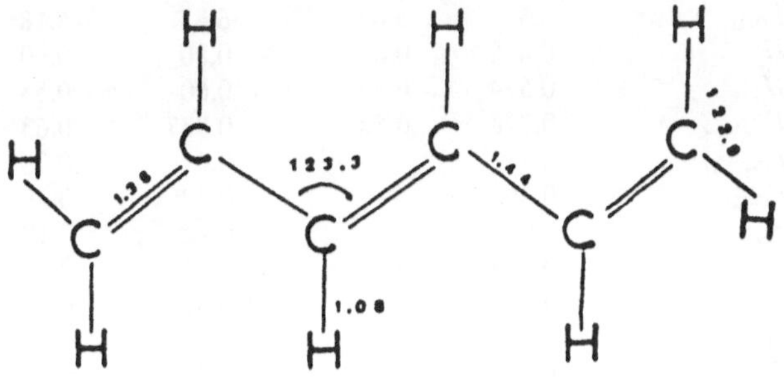

FIGURE 9.5. Bond lengths and angles used for *trans*-1,3,5-hexatriene. Bond lengths are given in Å.

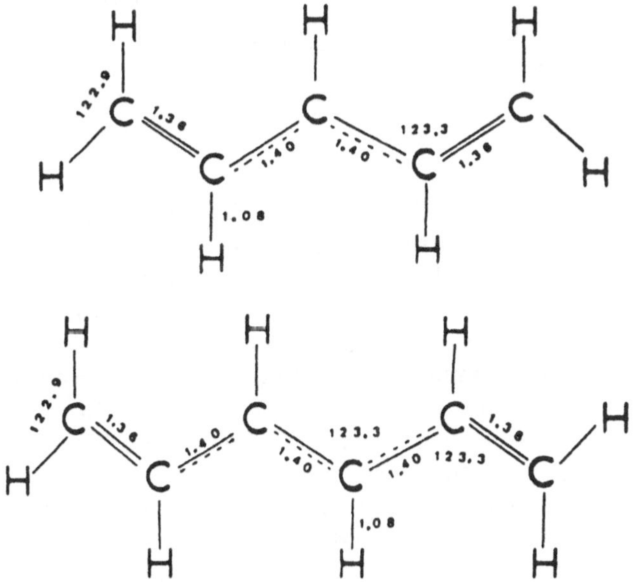

FIGURE 9.6. Cluster model for a soliton (top) and polaron (bottom) defect. Bond lengths are given in Å.

TABLE 9.7. Most Important Force Constants for Butadiene and Hexatriene[a]

Type	1,3-Butadiene		1,3,5-Hexatriene	
	MNDO	Exp.[b]	Unscaled	Scaled[c]
$F_{C=C}$	11.9	8.15	11.083	7.53
F_{C-C}	7.4	6.07	8.46	6.94
F_{C-H}	6.3	5.07	6.48	5.18
H_{α}	0.828	0.61	0.80	0.60
H_{β}	0.574	0.53	0.60	0.55
$F_{C=C/C-C}$	0.778	0.58	0.837	0.63
$F_{C=C/CH}$	0.41	—	0.41	0.28
$F_{C-C/CH}$	0.36	—	0.37	0.29
$F_{\alpha\alpha}$	0.08	—	0.12	0.10
$F_{\beta/\beta}$	0.035	—	0.038	0.0
$F_{C=C/\alpha}$	0.33	—	+0.32	0.23
$F_{C-C/\alpha}$	0.33	—	+0.33	+0.25
$F_{C-H/\alpha}$	−0.26	—	−0.26	−0.20
$F_{C=C/\beta}$	−0.31	—	−0.31	−0.25
$F_{C-C/\beta}$	−0.32	—	−0.32	−0.29
$F_{C-H/\beta}$	0.0	—	0.0	0.0

[a] For butadiene the MNDO and experimental values are given, for hexatriene the MNDO and scaled MNDO force constants are shown. The force constants for stretching and stretching–stretching coupling are given in mdyn/Å, for bending in mdyn Å/rad², and for stretching–bending coupling in mdyn/rad; α denotes the deformation, β the rocking force constant.[3]
[b] From Karpfen and Petkov.[36]
[c] As suggested in Pulay *et al.*[42]

are in good agreement with experiment. The same scale factor was later transferred to 1,3,5-hexatriene. The variations in the force constants on going from butadiene to hexatriene show the expected trend: the double-bond force constant decreases while the single-bond value increases.

The calculations have been carried out for neutral and singly charged clusters, assuming complete charge transfer of one electron from the donor dopant to the polymer backbone. The UHF treatment was used for the open-shell cases, as mentioned before. Dewar[43] noted that the overestimation of electron correlation can be reduced with a correction factor. In the calculation cited this was done by using a scale factor for the UHF-MNDO force constants that was the gometric mean of RHF scale factors for C=C and C–C stretching, or for the deformation and rocking-bending force constants.

The most important force constants are given in Table 9.8. As expected important variations can be observed, especially when the defect creates an unpaired electron. The stretching force constant for the C=C double bond in the neighborhood of the defect site decreases for the neutral soliton and charged polaron defects, while in the other cases the change is less pronounced. In all cases the force constants are lower with respect to the normal C=C bond. This leads to the conclusion that both the soliton and polaron defects are not localized in only two or three bonds. At the defect site, where the C–C bond length has been taken in

TABLE 9.8. Unscaled and Scaled MNDO Force Constants Calculated for the Soliton and Polaron Defect Modes[a]

	Soliton				Polaron			
	Charge = 0		Charge = − 1		Charge = 0		Charge = − 1	
Type	Unscaled	Scaled	Unscaled	Scaled	Unscaled	Scaled	Unscaled	Scaled[b]
$K_{C=C}$	9.24	6.90	10.86	7.39	10.9	7.4	9.96	7.44
$K_{-C\underline{\cdots}}$	8.75	6.56	9.51	7.13	9.96	7.44	9.46	7.01
H_{α_1}	0.68	0.56	0.833	0.62	0.92	0.69	0.68	0.58
H_{β_1}	0.57	0.47	0.576	0.53	0.58	0.53	0.57	0.44
H_{α_2}	0.62	0.51	0.727	0.55	0.60	0.50	0.60	0.50
H_{β_2}	0.60	0.50	0.615	0.56	0.62	0.55	0.60	0.51
$F_{C=C/C=C}$	0.27	0.20	0.11	0.075	0.008	0.0	0.16	0.12
$F_{C=C/C\underline{\cdots}C}$	1.69	1.24	1.15	0.86	0.94	0.70	1.33	1.0
$F_{C\underline{\cdots}C/C\underline{\cdots}C}$	1.06	0.78	1.09	0.81	− 0.09	− 0.07	0.073	0.059
$F_{C\underline{\cdots}C/C-H}$	0.41	0.29	0.41	0.29	0.38	0.27	0.40	0.28
F_{α_1/α_2}	0.075	0.06	0.019	− 0.014	− 0.14	0.11	0.11	0.09
F_{β_1/β_2}	0.040	0.033	0.023	− 0.021	− 0.019	0.017	− 0.023	− 0.021
$F_{C=C/\alpha_1}$	0.28	0.22	0.394	0.29	0.37	0.29	0.44	0.34
$F_{C=C/\beta_1}$	0.29	0.23	0.26	− 0.21	0.27	0.22	0.273	0.22
$F_{C\underline{\cdots}C/\alpha_1}$	0.32	0.26	0.35	0.27	0.39	0.31	0.30	0.24
$F_{C\underline{\cdots}C/\alpha_2}$	0.36	0.29	0.36	0.28	0.20	0.16	0.41	0.32
$F_{C\underline{\cdots}C/\beta_1}$	− 0.35	− 0.28	0.305	− 0.26	− 0.34	− 0.27	− 0.32	− 0.26
$F_{C\underline{\cdots}C/\beta_2}$	0.31	0.25	− 0.33	− 0.28	0.31	0.24	0.32	0.26
$OOP(CH)_3$	0.50	0.60	0.586	0.686	0.80	0.90	0.50	0.60
$OOP(CH)_2$	0.67	0.804	0.68	0.81	0.75	0.85	0.56	0.66
$K_{=C-H}$	6.28	5.0	6.25	5.0	6.51	5.20	6.46	5.17
$K_{\underline{\cdots}C-H}$	6.48	5.18	6.44	5.15	6.48	5.18	6.44	5.15
$K_{-C\underline{\cdots}C\underline{\cdots}}$	—	—	—	—	9.38	7.26	9.75	7.40
$F_{-C\underline{\cdots}C\underline{\cdots}/C\underline{\cdots}C}$	—	—	—	—	0.90	0.69	0.80	0.62
$F_{C\underline{\cdots}C/C=C}$	—	—	—	—	− 0.091	− 0.21[b]	0.0073	0.3[b]

[a] The units are the same as in Table 9.7. OOP stands for out-of-plane. $(CH)_i$, $i = 2, 3$ denotes the ith C–H group in Figure 9.8.[3]

[b] As suggested in Pulay *et al.*[42]

this simple model as the average of the short and long bond, different behavior for the soliton and polaron cases has been found. For the neutral soliton model the MNDO method predicts a very low force constant, comparable to the single-bond case, while in the charged soliton as well as in both polaron models these force constants are close to the arithmetic mean of the single- and double-bond force constants.

In the neutral polaron defect the C–C stretching force constants decrease monotonously from the center of the defect on, while in the charged polaron case a small alternation has been found. The general trend in the change of the bending force constant is a decrease in the deformation bending constant and an increase in the rocking force constant. It is also interesting to observe the increase in the coupling force constants between C–C bonds, especially in the neutral soliton cluster. A very large coupling is predicted for the double and intermediate bonds and between the two intermediate C–C bonds. Long-range coupling constants are increased.

Finally, out-of-plane force constants should be discussed briefly. Until now only one recent study[(44)] included also the out-of-plane motions of the hydrogen and carbon atoms. In the work of Peluso *et al.*[(3)] these motions were included because they may be useful in the interpretation of additional new bands in the infrared spectrum of doped PA. The scale factor for the out-of-plane C–H bending force constant has been obtained by comparing the MNDO results and experimental frequencies; symmetry decoupling of these modes from the in-plane modes makes this procedure possible. Calculations confirmed the conjecture: a large lowering of out-of-plane force constants was found for both soliton models and for the charged polaron cluster.

The force constant matrix $\mathbf{F}_R(k)$ for an infinite PA chain in internal coordinates has been constructed from the previously discussed force constants for hexatriene. No changes in the long-range coupling force constants have been introduced in order to take into account electron–phonon coupling, as suggested by Mele and Rice[(45)] and Zannoni and Zerbi.[(46)] The suggestion of Gavin and Rice[(47)] was followed, according to which these interaction force constants do not decrease as the bond separation increases but retain between 30% and 50% of the nearest-neighbor values. The calculation has been carried out in the second-neighbor-interactions approximation of the CO method.

The phonon dispersion curves and corresponding density of states, calculated from the force constant matrix $\mathbf{F}_R(k)$ with the help of methods described in Section 9.1, are shown in Figure 9.7 and atomic displacements of normal modes for $q = 0$ are sketched in Figure 9.8. The Raman spectrum of all-*trans*-PA, measured by Shirakawa *et al.*,[(48)] shows two

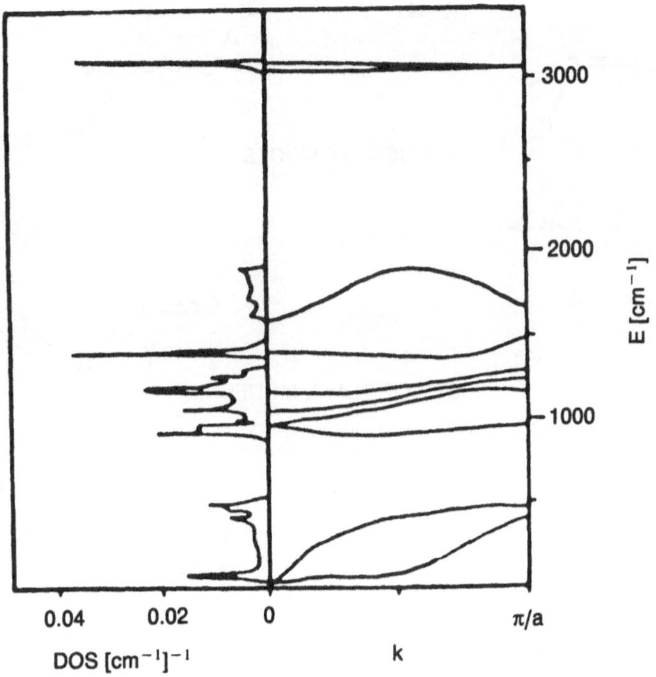

FIGURE 9.7. Density of states (DOS) and phonon dispersion curves for *trans*-polyacetylene.

strong bands at 1474 and 1080 cm^{-1} and a weak band at 1010 cm^{-1} (see Table 9.9). The latter has been ascribed to the C–H out-of-plane motion. The bands at 1474 and 1080 cm^{-1} were tentatively assigned to the C–C double- and single-bond stretching motions, respectively. However, this interpretation is still the subject of discussion. The results obtained are in agreement with those of Zannoni and Zerbi[46] and Inagaki.[47] The mode at 1100 cm^{-1} — calculation gave 1142 cm^{-1} (all values are given, as usual, for $q = 0$) — should be ascribed mainly to the C–C stretching motion, coupled to C–H in-plane bending, while Mele and Rice[45] and Schügerl and Kuzmany[49] find the single-bond stretching mode at 1200 cm^{-1} and ascribe the 1100 cm^{-1} mode to C–H in-plane bending with some amount of C=C and C–C stretching motions. In the present calculation a Raman-active C–H in-plane bending motion at a frequency of 1028 cm^{-1} was found, in very good agreement with the value of 1021 cm^{-1} calculated by Inagaki,[41] while others[49] have found a value between 1200 and 1290 cm^{-1}. The band at 1474 cm^{-1} can be ascribed

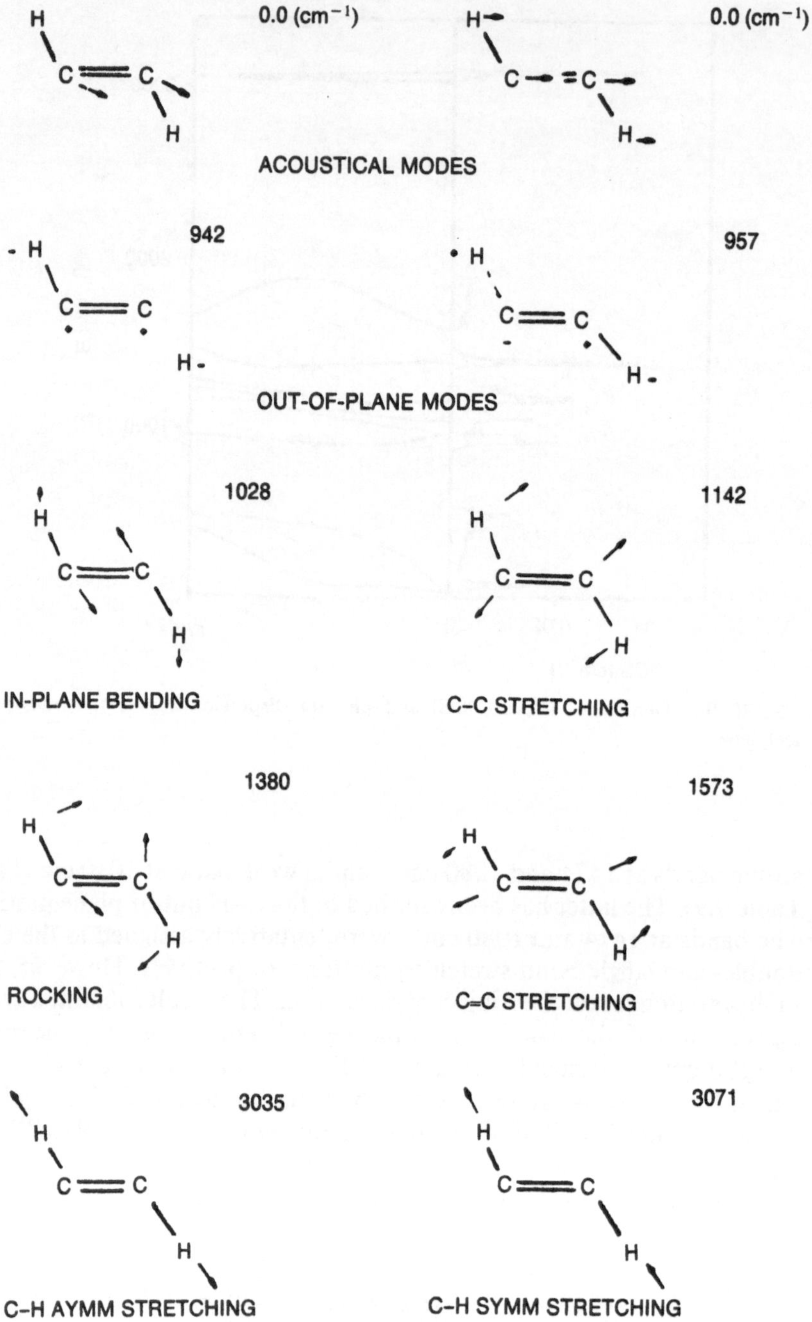

FIGURE 9.8. Atomic displacements of normal modes for selected frequencies in periodic *trans*-polyacetylene. The numbers indicate the frequency and are given in cm^{-1}.

TABLE 9.9. Observed Frequencies (in cm^{-1}) and Tentative Assignments in Doped and Undoped PA[a]

	Undoped tPA		Assignments
Raman[b]			
	1474	s	C=C stretching
	1291[d]	v v w	CCH in-plane bending
	1080	s	C–C stretching
	1016	w	C–H out-of-plane
Infrared[c]			
	1292	v w	CH in-plane deformation
	1015	v s	CH out-of-plane
	3013	m	CH stretching
	Doped tPA		**Assignments**
Infrared[d]			
	888	(very broad) s	CH in-plane deformation
	1288	s	C⋯C stretching +
	1397	s	C⋯C–H deformation

[a] s denotes strong, w weak, m medium, and v very.
[b] After Shirakawa *et al.*[48b]
[c] After Shirakawa and Ikeda.[48a]
[d] After Harada *et al.*[48c]

without doubt to the C=C stretching vibration weakly coupled with the C–H bending motion.

The infrared spectrum of polyacetylene exhibits three peaks at 3013, 1292, and 1015 cm^{-1}[48] (see Table 9.9). The peak at 3013 cm^{-1} is assigned to the C–H stretching vibration. Shirakawa and Ikeda assigned the mode at 1292 cm^{-1} to the *trans*-C–H in-plane deformation and the band at 1015 cm^{-1} to the out-of-plane C–H deformation. The results for these two bands are in good agreement with this interpretation.

For the out-of-plane motion, two bands between 950 cm^{-1} (infrared) and 1000 cm^{-1} (Raman) have been found in good agreement with the experimental results obtained by Lunde and Zechmeister[50] for several conjugated systems.

The experimental spectra of PA doped both by donor and acceptor species possess the following main features (see Table 9.9). The Raman spectrum does not exhibit additional peaks compared with undoped PA. In the infrared spectrum, on the other hand, at least two additional bands have been detected. The first, very broad band falls between 800 and 900 cm^{-1}, the second, much narrower band lies around 1300 cm^{-1}.[51] While there is agreement on the nature of the 1300 cm^{-1} band, the origin of the broad band starting at 800 cm^{-1} has been a matter of discussion for quite some time.[45,51,52] Before introducing the soliton concept Heeger and co-workers[51] suggested that this band is due to ionization of the dopant state in the semiconductor gap. Later, the isotopic shift of this band to lower frequencies found by Harada *et al.*[48c] demonstrated beyond doubt the vibrational origin of this band. These authors assigned the band to the in-plane C–C–H bending motion which, in undoped PA, is at 1010 cm^{-1}, but the reason for its large width could still not be explained.

The theoretical investigation of doped PA by Zannoni and Zerbi[40] has given a few peaks in this region. They ascribed the large bandwidth to the overlap of several molecular-like absorptions caused by the large dimension of the defect or by the presence of a distribution of defects of the same type but with slightly different electronic and geometrical characteristics. It is difficult, however, to understand the formation of defects with different characteristics in very lightly doped samples of PA which still show this very broad band.

In the band around 1300 cm^{-1} Harada *et al.*[48c] have found two peaks at 1288 cm^{-1} and 1397 cm^{-1}. The widths of these peaks are comparable to those of undoped PA. The band at 1397 cm^{-1} has been ascribed to the C–C stretching at the defect site. This assignment is in agreement with both the soliton and polaron models, which predict bond relaxation in the middle of the defect and therefore a shift of the C=C double-bond stretching vibration at 1474 cm^{-1} to lower frequencies. The other band near 1288 cm^{-1} has been ascribed to the C–H in-plane deformation which, in undoped PA, lies at 1290 cm^{-1}.

The eigenvalue spectrum of four different finite PA chains has been calculated using the negative-factor counting method in its matrix block form (see elsewhere[9] and Section 4.4). Each consists of 25 C_2H_2 units and contains three defects, either neutral (charged) solitons or neutral (charged) polarons, on positions 8, 18, and 23. These position numbers were generated by a random number generator. In Figure 9.9 the DOS for the chains containing neutral or charged solitons are shown, and in Figure 9.10 the DOS for the corresponding polaron cases are plotted. Selected normal modes for the respective neutral and charged, soliton and polaron were also calculated. We do not discuss them here, but refer the reader to the original paper.[3]

First of all, it should be pointed out that no large differences were found between the four cases. The calculated spectra for the charged and neutral soliton as well as for the corresponding polaron defects are generally in agreement with experimental trends and could at first sight explain the infrared spectrum of doped PA.

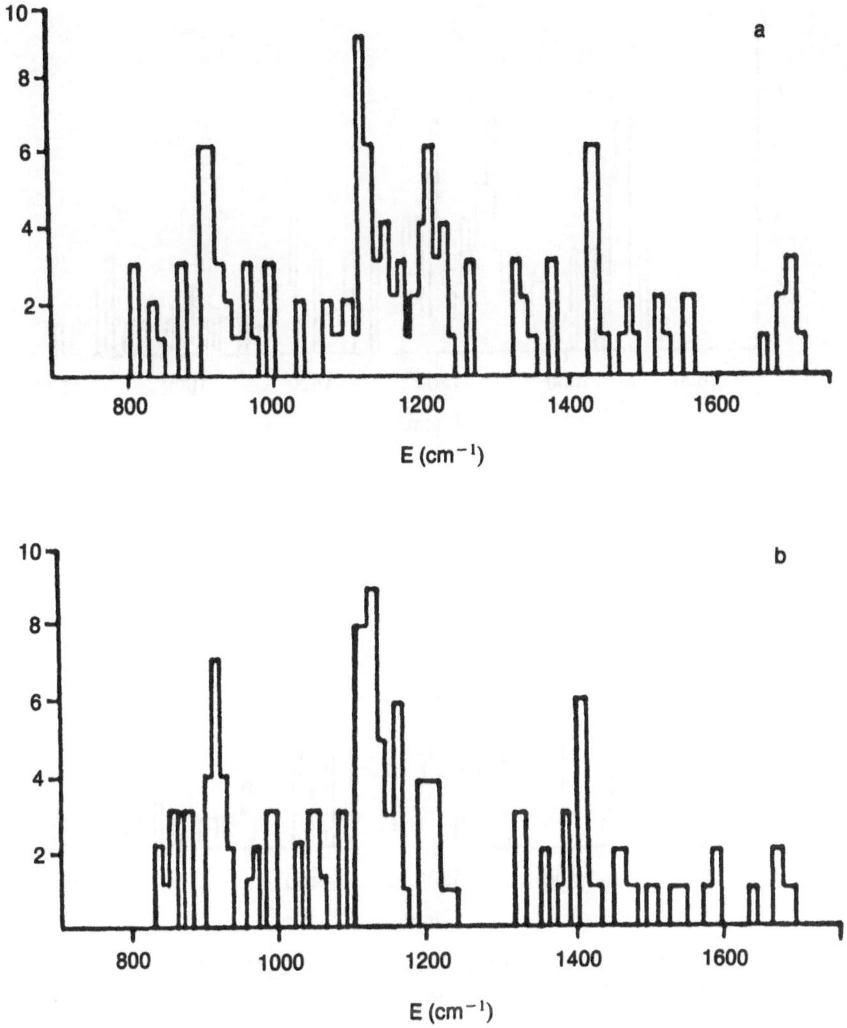

FIGURE 9.9. Phonon density-of-state histogram of a $C_{50}H_{52}$ chain containing three neutral (a) or three negatively charged (b) solitons.

For the band located approximately around 1300 cm^{-1}, calculations yield the same assignment as that given by Harada *et al.*[(48c)] Different behavior, however, has been found for the band at 800 cm^{-1}. In all cases in this range the presence of peaks has been found that have to be assigned to out-of-plane C–H motions and to in-plane C–C–H bending

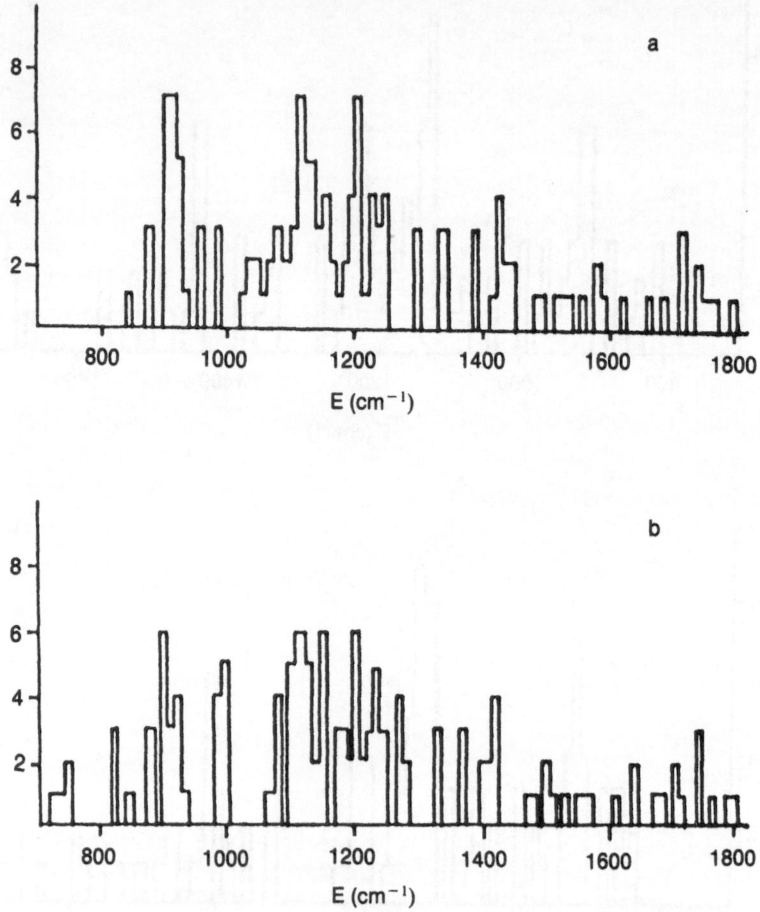

FIGURE 9.10. Phonon density-of-state histogram of a $C_{50}H_{52}$ chain containing three neutral (a) or three negatively charged (b) polarons.

motions. Harada *et al.* have discussed the possibility of assigning this band to C–H out-of-plane motions on the basis of deuteration and polarization experiments. In fact, the band at 888 cm^{-1} shifts to 790 cm^{-1} in the case of deuteration, while the C–D out-of-plane bending peak is located at 747 cm^{-1} in undoped PA. On the other hand, studies on the polarization dependence of these adsorption bands have shown that the 800–900 cm^{-1} absorption is much more strongly polarized along the chain, although a perpendicular polarization has also been observed. Therefore the possibility that the out-of-plane C–H bending deformation lies in the same region as the in-plane C–C–H bending motions is not ruled out by the experimental results. The theoretical results strongly suggest that the broad band between 800 and 900 cm^{-1} is due to the presence of these two motions. This interpretation was also put forward by Zannoni and Zerbi[(40)] on the basis of preliminary calculations for the out-of-plane vibrations.

Finally, the differences in the spectra and normal modes of the four chains investigated in this study should be discussed. In the case of the neutral polaron mode, two peaks at 1290 and 1380 cm^{-1} have been found that are in excellent agreement with experiment. In the range of 800–900 cm^{-1}, however, only a narrow band near 870 cm^{-1} (ascribed to the C–H out-of-plane bending) and one peak at 840 cm^{-1} (in-plane bending) are obtained. Therefore, it is difficult to explain the large width of this band. In the case of charged soliton defects more peaks are found, but the majority of them must be assigned to out-of-plane bending while experiments show the dominant role of the in-plane motions. The spectrum most consistent with experimental results is obtained for chains with either neutral solitons or charged polarons.

Calculation of the vibrational spectrum of doped PA, in which electron–electron interaction is included for the first time in a self-consistent way, shows that the infrared spectrum can be explained in terms of defects which invoke a local relaxation of the bond alternation. The interpretation is not uniquely in favor of the soliton picture. The polaron defect can serve as an alternative explanation. The broad band between 800 and 900 cm^{-1} results from in-plane C–C–H bending and C–H out-of-plane bending motions involving relaxed bond lengths of the defect site.

9.3. TRANSPORT PROPERTIES OF POLYMERS

Both the theoretical and experimental determination of charge transport, even in a periodic polymer, is a rather formidable task. In the case of the theory, one has to start with the dispersion curves of the different normal modes (see Sections 9.1 and 9.2) and, after calculating scattering of the electrons on these phonons, one has to conduct several rather complicated calculations to arrive at reliable values of the quantities (such as mean free path, mobility, and specific conductivity) which characterize a coherent (Bloch-type) electronic charge transport. On the experimental side, the samples for these measurements (especially in the case of biopolymers) are usually insufficiently defined (unknown types and quantities of impurities, even unknown sequences in many cases of biopolymers, a not-well-known 3D structure of many macroscopic polymer samples, unknown structural defects, etc.) to obtain reliable results, even with high-frequency ac investigations. Furthermore, there are hardly any polymers for which serious charge-transport calculations have been performed with simultaneously available reliable experimental results. In this situation one is restricted to a description of the different theoretical methods and their few applications and an examination of whether the results obtained exhibit any consistency at the theoretical level. On the other hand, references will be made to measurements on similar systems (for instance, calculations on periodic DNA chains compared with measurements on nonperiodic DNA B, calculations on periodic polypeptide chains compared with experiments on different native proteins, etc.) to see whether there is at least order-of-magnitude agreement with theory.*

In the case of nonperiodic polymers, like randomly doped polyacetylene or other organic polymers, biopolymers, etc., one cannot of course expect Bloch-type conduction because, depending on the degree of disorder, most (or all) states are localized (see Section 4.5). In such systems only hopping conduction can occur. For such cases an elementary theory for the calculation of the elementary hopping probabilities (primary jump rates) and its application to a four-component aperiodic polypeptide chain will be presented. There are no calculations of hopping conductivities for aperiodic polymers in the literature (and no publications of corresponding measurements), so the procedure for deducing from primary jump rates (using a stochastic model) to quantities describing charge transport in disordered polymers (giving, of course, relevant references) will be outlined only very briefly.

One could ask whether it is worthwhile discussing in this book the theory of charge transport in polymers if only very few serious calculations are available and there are no very reliable experimental measurements on the same systems, which hinders comparison of the results. The answer is yes, because the theory itself is sufficiently developed for

* The only exception seems to be the K^+TCNQ^- stack, for which there is an approximate theoretical calculation of its transport properties which agrees reasonably well with corresponding experiments.

different investigators to find more accurate and numerous applications in the future. At the same time, it is very probable that the necessary material science development will take place in the near future to produce well-defined polymeric samples on which reliable transport property-measurements can be performed.

9.3.1. The Theory of Bloch-Type Electric Conduction in Polymers and Its Applications

9.3.1.1. The Deformation Potential Approximation and Its Applications to Different Polymers

a. Basic Ideas of the Deformation Potential Approximation. If the widths of the conduction and valence bands of a polymer are order(s) of magnitude larger than the thermal energy at room temperature ($k_B T = 0.026$ eV at 300 K), the deformation potential approximation can be applied to obtain the transport properties of a crystal (polymer). On the other hand, if this is not the case, this simple approximation breaks down. In other words, if $\Delta E/k_B T \ll 1$, one has to take into account only electrons at the upper edge of the valence band or at the lower edge of the conduction band, in the scattering of electrons on phonons (especially if the phonon bands have also a much smaller energy than the electronic bands). However, if the criterion $\Delta E/k_B T \ll 1$ is not fulfilled, points differing from the band edges [eventually, levels in the whole Brillouin zone (corresponding to the first Brillouin zone and to Umklapp processes)] must be taken into account in the scattering processes [including transitions due to larger energy (short wavelength) phonons)].

The deformation potential method was first derived by Bardeen and Shockley.[53] According to their theory (which in its more general form is applicable also to shear strains,[54] see Section 10.3), if we expand a quasi-one-dimensional polymer (or a crystal) by a dilatation $\Delta = \delta a/a$ (in a three-dimensional crystal $\Delta = \delta V/V$), where a (V) is the equilibrium elementary translation (or the volume of the unit cell, respectively) and δa (δV) its change, we obtain bands differing slightly from those corresponding to the normal atomic distances. The shifts in the energy of the band extrema will in general be linear in the dilatation Δ. Therefore, one can write for the valence band $\delta\varepsilon_{v,u} = D_v \Delta$ and for the conduction band $\delta\varepsilon_{c,l} = D_c \Delta$, where $\delta\varepsilon_{v,u}$ and $\delta\varepsilon_{c,l}$ are the respective shifts of the upper edge of the valence band and of the lower edge of the conduction band; D_v and D_c are the deformation potential constants pertaining to the valence and conduction bands, respectively. In the case of a long-wavelength lattice vibration the dilatation varies with position and it can be assumed that there is an effective potential seen by the electrons that also varies with position: $\delta V_{c,v}(\mathbf{r}) = \Delta_{c,v}\Delta(\mathbf{r})$. In order to treat the electron–lattice interaction it is convenient to expand the displacements $\mathbf{x}_j$ of the lattice site at $\mathbf{R}_j$ in terms of normal coordinates:

$$\mathbf{X}_j = \sum_{\mathbf{q}} \mathbf{u}_{\mathbf{q}} \exp(i\mathbf{q}\cdot\mathbf{R}_j) \tag{9.19}$$

We shall always be interested in longitudinal vibrations of quasi-one-dimensional systems, so the vector notation can mostly be omitted in subsequent discussions if one refers to the component of a vector parallel to the chain axis. The phonon amplitudes u_q can be expressed in terms of phonon creation and annihilation operators as[14]

$$u_q = [\hbar/2MG\omega(q)]^{1/2}(\hat{a}_q + \hat{a}^{+}_{-q}) \tag{9.20}$$

where M is the mass of the lattice site, G is the number of sites, and $\omega(q)$ is the dispersion function of the phonon frequency (see Section 9.1). Further, since phonons are bosons (their spin is an even multiple of $\frac{1}{2}\hbar$), their creation and annihilation operators $\hat{a}_q^+$ and $\hat{a}_q$ satisfy (in contrast to the fermions: electrons, proteins, neutrons, etc., which possess spin with an odd multiple of $\frac{1}{2}\hbar$) the commutation relations

$$[\hat{a}_q, \hat{a}_{q'}] \equiv \hat{a}_q\hat{a}_{q'} - \hat{a}_{q'}\hat{a}_q = 0 \tag{9.21a}$$

$$[\hat{a}_q^+, \hat{a}_{q'}^+] = 0 \tag{9.21b}$$

$$[\hat{a}_q^+, \hat{a}_{q'}] = \delta_{q,q'} \tag{9.21c}$$

[Compare with the case of anticommutation relations valid for fermions; see equations (8.4) and (8.6) in Section 8.1.][55]

Since the dilatation is given by $\Delta = \operatorname{div}(x)$, the potential seen by an electron at position r due to the instantaneous displacements of the lattice with amplitude u_q is given by

$$V(r) = D\sum_q [\hbar/2MG\omega(q)]^{1/2} iq(\hat{a}_q + \hat{a}^{+}_{-q})\exp(iqr)$$

Assuming nearly-free electrons with plane-wave states $\psi(k) = G^{-1/2}\exp(ikr)$ and energies $E(k) = \hbar^2k^2/2m^*$, and introducing electron creation and annihilation operators $\hat{c}_k^+$ and $\hat{c}_k$, which satisfy anticommutation relations (8.6),[55] the Hamiltonian of the electron–phonon interaction becomes[55]

$$\hat{H}_{\mathrm{el,ph}} = \sum_k \sum_q V_q(\hat{a}_q + \hat{a}^{+}_{-q})\hat{c}^{+}_{k+q}\hat{c}_k \tag{9.22}$$

where matrix element V_q is given by $V_q = iD[\hbar/2MG\omega(q)]^{1/2}q$. It can be seen from equation (9.22) that the interaction couples only states in which a single electron has changed its wave number. Furthermore, if the starting phonon state is characterized by $|n_q, n_{-q}\rangle$, where n_q is the number of phonons with momentum q, one obtains[55]

$$(\hat{a}_q + \hat{a}^{+}_{-q})|n_q, n_{-q}\rangle = n_q^{1/2}|n_{q-1}, n_{-q}\rangle + (n_{-q}+1)^{1/2}|n_q, n_{-q+1}\rangle \tag{9.23}$$

The Golden Rule of time-dependent perturbation theory[56] now yields, for the probability of scattering per unit time from state $\psi(k)$ to $\psi(k+q)$,

$$P(k, k+q) = \frac{2\pi}{\hbar}|V_q|^2\{n_q\delta[E(k+q) - E(k) - \hbar\omega(q)] + (n_q+1)\delta[E(k+q) - E(k) + \hbar\omega - q)]\} \tag{9.24}$$

This expression can then be used to determine the different transport properties of a polymer using relaxation-time approximation for the solution of Boltzmann's transport equation.

b. The Solution of Boltzmann's Transport Equation in the Relaxation Time Approximation Using a Deformation Potential to Compute Scattering Processes.[14,59–61] When the macroscopic response of the electrons is linearly related to the electric field (the equation $\mathbf{j} = \boldsymbol{\sigma}\mathbf{E}$ holds, where $\mathbf{j}$ is the electric current density, $\boldsymbol{\sigma}$ the conductivity tensor, and $\mathbf{E}$ the field strength), the Boltzmann equation gives a fairly good description of charge transport with the exception of very high magnetic field interband transitions due to very high energy phonons and other unusual situations.[57] In these latter cases the more general (but also more complicated) Kubo approach must be applied.[58] However, since such phenomena were not treated theoretically in the case of polymers, only the Boltzmann equation and its approximate solution will be discussed here.

In the transport theory of solids one is concerned with average effects produced by many electrons. Each of them is supposed to follow a trajectory in six-dimensional phase space $(\mathbf{k}, \mathbf{r})$ and is determined by the dynamical equations[14]

$$\mathbf{v}(\mathbf{k}) = \hbar^{-1}\nabla_{\mathbf{k}}E(\mathbf{k}), \quad \hbar\dot{\mathbf{k}} = -e(\mathbf{E} + c^{-1}[\mathbf{v}(\mathbf{k}) \times \mathbf{B}]) \tag{9.25}$$

where $E(\mathbf{k})$ is the band energy, and $\mathbf{E}$ and $\mathbf{B}$ are the electric and magnetic field vectors, respectively. The gas of conduction electrons (or holes) can be characterized by a distribution function $f(\mathbf{k}, \mathbf{r}, t)$, which is equal to the probability of finding an electron with wave vector $\mathbf{k}$ and given spin orientation in the unit volume element around $\mathbf{r}$ at time t. The $\mathbf{r}$ dependence can be omitted when the solid is homogeneous (in the absence of internal inhomogeneities this means isothermal conduction). For the problems considered here, $f(\mathbf{k}, \mathbf{r}, t)$ will be the same for both spins. Hence, since the density of states in $\mathbf{k}$ space per unit volume of $\mathbf{r}$ space is $1/(2\pi)^3$, the density of electrons in $(\mathbf{k}, \mathbf{r})$ space is $(4\pi^3)^{-1}f(\mathbf{k}, \mathbf{r}, t)$.

The electric current density $\mathbf{j}$ (the flux of charge measured relative to the Fermi energy ε_F) can be defined in terms of $f(\mathbf{k}, \mathbf{r}, t)$:

$$\mathbf{j} = \frac{e}{4\pi^3}\int f(\mathbf{k}, \mathbf{r}, t)\mathbf{v}(\mathbf{k})d\mathbf{k} \tag{9.26}$$

In thermal equilibrium (no electric field, no thermal gradient) the distribution function is the well-known Fermi–Dirac function

$$f^0(\mathbf{k}) = \{\exp\{(1/k_BT)[E(\mathbf{k}) - \varepsilon_F]\} + 1\}^{-1}$$

which can be very well approximated, however, in most polymers and molecular crystals by the classical Maxwell–Boltzmann distribution (nondegenerate semiconductors). When the thermal equilibrium is not strongly perturbed one can assume that the disturbed distribution is still sufficiently close to the equilibrium one, so that one can write

$$f(\mathbf{k}, \mathbf{r}) = f^0(\mathbf{k}, \mathbf{r}) + f^1(\mathbf{k}, \mathbf{r}), \qquad f^1 \ll f^0 \tag{9.27}$$

In transport calculations, as first approximation one can linearize in the external disturbances represented by $f^1(\mathbf{k}, \mathbf{r})$, i.e., one can perform a linear approximation in the field strengths $\mathbf{E}$ and $\mathbf{B}$ (small fields). Since the equilibrium distribution does not carry any net flow, one must calculate the true distribution at least to first order in the driving forces. From statistical mechanics one knows that phase points move in the $(\mathbf{k}, \mathbf{r})$ space like an incompressible fluid, thus the total time derivative of

$f(\mathbf{k}, \mathbf{r}, t)$, namely

$$\left(\frac{df}{dt}\right)_{\text{drift}} = \frac{\partial f}{\partial t} + \mathbf{v}(\mathbf{k})\nabla_{\mathbf{r}} f + \dot{\mathbf{k}}\nabla_{\mathbf{k}} f \tag{9.28}$$

ought to be zero.[61] Since, however, there are always imperfections in a real solid which scatter the electrons, the above drift term will not be zero but, instead, equal to the negative rate at which f increases as a result of these collision processes. On writing down the equation of this balance one obtains the general Boltzmann equation

$$\frac{\partial f}{\partial t} + \mathbf{v}\nabla_{\mathbf{r}} f + \dot{\mathbf{k}}\nabla_{\mathbf{k}} f + \left(\frac{\partial f}{\partial t}\right)_{\text{coll}} = 0 \tag{9.29}$$

To derive the appropriate form of $(\partial f/\partial t)_{\text{coll}}$ one has to assume that the effect of the various collision events is to induce transition from an occupied state $\mathbf{k}$ to an empty state $\mathbf{k}'$. If $P(\mathbf{k}, \mathbf{k}')$ is the probability per unit time that an electron known to be occupying state $\mathbf{k}$ will be scattered into unoccupied state $\mathbf{k}'$, then Fermi statistics yields

$$\left(\frac{\partial f}{\partial t}\right)_{\text{coll}} = \left(\frac{1}{4\pi^3}\right) \int d\mathbf{k}\{ f(\mathbf{k}')[1 - f(\mathbf{k})P(\mathbf{k}', \mathbf{k}) - f(\mathbf{k})[1 - f(\mathbf{k}')]P(\mathbf{k}, \mathbf{k}')\} \tag{9.30}$$

In this equation variables $\mathbf{r}$ and t have been suppressed and integration is over the first Brillouin zone. Evaluation of transition probabilities $P(\mathbf{k}, \mathbf{k}')$ generally forms the most tedious part of any transport calculation, though if the relatively simple deformation potential aproximation can be applied its calculation (as we have seen previously) is not so difficult. In thermal equilibrium the collisions alone do not change the total density of representative points in the phase space and equation (9.29) vanishes. Thus we arrive at the principle of detailed balance:

$$f^0(\mathbf{k}')[1 - f^0(\mathbf{k})]P(\mathbf{k}, \mathbf{k}') = f^0(\mathbf{k})[1 - f^0(\mathbf{k}')]P(\mathbf{k}, \mathbf{k}') \tag{9.31}$$

In studying now the effect of applied fields and/or spatial gradients of the thermodynamic parameters, one can make the basic assumption that $P(\mathbf{k}, \mathbf{k}')$ is unaffected by these perturbations and therefore continues to satisfy condition (9.31).

First one should note that, provided equation (9.31) remains valid, application of a magnetic field has no effect on the uniform steady-state distribution since $f^0(\mathbf{k})$ is independent of both $\mathbf{r}$ and $\mathbf{t}$, and equations (9.29) and (9.25) give

$$-(e/\hbar c)[\mathbf{v}(\mathbf{k}) \times \mathbf{B}]df^0(\mathbf{k})/dE(\mathbf{k})\nabla_{\mathbf{k}}E(\mathbf{k}) = -(e/c)[\mathbf{v}(\mathbf{k}) \times \mathbf{B}]\mathbf{v}(\mathbf{k})df^0(\mathbf{k})/dE(\mathbf{k}) = 0$$

This shows that the Lorentz force is not a thermodynamical force; the magnetic field itself does not give rise to any net flow. As stated above, one assumes the field strengths and their spatial gradients to be first-order quantities while $f^0(\mathbf{k}, \mathbf{r})$ and $f^1(\mathbf{k}, \mathbf{r})$ in equation (9.27) will be assumed to be of zero and first order, respectively. Then in zero order the Boltzmann equation reduces to

$$-(e/\hbar c)[\mathbf{v}(\mathbf{k}) \times \mathbf{B}]\nabla_{\mathbf{k}} f^0(\mathbf{k}) = -(\partial f^0/\partial t)_{\text{coll}} \tag{9.32}$$

which can be seen to be satisfied by the Fermi–Dirac distribution even if T and ε_F are functions of $\mathbf{r}$, provided $P(\mathbf{k}, \mathbf{k}')$ satisfies equation (9.31).

Collecting now first-order terms on both sides of equation (9.29) one obtains the linearized Boltzmann equation on which all subsequent discussions will be based,[61]

$$\begin{aligned}\mathbf{v}(\mathbf{k})\nabla_{\mathbf{r}} f^0(\mathbf{k}) - (e/\hbar)\mathbf{E}\nabla_{\mathbf{k}} f^0(\mathbf{k}) - (e/\hbar c)[\mathbf{v}(\mathbf{k}) \times \mathbf{B}]\nabla_{\mathbf{k}} f^1(\mathbf{k})\\ = -(1/4\pi)^3 \int d\mathbf{k}\{ f^1(\mathbf{k}')[(1 - f^0(\mathbf{k}))P(\mathbf{k}', \mathbf{k}) + f^0(\mathbf{k})P(\mathbf{k}, \mathbf{k}')]\\ - f^1(\mathbf{k})[f^0(\mathbf{k}')P(\mathbf{k}', \mathbf{k}) + [1 - f^0(\mathbf{k}')]P(\mathbf{k}, \mathbf{k}')]\}\end{aligned} \tag{9.33}$$

It is customary to define the nonequilibrium part of the distribution function in the form

$$f^1(\mathbf{k}, \mathbf{r}) = -\phi(\mathbf{k}, \mathbf{r})df^0/dE(\mathbf{k}) = \phi(\mathbf{k}, \mathbf{r})(1/k_B T) f^0(\mathbf{k})[1 - f^0(\mathbf{k})] \tag{9.34}$$

The collision term on the right-hand side of equation (9.33) assumes the following simple form by substituting equation (9.34):

$$-(\partial f^1/\partial t)_{\text{coll}} = (1/4\pi^3 k_B T) \int d\mathbf{k} V(\mathbf{k}, \mathbf{k}')[\phi(\mathbf{k}) - \phi(\mathbf{k}')] \tag{9.35}$$

where

$$V(\mathbf{k}, \mathbf{k}') = f^0(\mathbf{k})[1 - f^0(\mathbf{k}')]P(\mathbf{k}, \mathbf{k}') \tag{9.36}$$

is the equilibrium transition rate between states **k** and **k**′, the actual number of transitions per unit time, when the system is in overall thermodynamic equilibrium. For later purposes one should note that equation (9.35) can be written in the very simple form

$$-(\partial f^1/\partial t)_{\text{coll}} = \hat{P}\phi(\mathbf{k}, \mathbf{r}) \tag{9.37}$$

where $\hat{P}$ is the integral operator of the scattering and possesses symmetric and positive kernel $V(\mathbf{k}, \mathbf{k}')$. Similarly, substitution of equation (9.34) in the left-band side of equation (9.32) results in a very transparent form of the linearized Boltzmann equation,

$$X(\mathbf{k}, \mathbf{r}) + \hat{M}\phi(\mathbf{k}, \mathbf{r}) = \hat{P}\phi(\mathbf{k}, \mathbf{r}) \tag{9.38}$$

where

$$X(\mathbf{k}, \mathbf{r}) = \{-e\mathbf{E} - \nabla\varepsilon_F + [E(\mathbf{k}) - \varepsilon_F T]\nabla(1/T)\}\mathbf{v}(\mathbf{k})df^0/dE(\mathbf{k}) \tag{9.39}$$

and, noting that $\nabla_{\mathbf{k}}(df^0/dE)$ is parallel to $\mathbf{v}(\mathbf{k})$, the magnetic operator $\hat{M}$ satisfies

$$\hat{M}\phi(\mathbf{k}, \mathbf{r}) = [df^0/dE(\mathbf{k})](e/\hbar c)[\mathbf{v}(\mathbf{k} \times \mathbf{B})]\nabla_{\mathbf{k}}\phi(\mathbf{k}, \mathbf{r}) \tag{9.40}$$

Since $\hat{M}$ is not associated with any flow, one may formally regard this drift term as a sort of collision term, as if the electrons were constantly "colliding" with the magnetic field.

It was shown earlier that the collision term of the Boltzmann equation involves a complicated integral over the unknown distribution function. In many cases, however, one can make certain assumptions which greatly simplify the problem. The most widely used simplification is the relaxation time approximation, which we now examine.

The collisions suffered by the electrons involve a change in their momenta and energies. From the standpoint of electrical conductivity the main point of interest is the change of **k**, as this determines the loss of forward drift. There are actually many cases in which the energy change during scattering is completely negligible besides the average energy of the conduction electrons. We shall see later that this elasticity of the scattering is a necessary condition for the existence of a relaxation time. Obvious examples of elastic scattering are collisions of electrons with structural defects. The electron–phonon interaction in most metals is also elastic at room tmeperatures as well as scattering of conduction electrons in semiconductors with long-wavelength phonons at all temperatures.

Now we consider a time-dependent but spatially homogeneous situation in the absence of applied fields. One can again write the distribution function in the form $f = f^0 + f^1$, where f^0 is the thermal equilibrium distribution and f^1 is a small perturbation. From equation (9.29) we obtain

$$\frac{\partial f^1}{\partial t} = -\left(\frac{\partial f^1}{\partial t}\right)_{\text{coll}} \tag{9.41}$$

One would expect that the system with any given perturbation $f^1 = f - f^0$ at $t = 0$ should decay to the distribution f^0 with a time constant characteristic of the collision processes. In the relaxation time approximation one therefore assumes that

$$\left(\frac{\partial f^1}{\partial t}\right)_{\text{coll}} = f^1(\mathbf{k})/\tau \tag{9.42}$$

which yields the solution of equation (9.41) in the form $f^1(\mathbf{k}, t) = f^1(\mathbf{k}, 0)e^{-t/\tau}$. The actual value of τ, if it exists at all, can be determined of course only by an investigation of the scattering process itself. For the moment we assume only that it does exist and depends on the energy of the electron, $\tau = \tau(E)$, alone. Equations (9.34) and (9.37) enable one to obtain from equation (9.42) that the scattering operator is also a scalar function of the energy:

$$\hat{P} = (df^0/dE)\tau(E)^{-1}$$

To proceed further with the solution of the linearized Boltzmann equation one assumes the magnetic field to be sufficiently weak so that the electrons undergo collisions long before they can complete a cyclotron orbit. Then one can regard the solution for the case $\mathbf{B} = \mathbf{0}$ as slightly perturbed by the addition of the magnetic field and hence write down a power-series expansion of $(\hat{P} - \hat{M})^{-1}$ in the form

$$(\hat{P} - \hat{M})^{-1} = \hat{P}^{-1} \sum_n (\hat{P}^{-1}\hat{M})^n \tag{9.43}$$

Introducing the operator $\hat{\Omega} = (df^0/dE)^{-1}\hat{M} = (e/\hbar c)[\mathbf{v}(\mathbf{k}) \times \mathbf{B}]$ and taking into account that the energy-dependent $\hat{P}$ commutes with $\hat{\Omega}$, one obtains the Jones–Zener expansion[62]

$$(\hat{P} - \hat{M})^{-1} = \frac{\tau}{df^0/dE}[1 + \tau\hat{\Omega} + (\tau\hat{\Omega})^2 + \cdots] \tag{9.44}$$

It only remains to substitute this expansion into the expression for σ_{ij} at constant temperature,[61]

$$\sigma_{ij} = -(e^2/4\pi^3)\int d\mathbf{k}[(df^0/dE)v_i(\mathbf{k})(\hat{P}-\hat{M})(df^0/dE)v_j(\mathbf{k})] \quad (9.45)$$

Equations (9.35) and (9.42) yield

$$\frac{1}{\tau} = \frac{1}{4\pi^3 k_B T(df^0/dE)}\int d\mathbf{k}' V(\mathbf{k},\mathbf{k}')\left[1-\frac{\phi(\mathbf{k}')}{\phi(\mathbf{k})}\right] \quad (9.46)$$

If it is assumed that τ depends only on $E(\mathbf{k})$, i.e., the constant-energy surfaces are spherical (otherwise there would be no reason to expect τ to be independent of the orientation of $\mathbf{k}$), and that the scattering processes are elastic, then the relaxation time can be obtained from equation (9.46) in the form[61]

$$\tau^{-1} = [4\pi^3 k_B T(df^0/dE)]\int d\mathbf{k}' V(\mathbf{k},\mathbf{k}')(1-\cos\theta) \quad (9.47)$$

where θ is the angle between vectors $\mathbf{k}$ and $\mathbf{k}'$.

A similar result can be obtained without the cosine factor for the so-called velocity-randomizing collisions for which $V(\mathbf{k},\mathbf{k}') = V(\mathbf{k},-\mathbf{k}')$. Except for the two cases discussed above, namely elastic and velocity-randomizing collisions in a spherically symmetric band, the introduction of single energy-dependent relaxation time is always an uncertain approximation. It is true, however, that in many cases the formal use of equation (9.46) and subsequent averaging of τ^{-1} over an energy surface yields meaningful results when substituted into the transport expressions.[60] At this point two brief remarks are in order. First, even if a correct relaxation time cannot be calculated, its approximate form may provide a very useful guide to the variational solution of the Boltzmann equation.[14] Second, it should be stressed that the possible range of validity of the Jones–Zener method[62] is much wider than the relaxation-time model. In fact, equation (9.43) suggests that it can be used for any scattering mechanism. If the operator $\hat{P}$ can be inverted by means of a variational calculation, then the power-series expansion of $(\hat{P}+\hat{M})^{-1}$ can be employed to simply account for a small perturbation due to a magnetic field.

If one substitutes equation (9.24) obtained in the deformation potential approximation into expression (9.47) for relaxation time, one obtains the latter quantity[63] in the form

$$\tau(E) = \frac{\pi^2 M v_s^2 \hbar^4 E^{-1/2}}{4\sqrt{2}(m^*)^{3/2} D^2 k_B T} \quad (9.48)$$

Here, the simple dispersion $\omega(q) = v_s q$ was used for the acoustic phonons (if whole molecules vibrate with respect to each other; like the change of the stacking distance in a stack one speaks of acoustic phonons, because they have a long wavelength comparable to those of acoustic waves) and the sound velocity v_s is determined by the relation $v_s = (c_l/\rho)^{1/2}$, where c_l is the longitudinal elastic constant and ρ is the mass density. One should remark that this very simple linear dispersion relation $\omega(q) = v_s q$ is not necessarily correct. With the help of the FG method described in Section 9.1 one can obtain more accurate dispersion curves. Equation (9.48) can now be used to calculate the charge carrier mobilities and free paths, defined in this case by $\mu = |e|\langle\tau/m^*\rangle$ and $\lambda = \langle\tau v\rangle$, respectively, where

$$\left\langle\frac{\tau}{m^*}\right\rangle = \frac{\int E^{3/2}\tau(E)m^*(E)^{-1}\exp(-E/k_BT)dE}{\int E^{3/2}\exp(-E/k_BT)dE} \tag{9.49}$$

and m^* is in the harmonic $[E(k) = k^2/2m^*]$ approximation,

$$m^* = \hbar^2\left[\frac{\partial^2 E(k)}{\partial k^2}\right]^{-1} \tag{9.50}$$

One obtains in this way,[63] for the quasi-one-dimensional case,

$$\mu = \frac{2|e|c_l\hbar^4}{3\pi D^2(m^*)^{3/2}(k_BT)^{1/2}}, \quad \lambda = \frac{\pi\hbar^4 c_l}{(m^*)^2 D^2 k_BT} \tag{9.51}$$

Similarly, the preexponential factor σ_0 of the electrical conductivity defined by $\sigma = \sigma_0 \exp(-\Delta E(2k_BT)$, where ΔE is the band gap, and assuming that both electrons and holes are present in a semiconductor, is

$$\sigma_0 = \frac{2e^2\hbar c_l}{\pi}\left(\frac{1}{m_e^* D_c^2} + \frac{1}{m_h^* D_v^2}\right) \tag{9.52}$$

if one applies the simple relation $\sigma(T) = |e|[n_e(T)\mu_e(T) + n_h(T)\mu_h(T)]$. Here $\mu_e(T)$ and $\mu_h(T)$ denote the electron and hole mobility, respectively, while $m_e(T)$ and $m_h(T)$ are the number of free electrons and holes, respectively, at temperature T. The latter can be expressed in the one-dimensional case as follows[36]:

$$n_e = e^{-(E_{c,l} - \varepsilon_F)/k_BT}\,\frac{1}{2^{3/2}\pi}\,\frac{m_e^{*1/2}}{\hbar}(k_BT)^{1/2} \tag{9.53a}$$

and

$$n_h = e^{-(\epsilon_F - E_{v,u})/k_B T} \frac{1}{2^{3/2}\pi} \frac{m_h^{*1/2}}{\hbar} (k_B T)^{1/2} \tag{9.53b}$$

c. Application to Different Polymers. The results of some calculations performed on different polymeric systems are shown in Table 9.10. In one case an adenine stack was investigated in the same conformation as in DNA B.[63] The values of D (the shifts of the band edges) were calculated by a perturbative method[63] using the simple tight-binding approximation. According to the results the mobility and free-path values of both electrons and holes in this polymer fall into the region where the Bloch-type delocalized description of the transport process is reasonable.[64] A further application of the same method[65] to other periodic DNA models with much narrower bands (0.003–0.060 eV) has shown, however, that for such systems the deformation-potential approximation breaks down, in accordance with our earlier considerations.

Other systems in Table 9.10 are polyglycine with a geometrical structure of the polypeptide backbones (polyglycine main chain) and in a hydrogen-bonded configuration (details are given elsewhere[66]). The energy band structures of both systems have been calculated by the *ab initio* SCF LCAO crystal-orbital method.[67] To obtain the deformation-potential constants the calculations were repeated by distorting the lattice in both directions and the derivatives of the appropriate band edges were computed numerically. The results show that the electrical conduction in proteins can be expected to be very anisotropic, since the main-chain direction exhibits transport properties similar to good semiconductors while along the hydrogen bonds the mean charge-carrier free paths are comparable with the lattice constants; thus the existence of a delocalized (band-type) transport becomes very questionable. We note that the measured values of σ for different polypeptides scatter in the range 10^4–10^6 Ω^{-1} cm^{-1},[68] thus supporting a conduction mechanism with charge-carrier delocalization along the main chains.

The last system investigated by the deformation-potential method is a member of the large family of charge-transfer molecular crystals containing as anion the tetracyanoquinodimethane (TCNQ) molecule. The K^+TCNQ^- salt as well as the other TCNQ crystals consist of columns of stacked TCNQ molecules coordinated by similar columns of the cations.[69] The energy band structures of the stack have been calculated earlier with the help of Pariser–Parr–Pople parametrization of the unrestricted Hartree–Fock crystal-orbital method.[70] The last two columns of Table 9.10 show that this material behaves again as a normal

TABLE 9.10. Transport Properties of Some Wide-Band Polymers Calculated by the Deformation Potential Method at $T = 300$ K

	Adenine stack[a]		Thymine stack[a]		Polyglycine (main chain)[b]		Polyglycine (H bonds)[b]		K^+TCNQ^-[c]	
Polymeric system	electrons	holes	electrons	holes	electrons	holes	electrons	holes	electrons	holes
Bandwidth (eV)	0.246	0.320	0073	0.276	1.376	2.098	0.141	0.291	1.163	0.979
Effective mass (in m_e)	3.4	2.6	115	3.0	1.049	0.688	4.834	2.342	0.665	0.778
Deformation potentials (in eV)	0.448	0.352			2.487	1.924	4.764	4.102	1.846	1.376
Drift mobility (in $cm^2 V^{-1} s^{-1}$)	156	485			878	4210	1.03	8.50	1034	1258
Mean free path (in Å)	265	725			987	2650	1.64	9.43	608	801
Preexponential factor of the conductivity (in $\Omega^{-1} cm^{-1}$)	9.73×10^3	2.04×10^3	1.16×10^3	3.06×10^3	3.97×10^3	1.01×10^4	44.1	123	2.25×10^3	3.46×10^3

[a] After Suhai[61] and Beleznay *et al.*[63]
[b] After Suhai and Ladik.[67]
[c] After Suhai.[71]

semiconductor with relatively large mobilities and free paths. Experiments confirm this prediction; K^+TCNQ^- has been shown to be a semiconductor in the whole temperature region investigated. Its electrical conductivity follows the law $\sigma = \sigma_0 \exp(-\Delta E/k_B T)$, and close agreement between the room-temperature microwave conductivity (3×10^{-4} Ω^{-1} cm^{-1}[72]) and the corresponding dc value (1×10^{-4} Ω^{-1} cm^{-1}[73]) makes it probable that the measured single particle gap may be regarded as the intrinsic value and not due to other, disordered-limited transport processes. For a calculated gap of $\Delta E = 0.829$ eV[70] the values in Table 9.10 yield for the room-temperature conductivity $\sigma_{T=300\,K} = 7.29 \times 10^{-4}$ Ω^{-1} cm^{-1}, in rather good agreement with the experimental ac value.

9.3.2. Calculation of Bloch Conduction for Narrow-Band Polymers

9.3.2.1. Calculation of Electron–Phonon Interaction and Solution of the Boltzmann Equation in the General Case

a. General Treatment of Electrons Scattered on Phonons in Polymers. In the case of narrow-band semiconductors the simple deformation-potential approximation described in Section 9.3.1 cannot be applied, because if the electronic bandwidths are of the order of the thermal energy, scattering processes throughout the whole valence and/or conduction band must be taken into account in order to determine the transition probabilities (9.24). Moreover, in narrow-band situations even the effective mass m^* becomes a sensitive function of k, i.e., $m^*(k)$. For these reasons one first requires a more general method for computing the electron–phonon interaction matrix elements.

b. Interaction of Electrons with Long-Wavelength (Acoustic) Phonon Modes.* In the case of narrow electronic bands the wave functions are localized. Therefore, in the usual procedures for the calculation of electron–phonon interaction matrix elements only the change in the local potentials due to the lattice motion is taken into account, while the wave functions themselves are constrained at the original lattice sites. Various authors[74–76] have pointed out that this leads to physically unreasonable, spurious scattering terms in the tight-binding limit. To overcome this difficulty Whitfield[75] proposed the use of a "deformed Bloch" representation in which the local displacements are built into the starting wave function and the perturbing Hamiltonian is expressed directly in terms of

* In reality, one should calculate the corresponding electron–phonon interaction matrix elements to obtain more accurate results for the primary jump rates.

the relative displacements. While this method is physically more correct, the resulting matrix elements are rather complicated. Another solution of this problem has been suggested by Friedman,[76] who expressed the interaction constants directly with the gradients of the characteristic molecular overlap integrals (within the framework of the simple tight-binding LCMO theory). Here, this latter proposal will be followed and for the derivation of the matrix elements of the electron–acoustic phonon scattering the *ab initio* SCF LCAO crystal-orbital method[77] and time-dependent perturbation theory will be used.[61]

The total wave function of the complete (electron–phonon) system can be written using the above-mentioned method as a linear combination of the instantaneous local atomic wave functions:

$$\psi_n(\mathbf{r}, \ldots, \mathbf{R}_s, \ldots, t) = G^{-1/2} \sum_{s=1}^{G} \sum_{l=1}^{g} d_{sm}(\ldots, \mathbf{R}_s, \ldots, t) c_{lm} \phi_l^s(\mathbf{r} - \mathbf{R}_s - \mathbf{R}_{sl}) \tag{9.54}$$

where m is the electronic band index (no interband scattering will be considered), $\mathbf{R}_s$ is the instantaneous lattice position ($\mathbf{R}_s = \mathbf{R}_s^0 + \mathbf{x}_s$), $\mathbf{R}_{sl}$ is the position of the atomic orbital ϕ_l^s at site s, G and g are respectively the number of lattice sites and atomic orbitals per site, while finally d_{sm} and c_{lm} are expansion coefficients whose physical meaning will be defined below. In the case of one-dimensional systems the component of the vectors along the chain direction will be used whenever possible. The Hamiltonian of the system can be written as

$$\hat{H} = \hat{H}_e + \hat{H}_L \tag{9.55}$$

where the electronic part is an effective one-electron Hamiltonian, $\hat{H}_e = \hat{H}_e^{\text{eff}}$, which will be specified later for different approximations; the lattice is treated in the harmonic approximation by

$$\hat{H}_L = -\frac{\hbar^2}{2M} \sum_{h=1}^{G} \nabla_{x_h}^2 + V_L(\ldots, \mathbf{R}_s, \ldots) \tag{9.56}$$

where M is the mass of the lattice site again. The total wave function now obeys the time-dependent Schrödinger equation

$$i\hbar \partial\psi_m/\partial t = \hat{H}\psi_m \tag{9.57}$$

On substituting the expression for ψ_m from equation (9.54) into equation

(9.57), multiplying on the left by ϕ_i^{t*}, and integrating over **r** one obtains

$$i\hbar \sum_s \sum_l (\partial d_{sm}/\partial t) c_{lm} \langle \phi_i^t \mid \phi_l^s \rangle$$
$$= \sum_s \sum_l d_{sm} c_{lm} \langle \phi_i^t \mid \hat{H}_e^{\text{eff}} \mid \phi_l^s \rangle + \sum_s \sum_l (V_L d_{sm}) c_{lm} \langle \phi_i^t \mid \phi_l^s \rangle$$
$$- \frac{\hbar^2}{2M} \sum_s \sum_l \sum_h (\nabla_{x_h} d_{sm}) c_{lm} \langle \phi_i^t \mid \nabla_{x_h} \mid \phi_l^s \rangle$$
$$- \frac{\hbar^2}{2M} \sum_s \sum_l \sum_h d_{sm} c_{lm} \langle \phi_i^t \mid \nabla_{x_h}^2 \mid \phi_l^s \rangle \qquad (9.58)$$

This equation can be reduced to a form which is easier to handle with the help of the following approximations:

1. It can be easily shown that the last term on the right-hand side is proportional to $(m_e/M)E_{\text{kin}}^{\text{el}}$, which can be neglected as small.
2. The electronic matrix elements can be expanded about their equilibrium values with respect to first-order relative displacements:

$$H_{il}^{ts} = \langle \phi_i^t \mid \hat{H}_e^{\text{eff}} \mid \phi_l^s \rangle = (H_{il}^{ts})_0 + (x_t - x_s)(\nabla H_{il}^{ts})_0 \qquad (9.59)$$

   i.e., only one-phonon processes will be allowed.
3. One can introduce the zero differential overlap (see Chapter 2) approximation (that is, one can set $\langle \phi_i^t \mid \phi_l^s \rangle = S_{il}^{ts} = \delta_{il}\delta_{ts}$), since the subsequent actual applications were based on π-electron methods.[78]

It should be emphasized, however, that this procedure could equally well be applied also within the framework of an *ab initio* crystal-orbital scheme. With these approximations and setting for a moment all lattice displacements and velocities equal to zero, we get the Schrödinger equation of the system with complete translational symmetry. It can be easily verified by some algebra that the solutions of this zero-order equation can be written in the usual Bloch form

$$d_{tm}^{k\{n\}}(t) c_{lm}(k) = e^{ikR_t^0} \chi^{\{n\}} e^{-(it/\hbar)[E_m(k) + E^{\{n\}}]} c_{lm}(k) \qquad (9.60)$$

where $\{n\}$ represents the totality of vibrational quantum numbers, and the vibrational wave function $\chi^{\{n\}}$ is a product of harmonic-oscillator eigenfunctions which satisfy the equation

$$\hat{H}_L \chi^{\{n\}} = E^{\{n\}} \chi^{\{n\}} \tag{9.61}$$

Substitution of equation (9.60) into the zeroth-order form [$x_s = 0$, $(\hbar/iM)\nabla_s = 0$] of equation (9.58) yields the matrix eigenvalue equation

$$\sum_s \sum_l e^{ikR_s^0} (H_{il}^{0s})_0 c_{lm}(k) = E_m(k) c_{im}(k) \qquad (i = 1, \ldots, g) \tag{9.62}$$

which is identical with the eigenvalue problem solved by determining the electronic band structure of the solid.[77] The LCAO coefficients $c_{lm}(k)$ completely determine the lattice Bloch functions having the form [see equation (1.66) in Section 1.2]

$$\psi_m(k) = G^{-1/2} \sum_s \sum_l e^{ikR_s^0} c_{lm}(k) \phi_l^s \tag{9.63}$$

From the preceding discussion it in evident that the effective one-electron Hamiltonian in equation (9.55) must be the same as that applied to the band structure calculation. A simple Hückel-type Hamiltonian $\hat{H}_e^{\text{eff}}$ was used in an earlier application.[78]

The next step is to take into account the two terms on the right-hand side of equation (9.58) that were neglected by looking for a zeroth-order solution. These terms arise from the part proportional to $(x_t - x_s)$ of the first term on the right-hand side of equation (9.58) and from the original third term on the right-hand side. These two terms are the sources of the electron–phonon scattering. Their physical meaning is also rather apparent: the first results from the variation in the one- and two-electron integrals which contain atomic functions at different sites, while the second describes the tendency of the moving lattice to drag the electron with it.[75] According to the usual technique of time-dependent perturbation theory, one can expand the solutions of the complete equation of motion (9.57) in terms of the zeroth-order solutions (9.60):

$$d_{tm} c_{im} = \sum_{k'\{n'\}} a^{k'\{n'\}}(t) d_{tm}^{k'\{n'\}} c_{im}(k') \exp\{-(it/\hbar)[E_m(k') + E^{\{n'\}}]\} \tag{9.64}$$

Substituting this expression into equation (9.57), multiplying on the left by $d_{tm}^{k\{n\}*} c_{im}^*(k)$, summing over indices t and i, and integrating over the vibrational coordinates, one obtains

$$i\hbar \frac{\partial a^{k\{n\}}}{\partial t} = \sum_{k',\{n'\}} a^{k',\{n'\}} \langle k, \{n\} \,|\, V \,|\, k', \{n'\} \rangle e^{-it/\hbar[E_n(k') + E_n(k)]} \tag{9.65}$$

where the matrix element is given by

$$\langle k, \{n\} | V | k', \{n'\} \rangle = G^{-1} \sum_s \sum_l \sum_t \sum_i e^{i(k'R_s^0 - kR_i^0)} c_{im}^*(k) c_{lm}(k') \times \left[(\nabla H_{il}^{ts})_0 \langle \chi^{\{n\}} | x_s - x_t | \chi^{\{n'\}} \rangle - \frac{\hbar^2}{M} \langle \chi^{\{n\}} | \nabla_{x_s} | \chi^{\{n'\}} \rangle \times \langle \phi_i^t | \nabla_{x_s} | \phi_l^s \rangle \right] \tag{9.66}$$

Using now expression (9.20) for the lattice displacements and its analogue

$$\nabla_{x_s} = \sum_q \left(\frac{M\omega(q)}{2\hbar G} \right)^{1/2} (\hat{a}_q - \hat{a}_{-q}^+) e^{iqR_s} \tag{9.67}$$

for the lattice velocities, and introducing the interaction constants

$$V_m^u(k, k') = \sum_i \sum_l c_{im}^*(k) c_{lm}(k') (\nabla H_{il}^{0u})_0 \tag{9.68a}$$

and

$$W_m^u(k, k') = \sum_i \sum_l c_{im}^*(k) c_{lm}(k') \langle \phi_i^0 | \nabla_{x_u} | \phi_l^u \rangle \tag{9.68b}$$

we can write the final expressions for the scattering matrix elements in the compact form

$$\langle k, \{n\} | V | k', \{n'\} \rangle = \sum_{\pm} \sum_u \sum_q \left[\left(\frac{\hbar}{2MG\omega(q)} \right)^{1/2} (1 - e^{\mp iqR_u^0}) V_m^u(k, k') \mp \left(\frac{\hbar^3 \omega(q)}{2MG} \right)^{1/2} W_m^u(k, k') e^{ikR_u^0} \right] \times \delta_{k, k'+K \pm q} \delta_{n_q, n_q' \mp 1} (n_q' + \tfrac{1}{2} \pm \tfrac{1}{2}) \tag{9.69}$$

The plus sign in this expression refers to scattering processes in which the electron emits one phonon of wave number q, while the minus sign indicates the absorption of a phonon with wave number q. Furthermore, K denotes an arbitrary lattice vector in the reciprocal space.

Finally, one should note that in the case of the interaction of electrons (holes) (which populate narrow bands) with short wavelength (optical) phonons the theory becomes still more complicated because, in such cases, the electron–phonon interaction becomes very strong. This is due to the fact that the widths of the electronic and vibrational bands are about the same and therefore the electronic localization time at a site, $\tau_{\text{loc}}^{\text{el}} \sim \hbar/\Delta E$, where ΔE is the electronic bandwidth, is comparable or even longer than $\tau_{\text{opt}} = 2\pi/\omega_{\text{opt}}$, the period of the intermolecular (mostly bond-stretching) type vibrations. Therefore, the strong electron–phonon interaction cannot be treated as a small perturbation but must be built into the zero-order Hamiltonian. For such a situation there exists only a single calculation in the literature (see Suhai[61] and references cited therein) on a TCNQ stack, therefore we shall not discuss this rather complicated theory but refer the reader to Suhai's work.

c. Solution of the Boltzmann Equation for Narrow-Band Semiconductors. In the case of narrow-band semiconductors one cannot use the simple relaxation-time approximation for the solution of the Boltzmann transport equation, but one must rather apply other methods. The most general and most frequently used method is the variational one, which requires that the solution $\phi(\mathbf{k}, \mathbf{r})$ of integral equation (9.38) given by

$$X(\mathbf{k}, \mathbf{r}) + \hat{M}\phi(\mathbf{k}, \mathbf{r}) = \hat{P}\phi(\mathbf{k}, \mathbf{r})$$

with respective definitions (9.39) and (9.40) for $X(k, \mathbf{r})$ and $\hat{M}\phi(\mathbf{k}, \mathbf{r})$ satisfies the condition[79]

$$\frac{\langle \phi \,|\, \hat{P}\phi \rangle}{|\langle \phi \,|\, X \,|\, \rangle^2} = \min \tag{9.70}$$

In practice one has to define a trial function ϕ with a certain number of arbitrary parameters, which have to be varied until the functional attains its minimum. A frequently used ansatz for ϕ is, in the one-dimensional case,[80]

$$\phi(k', \mathbf{r}) = \text{const} \times \sin(k'a)\tau_{\text{appr}}(k') \tag{9.71}$$

where the k-dependent approximate relaxation time $\tau_{\text{appr}}(k')$ is, instead of equation (9.47),[78]

$$\frac{1}{\tau_{\text{appr}}(k')} = 2 \sum_{k} \sum_{q} [W_{\text{E}}^{q}(k' \,|\, k) + W_{\text{A}}^{q}(k' \,|\, k)] \tag{9.72}$$

Here $W_E^q(k' \mid k)$ and $W_A^q(k' \mid k)$ are the scattering matrix elements (9.69) for the emission or absorption of a phonon with momentum q, corresponding to the + or − signs in equation (9.69), respectively.

The definition of this approximate relaxation time makes it possible (1) to evaluate the approximate mean free path $\tilde{\Lambda}_{appr}$ (which gives very important information about the consistency of the band model) and (2) to a great extent simplify the variational calculation of mobility.

In evaluating $\tau_{appr}(k')$, the approximate expression $E_k = \varepsilon[1 - \cos(ka)]$ (ε denotes the half bandwidth) and the equation $\hbar\omega(q) = \hbar\omega_{max} |\sin[(q/2a)]|$ have been used,[78] though it should be pointed out again that one could use the correct phonon dispersion curves $\omega(q)$ determined by the GE method described in Section 9.1. The maximum phonon frequency can be determined in one dimension by $\omega_{max} = (2/q)(c_l/\rho)^{1/2}$, where ρ is the mass density and c_l is the interbase elastic constant. (For this one can use the corresponding value found for graphite[82] because the distance between the different layers in graphite is the same as between the nucleotide base pairs in DNA.) The phonon system has been assumed to be in equilibrium described by the Bose–Einstein distribution. If the sums in expression (9.72) are transformed into integrals containing δ-functions, they can be evaluated using the relationship

$$\int g(q)\delta\{f(q)\}dq = \sum_i [g(q_i)/f'(q_i)] \tag{9.73}$$

where q_i denotes the roots of function $f(q)$ and the sum is taken over all of them. The conservation of momentum and of energy determines the possible k values for each type of scattering. We denote the principal value of the function $(1/a)\arcsin[(\hbar\omega_{max})/(2\varepsilon)]$ by Ω. Normal scattering due to absorption of a phonon is found to be possible for $\Omega - (\pi/2a) \leqslant k' < \Omega$ and $(\pi/a) - 2\Omega \leqslant k' < (\pi/a) - \Omega$ while Umklapp scattering (scattering from the first Brillouin zone) of this type occurs in the region $(\pi/2a) - \Omega \leqslant k' < (\pi/a) - 2\Omega$. It should be noted that for broad bands ($2\varepsilon \gg \hbar\omega_{max}$) one gets back the usual "horizontal" scattering processes, inelasticity becomes important only for narrow energy bands.

The above-defined approximate relaxation times can be used to evaluate the mean free path of the charge carriers, averaging over the states of a band using Boltzmann statistics,

$$\tilde{\Lambda}_{appr} = \frac{\int v(k)\tau_{appr}(k)\exp[-E(k)/(k_B T)]dk}{\int \exp[-E(k)/(k_B T)]dk} \tag{9.74}$$

The mobility is calculated by applying the variational method of transport theory, which gives the following expression for the electrical conductivity (which can be obtained by changing the appropriate expression for metals[80] to the case of nondegenerate semiconductors):

$$\sigma = \frac{(2/k_B T)\left\{\sum_k ev(k)\phi(k) f^0[E(k)]\right\}^2}{\sum_{k'}\sum_k\sum_q [\phi(k') - \phi(k)]^2 f^0[E(k)][W_E^q(k' \mid k) + W_A^q(k' \mid k)]} \tag{9.75}$$

For the trial function $\phi(k)$ one can employ (instead of performing a complete variational calculation) the ansatz (9.71), where the constant in equation (9.71) cancels in expression (9.75). The equilibrium charge-carrier distribution $f^0[E(k)]$ may be written as

$$f^0[E(k)] = F^0 \exp[-E(k)/(k_B T)] \tag{9.76}$$

Since one is interested only in mobility values, the constant F^0 will also cancel from the mobility results.

During the evaluation of σ it can be shown that the regions of possible scatterings are the same as before. The conductivity values have been determined by numerical integration. Using the well-known expression $\mu = \sigma/ne$, where the charge-carrier density is given by $n = \Sigma_k f^0[E(k)]$, one finally obtains the mobilities.

d. Application to Different Periodic Nucleotide Base and Base Pair Stacks. Numerical results for bandwidths, mean free paths at 300 K, and mobilities at 180 and 300 K, calculated with the aid of the aforementioned method, are given in Table 9.11. It can be seen that the periodic nucleotide base stacks can be divided into two different groups from the viewpoint of their transport properties. The group of the periodic single base stacks and of the poly(A-T), poly(G-C) systems has bandwidths of 0.1–0.3 eV and corresponding mobility values lie between 30 and 100 $cm^2 V^{-1} s^{-1}$, while the bandwidths of the more complicated poly(A-T/G-C)-type systems are much smaller (0.03–0.61 eV) and also their mobilities are smaller by one order of magnitude. All the mobility values follow the T^{-n} law, but with $n = 1.5$–2 for the former group and $n = 2$–2.5 for the latter. At the same time, in broad-band models the carrier free paths are 20–50 Å at 300 K, hence the band model seems to be applicable to them. For the second group, however, the $\tilde{\lambda}_{appr}$ values are of the order of the lattice distance ($a = 3.36$ Å), where the delocalized model becomes inadequate.

TABLE 9.11. Characteristic Quantities of Charge Carrier Transport in Periodic DNA Nucleotide Base and Base Pair Stacks (after Suhai[78])

Periodic stack		δE (eV)	$\tilde{\lambda}^{300\,K}_{appr}$ (Å)	$\mu^{180\,K}$	$\mu^{300\,K}$
polyT	electron	0.072	12	189	30
	hole	0.274	42	297	82
polyA	electron	0.244	32	227	65
	hole	0.318	54	352	94
polyC	electron	0.142	15	183	54
	hole	0.310	41	270	87
poly(A–T)	electron	0.268	25	142	40
	hole	0.254	32	150	43
poly(G–C)	electron	0.142	18	146	42
	hole	0.268	27	254	77
poly(T/C)[a]	hole	0.056	10	82	16
poly(T/A)	hole	0.030	6	78	12
poly(G/A)	hole	0.040	6.8	64	9.5
poly(A–T/G–C)	electron	0.046	8.5	45	7
	hole	0.053	8	63	10
poly(A–T/T–A)	hole	0.037	5.2	82	13
poly(G–C/C–G)	electron	0.027	5	24	4
	hole	0.020	3.5	28	4

[a] The notation poly(T/C), etc., denotes the periodic sequence TCTCTCTC... in a single stack.

Unfortunately, no direct experimental data are available for the drift mobility in DNA. (It should be noted here that only the high-frequency ac measurements can give information about the real charge-carrier dynamics, because in dc measurements the current also flows through the junctions between the macromolecules and between the electrodes and macromolecules and this has a very strong disturbing effect.) There seems, however, to be a possibility of estimating the order of magnitude of the mobility from the high-frequency ac conductivity measurements of O'Konski *et al.*[81] They observed saturation of the conductivity at about 10^8 cps. If the conductivity is assumed to be electronic, this saturation must be the consequence of the fact that at such high frequencies the charge carriers have no time to accumulate at the ends of the macromolecular chains. Assuming the average number of base pairs in one chain to be 10^6 and using the relation $l = \mu F t$ (where l is the length of the individual chains, F is the field strength, and t is the period of the external field) one obtains from both calculations and measurements mobility values of 10^2 cm^2 V^{-1} s^{-1}. This estimate is uncertain, because we do not know l well. (In DNA of higher organisms the number of base pairs is 10^8–10^9, but in the course of sample preparation the macromolecules might have broken.) It may be accepted only for orientation. To be able to compare more seriously experimental and theoretical results, both the calculations and measurements must be refined. On the theoretical side it will be necessary to take into account also the effect of intramolecular vibrations, uncorrelated hopping-type transitions, the presence of various impurities (which can very strongly influence both the band structure and the transport process), and, last but not least, the aperiodic nature of real DNA. The transport property experiments will become reliable only when it will be possible to produce well-defined samples with known sequences and structure, with at least approximately homogeneous chain lengths and with known, very low concentrations of known types of impurities. Of course, to produce such samples suitable for solid-state physical measurements, major developments are necessary in biomaterial science.

9.3.3. Theory of Hopping Conduction in Very Narrow Band Polymers and in Disordered Polymers with Applications

9.3.3.1. Hopping Conduction in the Very Narrow Band Case

If a periodic polymer chain has a very narrow bandwidth [like the so-called narrow-band (widths of order 10^{-2} eV) periodic nucleotide base stacks or base pair stacks in Table 9.11) or a polymer consists of a nonperiodic sequence of different types of units, then the electronic states become localized molecular states and coherent Bloch-type conduction is no longer possible.

In the first case, the time τ_{loc} during which an electron remains at a certain site becomes larger than or comparable to the time τ_{opt} of vibration of an optical (intramolecular short-wavelength vibrations) phonon, and therefore the transport of electrons can take place only through an uncorrelated "hopping"-type motion between these localized states. This means that during the application of perturbation theory, the zeroth-order Hamiltonian contains the molecular potential plus the electron–vibration interaction and one must treat as a perturbation the intermolecular potential, which earlier gave rise to the formation of Bloch functions. The transport thus becomes a diffusion process and the mobility is determined by the Einstein relation $\mu = ed/k_B T$, where the diffusion constant d is given by $d = Wa^2$ in the one-dimensional case. The transition probability W per unit time between two neighboring localized states can be determined again by the Golden Rule. Here, only the main results of such a calculation performed for the above-mentioned very-narrow-band periodic DNA models[61,82] can be summarized as follows:

1. The polaron binding energies lie between 0.180 and 0.260 eV; the resulting band narrowing factor is about 0.4–0.2.
2. The hopping mobilities are 10^{-1}–10^{-2} $cm^2 V^{-1} s^{-1}$, nearly two orders of magnitude smaller than in the band case.
3. The temperature dependence of the mobility depends on the strength of the electron–phonon interaction. For the obtained polaron binding energies it is proportional to T^{-1}; temperature-activated properties would actually be observable only for much stronger coupling.

9.3.3.2. Hopping Conduction in Disordered Polymers

The situation is somewhat different for a disordered multicomponent chain. In this case all the electronic states are also localized on the molecular residues, however, calculations of four-component nonperiodic DNA single strands[83] and four-, five-, and six-component aperiodic polypeptide chains[84] (for which random sequences were constructed with the help of a Monte Carlo program) have indicated extensive broadening of the electronic DOSs compared to periodic sequences of the same 1:1:1:1, etc., composition in both the valence- and conduction-band regions. This is demonstrated in Figure 9.11 by the DOS distribution in the valence-band region for a random sequence comprising 1:1:1:1 of a glycine (gly), serine (ser), cysteine (cys), asparagine (asn) chain of 300 units, obtained with the help of the matrix block NFC method (see Section 4.4). For comparison, Figure 9.12 presents thc DOS in the valence-band region for the same system if the four different units form a periodic sequence. The broadening effect of the aperiodicity on the DOS is seen to be quite dramatic and, moreover, there are only very few, and narrow, gaps in the DOS curve. Figure 9.13 shows the same, very large broadening of allowed states in the above system for the conduction-band region if the sequence is random.

The explanation of this phenomenon is that, for a periodic sequence of a multicomponent chain, each type of unit experiences the same potential (of course, this potential is different for the various types of units). On the other hand, in a random sequence each unit (even of the

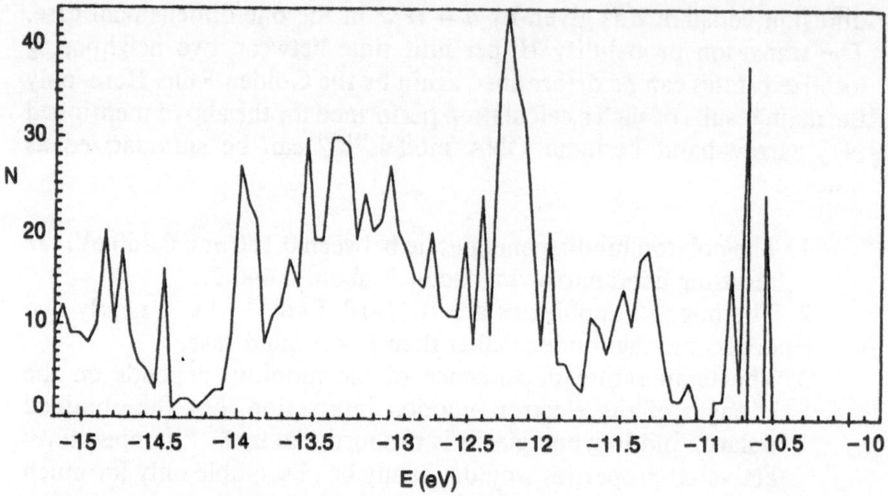

FIGURE 9.11. DOS distribution for the valence-band region of a peptide chain of 300 units in a random sequence, having a composition of 1:1:1:1 composition of gly:ser:cys:asn; N is the total number of states per histogram interval (0.05 eV) and energy E is given in eV.

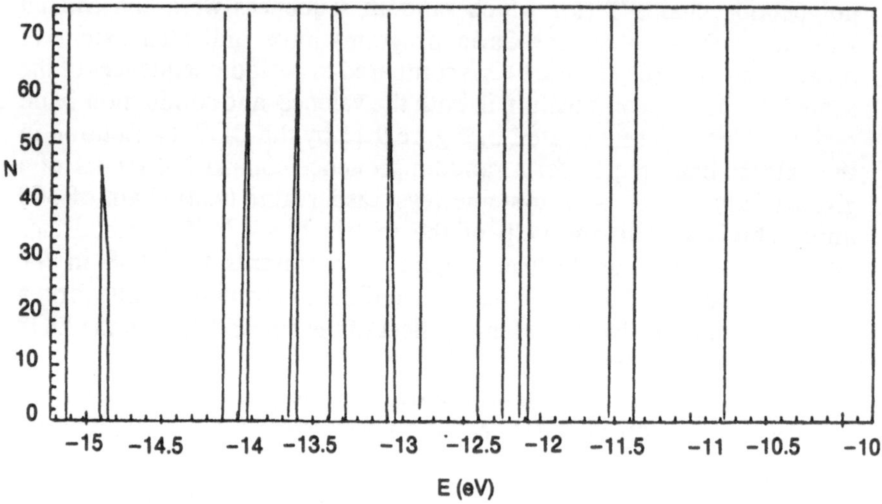

FIGURE 9.12. DOS distribution for the valence-band region of a periodic poly(gly, ser, asn, cys) chain; N is the total number of states per histogram interval (0.05 eV) and energy E is given in eV.

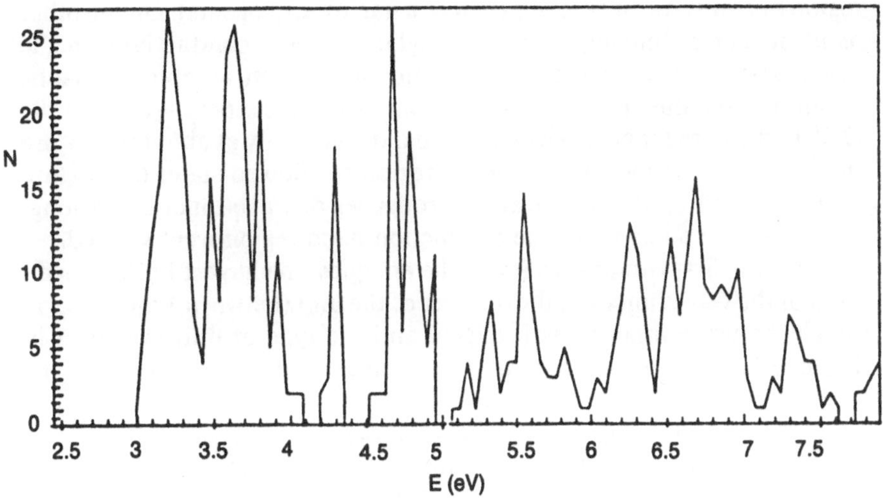

FIGURE 9.13. DOS distribution for the conduction-band region of the random gly-ser-cys-asn chain with 300 units and 1:1:1:1 composition; N is the total number of states per histogram interval (0.05 eV) and energy E is given in eV.

same type) experiences many different potentials acting on it because, owing to the randomness, each unit can have many different neighboring units. In our experience, this explanation requires that the number n of different types of unit should be at least 3, but the broadening effect on the allowed regions of state becomes really strong only if $n \geqslant 4$.

In the case of the four-component gly-ser-cys-asn system with random sequence and $N = 300$ units of chain length, the localization properties of the wave functions of the first 56 lowest levels in the conduction-band region were investigated using the inverse-iteration technique (see Section 4.5). The reason for this choice was that, most probably, *in vivo* there is a charge transfer from the phosphate group of DNA (which carries 1.05 e negative charge instead of 1.0 e due to charge shift from the sugar part of a polynucleotide to the phosphate group, as shown by detailed calculations[85]) and the positively charged guanidium end-groups of arginine, protonated lysine residues, or histidine residues in their cationic form. [In a nucleic acid–protein complex (so-called nucleo-protein), the protein molecules histones play a key role and arginine residues occur rather frequently in them]. If it is assumed that, on the average, there is 0.1 e charge transfer per unit cell from DNA to the polypeptide chain of length 300 residues, the first 15 levels will be filled and, as is evident from Figure 9.13, there is a more than 1-eV-broad region of allowed empty levels (before a narrow gap occurs). On the other hand, it seems difficult to achieve higher hopping conductivity in the valence-band region even with the aid of electron acceptor dopants, because at the upper edge of this region only narrow peaks occur in the DOS (which are separated by narrow bands) and only at about 0.5 eV are there lower energy levels in a broad region of allowed states (see Figure 9.11). Therefore, only localization properties of wave functions belonging to the first 56 levels of the conduction-band region were studied.

For the first twenty-five levels the energies, the largest LCAO coefficient of the wave function, the number of the unit (between 1 and 300) on which the wave function is localized and the type of unit are given in Table 9.12. If one assumes a 0.1 *e* charge transfer per unit to the

TABLE 9.12. Orbital Energy ε_i for the Lowest Twenty-Five Unoccupied States of the gly-ser-cys-asn Random Chain Whose DOS Distribution is Shown in Figure 9.13[a]

i	ε_i (eV)	$\lvert c_{i\,\max}\rvert$	N_{unit}	Type
1	2.98441	0.939	252	ser
2	2.99009	0.755	207	asn
3	2.99577	0.687	101	asn
4	3.00145	0.797	63	ser
5	3.00713	0.871	63	ser
6	3.01280	0.902	24	ser
7	3.01848	0.693	192	gly
8	3.02416	0.890	192	gly
9	3.02984	0.802	193	gly
10	3.03552	0.873	108	gly
11	3.04119	0.792	227	asn
12	4.04688	0.896	193	gly
13	3.05255	0.893	249	ser
14	3.05823	0.775	210	ser
15	3.06391	1.230	81	asn
16	3.06959	0.762	149	gly
17	3.07527	0.647	251	asn
18	3.08085	0.901	108	gly
19	3.08662	0.708	63	ser
20	3.09231	0.902	108	gly
21	3.09799	0.901	108	gly
22	3.10366	0.749	134	asn
23	3.10934	0.795	192	gly
24	3.11502	0.904	192	gly
25	3.12070	0.905	192	gly

[a] ε_1 denotes the lowest energy level, $c_{i\,\max}$ the largest LCAO coefficient of the one-electron wave function, N_{unit} is number of unit (between 1 and 300) on which the wave function is localized, and type denotes the type of unit.

polypeptide chain, then the Fermi energy lies at 2.06391 eV (eigenvalue number 15 in Table 9.12). When first excitations from these fifteen occupied states to the next ten empty ones are considered, one first-neighbor pair (193, 192) is found in this group, namely, the excitation from level 12 or 23. If higher excitations are also considered, or we take into account the fact that some of the occupied levels become empty owing to excitations, then the following excitations between first neighbors can be found (the numbers of the participating unit cells are given in parentheses): excitations from level 7 to 9 (192, 193), from 8 to 9 (192, 193), from 27 to 31 and 32 (98, 97), from 16 to 51 (149, 150), from 12 to 37 (193, 194), and from 37 to 44 and 45 (194, 193). Possible excitations between second-neighbor pairs occur from level 13 to 17 (249, 251) and from 26 to 37 (192, 194). It can also be seen from Table 9.12 that excitations within the same unit are also possible. These excitations can help to promote an electron to a higher level, from which it can be excited in a subsequent step to an orbital which is localized on a different site. Information on the localization properties of the one-electron functions around the Fermi energy allows one to discuss the consequences for conductivity in terms of the hopping model.

A pure hopping model is considered here, that is, the conduction is due to electrons in localized states with energies near the Fermi energy. The primary jump rate $W_{i\to j}$, i.e., the number of jumps from one localized state i to another j per unit time, is given to an acceptable approximation by*[86]

$$W_{i\to j} \approx \nu_{\text{phonon}}[\exp(-\alpha|\mathbf{R}_i - \mathbf{R}_j|)]^2 \exp(-\Delta E_{ij} \mid k_B T) \qquad (9.77)$$

where $\Delta E_{ij} = \varepsilon_j - \varepsilon_i$, $\mathbf{R}_i$ and ε_i are the position vector and energy of the electron localized at site i, and $\exp(-\alpha|\mathbf{R}_i - \mathbf{R}_j|)$ is the overlap integral between two exponentially localized wave functions.[86] This overlap integral can take the form of an LCAO expression and be approximated by the dominant terms. In this way [see equation (20) of the paper by Ladik *et al.*[88]] one obtains

$$W_{A\to B} = \nu_{\text{phonon}} \left(\sum_{r,s} c_{i,r} c_{j,s} \langle \chi_r \mid \chi_s \rangle \right)^2 \exp(-\Delta E_{ij} \mid k_B T) \qquad (9.78)$$

where ΔE_{ij} is again given by $\Delta E_{ij} = \varepsilon_j - \varepsilon_i$, $c_{i,r}$ and $c_{j,s}$ are the dominant

* In reality, one should calculate the corresponding electron–phonon interaction matrix elements to obtain more accurate results for the primary jump rates.

LCAO coefficients of wave functions ψ_i (ψ_j) localized at site A (B) and correspond to energies ε_i and ε_j, respectively.

The phonon frequencies ν_{phonon} in proteins have the following characteristic values[89]: at acoustic modes they lie between 10^{11} and 10^{12} s^{-1} ($\sim$0.0004–0.004 eV); in a polypeptide chain the acoustic modes are generated by vibrations, which cause changes in the relative distances and orientations of the side chains. The upper limit is given by the high-frequency intramolecular vibrational modes of hydrogen atoms with a frequency of 10^{14} s^{-1} ($\approx$0.4 eV). At body temperatures ($k_B T = k_B$ 310 K $\approx$ 0.03 eV), the acoustic modes will be most active in the scattering of electrons. Therefore for the subsequent considerations $\nu_{\text{phonon}} = 10^{12}$ s^{-1} has been taken as the characteristic value.

Table 9.13 presents typical primary jump rates, calculated from equation (9.78). The first two values (5×10^7 and 7×10^8 s^{-1}) refer to hopping between first neighbors. The value of the square of the overlap integral is in both cases approximately 7×10^{-4}; the order-of-magnitude difference is determined by the Boltzmann factor. The characteristic value for second-neighbor hopping is about 10^5 jumps per second. Here, the overlap integral factor is about 10^{-6}. The last three entries give primary jump rates between orbitals localized in the same unit cell. The rates for these local excitations (which can be important intermediate steps in the promotion of an electron to a level from which it can more easily reach a neighboring cell due to a larger Boltzmann factor) are between 5×10^{11} and 9×10^{11} s^{-1}. The small differences are due to discrepancies in the overlap between the initial and final states.

One should note that if one electron is to be transported from one end of the the chain to the other and it is assumed that each of the wave

TABLE 9.13. Characteristic Primary Jump Rates $W_{A \to B}$ (in s^{-1}) Calculated from Equation (9.78)[a]

A	B	i	j	ΔE_{ij}	$P_{A \to B}$
193	192	12	23	0.062	5×10^7
192	193	8	9	0.006	7×10^8
192	194	26	37	0.062	1×10^5
192	192	7	8	0.006	7×10^{11}
108	108	18	20	0.011	7×10^{11}
108	108	20	21	0.006	9×10^{11}

[a] A, B denote the localization sites, i, j the numbering of the energy levels (1 denotes the lowest one), ΔE_{ij} (in eV) the energy difference entering the Boltzmann factor, ν_{phonon} was chosen to be 10^{12} s^{-1}, $k_B T = 2.67 \times 10^{-2}$ eV for $T = 310$ K.

functions belonging to the first 300 levels in the conduction-band region is localized on a different amino-acid residue, then the hopping electron must be excited from the bottom of the conduction-band region by about 1 eV until it reaches the end of the chain. (This can be seen from Figure 9.13, where the average total DOS is 15 and the step length in the histogram is 0.05 eV; 15/0.05 = 300, that is, there are 300 states in an energy region of approximately 1 eV.) This would lead to a Boltzmann factor $e^{-1/0.03} = e^{-33} \approx 10^{-15}$. On the other hand, it can be seen from Table 9.12 that the wave function of the bottom level of the conduction-band region is not localized at the end of the chain but at amino acid 252, and the subsequent levels have wave functions that are distributed randomly. Therefore, a 1-eV excitation from one site to a neighboring one is highly improbable and a more realistic estimate of the average excitation energy to reach a neighboring site is about 0.5 eV. Further, one should not forget that in both the β-pleated sheet and α-helix structure an electron can always jump from a site in one chain to another in a neighboring chain, if a jump to a first (or second) neighbor in its own chain requires too much energy. Therefore, it seems that a conservative estimate for the averge jumping energy is $\Delta E_{ij} \approx 0.25$ eV, which provides a Boltzmann factor of $e^{-0.25/0.03} \approx e^{-8.3} \approx 10^{-3.6}$. The water and ion environment and the protonation of some side chains probably provide extra levels, which further increase this average Boltzmann factor in real proteins in solution. A value of 10^{-3} can be assumed which decreases the Boltzmann factor only by two orders of magnitude (the Boltzmann factor corresponding to 0.06 is $e^{-0.06/0.03} = e^{-2} \approx 10^{-1}$). Hence the physically most interesting, first and third $P_{A\text{-}B}$ values given in Table 9.13 must be decreased only by this value.

One can compare the primary jump rates, which were calculated for proteins using *ab initio* one-electron wave functions, with those used in hopping models for other noncrystalline solids. For the contribution of hopping transport by carriers with energies near the Fermi level to the ac conductivity in noncrystalline semiconductors, Mott and Davis[90] chose the following values for the quantities which enter equation (9.77): $\nu_{phonon} = 10^{12}\ s^{-1}$, $\alpha^{-1} = 8$ Å, $R = |\mathbf{R}_i - \mathbf{R}_j| = 60$ Å, $\Delta E_{ij} = 0.1$ eV, $T =$ 300 K. This yields for the square of the overlap integral $e^{-15} \approx 3 \times 10^{-7}$, for the Boltzmann factor about 3×10^{-2}, and for the primary jump rate $W_{i \to j} \approx 10^4$, while we have obtained for proteins (using already the estimated 0.25 eV averaged excitation energy of first-neighbor jumps) the value 5×10^5. In another numerical study of a general hopping transport theory,[91] which should be applicable to photoexcited carriers in amorphous silicon and chalcogenides, the following numbers were used: $\nu_{phonon} = 10^{13}\ s^{-1}$ and $2\alpha R = 10$. This yields for the overlap contribution

$e^{-10} \approx 5 \times 10^{-5}$ and, assuming a Boltzmann factor of the order of unity, a primary jump rate of about 5×10^8. This is about two or three orders of magnitude larger than the calculated first-neighbor jump rates (with the average excitation energy) for electrons in localized states in proteins.

It should be noted, however, that in this second calculation a tenfold larger phonon frequency was chosen than the one that is realistic in the case of proteins, and a Boltzmann factor of the order of unity seems to be rather overestimated. These two facts certainly account for at least two orders of magnitude in the difference of three orders of magnitude in the primary jump rates between their and our calculations.

Finally, we note that after determining the primary jump rates between different sites of a multicomponent disordered polypeptide, these data can be used as input in a stochastic (random walk) theory of hopping conductivity in random media. Since, however, this theory has not yet been applied to any real polymers, we do not describe this method here but refer to the appropriate papers.[92]

REFERENCES

1. E. Wilson, Jr., J. C. Decines, and P. C. Cross, *Molecular Vibrations*, MacGraw-Hill Book Company, New York (1955).
2. P. Pulay, in: *Applications of Electronic Structure Theory, Modern Theoretical Chemistry* (H. F. Schaefer III, ed.), Vol. 4, p. 153, Plenum Press, New York (1977).
3. A. Peluso, M. Seel, and J. Ladik, *Can. J. Chem.* **63**, 1553 (1985).
4. P. Bosi, G. Zerbi, and E. Clementi, *J. Chem. Phys.* **66**, 3376 (1977).
5. M. Tasumi and T. Shimanouchi, *J. Chem. Phys.* **43**, 1245 (1965); M. Tasumi and S. Krimm, *J. Chem. Phys.* **46**, 755 (1967).
6. L. Piseri and G. Zerbi, *J. Mol. Spectrosc.* **26**, 254 (1961).
7. See, for instance: C. Kittel, *Quantum Theory of Solids*, Wiley, New York–London (1963).
8. A. Peluso, M. Seel, and J. Ladik, *Solid State Commun.* **53**, 893 (1985).
9. R. S. Day and F. Martino, *Chem. Phys. Lett.* **84**, 86 (1981); J. Ladik, *Int. J. Quantum Chem.* **23**, 1073 (1983).
10. G. Zannoni and G. Zerbi, *Solid State Commun.* **47**, 213 (1983).
11. P. Dean, *Rev. Mod. Phys.* **44**, 127 (1972).
12. J. Delhalle, personal communication.
13. J. H. Wilkinson, *The Algebraic Eigenvalue Problem*, p. 633, Clarendon Press, Oxford (1965).
14. J. Ziman, *Electrons and Phonons*, Clarendon Press, Oxford (1972); W. Jones and N. March, *Theoretical Solid State Physics*, Vol. 1, pp. 195, 210, 277, 281–285, Dover Publ. Inc., New York (1985).
15. P. E. Van Camp, V. E. Van Doren, and J. T. Devreese, *Solid State Commun.* **42**, 1224 (1979).
16. M. J. S. Dewar, Y. Yamaguchi, and S. H. Such, *Chem. Phys.* **43**, 145 (1979).
17. H. Teramae, T. Yamabe, and A. Imamura, *J. Chem. Phys.* **81**, 3564 (1984).

18. C. E. Blom, *Lecture Notes in Physics*, No. 113, p. 233, Springer-Verlag, Berlin–Heidelberg–New York (1980).
19. J. Shirakawa and S. Ikeda, *Polym. J.* **2**, 231 (1971); F. B. Schügerl and H. Kuzmany, *J. Chem. Phys.* **74**, 953 (1981).
20. D. Wöhrle, *Tetrahedron Lett.* 1969 (1971); *Makromol. Chem.* **175**, 1751 (1974).
21. A. Karpfen, *J. Chem. Phys.* **75**, 238 (1981).
22. S. Huzinaga, *J. Chem. Phys.* **42**, 1293 (1965); *Approximate Atomic Functions I*, University of Alberta, Alberta (1971).
23. H. Schachtschneider and R. G. Snyder, *Spectrochim. Acta* **19**, 117, 865 (1963); *ibid.* **19**, 865 (1963).
24. A. Beyer and A. Karpfen, *Chem. Phys.* **64**, 343 (1982).
25. L. Piela and J. Delhalle, *Int. J. Quantum Chem.* **13**, 605 (1978); J. Delhalle, J. Piela, J.-L. Brédas, and J.-M. André, *Phys. Rev. B* **22**, 6254 (1980).
26. A. Karpfen, *Int. J. Quantum Chem.* **19**, 1207 (1981).
27. A. Imamura and H. Fujita, *J. Chem. Phys.* **61**, 1115 (1974); I. I. Ukrainsky, *Theor. Chim. Acta (Berlin)* **38**, 139 (1975); A. Blumen and C. Merkel, *Phys. Status Solidi B* **83**, 425 (1977).
28. L. A. Curtiss and J. A. Pople, *J. Mol. Spectrosc.* **61**, 1 (1976).
29. H. Lischka, *J. Am. Chem. Soc.* **96**, 4761 (1974).
30. A. Axmann, W. Bien, P. Barsch, F. Hoszfeld, and H. Stiller, *Discuss. Faraday Soc.* 7, 69 (1969); R. Turbino and G. Zerbi, *J. Chem. Phys.* **51**, 4509 (1969); A. Anderson, B. H. Terris, and W. S. Tse, *J. Raman Spectrosc.* **10**, 148 (1981).
31. A. Zunger, *J. Chem. Phys.* **63**, 1713 (1975); A. Karpfen and P. Schuster, *Chem. Phys. Lett.* **44**, 459 (1976).
32. J. S. Kittelberger and D. F. Hornig, *J. Chem. Phys.* **46**, 3099 (1967).
33. A. Anderson, B. H. Terrie, and W. S. Tse, *Chem. Phys. Lett.* **70**, 300 (1980).
34. W. Kusmierczuk and A. Witkowski, *Chem. Phys. Lett.* **81**, 558 (1981).
35. D. Hodzi and S. Bratož, in: *The Hydrogen Bond Recent; Developments in Theory and Experiments* (P. Schuster, G. Zundel, and C. Sándorfy, eds.), Vol. 2, p. 565, North-Holland Publishing Company, Amsterdam (1976).
36. A. Karpfen and J. Petkov, *Solid State Commun.* **29**, 251 (1979); *Theor. Chim. Acta (Berlin)* **53**, 65 (1979).
37. T. Shimanouchi, *Phys. Chem.* **4**, 233 (1970).
38. M. J. S. Dewar and W. Thiel, *J. Am. Chem. Soc.* **99**, 4899, 4907 (1977); QCPE program number 353, Indiana University, Bloomington, Indiana; M. J. S. Dewar, G. P. Ford, M. L. McKee, H. S. Rzepa, W. Thiel, and Y. Yamagouchi, *J. Mol. Struct.* **43**, 135 (1978).
39. S. Suhai, *Chem. Phys. Lett.* **96**, 619 (1983).
40. G. Zannoni and G. Zerbi, *J. Mol. Struct.* **100**, 505 (1983).
41. F. Inagaki, M. Tasumi, and T. Miyazawa, *J. Mol. Spectrosc.* **50**, 286 (1974).
42. F. Török and P. Pulay, *J. Mol. Struct.* **32**, 93 (1970); K. Kozmutza and P. Pulay, *Theor. Chim. Acta (Berlin)* **37**, 67 (1979).
43. M. J. S. Dewar and S. Olivella, *J. Am. Chem. Soc.* **100**, 5290 (1978).
44. M. Damjanovich and M. Vujicu, *Phys. Rev. B* **28**, 1997 (1983).
45. E. J. Mele and M. J. Rice, *Phys. Rev. Lett.* **45**, 926 (1980).
46. G. Zannoni and G. Zerbi, *Chem. Phys. Lett.* **85**, 50 (1982); *J. Mol. Struct.* **100**, 485 (1983).
47. R. M. Gavin, Jr. and S. A. Rice, *J. Chem. Phys.* **55**, 2675 (1971).
48. (a) J. Shirakawa and S. Ikeda, *Polym. J.* **2**, 231 (1971). (b) J. Shirakawa, T. Ito,

AND S. IKEDA, *Polym. J.* **4**, 460 (1973). (c) I. HARADA, Y. FURUKAWA, M. TASUMI, H. SHIRAKAWA, AND S. IKEDA, *J. Chem. Phys.* **73**, 474 (1980).
49. F. B. SCHÜGERL AND H. KUZMANY, *J. Chem. Phys.* **74**, 953 (1981).
50. K. LUNDE AND L. ZECHMEISTER, *Acta Chim. Scand.* **8**, 1421 (1954).
51. C. R. FINCHER, JR., M. OZAKI, A. J. HEEGER, AND A. G. MACDIARMID, *Phys. Rev. B* **19**, 4140 (1979).
52. B. HOROVITZ, *Phys. Rev. Lett.* **47**, 1491 (1981).
53. J. BARDEEN AND W. SHOCKLEY, *Phys. Rev.* **80**, 72 (1950).
54. H. BROOKS, *Adv. Electron.* **7**, 85 (1955); W. P. DUNKE, *Phys. Rev.* **101**, 531 (1956); C. HERRING AND E. VOGT, *Phys. Rev.* **101**, 944 (1956).
55. See, for instance: A. L. FETTER AND J. D. VALECKA, *Quantum Theory of Many Particle Systems*, McGraw-Hill Book Company, New York (1971).
56. See, for instance: L. I. SCHIFF, *Quantum Mechanics*, McGraw-Hill Book Company, New York (1959).
57. R. PEIERLS, in: *Transport Phenomena* (G. Kirczenow and J. Marro, eds.), Springer-Verlag, Berlin–New York–Heidelberg (1974).
58. R. KUBO, *J. Phys. Soc. Jpn.* **12**, 570 (1957).
59. A. H. WILSON, *The Theory of Metals*, Cambridge University Press, Cambridge, England (1954).
60. A. C. SMITH, J. E. JANEK, AND R. B. ADLER, *Electronic Conduction in Solids*, McGraw-Hill Book Company, New York (1967).
61. S. SUHAI, in: *Quantum Theory of Polymers* (J.-M. André, J. Ladik, and J. Delhalle, eds.), p. 335, D. Reidel Publ. Co., Dordrecht–Boston (1978).
62. H. JONES AND C. ZENER, *Proc. R. Soc. London, Ser. A* **144**, 101 (1934).
63. F. BELEZNAY, G. BICZÓ, AND J. LADIK, *Acta Phys. Acad. Sci. Hung.* **18**, 213 (1965).
64. S. H. GLARUM, *J. Phys. Chem. Solids* **24**, 1577 (1963).
65. J. LADIK, G. BICZÓ, AND G. ELEK, *J. Chem. Phys.* **44**, 483 (1966).
66. S. SUHAI, *Theor. Chim. Acta (Berlin)* **34**, 157 (1974).
67. S. SUHAI AND J. LADIK, unpublished results.
68. D. D. ELEY AND D. I. SPIVEY, *Trans. Faraday Soc.* **56**, 1432 (1961).
69. R. G. ANDERSON AND C. J. FRITSCHE, *Proc. 2nd Natl. Meeting of the Soc. of Applied Spectroscopy*, paper 111, San Diego (1963).
70. S. SUHAI, *Solid State Commun.* **21**, 117 (1977).
71. S. SUHAI, unpublished results.
72. S. K. KHANNA, A. A. BRIGHT, A. F. GARITO, AND A. J. HEEGER, *Phys. Rev. B* **10**, 2139 (1974).
73. R. G. KEPLER, *J. Cjem. Phys.* **39**, 3528 (1963).
74. C. HERRING, *Proc. Int. Conf. Semiconduct. Prague*, p. 60 (1961).
75. G. D. WHITEFIELD, *Phys. Rev.* **121**, 720 (1961).
76. L. FRIEDMAN, *Phys. Rev.* **140**, 1649 (1969).
77. G. DEL RE, J. LADIK, AND G. BICZÓ, *Phys. Rev.* **155**, 997 (1967).
78. S. SUHAI, *J. Chem. Phys.* **57**, 5599 (1972).
79. See Ref. 14 of Chapter 7.
80. F. J. BLATT, *Physics of Electronic Conduction in Solids*, pp. 135 and 186, McGraw-Hill Book Company, New York (1968).
81. C. T. O'KONSKI, P. MOSER, AND M. SHIRAI, *Biopolymers Symp.* **1**, 479 (1964).
82. S. SUHAI, unpublished results.
83. A. K. BAKHSHI, P. OTTO, J. LADIK, AND M. SEEL, *Chem. Phys.* **108**, 215 (1986).
84. A. K. BAKHSHI, J. LADIK, M. SEEL, AND P. OTTO, *Chem. Phys.* **108**, 233 (1986).
85. See Refs. 78–80 of Chapter 2.

86. N. F. Mott and E. A. Davis, *Electronic Processes in Non-Crystalline Materials*, Clarendon Press, Oxford (1971); M. Grünewald, B. Pohlmann, B. Mohagvan, and D. Würtz, *Philos. Mag. B* **49**, 341 (1984); M. Grünewald, B. Mohagvan, B. Pohlmann, and D. Würtz, *Phys. Rev. B* **32**, 8191 (1985).
87. J. M. Ziman, *Models of Disorder*, Cambridge University Press, Cambridge, England (1979).
88. J. Ladik, M. Seel, P. Otto, and A. K. Bakhshi, *Chem. Phys.* **108**, 203 (1986).
89. F. Parak, in: *Structure and Motion: Membranes, Nucleic Acids and Proteins* (E. Clementi, G. Corongiu, M. H. Sarma, and R. H. Sarma, eds.), p. 243, Adenine Press, New York (1985).
90. N. F. Mott and E. A. Davis, *Electronic Processes in Non-Crystalline Materials*, p. 213, Clarendon Press, Oxford (1971).
91. M. Silver, G. Schoenherr, and G. Baessler, *Phys. Rev. Lett.* **48**, 352 (1982).
92. H. Scher and M. Lax, *Phys. Rev. B* **7**, 4491, 4502 (1973); T. Odagaki and M. Lax, *Phys. Rev. B* **24**, 5284 (1981); T. Odagaki and M. Lax, *Phys. Rev. B* **26**, 6480 (1982); T. Odagaki and M. Lax, *Phys. Rev. Lett.* **45**, 847 (1980). For a review see: F. Martino, in: *Quantum Chemistry of Polymers; Solid State Aspects* (J. Ladik, J.-M. André, and M. Seel, eds.), p. 279, D. Reidel Publ. Co., Dordrecht–Boston–Lancaster (1984).

Chapter 10

Magnetic, Electric, and Mechanical Properties of Polymers

10.1. HARTREE–FOCK EQUATIONS FOR PERIODIC POLYMER CHAINS IN A MAGNETIC FIELD

10.1.1. Static Magnetic Field

If we neglect interaction of the electronic and/or nuclear spins with an external magnetic field (which can be treated correctly only within the framework of a relativistic CO theory; see Section 1.5), the one-electron Hamiltonian of a system in the presence of a magnetic field **H** can be written in the form

$$\hat{\tilde{H}} = \frac{\hat{\mathscr{P}}^2}{2m} + V(\mathbf{r}) \tag{10.1}$$

where

$$\hat{\mathscr{P}} = \hat{\mathbf{p}} - \frac{e}{c}\mathbf{A}; \qquad \mathbf{H}(\mathbf{r}) = \operatorname{curl} \mathbf{A}(\mathbf{r}) \tag{10.2}$$

In the Hartree–Fock case one can then write

$$\hat{F} = \hat{\tilde{H}}^N + \sum_j (2\hat{J}_j - \hat{K}_j) \tag{10.3}$$

with

$$\hat{\tilde{H}}^N = \frac{1}{2m}\left(\hat{\mathbf{p}} - \frac{e}{c}\mathbf{A}\right)^2 - \sum_{q_1=-N}^{N} \sum_{\alpha=1}^{M} \frac{Z_\alpha}{|\mathbf{r} - \mathbf{R}_\alpha^{q_1}|} = \frac{\hat{\mathscr{P}}^2}{2m} + \mathrm{EN} \tag{10.4}$$

for a linear chain of $2N+1$ unit cells and M atoms in the cell, where the abbreviation EN denotes all the electron–nuclear interaction terms in the chain. In equation (10.3) summation over j indicates, as before, summation over all filled bands and integration in k-space over the first Brillouin zone.

If equation (10.4) operates on a wave function φ_i, one obtains

$$\begin{aligned}\tilde{\hat{H}}^N\varphi_i &= \frac{1}{2}\left(\hat{\mathbf{p}}-\frac{1}{c}\mathbf{A}\right)^2\varphi_i + \text{EN}\,\varphi_i\\ &= \frac{1}{2}\left(\frac{1}{i}\nabla-\frac{1}{c}\mathbf{A}\right)^2\varphi_i + \text{EN}\,\varphi_i\\ &= -\frac{1}{2}\left\{\Delta\varphi_i+\frac{1}{ic}[\nabla(\mathbf{A})\varphi_i+\mathbf{A}\nabla\varphi_i]\right\}+\frac{1}{2c^2}A^2\varphi_i+\text{EN}\,\varphi_i\\ &= -\frac{1}{2}\left[\Delta\varphi_i+\frac{1}{ic}(\text{div}\,\mathbf{A}\varphi_i+\mathbf{A}\,\text{grad}\,\varphi_i)\right]+\frac{1}{2c^2}A^2\varphi_i+\text{EN}\,\varphi_i\end{aligned} \tag{10.5}$$

where atomic units have been applied everywhere. If the field is not very strong, one can usually neglect the term containing A^2.

10.1.1.1. Static Homogeneous Magnetic Field

In the simplest case $\mathbf{H} = \mathbf{const}$. For instance, if $\mathbf{H} = H_z\mathbf{k}$ where z is the direction of the long axis of the chain, one obtains

$$\begin{gathered}H_z = (\text{curl}\,\mathbf{A})_z = \frac{\partial A_y}{\partial x}-\frac{\partial A_x}{\partial y}\\ A_x=0,\quad A_y = xH_z,\quad A_z=0\end{gathered} \tag{10.6}$$

With allowance for equations (10.5) and (10.6) the matrix elements of $\tilde{\hat{H}}^N$ in an LCAO approximation can then be expressed as

$$\begin{aligned}(\tilde{\hat{H}}^N)^{0,q}_{r,s} &= \langle\chi^0_r\,|\,\tilde{\hat{H}}^N\,|\,\chi^q_s\rangle\\ &= \left\langle\chi^0_r\left|-\tfrac{1}{2}\Delta\underbrace{-\frac{1}{2ic}\frac{\partial}{\partial y}(xH_z)}_{0}-\frac{1}{2ic}xH_z\frac{\partial}{\partial y}+\frac{1}{2c^2}A^2+\text{EN}\right|\chi^q_s\right\rangle\\ &\approx\left\langle\chi^0_r\left|-\tfrac{1}{2}\Delta-\frac{1}{2ic}xH_z\frac{\partial}{\partial y}\right|\chi^q_s\right\rangle\\ &\quad+\left\langle\chi^0_r\left|\sum_{q_1}\sum_{\alpha}\frac{Z_\alpha}{|\mathbf{r}-\mathbf{R}^{q_1}_\alpha|}\right|\chi^q_s\right\rangle\end{aligned} \tag{10.7}$$

If **H** remains a constant vector but points in a general direction, then

$$\mathbf{H} = H_x\mathbf{i} + H_y\mathbf{j} + H_z\mathbf{k} \tag{10.8}$$

Hence the equations

$$H_x = \frac{\partial A_z}{\partial y} - \frac{\partial A_y}{\partial z}$$
$$H_y = \frac{\partial A_x}{\partial z} - \frac{\partial A_z}{\partial x}$$
$$H_z = \frac{\partial A_y}{\partial x} - \frac{\partial A_x}{\partial y} \tag{10.9}$$

yield for the components of **A**

$$A_x = zH_y, \quad A_y = xH_z, \quad A_z = yH_x \tag{10.10}$$

These vector potential components enable one to obtain the following expression for the elements of the one-electron part of the Fock matrix (the two-electron part remains unchanged):

$$\langle \chi_r^0 \mid \tilde{\hat{H}}^N \mid \chi_s^0 \rangle = \left\langle \chi_r^0 \right| -\tfrac{1}{2}\Delta \underbrace{- \frac{1}{2ic}\operatorname{div}\mathbf{A}}_{0} - \frac{1}{2ic}\left(zH_y \frac{\partial}{\partial x} + xH_z \frac{\partial}{\partial y} + yH_z \frac{\partial}{\partial z} \right) + \frac{1}{2c^2}A^2 + \mathrm{EN} \left| \chi_s^q \right\rangle \tag{10.11}$$

10.1.1.2. Static Inhomogeneous Magnetic Field

a. Inhomogeneous Magnetic Field without Perturbation of the Periodic Symmetry. An inhomogeneous magnetic field does not perturb the periodic symmetry of a linear chain (with z as main axis direction) if (1) $H_z = 0$ and (2) H_x and H_y do not depend on z,

$$H_x(x, y) \neq 0, \quad H_y(x, y) \neq 0, \quad H_z = 0 \tag{10.12}$$

A very simple example which fulfills both conditions is $H_x = xH_0$, $H_y = -yH_0$, and $H_z = 0$. From the equation $\mathbf{H} = \operatorname{curl}\mathbf{A}$ one can then show that $A_x = xH_0$, $A_y = yH_0$, and $A_z = xyH_0$, where H_0 is constant.

Substitution of these expressions into equation (10.5) immediately reduces the matrix elements to the form

$$\langle \chi_r^0 \,|\, \tilde{H}^N \,|\, \chi_s^q \rangle$$
$$= \left\langle \chi_r^0 \right| -\tfrac{1}{2}\Delta - \frac{1}{2ic}(H_0 + H_0) - \frac{H_0}{2ic}\left(x\frac{\partial}{\partial x} + y\frac{\partial}{\partial y} + xy\frac{\partial}{\partial z}\right)$$
$$+ \frac{1}{2c^2}A^2 + \mathrm{EN}\left| \chi_s^q \right\rangle \qquad (10.13)$$

b. Symmetry-Breaking Inhomogeneous Magnetic Field. In the general case, even if the magnetic field is static

$$H_x(x, y, z) \neq 0, \quad H_y(x, y, z) \neq 0, \quad H_z(x, y, z) \neq 0 \qquad (10.14)$$

Of course, in this case different segments of a linear chain are affected by different magnetic field strengths and therefore the original periodic symmetry is broken. If, however, the field strength along the chain is only a slowly varying function of z, one can try to define different average field strengths for different segments (20–30 units) of the chain. If this still remains a resonable approximation, one must treat the interfaces between these different segments (see Figure 10.1). For this purpose one can apply a combination of the one-particle Green matrix equation (the Dyson equation; see Section 4.1) with the mutually consistent field (MCF) procedure (see Section 6.2).

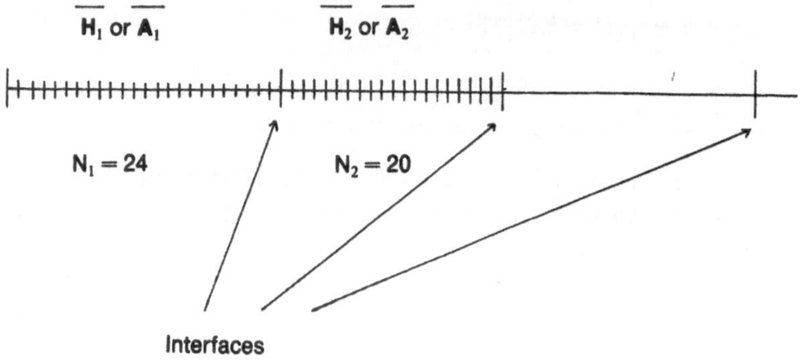

FIGURE 10.1. A linear chain with different segments to which correspond different average magnetic field strengths $\bar{\mathbf{H}}_i$. The number of units in the segments is not necessarily equal (in this example $N_1 = 24$, $N_2 = 20$). The interfaces between the different segments are indicated by arrows.

On applying the above approximation we can set

$$\tilde{\hat{H}}_1^N = -\tfrac{1}{2}\Delta \underbrace{-\frac{1}{2ic}\operatorname{div}\bar{\mathbf{A}}_1}_{0} - \frac{1}{2ic}\bar{\mathbf{A}}_1 \operatorname{grad} + \frac{1}{2c^2}\bar{A}_1^2 - \sum_{q_1=1}^{N_1}\sum_{\alpha=1}^{M}\frac{Z_\alpha}{|\mathbf{r}-\mathbf{R}_\alpha^{q_1}|}$$

$$\overline{\mathbf{H}}_1 = \operatorname{curl}\bar{\mathbf{A}}_1 \tag{10.15a}$$

and

$$\tilde{\hat{H}}_2^N = -\tfrac{1}{2}\Delta \underbrace{-\frac{1}{2ic}\operatorname{div}\bar{\mathbf{A}}_2}_{0} - \frac{1}{2ic}\bar{\mathbf{A}}_2 \operatorname{grad} + \frac{1}{2c^2}\bar{A}_2^2 - \sum_{q_1=1}^{N_2}\sum_{\alpha=1}^{M}\frac{Z_\alpha}{|\mathbf{r}-\mathbf{R}_\alpha^{q_1}|}$$

$$\overline{\mathbf{H}}_2 = \operatorname{curl}\bar{\mathbf{A}}_2 \tag{10.15b}$$

for the one-electron parts of two neighboring segments with average vector potentials $\bar{\mathbf{A}}_1$ and $\bar{\mathbf{A}}_2$, respectively. The corresponding Fock operators and matrices of the noninteracting segments will then be

$$\hat{F}_1^{[0]} = \tilde{\hat{H}}_1^N + \hat{W}^{[0]}, \quad \mathbf{F}_1^{[0]} = \tilde{\mathbf{H}}_1^N + \mathbf{W}^{[0]} \tag{10.16a}$$

and

$$\hat{F}_2^{[0]} = \tilde{\hat{H}}_2^N + \hat{W}^{[0]}, \quad \mathbf{F}_2^{[0]} = \tilde{\mathbf{H}}_2^N + \mathbf{W}^{[0]} \tag{10.16b}$$

respectively. According to Section 1.1, the Fock matrices provide the starting eigenvalues and eigenvectors, $\varepsilon_{1,l}^{[0]}(k)$, $\mathbf{d}_{1,l}^{[0]}(k)$ and $\varepsilon_{2,l}^{[0]}(k)$, $\mathbf{d}_{2,l}^{[0]}(k)$, respectively, for both systems. (Superscript [0] indicates the starting matrices of segments 1 and 2.)

After the Fock matrices $\mathbf{F}_i^{[0]}(k)$ $(i = 1, 2)$ have been obtained for both unperturbed segments, the corresponding Green matrices can be expressed as

$$\mathbf{G}_1^{0\,[0]}(z, k) = \frac{1}{z\mathbf{S}(k) - \mathbf{F}_1^{[0]}(k)} \tag{10.17a}$$

and

$$\mathbf{G}_2^{0\,[0]}(z, k) = \frac{1}{z\mathbf{S}(k) - \mathbf{F}_2^{[0]}(k)} \tag{10.17b}$$

where $z = E + \varepsilon i$, $\varepsilon \geqslant 0$. These Green matrices can be substituted into the Dyson equation to yield

$$\mathbf{G} = \mathbf{G}^0 + \mathbf{G}^0\mathbf{V}\mathbf{G} \tag{10.18a}$$

and

$$\mathbf{G} = (\mathbf{1} - \mathbf{G}^0\mathbf{V})^{-1}\mathbf{G}^0 \tag{10.18b}$$

respectively. Here, initially,

$$\mathbf{V}_1^{[0]} = \tilde{\mathbf{H}}_2^N(\bar{\mathbf{H}}_2) - \tilde{\mathbf{H}}_1^N(\bar{\mathbf{H}}_1) \tag{10.19a}$$

and

$$\mathbf{V}_2^{[0]} = \tilde{\mathbf{H}}_1^N(\bar{\mathbf{H}}_1) - \tilde{\mathbf{H}}_2(\bar{\mathbf{H}}_2^N) \tag{10.19b}$$

are the differences in the one-electron parts of the Fock matrices in the two segments owing to the different average fields $\bar{\mathbf{H}}_2$ and $\bar{\mathbf{H}}_1$, respectively. The dimensions of the nonzero parts of matrices $\mathbf{V}_1^{[0]}$ and $\mathbf{V}_2^{[0]}$ is $m(\tilde{N}_1 \times \tilde{N}_1)$ and $m(\tilde{N}_2 \times \tilde{N}_2)$, where $\tilde{N}_1 \ll N_1$ and $\tilde{N}_2 \ll N_2$ are the number of cells, respectively, for which, in one segment, the effect of the other segment with a different average field strength is taken into account while m is the number of orbitals in the unit cell. (Since $\tilde{N}_1$ is not necessarily equal to $\tilde{N}_2$, the dimensions of the nonzero parts of $\mathbf{V}_1^{[0]}$ and $\mathbf{V}_2^{[0]}$ is also generally not equal, therefore $\mathbf{V}_1^{[0]} \neq -\mathbf{V}_2^{[0]}$.) With these definitions of matrices $\mathbf{V}_1^{[0]}$ and $\mathbf{V}_2^{[0]}$ equation (10.18b) enables us to write

$$\mathbf{G}_1^{[0]} = (\mathbf{1} - \mathbf{G}_1^{0\,[0]}\mathbf{V}_1^{[0]})^{-1}\mathbf{G}_1^{0\,[0]} \tag{10.20a}$$

and

$$\mathbf{G}_2^{[0]} = (\mathbf{1} - \mathbf{G}_2^{0\,[0]}\mathbf{V}_2^{[0]})^{-1}\mathbf{G}_2^{0\,[0]} \tag{10.20b}$$

The forms of $\mathbf{V}_1^{[0]}$ and $\mathbf{V}_2^{[0]}$ are, respectively,

$$\mathbf{V}_1^{[0]} = \left[\begin{array}{cc} \begin{array}{c} \times\times\cdots\times \\ \vdots \\ \times\times\cdots\times \end{array} \;\; \tilde{N}_1 m & \mathbf{0} \\ \mathbf{0} & \mathbf{0} \end{array}\right] \updownarrow N_1 m \quad \text{and}$$

$$\mathbf{V}_2^{[0]} = \left[\begin{array}{cc} \begin{array}{c} \times\times\cdots\times \\ \vdots \\ \times\times\cdots\times \end{array} \;\; \tilde{N}_2 m & \mathbf{0} \\ \mathbf{0} & \mathbf{0} \end{array}\right] \updownarrow N_2 m \tag{10.21}$$

Hence one can also partition the corresponding Green matrices $\mathbf{G}_i^0$ $(i = 1, 2)$ as

$$\mathbf{G}_i^{(0)} = \begin{pmatrix} \mathbf{G}_{i,00}^0 & \mathbf{G}_{0.B_i}^0 \\ \mathbf{G}_{B_i,0}^0 & \mathbf{G}_{B_i,B_i}^0 \end{pmatrix} \qquad (i = 1, 2) \tag{10.22}$$

where $\mathbf{G}_{i,00}^0$ denotes that part of the Green matrix $\mathbf{G}_i^0$ which corresponds to the nonzero parts of $\mathbf{V}_i^{[0]}$.

After these preparations one can develop the following numerical procedure to obtain a mutually consistent solution for the cells in the neighborhood of the interface.

1. For the actual calculations we need only the submatrices $\mathbf{G}_{i,00}^0$, so we can write for their elements [see equation (4.109) in Section 4.6.1 and work of Del Re and Ladik[1] and Seel *et al.*[2]]

$$[\mathbf{G}_{i,00}^{0\,[0]}(z, k)]_{a,p,bq} = \frac{1}{2\pi} \sum_{l=1}^{n_i^*} \int \frac{d_{i,l;a}^{[0]}(k) d_{i,l;b}^{[0]}(k)}{z - \varepsilon_{i,l}^{[0]}(k)} e^{\tilde{\imath}k(p-q)} dk$$

$$(i = 1, 2), \qquad \tilde{\imath} = \sqrt{-1} \tag{10.23}$$

where a and b are orbital indices and p and q are cell indices. In this way one can construct the starting matrices $\mathbf{G}_{1,00}^{0\,[0]}$ and $\mathbf{G}_{2,00}^{0\,[0]}$.

2. Starting values for $\mathbf{V}_1$ and $\mathbf{V}_2$ ($\mathbf{V}_1^{[0]}$ and $\mathbf{V}_2^{[0]}$) can be obtained by substituting for $\mathbf{H}_i(q)$ in the qth cell from the interface in segments 1 and 2, respectively,

$$\mathbf{H}_2(q) = [\exp(-\alpha_2 q a_z)]\bar{\mathbf{H}}_2 \tag{10.24a}$$

and

$$\mathbf{H}_1(q) = [\exp(-\alpha_1 q a_z)]\bar{\mathbf{H}}_1 \tag{10.24b}$$

where α_i are positive parameters whose values depend on the difference between $\bar{\mathbf{H}}_2$ and $\bar{\mathbf{H}}_1$. To start the iterations one begins with $\mathbf{P}_{i,00}^{[0]}$ $(i = 1, 2)$, charge-bond-order matrix blocks, which can be computed in the standard way (see Section 1.1) with the aid of band structure calculations by assuming an infinite chain in the constant magnetic fields $\bar{\mathbf{H}}_1$ and $\bar{\mathbf{H}}_2$, respectively. [We note that the eigenvalues and eigenvector components in equation (10.23) should be determined from such band structure calculations in the presence of these homogeneous magnetic fields.)

3. The starting quantities $\mathbf{G}_{i,00}^{0\,[0]}(z, k)$ and $\mathbf{V}_{i.00}^{[0]}$ enable us to solve the Dyson equations

$$\mathbf{G}_{1,00}^{[0]}(z, k) = (\mathbf{1}_{00} - \mathbf{G}_{1,00}^{0\,[0]}\mathbf{V}_{1,00}^{[0]})^{-1}\mathbf{G}_{1,00}^{0\,[0]} \tag{10.25a}$$

and

$$\mathbf{G}_{2,00}^{[0]}(z,k) = (\mathbf{1}_{00} - \mathbf{G}_{2,00}^{0\,[0]}\mathbf{V}_{2,00}^{[0]})^{-1}\mathbf{G}_{2,00}^{0\,[0]} \tag{10.25b}$$

For this purpose one must select a dense mesh of E within the bulk bands and for the extra states in the band gaps one can use the simple criteria [see equation (4.114) in Section 4.6.1]

$$D_1(E) = \det[\mathbf{1}_{00} - \mathbf{G}_{1,00}^{0\,[0]}\mathbf{V}_{1,00}^{[0]}] = 0 \tag{10.26a}$$

and

$$D_2(E) = \det[\mathbf{1}_{00} - \mathbf{G}_{2,00}^{0\,[0]}\mathbf{V}_{2,00}^{[0]}] = 0 \tag{10.26b}$$

4. Finally, one computes the new $\mathbf{P}_i^{[1]}$ matrices for the next iteration step from the equation

$$\mathbf{P}_i^{[1]} = -\frac{1}{\pi}\int_{-\infty}^{E_F} \operatorname{Im}\mathbf{G}_i(E)dE \tag{10.27}$$

For this purpose we take into account both the bound (gap) state and the band region contributions. The former can be computed using equation (4.121) or (4.126a) together with the normalization condition (4.129) (see Section 4.6.1 and work of Seel *et al.*[1,3]). The band state contributions can be obtained directly from equation (10.27) using a nonequidistant dense mesh (using more points near the band edges). After obtaining matrices $\mathbf{P}_i^{[1]}$ one can construct the two-electron parts $\mathbf{W}_i^{[1]}$ of the Fock matrices and the Fock matrices $\mathbf{F}_i^{[1]}$ themselves in the next iteration step. One should point out that, owing to equations (10.25) and (10.27), the effect of one segment of the chain with a different average magnetic field strength may cause a change also in the two-electron part of the Fock matrix in the other segment of the chain, and *vice versa*. Therefore, generally $\mathbf{W}_i^{[t]} \neq \mathbf{W}_i^{[0]}$, $\mathbf{F}_i^{[t]} \neq \mathbf{F}_i^{[0]}$, $\mathbf{V}_i^{[t]} \neq \mathbf{V}_i^{[0]}$ and $\tilde{N}_i^{[t]} \neq \tilde{N}_i^{[0]}$ $(i = 1, 2)$ where $[t]$ means the tth iteration step. In other words, though initially one restricts the effect of one segment of the chain on the other to the one-electron parts of the Fock operator, during the iterations also their two-electron parts will be affected and, furthermore, the number of cells $\tilde{N}_i$ counted from the interface of the two segments may change during the iterations. Therefore one can write

$$\mathbf{V}_1^{[t]} = \tilde{\mathbf{H}}_2^{N\,[0]}(\mathbf{H}_2) - \tilde{\mathbf{H}}_1^{N}(\bar{\mathbf{H}}_1) + \mathbf{W}_2^{[t]} - \mathbf{W}_1^{[t]} \qquad (\mathbf{W}_2^{[0]} = \mathbf{W}_1^{[0]} = \mathbf{W}^{[0]}) \tag{10.28}$$

and a similar expression for $\mathbf{V}_2^{[t]}$.

The iteration procedure can be summarized by the following sketch:

$$\begin{aligned}
&\mathbf{P}^{[0]} \rightarrow \mathbf{W}^{[0]},\\
&\tilde{\mathbf{H}}_i^N \rightarrow \mathbf{F}_i^{[0]}(k) \rightarrow \mathbf{d}_{i,l}^{[0]}(k),\\
&\varepsilon_{i,l}^{[0]}(k) \rightarrow \mathbf{G}_i^{0\,[0]}(z,k),\\
&\mathbf{V}_i^{[0]} \rightarrow \mathbf{G}_i^{[0]} \rightarrow \mathbf{P}_i^{[1]} \rightarrow \mathbf{W}_i^{[1]} \rightarrow \mathbf{F}_i^{[1]}(k) \rightarrow \mathbf{d}_{i,l}^{[1]}(k),\\
&\varepsilon_{i,l}^{[1]}(k) \rightarrow \mathbf{G}_i^{0\,[1]}(z,k),\\
&\mathbf{V}_i^{[1]} \rightarrow \mathbf{G}_i^{[1]}(z,k) \rightarrow \cdots \rightarrow \mathbf{P}_i^{[\mathrm{MCF}]} \rightarrow \mathbf{W}_i^{[\mathrm{MCF}]} \rightarrow \mathbf{F}_i(k)^{[\mathrm{MCF}]} \rightarrow \mathbf{G}_i^{0\,[\mathrm{MCF}]}(z,k),\\
&\mathbf{V}_i^{[\mathrm{MCF}]} \rightarrow \mathbf{G}_i^{[\mathrm{MCF}]}(z,k) \rightarrow \mathbf{P}_i^{[\mathrm{MCF}]} \rightarrow \mathbf{F}_i^{[\mathrm{MCF}]}(k) \rightarrow \mathbf{d}_{i,l}^{[\mathrm{MCF}]}(k),\\
&\varepsilon_{i,l}^{[\mathrm{MCF}]}(k) \qquad (i = 1, 2)
\end{aligned} \tag{10.29}$$

In this iteration scheme the superscript [MCF] denotes the converged (mutually consistent) solutions for both segments, in which case all the elements of both charge-bond-order matrices do not differ more than a prescribed threshold in two subsequent iteration steps.

It should be noted that the above procedure could be easily modified for the case in which two different polymer chains are coupled. Here, of course, the two-electron part of the Fock operator also differs from the very beginning and the overlap matrices are different:

$$\mathbf{V}_1^{[0]}(k) = \mathbf{H}_2^N - \mathbf{H}_1^N + \mathbf{W}_2^{[0]} - \mathbf{W}_1^{[0]} - z[\mathbf{S}_2(k) - \mathbf{S}_1(k)] \tag{10.30}$$

with a similar expression for $\mathbf{V}_2^{[0]}(k)$.

In the case of two interacting chains, or one chain with various segments corresponding to different average magnetic fields, if one performs the above iterative procedure, one expects not an abrupt change in the bands at the interface (which would occur without allowing the two chains to interact) but a continuous variation in the band limits around the interface (see Figure 10.2).

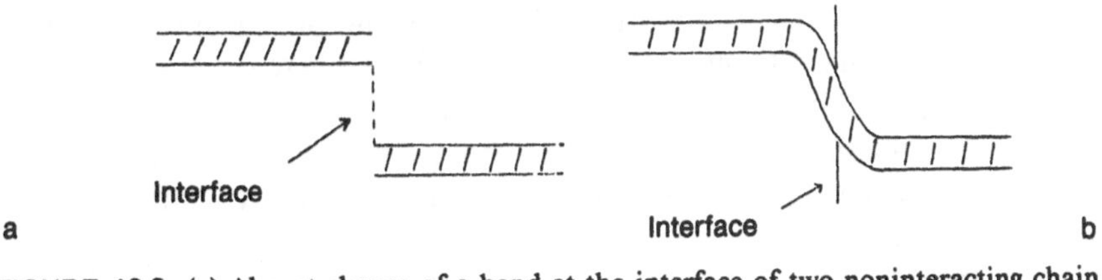

FIGURE 10.2. (a) Abrupt change of a band at the interface of two noninteracting chain segments. (b) Continuous variation in a band ("bending band") at the interface of two interacting chain segments.

10.1.2. Some General Remarks about the Theory of the Effects of Magnetic Fields on Polymers

In the case of time-dependent (nonstationary) magnetic fields, one cannot derive Hartree–Fock equations by varying the expectation value of the total Hamiltonian (the total energy is no longer constant) but one must vary the more complicated functional[4,5]

$$\tilde{J} = \frac{1}{T}\int_0^T dt \frac{\left\langle \Phi \middle| \hat{H} - i\frac{\partial}{\partial t} \middle| \Phi \right\rangle}{\langle \Phi \mid \Phi \rangle} \tag{10.31}$$

where Φ is a Slater determinant constructed from time-dependent one-electron functions and possesses a time-dependent exponential prefactor:

$$\Phi = (\mathbf{r}_1, \mathbf{r}_2, \ldots, \mathbf{r}_n, t) = e^{-iW_0 t}\hat{A} \prod_{i=1}^{n} \Phi_i(\mathbf{r}_i, t) \tag{10.32}$$

Further, W_0 is the magnetic-field-free total energy of the system, n is the number of electrons, and $\hat{A}$ is the antisymmetrizer operator. In formulating sketch (10.29) it was assumed that the magnetic field strength has a constant period $T = 2\pi/\omega$, which allows one to form the time average in equation (10.31). If this were not the case, the problem would become extremely difficult.[4,5]

The variation $\delta\tilde{J} = 0$ yields* the so-called time-dependent coupled Hartree–Fock equations.[6,7] When the field does not break the periodic symmetry of the polymer, these equations are already formulated for a polymer.[8] Since, however, the different cases of the effects of stationary magnetic fields on the band structure of a polymer have not yet been programmed, and therefore no calculations are available for this problem, we shall not give here the rather complicated equations obtained for the case of nonstationary magnetic fields.

Finally, we note that when band-structure (and wave function) calculations will be available for polymers in the presence of stationary and nonstationary magnetic fields, it will be possible to compute both the current induced by the magnetic field and the static and frequency-dependent magnetic susceptibilities. Since, however, no such calculations for polymers (even in a very approximate form) are available in the

* Since the operator $\hat{H} - i\partial/\partial t$ is not bounded, varying the functional $\tilde{J}$ generally yields only a stationary condition and not necessarily a minimum. However, this does not cause any problem because $\hat{H} - i\partial/\partial t$ is not an energy operator.[4]

literature, we do not give here the necessary formalism but refer to the solid-state physical literature.[9] Nevertheless, a great deal of experimental work has been conducted on the effect of magnetic fields on different organic polymers, especially biopolymers (a review is given elsewhere[10]). Hence it is rather urgent that the theory be developed further so that the various interesting experimental results obtained by numerous investigators[10] can be interpreted.

10.2. ELECTRIC POLARIZABILITIES OF POLYMERS

10.2.1. Theoretical Methods

Calculation of the polarizability tensors, defined by the equation

$$\boldsymbol{\mu}^{\text{ind}} = \boldsymbol{\alpha}\mathbf{E} \tag{10.33}$$

for polymers, as well as determination of their hyperpolarizability tensors* given by the equation

$$\begin{aligned}\boldsymbol{\mu}^{\text{ind}\prime} &= \boldsymbol{\alpha}\mathbf{E} + \tfrac{1}{2}\boldsymbol{\beta}\mathbf{E}^2 + \frac{1}{3!}\gamma E^2\mathbf{E} + \cdots \\ &= \sum_{t=1}^{n} \frac{1}{t!}\boldsymbol{\alpha}_t\mathbf{E}^t\end{aligned} \tag{10.34}$$

possess great importance in finding new organic electro-optic and nonlinear optical materials. For high-electric-field lasers one must set $n = 4$ or 5 in order to describe the relevant phenomena.

In principle, these calculations could be performed by solving the Hartree–Fock equations with the modified Fock operator,

$$\hat{F}' = \hat{F} - |e|(\mathbf{rE}) \tag{10.35}$$

where $\hat{F}$ is the Fock operator of the system in the absence of the electric field and $-|e|(\mathbf{rE}) = \boldsymbol{\mu}^{\text{ind}}\mathbf{E}$ is the perturbation caused by it.

In the case of long (infinite) chains, the difficulty arises that the component along the long (say z) axis of

$$-|e|(\mathbf{rE})_z = -|e|zE_z \tag{10.36}$$

* The elements of hyperpolarizability tensor $\boldsymbol{\beta}$ and of second-order hyperpolarizability tensor γ have 3 or 4 indices, respectively.

goes to infinity as $z \to \infty$. Some authors[11,12] avoided this problem by attempting to introduce zE_z in the form shown in Figure 10.3. This seems, however, to be a rather artificial way of circumventing this essentially still unsolved difficulty.

The first calculations of the π-electron polarizabilities of long-chain molecules employed the free-electron model and simple Hückel theory.[11,12] In both cases the computed polarizability tensor elements were proportional to the cube of the length of the chain, $\alpha_{k,l} \sim l^3$. In these calculations (and in the calculation of $\boldsymbol{\beta}$ by Zyss,[13] who showed also a definite dependence of the elements of $\boldsymbol{\beta}$ on chain length), no electron–electron interaction was taken into account. In order to take Coulomb interaction at least partially into account one must perform Hartree–Fock calculations on the *ab initio* level. Such calculations would be expected to affect (most probably decrease somewhat) the dependence on l.

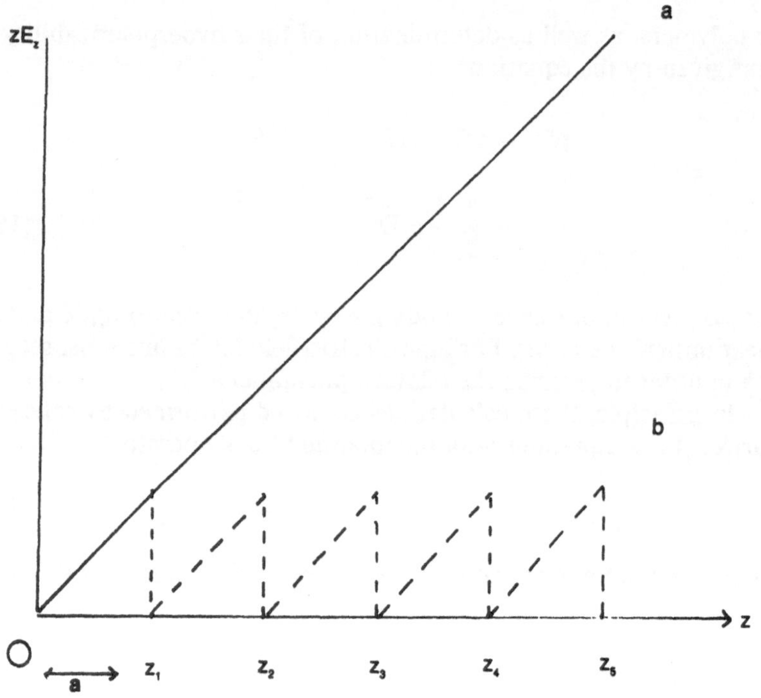

FIGURE 10.3. (a) The term zE_z at constant E_z. (b) The term zE_z by choosing its form on the basis of the argument that one can choose the reference (0) cell anywhere in an infinite chain, where a is the elementary translation in the periodic quasi-one-dimensional polymer.

Unfortunately, the elements of the polarizability and hyperpolarizability tensors are very sensitive to the choice of basis set and therefore high-quality basis sets are needed to obtain reliable results. It has been shown, however, on a series of large hydrocarbons, that if a minimal basis set is used one can obtain rather good agreement with experiment (in cases for which experimental data are available) if one scales the minimal basis results with respect to results obtained with better (double-ζ) basis sets computed for the first members of the hydrocarbon series.[14] This is true, however, only for the physically interesting longitudinal component of $\boldsymbol{\alpha}$, namely α_{zz}, and not for the other elements, owing to the lack of flexibility of the basis sets in the other two directions.

Direct HF CO calculations on an infinite chain in the presence of an electric field are not feasible because, as mentioned above, the perturbation term $-|e|(\mathbf{rE})$ is unbounded. Therefore, finite chain length *ab initio* calculations have recently been performed for different unsaturated hydrocarbon series.[15,16] In the course of the work, STO-3G minimal and in some cases split valence 4-31G basis sets have been used to solve the modified HF equations,

$$\hat{F}'(\mathbf{E})\psi_m(\mathbf{E}) = \varepsilon_m(\mathbf{E})\psi_m(\mathbf{E}) \tag{10.37}$$

The resulting one-electron wave functions were used to compute the induced dipole moments of the chains,

$$\boldsymbol{\mu}^{\mathrm{ind}\prime} = 2 \sum_{m=1}^{n'} \langle \psi_m(\mathbf{E}) \,|\, \hat{\boldsymbol{\mu}} \,|\, \psi_m(\mathbf{E}) \rangle \tag{10.38}$$

where n' is the number of doubly filled orbitals and $\hat{\boldsymbol{\mu}}$ the dipole moment operator.

If equation (10.34), or the corresponding equation for the components of $\boldsymbol{\mu}^{\mathrm{ind}\prime}$

$$\mu_i^{\mathrm{ind}\prime} = \sum_{j=1}^{3} \alpha_{ij} E_j + \frac{1}{2} \sum_{j,k=1}^{3} \beta_{ijk} E_j E_k + \frac{1}{3!} \sum_{j,k,l=1}^{3} \gamma_{yijkl} E_j E_k E_l \qquad (i = 1, 2, 3) \tag{10.39}$$

is taken into account, the elements of the polarizability and the different hyperpolarizability tensors can be obtained in the form

$$\alpha_{ij} = \left.\frac{\partial \mu_i^{\mathrm{ind}\prime}}{\partial E_j}\right|_{\mathbf{E}=\mathbf{0}} \tag{10.40a}$$

$$\beta_{ijk} = \frac{1}{2} \frac{\partial^2 \mu_i^{\text{ind}\prime}}{\partial E_j \partial E_k}\bigg|_{\mathbf{E}=0} \tag{10.40b}$$

$$\gamma_{ijkl} = \frac{1}{3!} \frac{\partial^3 \mu_i^{\text{ind}}}{\partial E_j \partial E_k \partial E_l}\bigg|_{\mathbf{E}=0} \tag{10.40c}$$

In order to calculate α_{zz}, numerical differentiation in equation (10.40a) was applied[15,16] because it is very difficult to express analytically $\mu_i^{\text{ind}\prime}$ as a function of E from equation (10.38). This numerical approach, however, requires a much higher accuracy for the total HF energy than in other applications. Therefore, the convergence threshold for the density matrix elements was set at 10^{-5}–10^{-6}. Having determined $\mu_i^{\text{ind}\prime}$ at different field strengths, equation (10.40a) can be approximated by the numerical difference quotient for α_{zz}:

$$\alpha_{zz} = \frac{[\mu_z^{\text{ind}\prime}(\mathbf{E}) - \mu_z^{\text{ind}\prime}(-E_z)]}{2E_z} \tag{10.41}$$

In actual calculations four values of the external field have been applied: $E_z = -0.002$, -0.001, 0.001, and 0.002 au.[15] The numerical stability of the result was checked by the Richardson extrapolation procedure[17] coupled with the Romberg algorithm[18] using the above four values of E_z. In this way the values obtained for α_{zz} can be improved, and one can also detect numerical instabilities which in most cases are caused by insufficiently accurate calculated values of $\mu_z^{\text{ind}\prime}(E_z)$.

Another way of obtaining the polarizability tensor elements would be to consider $-|e|(\mathbf{rE})$ or, for the total Hamiltonian, $-|e|\sum_{i=1}^{n}(\mathbf{r}_i\mathbf{E})$ as a perturbation and obtain the elements of $\boldsymbol{\alpha}$, $\boldsymbol{\beta}$, $\boldsymbol{\gamma}$, etc., at different orders of perturbation theory. However, such an approach seems rather tedious. The perturbation theory would, of course, provide only $\boldsymbol{\mu}^{\text{ind}\prime}$ at different orders and equations (10.40) and (10.41) would still have to be applied.

10.2.2. Numerical Applications

Bodart *et al.*[15] first performed both STO-3G and 4-31G *ab initio* calculations for α_{zz} of acetylene and ethylene. The 4-31G values (27.9280 and 32.4075 au) agree quite well with experiment (30.6 and 36.4 au, respectively[19]) while the STO-3G values are far too small (18.1793 and 19.5983 au, respectively). Their next step was to calculate four members of the ethylene and acetylene series using both the STO-3G and 4-31G bases. Table 10.1 presents only their 4-31G results together with the scaling factor (see the previous section) $f = \alpha_{zz}(\text{STO-3G})/\alpha_{zz}(\text{4-31G})$.

TABLE 10.1. Calculated Values of α_{zz} for the First Four Members of the Ethylene H–(CH═CH)$_n$–H] and Acetylene [H–(C≡C)$_n$–H] Series (in au) (after Bodart *et al.*[15])

Compound	α_{zz} (with 4-31G basis)	f
Ethylene	29.1906	0.60
1,3-*trans* Butadiene	73.3961	0.65
1,3,5-*trans* Hexatriene	152.6748	0.68
1,3,5,7-*trans* Octatetraene	246.1033	0.71
Acetylene	27.9280	0.65
1,3-Butadiyne	73.9655	0.69
1,3,5-Hexatriyene	131.3900	0.69
1,3,5,7-Octatetrayne	207.5293	0.70

In connection with Table 10.1, it should first be noted that the different 4-31G values of α_{zz} found for ethylene are due to the fact that the direction of the C═C bond lies at 30° to the z axis in this calculation, while the previously cited larger value of α_{zz} was obtained assuming that the C═C bond lies in the direction of the z axis. Further, the α_{zz} data exhibit a stronger than linear chain-length dependence, but one that is still much weaker than l^3 obtained in the early calculations without electron–electron interaction.[11,12] Finally, one should note that the scaling factor f varies from $n = 2$ to $n = 4$ between 0.65 and 0.71 in the ethylene series and between 0.65 and 0.70 in the acetylene series (already starting from $n = 1$). Since there is only a small difference in f in both cases if one proceeds from $n = 3$ to $n = 4$, it appears safe to use an f value of 0.70 for various other series of unsaturated hydrocarbons (or for further members of the already discussed two series) to scale the STO-3G minimal basis results.[15]

It should be further noted that the factor between the 4-31G and experimental α_{zz} values for acetylene and ethylene is 0.91 (for no other unsaturated hydrocarbons are experimental results available). Assuming that this value remains reasonably constant for other unsaturated hydrocarbons as well, one can define the combined factor $f' = 1.57 = 1/(0.7 \times 0.9)$ to predict from STO-3G anticipated experimental α_{zz} values: $\alpha_{zz}(\text{exp}) = \alpha_{zz}(\text{STO-3G}) \times 1.57$. Bodart *et al.*[15] employed this procedure to scale their STO-3G α_{zz} values both in the vinylacetylene series (proceeding to $n = 3$) and for a larger number of other unsaturated hydrocarbons (details are available elsewhere[15]). In a subsequent paper Delhalle *et al.*[16] also applied the same scaling procedure to the allene, diallene, and triallene series. They reported that the calculated α_{zz} values exhibit extreme sensitivity to the geometry used. This must be taken into account in order to perform a successful scaling procedure, which can be tested finally only when more experimental data become available.

One should point out, however, that the described procedures only pertain to an early stage in the theory of polarizabilities and hyperpolarizabilities of polymers. Further theoretical research is required to solve the problem of the unbounded perturbation operator in equation (10.35) which blocks direct CO calculations. Moreover, the Romberg algorithm[18] could be employed not only to test the accuracy of the numerical differentiation (10.41), but also to extrapolate the α_{zz} (and other polarizability and hyperpolarizability tensor element) values obtained for a few terms in a series of similar molecules for large n ($n \rightarrow \infty$). Our first results for the band structure of a $(\text{-H-H H-H-})_n$ chain have been very promising in this respect.[20]

In calculating α_{zz}, one should not assume only the oversimplified case $\mu_z^{\text{ind}\prime}(E_z)$, but instead one should use the more general $\mu_z^{\text{ind}\prime}(\mathbf{E})$ dependence on the basis of equation (10.39).

Finally, there remains the open question,[15] whether if one joins different chains α_{zz} decreases or increases (assuming the same length for a homogeneous chain or for two coupled chains). To answer this question the MCF–Green matrix technique for the treatment of interfaces (described in Section 10.1 for a chain with segments affected by different average magnetic field strengths) should be appropriately modified.

10.3. MECHANICAL PROPERTIES OF POLYMERS

Different mechanical properties of polymers, especially those which serve as structural materials, are of rapidly growing interest. Unfortunately, not very much theoretical work has yet been conducted in this field. The next section contains a summary of the existing material.

10.3.1. Theoretical Considerations

The bulk (Young's modulus) of a quasi-one-dimensional polymer chain is given by[21]

$$B = \tfrac{1}{3}(c_{11} + 2c_{12}) = \frac{1}{A}\left(\frac{\partial^2 E}{\partial a^2}\right)_{a=a_0} = \frac{1}{A}\left(\frac{\partial F_z}{\partial a}\right)_{a=a_0} \qquad (10.42)$$

where E is the total energy per unit cell, a_0 the elementary translation, A

the average cross section of the monomer (perpendicular to the direction of the long axis of the chain), and F_z denotes the z-component of the force.

For a two-dimensional polymeric system, one can define the shear modulus

$$C^1 = \tfrac{1}{2}(c_{11} - c_{12}) \sim \left(\frac{\partial^2 E}{\partial a \partial b}\right)_{\substack{a=a_0 \\ b=b_0}} \qquad (10.43)$$

where b is the lattice constant in the second direction. Equations (10.42) and (10.43) together give the longitudinal and shear elastic constants, c_{11} and c_{12}, respectively.[21]

If the elongation $\varepsilon = \delta a/a_0$ is very small, instead of recalculating the polymeric band structure at different lattice constants one can use perturbation theory (as in Beleznay *et al.*,[22] where besides the deformation potential the bulk modulus has also been computed). If, however, $\varepsilon \geqslant 0.05$, one has to recalculate the band structure at different values of a and the second derivative of the total energy per unit must be determined by a polynomial fit of the energy values obtained near equilibrium.[23] Furthermore, at larger elongations (e.g., at the dissociation of molecules) correlation effects obviously become more and more important. In the case of polyethylene, for the latter purpose its quasi-particle band structure has been determined at different lattice constants using the MP/2 many-body perturbation theory (MBPT) as described in Section 5.3. If $\varepsilon > 0.1$, it even becomes questionable whether a Slater determinant constructed from restricted HF orbitals is a good enough reference state for the MBPT calculations instead of a determinant constructed from UHF (DODS) orbitals.

It should also be noted that the results of the correlation calculations (as noted earlier in Chapter 5), especially the change in the total energy per unit cell including correlation effects, depend strongly on the choice of basis set. It can be estimated that MP/2 MBPT gives about 50% of the total valence-shell correlation applying a double ζ basis set (core–core correlations are not important for mechanical properties while the core–valence correlation itself comprises only about 2% of the valence-shell correlation and practically does not vary along an energy surface).[23] The inclusion of polarization functions increases the valence-shell correlation using MP/2 to about 70–75% of the total valence correlation. For this reason, if $\varepsilon > 0.2$ they must also be taken into account in the calculations.

An *ab initio* program has been developed in Erlangen for computing the band structure of a two-dimensional system with an arbitrary number of orbitals in the unit cell and arbitrary symmetry (applying a nonlocal exchange term), which is necessary to determine the shear modulus. This program has been tested for graphite and for a two-dimensional polyglycine antiparallel β-pleated sheet structure,[24] and is now being further developed to determine the correlated (QP) band structure of an arbitrary two-dimensional system.[25] Calculation of the bulk modulus in both directions and of the shear modulus of the polyglycine network is in progress, but no numerical results are yet available.

There are different possibilities for calculating the cohesion energy of a polymeric crystal. On the one hand, it is possible to determine the total energy per unit cell of a chain (including correlation) and add interactions with the other chains in the other directions. These latter calculations can be performed using either perturbation theory (including the dispersion term) or, better, the mutually consistent field (MCF) method (see Section 6.2) together with the empirical expression for the dispersion energy [equation (6.34) in Section 6.2]. In both cases the smallest interchain distance should be no less than 2.5 Å to obtain reasonable results. If the interchain distances are smaller, two-dimensional segments of the polymeric crystal should be selected for which two-dimensional direct CO calculations need to be undertaken with the aid of the above-mentioned programs. After the planned extension of this program has been completed to three dimensions, the total energy per unit cell of the three-dimensional polymeric crystal will be determined directly. Subtraction from this of the total energy of the monomer unit will yield the cohesion energy. One should note, however, that until now only interaction energies of alternating *trans*-polyacetylene (PA) chains have been computed by taking the interactions in both directions. For this purpose, since the general two-dimensional *ab initio* CO program was not yet available, Suhai[26] performed superchain calculations involving three chains together in the two different planes. Since, however, these minimal basis calculations did not contain correlation effects and the superchain calculations in this case do not provide satisfactorily the effects of the second and third dimensions in PA, his numerical results will not be discussed here.

10.3.2. Application to Polyethylene

The only system for which detailed investigations of the longitudinal Young modulus are available is polyethylene (PE). The most detailed study of this problem was performed by Suhai.[23] His first step in calculating the elastic modulus was to optimize the geometrical structure of PE at both the HF and correlated levels. In stretching the polymer he kept it planar and reoptimized its structural parameters for the values $\varepsilon = 0.02, 0.04, 0.06, 0.10, 0.15$, and 0.20, first in the case of HF/STO-3G. He found that the geometrical changes within the methylene group affect only to a very small extent the energetic changes taking place within the carbon chain (owing to the smallness of the corresponding coupling force constants). Therefore, in the case of the extended basis set as well as correlation energy calculations he fixed the coordinates of the methylene group at the appropriate HF/STO-3G value.

The theoretical values of the Young 'modulus of PE calculated at different levels of accuracy in his work are listed in Table 10.2 together with earlier values obtained by other (semiempirical or *ab initio*) quantum-mechanical methods. The value $A = 18.24$ Å^2 suggested by Bunn[30] was introduced for the cross-sectional area of the chain. Analysis of these results leads to the following observations: (1) The relatively high MNDO value reflects the well-known tendency of this method to overestimate force constants and vibrational frequencies. (2) As mentioned above, there is very good agreement between the MO cluster difference calculations[28] and the periodic CO calculations if the same basis set is used. However, extension of the basis, especially in the valence region, substantially reduces the value of the modulus from 417 to 339 GPa. A further significant reduction occurs owing to correlation effects, leading to a theoretically predicted value of 303 GPa.

In order to be able to compare these results with experiments, data obtained from very different measurement techniques are presented in Table 10.3.[23] The particularly large deviations between the values

TABLE 10.2. Theoretical Values of the Longitudinal Elastic Modulus of Polyethylene Obtained by Various Methods Using the Molecular Orbital (MO) Approach to Clusters or the Crystal Orbital (CO) Approach to Infinite Periodic Polymers[a]

Method of calculation	Modulus (GPa)	References
MNDO-CO (semiempirical)	493.5	27
HF/STO-3G MO (C_9H_{20}–C_7H_{16}) cluster	420 ± 30	28
MF/STO-3G CO	417 (406)[b]	23 (29)
HF/STO-3G CO + correlation (MP/2)	362	23
HF/6-31G CO	339 (345)	23 (29)
HF/6-31G CO + correlation (MP/2)	303	23

[a] Calculation with different atomic basis sets at the *ab initio* Hartree–Fock (HF) level and also including electron correlation effects with second-order Møller–Plesset (MP) perturbation theory are given.

[b] The values in parentheses are from work of Karpfen.[29]

TABLE 10.3. Experimental Values of the Longitudinal Elastic Modulus of Polyethylene Obtained with Different Types of Measurements

Experimental method	Modulus (GPa)	References
X-ray diffraction	235–255	31–33
Inelastic neutron scattering	239	34
Raman spectroscopy[a]	358, 290	35–36
IR vibrational analysis[b]	257–340	37–41

[a] Observations of the longitudinal accoustic mode.
[b] Application of different empirical force fields.

obtained by vibrational analysis are due to the very different force-field parameters applied in these studies. The spectroscopically derived moduli are calculated, however, by using force constants which do not contain anharmonicity corrections and assume an ideal tetrahedral geometry. These factors may cause some differences in the comparison with the theoretically derived values. The longitudinal acoustic mode (LAM), corresponding to accordion-like vibrations of planar chains, can best be compared with the deformations calculated in the theoretical work. Since X-ray experiments show a negative thermal expansion coefficient along the crystallographic c axis of PE,[42,43] and may be accounted for by torsional rotations about C–C bonds, the modulus obtained by Raman measurements for the LAM may be shifted to somewhat lower values than those belonging to the ideally planar chains.[44]

Suhai[23] computed the first derivative of the energy with respect to the elongation (the force acting due to deformation on the polymer) and observed a nearly linear increase until $\varepsilon \approx 0.05$, but then a clear deviation from Hooke's law is evident. This deviation is about 15% at $\varepsilon \approx 0.1$, but increases very sharply for larger values of ε. Since correlation fundamentally influences this deviation, one has to investigate this region more carefully, using better basis sets and a more flexible reference state for the perturbation theory or for other methods to account for the correlation.

Finally, we note that the longitudinal elastic modulus computed or measured in PE represents only an upper bound for the values which are measured in macroscopic samples. The reason is that both the calculated and measured moduli refer to a single PE chain (in Table 10.3, only the results of such experiments are given). In reality, however, due to impurities, chain breakages, and distortions, interactions between chains with different orientations (in a macroscopic sample, the chains do not have the same orientation), etc., in macroscopic PE samples, results about two orders of magnitude smaller have been obtained by different investigators.[45] One task for the future is to take into account theoretically, at least approximately, these effects in different polymeric crystals. In this way, one will be able to compare theoretical data on different mechanical properties of real polymers with experimental data on macroscopic polymer samples.

REFERENCES

1. G. Del Re and J. Ladik, *Chem. Phys.* **49**, 32 (1980).
2. M. Seel, G. Del Re, and J. Ladik, *J. Comput. Chem.* **3**, 451 (1982).
3. M. Seel, *Int. J. Quantum Chem.* **26**, 753 (1984).
4. P.-O. Löwdin and P. K. Mukherjee, *Chem. Phys. Lett.* **14**, 1 (1972).
5. P. W. Langhoff, S. T. Epstein, and M. Karplus, *Rev. Mod. Phys.* **44**, 602 (1962).
6. A. Dalgarno and G. Victor, *Proc. R. Soc., Ser. A* **291**, 285 (1966); S. Sengupta and M. Mukherji, *J. Chem. Phys.* **47**, 260 (1967); P. K. Mukherjee and R. K. Moitra, *J. Phys. B* **11**, 2813 (1978).
7. For a relativistic generalization of the time-dependent coupled Hartree–Fock equations see: J. Ladik, J. Čížek, and P. K. Mukherjee, in: *Relativistic Effects in Atoms, Molecules and Solids* (G. L. Malli, ed.), p. 305, Plenum Press, New York–London (1983).
8. J. Ladik (unpublished results).
9. See, for instance: W. Jones and N. H. March, *Theoretical Solid State Physics*, Vol. I, pp. 393–398, Dover Publ. Inc., New York (1985).
10. G. Maret and K. Dransfeld, in: *Strong and Ultrastrong Magnetic Fields and their Applications* (F. Herbach, ed.), Topics in Applied Physics, Vol. 57, p. 143, Springer-Verlag, Berlin–Heidelberg–New York–Tokyo (1985), and references cited therein.
11. P. L. Davies, *Trans. Faraday Soc.* **48**, 789 (1952).
12. K. C. Rustagi and J. Ducuing, *Opt. Commun.* **10**, 258 (1974); H. F. Hameka, *J. Chem. Phys.* **67**, 2935 (1977).
13. J. Zyss, *J. Chem. Phys.* **71**, 909 (1979).
14. A. Chablo and A. Hinchlife, *Chem. Phys. Lett.* **72**, 149 (1980).
15. V. P. Bodart, J. Delhalle, J.-M. André, and J. Zyss, *Can. J. Chem.* **63**, 1631 (1985).
16. J. Delhalle, V. P. Bodart, M. Dory, J.-M. André, and J. Zyss, *Int. J. Quantum Chem.* **S19**, 313 (1986).
17. See, for instance: H. Rutishauser, *Numer. Math.* **5**, 48 (1963).
18. W. Romberg, *Det. Kong. Norske Videnskabers Selskab. Forhandlinger* **28**, 7 (1955).
19. H. C. Allen and E. K. Phyler, *J. Am. Chem. Soc.* **80**, 2673 (1958).
20. C.-M. Liegener, F. Beleznay, and J. Ladik, *Phys. Lett.* (accepted).
21. R. W. Keyes, *IBM J. Res. Dev.* **5**, 266 (1961).
22. F. Beleznay, G. Biczó, and J. Ladik, *Acta Phys. Hung.* **18**, 213 (1965).
23. S. Suhai, *J. Polym. Sci., Polym. Phys. Ed.* **21**, 134 (1983).
24. O. Fleck, P. Otto, and J. Ladik, *Solid State Comm.* (accepted).
25. M. Vracko, C.-M. Liegener, and J. Ladik (to appear).

26. S. Suhai (unpublished results).
27. M. J. S. Dewar, Y. Yamaguchi, and S. Huck, *Chem. Phys.* **43**, 145 (1979).
28. A. L. Brower, J. R. Sabin, B. Crist, and M. A. Ratner, *Int. J. Quantum Chem.* **18**, 651 (1980).
29. A. Karpfen, *J. Chem. Phys.* **75**, 238 (1981).
30. C. W. Bunn, *Trans. Faraday Soc.* **35**, 482 (1939).
31. I. Sakurada, U. Nukushina, and T. Ito, *J. Polym. Sci.* **57**, 651 (1962).
32. I. Sakurada, T. Ito, and K. Nakamae, *J. Polym. Sci. C* **15**, 75 (1966).
33. J. Clements, R. Jakeways, and I. M. Ward, *Polymer* **19**, 639 (1978).
34. R. A. Feldkamp, G. Venkaterman, and J. S. King, in: *Neutron Inelastic Scattering, IAEA, Vienna*, Vol. 2, p. 159 (1968).
35. R. G. Schaufele and T. Schimanouchi, *J. Chem. Phys.* **42**, 2605 (1967).
36. G. R. Strobl and E. Eckel, *J. Polym. Sci., Polym. Phys. Ed.* **14**, 913 (1976).
37. T. Schimanouchi, A. Asahina and E. Enomoto, *J. Polym. Sci.* **59**, 99 (1962).
38. A. Odajima and T. Maeda, *J. Polym. Sci. C* **15**, 55 (1966).
39. G. Wobsner and S. Blasenberg, *Kolloid-Z. Z. Polym.* **241**, 985 (1970).
40. K. Tashiro, M. Kobayashi, and H. Tadokoro, *Macromolecules* **11**, 914 (1978).
41. J. Barham and A. Koller, *J. Polym. Sci., Polym. Lett. Ed.* **17**, 591 (1979).
42. G. T. Davies, P. K. Eby, and J. P. Coulson, *J. Appl. Phys.* **41**, 4366 (1970).
43. Y. Kobayashi and A. Keller, *Polymer* **11**, 114 (1970).
44. B. Crist, M. A. Ratner, A. L. Brower, and J. R. Sabin, *J. Appl. Phys.* **50**, 6047 (1980).
45. F. Schwarzl (personal communication).

Chapter 11

The Possible Role of Solid-State Physical Properties of Biopolymers in Their Biological Functions

11.1. MUTATION AND AGING

11.1.1. Theory of Point Mutation

Watson and Crick[1] have assumed that a point mutation (a single nucleotide-base substitution in the sequence of the bases in a DNA strand) occurs, if one of the nucleotide bases undergoes a tautomeric rearrangement via the intramolecular shifts of two protons in their hydrogen bonds (see Figure 11.1). The unusual tautomeric form of nucleotide bases (denoted by a star) is shown in Figure 11.2. If one writes the unusual and usual tautomeric forms of these bases side by side, one can easily see by inspecting the possibilities of hydrogen-bond formation that the normal A-T, G-C base pairing relations are replaced by the new relation

$$A^*\text{-}C,\quad A\text{-}C^*,\quad G^*\text{-}T,\quad G\text{-}T^* \tag{11.1}$$

Hence if, for instance, one has an A* instead of an A in one of the strands of a double helix at the instant of its replication, after a single duplication one obtains C instead of T in the complementary helix (see Figure 11.3) causing a point mutation.

In 1962 P.-O. Löwdin suggested that the double proton shifts in a base pair causing a point mutation can occur also between the two bases (*intermolecular* proton shifts[2]). His idea is supported by the fact that the

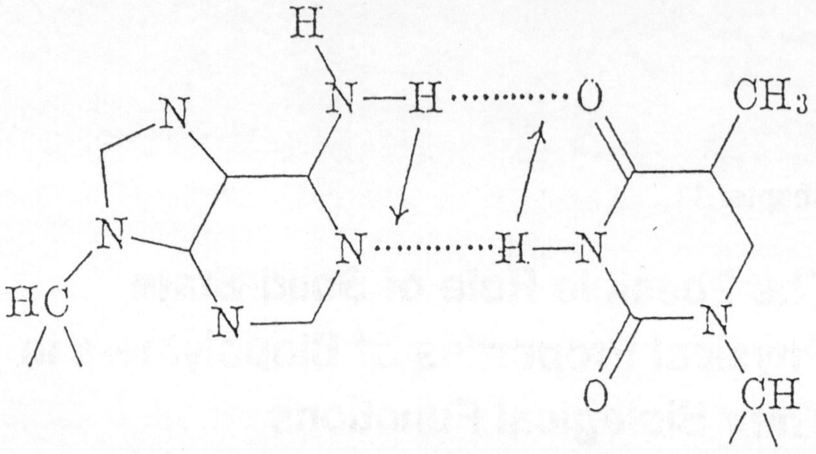

FIGURE 11.1. Tautomeric rearrangement of an A-T base pair (caused by two *intramolecular* proton shifts) which leads to a point mutation.

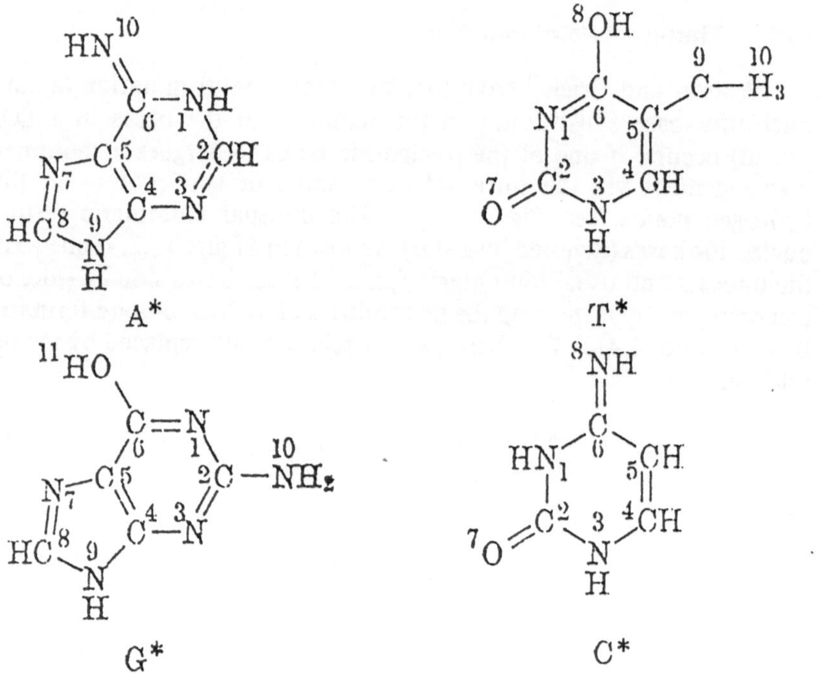

FIGURE 11.2. Chemical formulas of the unusual tautomeric forms of the DNA bases which can cause a point mutation.

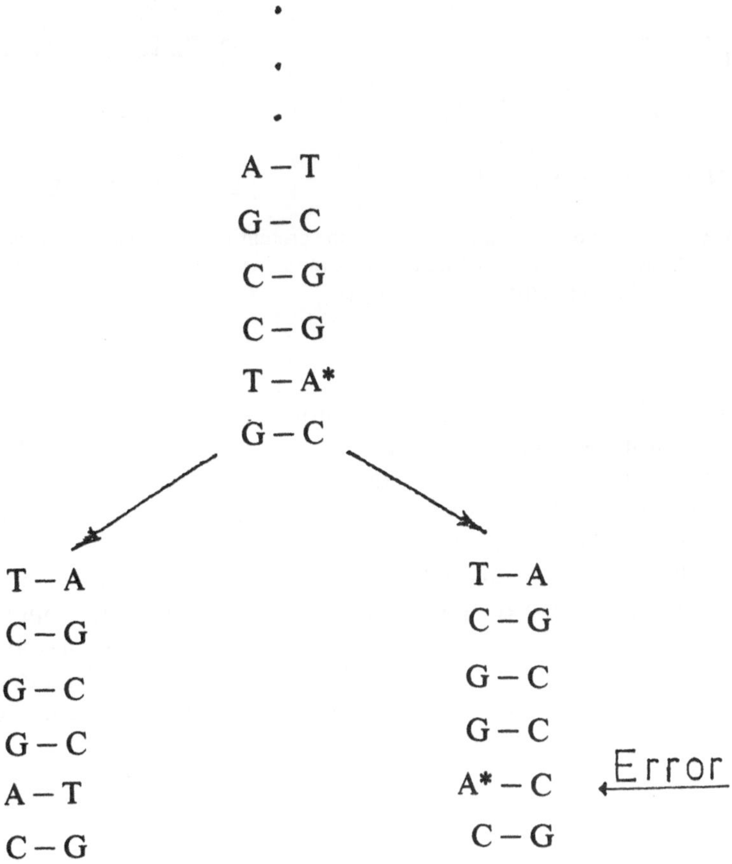

FIGURE 11.3. The Watson–Crick mechanism of point mutation (after Watson and Crick[1]).

potential barrier according to quantum-mechanical calculations for an intramolecular proton shift is essentially larger than for intermolecular proton shifts.[3] The Watson–Crick-type and Löwdin-type proton shifts in an A-T base pair are compared in Figure 11.4. When Löwdin's mechanism is valid, the base-pairing relations would be

$$A^*\text{-}T^*, \quad G^*\text{-}C^* \tag{11.2}$$

and lead to a simultaneous base substitution in both strands of a DNA double helix. In addition, one should note that both mechanisms lead to a Pyrimidine (T or C) → Py or Purine (A or G) → Pu based substitution, but

FIGURE 11.4. Tautomeric rearrangement of the protons in an A-T base pair (schematic): (a) after Watson and Crick[1] (intramolecular proton shifts) and (b) after Löwdin[2] (intermolecular shifts of the protons).

not a Py → Pu or Pu → Py base substitution (which occurs by the activation of a proto-oncogene; see below).

Löwdin has further assumed that in order to obtain the intermolecular double proton shift in a nucleotide base pair, the protons do not have to overcome the potential barriers, but the whole phenomenon can occur by simultaneous tunneling of both protons, if the first tunneling level is not much higher than the ground state of the proton in the double-well potential, that is, if the double-well potential is not highly asymmetric (see Figure 11.5). In order to study the double-well potential and hence the tunneling probabilities of the protons (or potential surface in the case

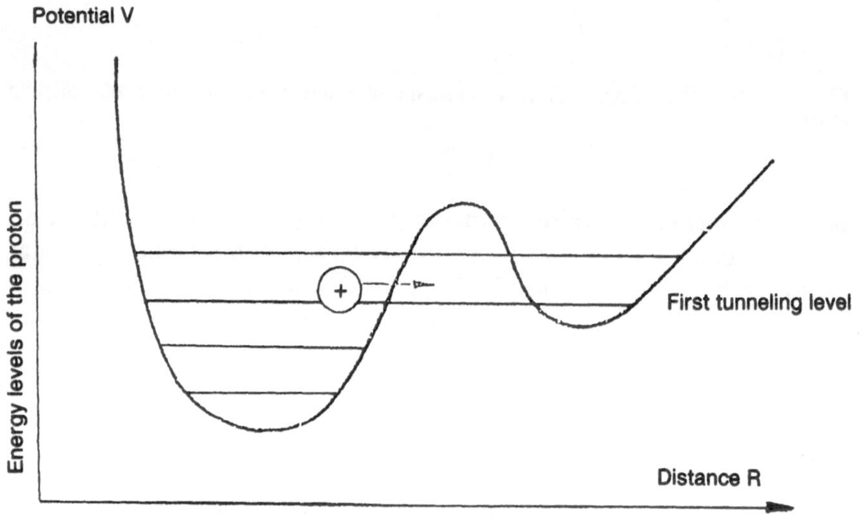

FIGURE 11.5. Tunnel effect in a double-minima potential function (schematic).

of simultaneous motion of the two protons in the hydrogen bonds), many calculations were conducted by the co-workers of Löwdin.[4] Most of these calculations employed the Pariser–Parr–Pople method, or empirical potential functions, and therefore their results cannot be considered as quantitative. Qualitatively, in all cases rather asymmetric potential wells have been obtained in the ground state of the base pairs (see Figure 11.6 for a G-C base pair) while in the first excited state of G-C the wells become rather symmetric (Figure 11.7).[5,6] This means that the tunneling probabilities are small in the ground state of the system, in agreement with the stability of the genetic code. On the other hand, in an appropriate excited state (in the first singlet excited state a charge transfer of 0.900*e* occurs from G to C[6]) the tunneling probability becomes rather large, which can explain the mutagenic effect of electromagnetic radiation.

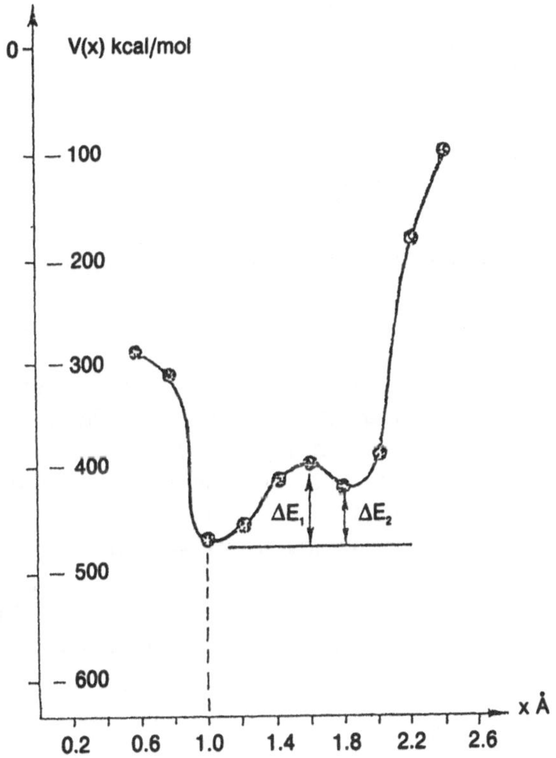

FIGURE 11.6. Potential-energy function of the N–H···N hydrogen bond in the ground state of a G–C base pair.

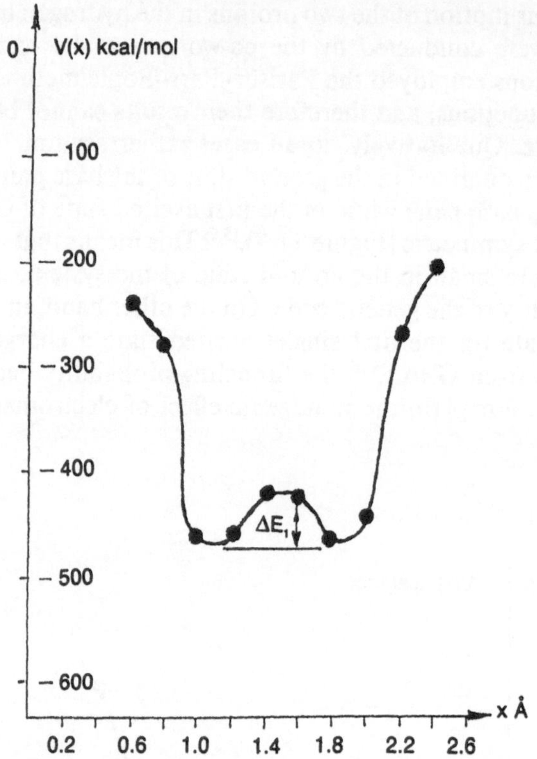

FIGURE 11.7. Potential-energy function of the N–H···N hydrogen bond in the first singlet excited state of the G-C base pair.

Quantitative data on the potential hypersurface of two simultaneously moving protons in the hydrogen bonds of the A-T and G-C base pairs can be obtained by performing an *ab initio* Hartree–Fock calculation with a good basis set and allowing for the major part of the correlation energy (for instance, with the help of MP/2 many-body perturbation theory). This already formidable undertaking is complicated by the fact that the motion of the protons in one base pair is certainly coupled to their motions at least in their first neighbors. Even if one had an AAA sequence, this would involve the simultaneous treatment of the motion of the protons in six hydrogen bonds. In addition, with each base pair at least two water molecules interact again through hydrogen bonds. The motion of the protons in these hydrogen bonds also influences the motion of the protons between the nucleotide bases in a base pair. Thus finally one is faced with at least 12 simultaneously moving protons. The quantitative treatment of this problem exceeds the present capabilities of the largest supercomputers. A further 5–10 years will probably pass until the problem of the proton shifts in DNA can be treated in a quantitative way. Despite these difficulties Löwdin's idea[2] seems quite interesting, and when computer technology has been developed sufficiently it will certainly be the subject of extensive research.

Finally, it is noteworthy that the tautomeric rearrangement of the nucleotide bases through proton shifts is not only interesting from the standpoint of point mutations. In all probability it also plays an important role in aging (see the next section) and also in the activation of oncogenes. In other words, the first human oncogene that was discovered becomes active through a single base substitution (the codon GGC has been changed to GTC at position 12 of the corresponding protein*)[7] and the activated oncogene can cause human EJ bladder carcinoma. One should observe that a

$$\begin{array}{ccc} \mathrm{G} & & \mathrm{G} \\ \mathrm{G} & \rightarrow & \mathrm{T} \\ \mathrm{C} & & \mathrm{C} \end{array} \tag{11.3}$$

base substitution is a Pu → Py type transition, which cannot be achieved either by the Watson–Crick-type or Löwdin-type tautomeric shifts. On the other hand, if one assumes the scheme

$$\begin{array}{c} \mathrm{G\text{-}C} \\ \mathrm{G\text{-}A} \\ \mathrm{C\text{-}G} \end{array} \begin{array}{c} \nearrow \\ \\ \searrow \end{array} \begin{array}{c} \mathrm{G\text{-}C} \\ \mathrm{G\text{-}C} \\ \mathrm{C\text{-}G} \\ \\ \mathrm{C\text{-}G} \\ \mathrm{A\text{-}T} \\ \mathrm{G\text{-}C} \end{array} \tag{11.4}$$

one can explain scheme (11.3) if, due to the perturbation of the normal conformation of DNA B (caused most probably by a chemical carcinogen), the unusual G-A base pair is temporarily formed in the double helix.[8] (Namely, A and G can form a base pair through hydrogen bonds, but the diameter of the double helix must be larger than usual in order to accommodate two Pu-type bases.)[9]

* The GGC → GTC sequence change has been determined by sequencing the important (protein-coding) parts of the gene in normal and tumor cells.[7]

11.1.2. Remarks about Aging

Many explain the process of aging by assuming that during a replication of DNA spontaneous point mutations occur with a certain small probability (see the previous section). If the organism is exposed to radiation or mutation-causing chemicals (the so-called mutagens) this probability can increase by several orders of magnitude. In the subsequent duplication cycles of the cells these mutation errors in the sequence of DNA accumulate. Consequently, the sequences (and hence also the conformations) of the proteins coded by the corresponding genes of DNA will also change and therefore the newly synthetized enzymes will be partially inactive, and/or exert their effects somewhat less precisely. This leads generally to a change in the rate of the biochemical reactions whose ensemble provides the basis of the self-regulation of the cells. The process called aging then takes place owing to the less effective self-regulation of the different kinds of cells and their interacting ensembles which form the different organs.

We note that if this mutation theory of aging is accepted, then the initiation of carcinogenesis can be regarded as a special case of aging. Namely, if at some spot of DNA a certain point mutation occurs, this can lead to the activation of an oncogene, which ultimately can cause the malignant transformation of the cell (see the previous and subsequent sections).

According to another theory of aging this is caused primarily by the stepwise increase in the viscosity of the cytoplasm.[10] This process could again change the rate of the biochemical reactions inside the cells and hence cause its less effective and precise self-regulation through the network of the resulting reactions. It remains, however, an open question whether the increase in viscosity is a consequence of the deterioration of the genetic information of DNA through point mutations, or is due to other causes.

There is a well-known fact which lends support to the genetic theory of aging. It is known from medicine that in a given human or animal organism the different organs are aging at quite different rates.* In other words, a 50-year-old man may have a lung like that of a 40-year-old person but with "60-year-old" intestines. One can go further and postulate that point mutations in a certain type of cell accelerate aging to a greater extent if they damage certain genes (one can call them *gerontogenes*), than if they occur at other parts of a DNA macromolecule. To

* This is true even when the organs are not exposed to especially strong external effects, such as the kidneys of an alcoholic or the liver of an overweight person.

prove this hypothesis would unfortunately require experiments extending over decades involving determination of the nucleotide base sequences of the active (protein-coding) parts of the DNA of different organs of a number of test persons at different ages (say once every ten years). After determining the biological age of the chosen organs in a test person at different physical ages, one could correlate the sequence changes found with the biological ages of the organs. In this way the gerontogenes (if they exist) could be pinpointed in the different types of cells. It is clear that such experiments impose extensive requirements on molecular biologists, but it is hoped that with the aid of more rapid base-sequencing techniques they will become possible within the next decade.

11.2. CARCINOGENESIS CAUSED BY CHEMICALS AND DIFFERENT RADIATIONS

11.2.1. Different Biochemical Mechanisms of Oncogene Activation through Chemical Carcinogens

After the discovery of oncoviruses and oncogenes in plants and animals in recent years about thirty human oncogenes have been discovered. These oncogenes exhibit the following four basic biochemical mechanisms (at least these four mechanisms are presently known). In one case a simple base substitution (the codon GGC has been changed to GTC; see Section 11.1.1) transforms a protooncogene to an active oncogene, which causes human EJ bladder carcinoma.(7) The change determined through sequencing of the important part of the gene in normal and tumor cells corresponds to a single glycine–valine substitution in the protein coded by this gene. The substitution of glycine by the more bulky valine changes strongly the conformation of the enzyme which is coded by this gene, as shown by calculations with the aid of empirical (Scheraga-type) potential functions.(11) This conformational change most probably inactivates the enzyme (oncoprotein), or changes its function, which obviously changes the regulation of the cells in which this process has occurred.

Another example of a single base substitution causing the transformation of a protooncogene to an active oncogene is the CAG→CTG codon transition.(12) This point mutation corresponds to a substitution of leucine for glutamine in position 61 of the protein coded by the protooncogene. The transformed gene (oncogene) was detected in a human lung carcinoma-derived cell line, HS 242, but proved to be transformed from the same protooncogene that can also be activated through the GGC→

GTC point mutation (but at position 12 instead of 61).[12] One should further observe that in both cases, at the end, an amino acid with a branched isopropyl group is substituted into the protein instead of the original amino acid, leading to the probable inactivation of an enzyme, although in this case leucine is not larger than glutamine.

In the second case, a gene occurring in normal cells, becomes overactivated (without any change in its sequence) by binding a new control element [the so-called long terminal repeat (LTR)] to its end. The slicing out of LTR (which is of viral origin) from another part of DNA and its insertion into DNA at the end of the oncogene happens in an enzymatic way.[13] The overproduction of the protein coded by the overactivated oncogene (by binding LTR to it) obviously again changes the regulation of the cells so much that this can lead to a malignant transformation.[13]

The third mechanism is similar to the second. In this case a whole, normally functioning gene or cluster of genes gets sliced out from one chromosome region and becomes inserted into another "more active" chromosomal region[14] (for instance, from chromosome 8 to 12). This DNA transposition (performed with the help of appropriate enzymes) again causes overproduction of the protein coded by this gene. This leads to changes in cell regulation,[14] which finally result in the development of a tumor.

A fourth mechanism, the strong multiplication of an oncogene (until it occurs a thousand times) in the same strand of DNA, was observed[15] at a more advanced stage of cell transformation. This process has the same effect as one gene overactivated to produce one thousand times the normal value of the protein molecule that is coded by this gene (which is the situation in the second and third mechanisms).

Oncogene activation through point mutations can be explained on the basis of short-range effects of a chemical carcinogen binding in the neighborhood of the site of point mutation [such as local conformational change, change in the charge distribution (including charge transfer), breakage and formation of new bonds, local changes in the vibrations, etc.]. However, this is not so in the case of the other mechanisms.

The transposition of the LTR sequence in DNA (the second mechanism) or the transposition of a whole gene to another chromosome is explained on the biochemical level by enzyme action.[13,14] But in order to understand on the physicochemical level how the binding of another DNA sequence (the LTR) to the end of a gene influences strongly its regulation, one must assume that DNA–protein interactions are essentially dependent on the sequences of the neighboring DNA segments. (Overactivation of an oncogene most probably means that it is blocked by a protein in a smaller fraction of time than under normal conditions.) In the same way, for the transposition of longer DNA sequences (with the third mechanism for a whole gene or clusters of them) into another location, one has to assume that the DNA–protein interactions of these DNA segments have been weakened to some extent (the DNA segments become deblocked) so that the necessary enzymatic reactions could take place. Since experiments reported[13,14] have been performed with whole cancer cells and the corresponding normal cells, they do not contain any information pertaining to the start of these processes.

The most plausible assumption is also found in the second and third mechanisms, namely, that they were initiated by carcinogens and/or radiations which possess not only local but also long-range effects. In a previous paper[16] a number of possibilities for long-range solid-state physical effects of carcinogen bindings on DNA–protein interactions have been reviewed: change of strength of the DNA–protein interaction in a longer sequence caused by charge transfer, which can influence strongly the polarization and dispersion forces between the two chains; long-range effect of carcinogen binding to DNA on its tertiary structure, which results again in a change of DNA–protein interaction; additional aperiodicity caused by carcinogen binding influencing DNA–protein interaction; soliton formation, etc.

The necessity of postulating long-range effects of carcinogens (even if the carcinogen binds to a site within an oncogene) can be shown on the basis of the following simple statistical consideration. One knows that nucleohistone is a protein which most frequently blocks a gene in DNA–protein (P) complexes. This protein is rather rich in the amino acid arginine (see Figure 11.8). The arginine molecule has an end group, the so-called guanidium group, which can form a hydrogen-bonded complex with the phosphate group of DNA (see Figure 11.9). Therefore one can model the interaction of an arginine side chain of a polypeptide chain with DNA through this guanidium–phosphate complex. Actually this complex also exists in a crystallized form and an X-ray investigation by Cotton and his co-workers[17] has provided an accurate geometrical structure for it. Knowing this, it was not difficult to perform a quantum-chemical calculation, which yielded a binding energy of about 2.5 eV, essentially of an electrostatic nature.[18] This is a rather large interaction energy, similar to the binding energies of normal chemical bonds.

We now assume that a nucleohistone molecule is 300 amino acids long and possesses only 5% (namely 15) arginine molecules (which is a very conservative estimate). This means that our nucleohistone molecule is strongly bound to DNA at 15 different places. We assume further that a carcinogen bound to DNA can have only local effects (say, at the base to which it is bound and at its first right and left neighbors). From the pitch height of a turn of an α-helix (5.4 Å) and from the number of amino acids in a turn (3.6), as well as from the height of a turn of DNA B (34 Å) and the number of base pairs in one turn (10), it can easily be calculated that about 20 (more precisely 19) amino acids correspond to 10 stacked bases in the DNA chain (if we assume an α-helix configuration for the protein chain). This means that a nucleohistone of 300 amino acids can block a segment of 150 base pairs of DNA B [a usual gene is, on the average, 1000 units long, but since in differentiated cells it can consist of different pieces, it can easily happen that the blocking–deblocking of a segment of 150 bases of it can regulate (deactivate or activate, respectively) the whole gene].

If effects on first neighbors are also taken into account, the probability that a carcinogen affects locally a strong arginine–phosphate bond is $1/(150/3) = 1/50$. The probability of affecting a second such bond would again be 1/50, and that of affecting both bonds simultaneously is

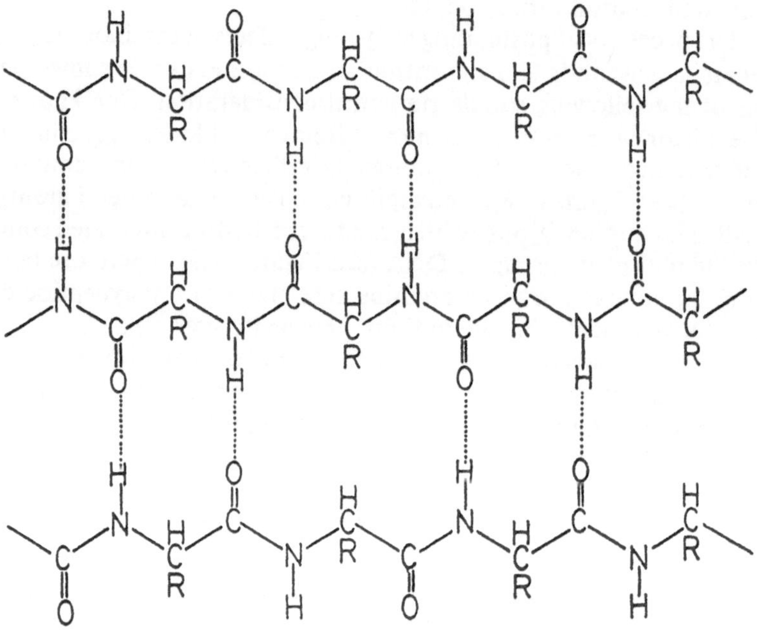

FIGURE 11.8. The β-pleated sheet structure of protein. Dotted lines indicate hydroge. bonds.

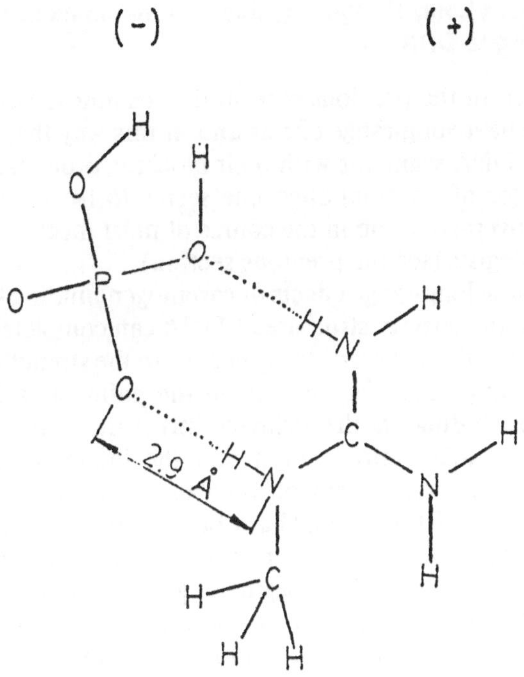

FIGURE 11.9. Chemical structure of the phosphate–guanidium complex; ΔE, the interaction energy, is 2.55 eV.

(1/50)·(1/49). Hence the overall probability of affecting all the 15 strong arginine–phosphate bonds at the same time would be

$$\frac{1}{50}\frac{1}{49}\frac{1}{48}\cdots\frac{1}{37}\frac{1}{36} \ll 10^{-15} \tag{11.5}$$

which is an extremely small number.

From this simple estimation we can conclude that it is extemely improbable to deblock even a part of an oncogene with the aid of carcinogens bound to it, if we assume only their local effects. On the other hand we have seen, with the exception of oncogene activation through point mutation, that all other methods of their activation presume their deblocking from their protein shield. The only way out of this dilemma seems to be to assume that carcinogens bound to DNA also exert long-range (usually solid-state physical) effects.

11.2.2. Different Long-Range Physical Mechanisms of Carcinogen Binding to DNA

It was seen in the previous section that chemical carcinogens most probably also have long-range effects and in this way they can affect the interaction of DNA segments with their blocking proteins. The deblocking of a part (or of a whole) oncogene seems to be the first step in its activation or overactivation in the course of most mechanisms found by molecular biologists (see the previous section).

Among these long-range effects of carcinogen binding the simplest is to assume that the tertiary structure of DNA can completely change (see Figure 11.10) and this causes a strong change in the strength of the DNA–P interaction somewhere far away from the point of carcinogen (C in Figure 11.10) binding in the primary structure. This can be easily explained if one takes into account that the interaction between two polymers is strongly dependent on their environment. In the example shown in Figure 11.10 the original back-bending of the DNA double helix (left side) ceases in consequence of the carcinogen bonding (right side). In this way the protein and DNA segment which is blocked by it has a completely different environment (including the change in the structure of the surrounding water molecules and K^+ ions) which can ultimately radically change the energy of the DNA–P interaction.[16]

Another long-range mechanism of carcinogen binding can be caused by charge transfer (CT) between the ultimate (metabolized) carcinogen and the nucleotide base or other DNA constituent to which it is binding. In other words, one can assume that due to CT from DNA to protein (*in vivo* this is rather probable, because the phosphate group has a negative charge and some amino-acid residues like arginine have positive charges) both DNA and proteins (even if they are disordered chains) have a

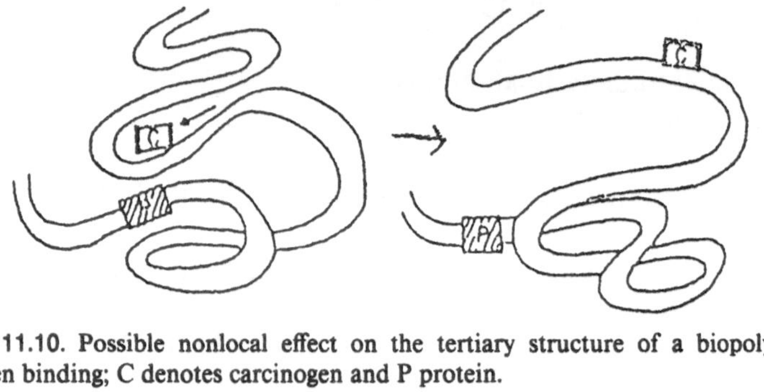

FIGURE 11.10. Possible nonlocal effect on the tertiary structure of a biopolymer of carcinogen binding; C denotes carcinogen and P protein.

partially filled valence-band region with very few gaps in the density-of-states curve.[19] Hence the Fermi level most probably falls within a broad region of allowed, very close lying energy levels.[19] One can show that the dispersion and polarization forces of such chains are orders of magnitude larger[20,21] than the case in which the valence-band region of even one chain is completely filled (this was noted by Forbes;[10] see Section 6.1). If now, due to CT between the carcinogen and a DNA constituent or an amino-acid residue in a polypeptide chain, the valence-band region of one of the chains becomes completely filled, the DNA–protein interaction becomes much weaker.

Further possibilities for long-range effects of chemical carcinogens are provided by the change or occurrence of different collective states in these complex polymers. These collective states can be vibrational or conformational solitons, Mott insulator states, Peierls instabilities, plasmon-type states, charge and spin density waves, excitonic insulator states, etc. Here only one example (which has been worked out in some detail), namely a conformational soliton caused by carcinogen binding, will be discussed.

To understand the formation of a conformational soliton due to carcinogen binding we assume that a bulky carcinogen, like the ultimate of 3,4-benzpyrene, is bound to a nucleotide base. Certainly, in the neighborhood of the attached carcinogen the structure of DNA, primarily the positions of the stacked base pairs, will become strongly distorted. With this conformational change a change in the electronic structure of DNA will be coupled, because the electronic interaction of the stacked base pairs is strongly dependent on their relative positions. In this way a nonlinear change (conformational change coupled with electronic structure change) takes place at the site and neighborhood of the carcinogen binding (see Figure 11.11).

In vitro, a covalently bound ultimate carcinogen would remain attached to DNA for an indefinitely long time. This is not the case, however, *in vivo* where, for instance, repair enzymes can remove the carcinogen within a few hours.[22] Until the carcinogen sits at its binding site, the above-described nonlinear change will remain localized to the neighborhood of this site. After removal of the carcinogen, however, it seems rather probable that the system will not relax immediately (the original conformation will not be restored instantaneously) because, for this to happen, bulky constituents of DNA (together with the water and ion structure surrounding them) have to be moved. On the other hand, it is well known that solitons have several orders of magnitude longer lifetimes than simple electronic excitations, and that they can travel as solitary waves in an extended system.[23] Therefore, one can postulate

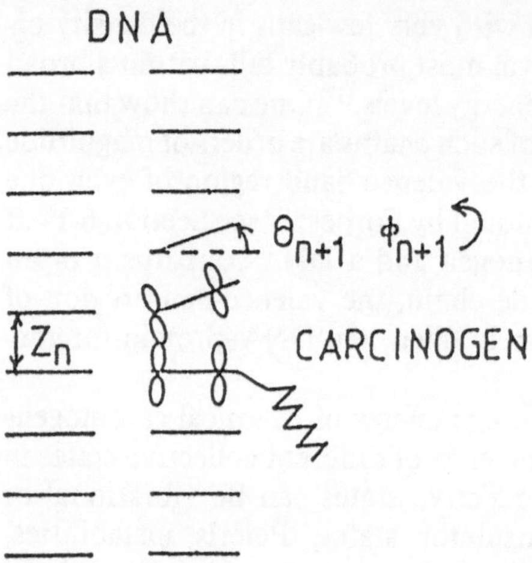

FIGURE 11.11. Distortion of a DNA stack due to binding of a bulky carcinogen to a nucleotide base. Z_n denotes the change in stacking distance at the nth unit, the distortion angle θ_{n+1} measures the deviation in the plane of the $(n + 1)$th nucleotide base from its original position (perpendicular to the main axis of the double helix), and ϕ_{n+1} is the deviation in the latter unit from the equilibrium value of rotation around the helix axis (36° in DNA B).

that after the removal of a carcinogen from DNA, the nonlinear (but previously local) change caused by its binding can travel through rather large distances along the chain (see Figure 11.12) causing a long-range effect, which may affect a larger segment of DNA and its interaction with a protein molecule.

To test this hypothesis one can start by writing down a Hamiltonian for the soliton. Generalizing the theory of Su, Schrieffer, and Heeger

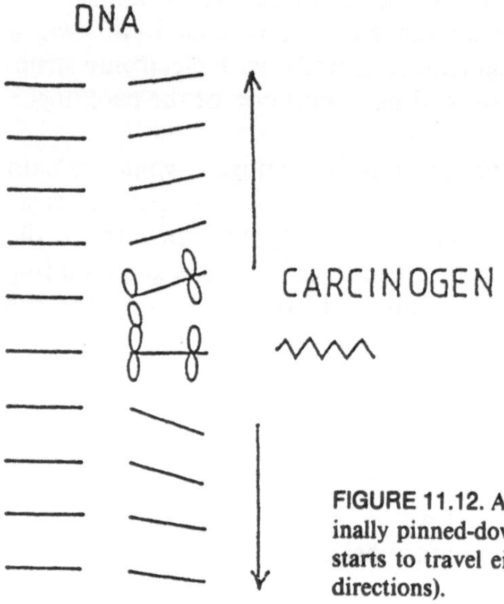

FIGURE 11.12. After removal of the carcinogen the originally pinned-down nonlinear change in the DNA stack starts to travel either upward or downward (or in both directions).

(SSH)[24] developed for the nonlinear change caused by a kink in polyacetylene, we can write

$$\tilde{H} = \tilde{H}_{el} + \tilde{H}_{conf} + \tilde{H}_{el\text{-}conf} \tag{11.6}$$

Here, as first (and rough) approximation for the description of the overlapping π electrons of the stacked bases, we take $\tilde{H}_{el}$ in the form (Hückel-type or tight-binding approximation)

$$\hat{H}_{el} = -t_0 \sum_{n,s,\alpha} (\hat{C}^{+}_{n+1,s,\alpha}\hat{C}_{n,s,\alpha} + \hat{C}^{+}_{n,s,\alpha}\hat{C}_{n+1,s,\alpha}) \tag{11.7}$$

where n is the site index (base), $s(\pm\frac{1}{2})$ denotes the spin, and $\alpha = 1, 2$ indicates the two strands in the DNA double helix. Further, the $\hat{C}^{+}$ and $\hat{C}$ are creation and annihilation operators, respectively, and t_0 is the hopping integral for the undistorted chain. One should note immediately that $\hat{H}_{el}$ is incomplete because it does not contain electron–electron interaction terms.

The conformational changes of the stacked based can be described by introducing three variables (in contrast to the one variable of the poyacetylene case[24]). We denote by $Z_{n,\alpha}$ the shift of the nth base in the αth chain from its equilibrium position along the Z axis [the main axis of the double helix; if Z_n is positive, Z_{n+1} will also be positive, however Z_{n-1} and Z_{n-2} will be negative (see Figure 11.11)] and by $\varphi_{n,\alpha}$ (following Krumhansl and Alexander[25]) the rotation of this base (again measured from its equilibrium position) in the plane perpendicular to the Z axis. Finally, the angle $\theta_{n,\alpha}$ measures the tilting of the base (thus θ is the angle between the plane of the displaced based and the plane perpendicular to the Z axis; see Figure 11.11). If we assume that the motion of the base described by these three variables can still be treated in the harmonic approximation, then we can write

$$\begin{aligned}\tilde{H}_{conf} = \frac{1}{2}\sum_{n,\alpha} [&K_1(Z_{n+1,\alpha} - Z_{n,\alpha})^2 + K_2(l_{n+1,\varphi,\alpha}\varphi_{n+1,\alpha} - l_{n,\varphi,\alpha}\varphi_{n,\alpha})^2 \\ &+ K_3(l_{n+1,\theta,\alpha}\theta_{n+1,\alpha} - l_{n,\theta,\alpha}\theta_{n,\alpha})^2 \\ &+ M_{n,\alpha}(\dot{Z}^2_{n,\alpha} + l_{n,\varphi,\alpha}\dot{\varphi}^2_{n,\alpha} + l_{n,\theta,\alpha}\theta^2_{n,\alpha})]\end{aligned} \tag{11.8}$$

Here K_1, K_2, and K_3 are the force constants corresponding to the three variables Z, φ, and θ, while $M_{n,\alpha}$ is the mass of the nth base in the αth strand (of course $M_{n,\alpha}$ has only four different values). In addition, $l_{n,\varphi,\alpha}$ is the distance of the center of mass of the nth unit from the z axis and $l_{n,\theta,\alpha}$

the corresponding distance from the y axis (which lies in the plane of the planar base pairs).

Finally, the conformational change can be coupled to the electronic structure change by introducing the modified hopping integral $\tilde{t}_{n+1,\alpha;n,\alpha}$ in the form

$$\tilde{t}_{n+1,\alpha;n,\alpha} = t_0 - [\beta_1(Z_{n+1,\alpha} - Z_{n,\alpha}) + \beta_2(l_{n+1,\varphi,\alpha}\varphi_{n+1,\alpha} - l_{n,\varphi,\alpha}\varphi_{n,\alpha}) + \beta_3(l_{n+1,\theta,\alpha}\theta_{n+1,\alpha} - l_{n,\theta,\alpha}\theta_{n,\alpha})] \tag{11.9}$$

where β_1, β_2, and β_3 are the electron-base displacement (phonon) coupling constants. We note that equation (11.9) is a straightforward generalization of equation (2.2) of SSH.[24] Expression (11.9) can be employed to express the nonlinear electron-base displacement coupling term as

$$\hat{H}_{\text{el-conf}} = \sum_{n,s,\alpha} (\tilde{t}_{n+1,\alpha;n,\alpha} - t_0)(\hat{C}^+_{n+1,s,\alpha}\hat{C}_{n,s,\alpha} + \hat{C}^+_{n,s,\alpha}\hat{C}_{n+1,s,\alpha}) \tag{11.10}$$

Following SSH[24] and Krumhansl and Alexander,[25] one can substitute the Hamiltonian (11.6) with definitions (11.7)–(11.10) (after taking expectation values of the electron operators with the resulting many-electron reference function) into the classical canonical equations of motion of Hamilton. For actual numerical calculations one must choose a set of values K_i and β_i ($i = 1, 2, 3$), or one can try to determine them on the basis of quantum-mechanical potential-surface calculations and computing the corresponding electron–phonon matrix elements.

However, in reality, the three variables describing the change in the positions of nucleotide bases are not independent but have to be coupled. Further, an at least approximate treatment of the electron–electron interaction is indispensable. Therefore, Hamiltonian (11.6) must be written in the form

$$\hat{H} = \hat{\tilde{H}} + \hat{H}' \tag{11.11}$$

where

$$\hat{H}' = \hat{H}'_{\text{el}} + \hat{H}'_{\text{conf}} + \hat{H}'_{\text{el-conf}} \tag{11.12}$$

Here

$$\hat{H}'_{\text{el}} = \sum_{n,m} \gamma_{n,m}(\hat{n}_m - 1)(\hat{n}_m - 1) \tag{11.13}$$

where

$$\hat{n}_m = \hat{C}^+_m\hat{C}_m$$

is the number operator at site m. Equation (11.13) describes the electron–electron interaction in the well-known Pariser–Parr–Pople (PPP) approximation. Further H'_{conf} can be written as

$$H'_{\text{conf}} = \frac{1}{2} \sum_{n,\alpha} [X_1(Z_{n+1,\alpha} - Z_{n,\alpha})(l_{n+1,\varphi,\alpha}\varphi_{n+1,\alpha} - l_{n,\varphi,\alpha}\varphi_{n,\alpha}) + X_2(Z_{n+1,\alpha} - Z_{n,\alpha})(l_{n+1,\theta,\alpha}\theta_{n+1,\alpha} - l_{n,\theta,\alpha}\theta_{n,\alpha}) + X_3(l_{n+1,\varphi,\alpha}\varphi_{n+1,\alpha} - l_{n,\varphi,\alpha}\varphi_{n,\alpha})(l_{n+1,\theta,\alpha}\theta_{n+1,\alpha} - l_{n,\theta,\alpha}\theta_{n,\alpha})] \tag{11.14}$$

where X_1 X_2, and X_3 are coupling constants between the different coordinates describing the distortion of the original position of a nucleotide base (these coupling constants could also be computed from the corresponding potential surface). Finally, $H'_{\text{el-conf}}$ can be easily constructed if one replaces $\tilde{t}_{n+1,\alpha;n,\alpha}$ in equation (11.10) by

$$t_{n+1,\alpha;n,\alpha} = \tilde{t}_{n+1,\alpha;n,\alpha} + t'_{n+1,\alpha;n,\alpha} \tag{11.15}$$

with

$$t'_{n+1,\alpha;n,\alpha} = -\{[\gamma_1(l_{n+1,\varphi,\alpha}\varphi_{n+1,\alpha} - l_{n,\varphi,\alpha}\varphi_{n,\alpha}) + \gamma_2(l_{n+1,\theta,\alpha}\theta_{n+1,\alpha} - l_{n,\theta,\alpha}\theta_{n,\alpha})](Z_{n+1,\alpha} - Z_{n,\alpha}) + \gamma_3(l_{n+1,\varphi,\alpha}\varphi_{n+1,\alpha} - l_{n,\varphi,\alpha}\varphi_{n,\alpha})(l_{n+1,\theta,\alpha}\theta_{n+1,\alpha} - l_{n,\theta,\alpha}\theta_{n,\alpha})\} \tag{11.16}$$

The additional electron–phonon coupling constants could be calculated again with the help of quantum mechanically computed electronic wave functions of the stack (band structures) using again a potential hypersurface to determine the wave functions of these coupled vibrations. Presently, however, to obtain an orientation one can treat these coupling constants as parameters substituting different values for them.

The dynamical problem can be solved, at least classically, by following the numerical iterative procedure of Su and Heeger[26] and Su.[27] This procedure involves determining the eigenvalues of the matrices $\mathbf{T}^{(r)}$, $r = 1, 2, 3$ (describing the electronic problem) whose elements are defined by the equations

$$T^{(1)}_{n+1,n} = t_0 - \beta_1(Z_{n+1} - Z_n) \tag{11.17a}$$

$$T^{(2)}_{n+1,n} = t_0 - \beta_2(l_{n+1,\varphi}\varphi_{n+1} - l_{n,\varphi}\varphi_n) \tag{11.17b}$$

$$T^{(3)}_{n+1,n} = t_0 - \beta_3(l_{n+1,\theta}\theta_{n+1} - l_{n,\theta}\theta_n) \tag{11.17c}$$

where subscript α has been omitted. If initial values $\{Z_n^{(0)}\}$, $\{\varphi_n^{(0)}\}$, and $\{\theta_n^{(0)}\}$ are attributed to all the three variables for all n, the classical potential can be expressed as

$$\begin{aligned}\tilde{V}^{(0)} &= \tilde{V}(\{Z_n^{(0)}\}, \{\varphi_n^{(0)}\}, \{\theta_n^{(0)}\}) \\ &= \frac{1}{2}\sum_n [K_1(Z_{n+1}^{(0)} - Z_n^{(0)})^2 \\ &\qquad + K_2(l_{n+1,\varphi}^{(0)}\varphi_{n+1}^{(0)} - l_{n,\varphi}^{(0)}\varphi_n^{(0)})^2 + K_3(l_{n+1,\theta}^{(0)}\theta_{n+1}^{(0)} - l_{n,\theta}^{(0)}\theta_n^{(0)})^2] \\ &\quad + \sum_n [A_1(Z_{n+1}^{(0)} - Z_n^{(0)}) + A_2(l_{n+1,\varphi}^{(0)}\varphi_{n+1}^{(0)} - l_{n,\varphi}^{(0)}\varphi_n^{(0)}) \\ &\qquad + A_3(l_{n+1,\theta}^{0}\theta_{n+1} - l_{n,\theta}^{(0)}\theta_n^{(0)})] \\ &\quad + \sum_{v,s} \hat{n}_{v,s}\tilde{\varepsilon}_{v,s}^{(r)(0)}, \qquad r = 1, 2, 3\end{aligned} \tag{11.18}$$

where s denotes the spin. In this equation the terms containing the constants A_1, A_2, and A_3 are necessary because otherwise (due to the lack of electron–electron interaction) the system would collapse. Their values can be determined by the conditions (in analogy to the polyacetylene case)

$$\frac{\delta\tilde{V}}{\delta\{Z_n\}} = 0, \quad \frac{\delta\tilde{V}}{\delta\{l_{n,\varphi}\varphi_n\}} = 0, \quad \frac{\delta V}{\delta\{l_{n,\theta}\theta_n\}} = 0$$

where the first equation describes uniform compression, the second uniform rotational deformation around the main axis, and the third a uniform tilting of the nucleotide bases.

Following again Su and Heeger[26] and Su,[27] the next step is to solve numerically the equations of motion (taking a compositionally averaged mass M of the four nucleotide bases)

$$\bar{M}\ddot{Z}_n = -\delta\tilde{V}/\delta Z_n \tag{11.19a}$$

$$\bar{M}l_{n,\varphi}\ddot{\varphi}_n = -\delta\tilde{V}/\delta\varphi_n \tag{11.19b}$$

$$\bar{M}l_{n,\theta}\ddot{\theta}_n = -\delta\tilde{V}/\delta\theta_n \tag{11.19c}$$

in order to obtain a new set of values of the variables $\{Z_n^{(1)}\}$, $\{\varphi_n^{(1)}\}$, and $\{\theta_n^{(1)}\}$ (for all n) which will describe the system at $t = \tau$ (one can choose a grid in time with steps τ). Substitution of these new values into the matrices $\mathbf{T}^{(r)}$ yields new eigenvalues $\tilde{\varepsilon}_{r,s}^{(r)}$ with the aid of which one can solve the dynamical problem again. In this way the time evolution of the stack can be followed until any required time.

This procedure was applied to the polyacetylene (PA) problem of 51 CH units with one kink. It was found that over a broad range of parameter values K_i and β_i it was possible to obtain a true soliton solution (with an infinite lifetime).[28]

One should point out, however, that in the aformentioned procedure we have used the simple Hamiltonian (11.6), which does not include couplings between the three different distortion variables of the DNA stack. Therefore the outlined method would describe three independent distortion problems, which differ from each other only in the values of the parameters β_i and K_i ($i = 1, 2, 3$) and in A_i. Therefore the existing polyacetylene kink program can be easily used to study the time dependence of these distortions of a DNA stack.

If one wants to study the dynamic behavior of the more realistic system described by the Hamiltonian defined through equations (11.11)–(11.16), there is only one matrix $\mathbf{T}$ describing the electronic problem whose elements are now given by

$$T_{n+1,n} = \sum_{r=1}^{3} T^{(r)}_{n+1,n} - 2t_0 + t'_{n+1,n} + \sum_{n,m} \gamma_{n,m}(\hat{n}_n - 1)(\hat{n}_m - 1) \qquad (11.20)$$

where $t'_{n+1,n}$ is defined by equation (11.16) and the last term in expression (11.20) was explained after equation (11.13). On solving the eigenvalue problem of matrix $\mathbf{T}$ with initial values $\{Z_n^{(0)}\}$, $\{\varphi_n^{(0)}\}$, and $\{\theta_n^{(0)}\}$, one obtains for the potential at $t = 0$

$$\begin{aligned} V^{(0)} = \frac{1}{2}\sum_n [&K_1(Z^{(0)}_{n+1} - Z^{(0)}_n)^2 \\ &+ K_2(l_{n+1,\varphi}\varphi_{n+1} - l_{n,\varphi}\varphi_n)^2 + K_3(l_{n+1,\theta}\theta_{n+1} - l_{n,\theta}\theta_n)^2] \\ &+ \sum_{v,s} \hat{n}_{v,s}(\varepsilon^{(0)}_{v,s} + T_{v,s}) + H'_{\text{conf}}[\{Z_n^{(0)}\}, \{\varphi_n^{(0)}\}, \{\theta_n^{(0)}\}] \end{aligned} \qquad (11.21)$$

Here, H'_{conf} is defined by equation (11.14) (using the initial values of the variables), and since the matrix $\mathbf{T}$ now contains electron–electron interaction [see equation (11.20)] the terms containing the constants A_i ($i = 1, 2, 3$) in equation (11.18) had to be omitted. In addition, $T_{v,s}$ denotes the one-electron integral, $T_{v,s} = \langle \phi_{v,s} | \hat{H}^N | \phi_{v,s} \rangle$, where $\phi_{v,s}$ is a one-electron spin orbital and $\hat{H}^N$ is the one-electron part of the Hamiltonian.

Having defined the starting potential $V^{(0)}$ in order to describe the time evolution of the system, one can proceed in the same way as above [see equation (11.11) and the subsequent remarks]. The modification of our program to include the more realistic Hamiltonian (11.11) of the problem has been executed. In this case too the soliton in PA had an infinite lifetime.[28]

The solutions of the classical equations of motion will then provide the time evolution of the solitary wave generated at the site of carcinogen binding. In this way one can learn about the lifetime and the range of the traveling nonlinear distortion caused by carcinogen binding and subsequent release as a function of the parameter values. With a realistic estimate of the parameter values one would most probably not obtain a soliton in the strict sense (infinite lifetime) but rather a solitary wave, which possesses a long enough lifetime to travel along the stack causing long-range interference with the DNA–protein interactions. In this way it can possibly initiate the activation of an oncogene. Obviously such results will have, besides their physical significance, profound biological importance.

Against the effect of a conformational solition on DNA–P interaction one could argue that if the solitary wave loses energy during its motion by breaking H bonds between DNA and protein, after a few units it would lose its energy. This would certainly be true if DNA were in vacuum. Since, however, DNA has a water and K^+-ion environment which interacts again with other molecules and macromolecules in the cell, one can regard DNA as being a macromolecule in a heat bath which can refuel the energy of the solitary wave. Therefore, the range of this wave in a DNA stack remains an open question.

The problem was attacked very recently by Hoffmann *et al.*[29] who investigated the soliton dynamics of a formamide

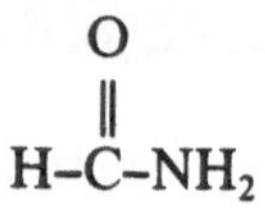

stack of 30 units.The same relative geometry of the stacked molecules was assumed as in a nucleotide base stack of DNA B. The calculations were performed using the simple SSH[24] Hamiltonian. According to the results obtained the soliton has an infinite lifetime also in this case, even if all three geometrical variables (Z, φ, and θ) were changed at $t = 0$ at the two ends of the stack (soliton–antisoliton system).[29]

Finally, it should be mentioned that in our model we have neglected, unlike Krumhansl and Alexander,[25] coupling of the distortions of the stacked bases with changes in the conformation of the sugar rings (sugar puckering), nonlinear terms in the motion of the bases and the sugar rings, and coupling to the environment (water molecules and ions).

11.2.3. Remarks on the Effect of UV and Particle Radiations and the Initiation of Carcinogenesis

UV and particle radiation has two main effects. If UV radiation is absorbed in the cytoplasm or particle radiation passes through it, usually organic radicals are formed which can easily bind to DNA and then act as chemical carcinogens. This event is more probable than direct hits on DNA.

On the other hand, if UV or particle radiations hit directly DNA then one has again both local and most probably long-range effects. In the case of UV radiation, bond breaking and forming (such as the formation of the photodimer of two T molecules), local electronic and vibrational excitations, ionization, etc., are local effects. However, the formation of excitons which can travel along the chain is one possible long-range effect. Also, breaking the extra bonds forming a thymine photodimer (for instance, owing to the action of a repair enzyme) can give rise to a conformational soliton for the same reasons as in the case of carcinogen binding to DNA and release from it (see the previous section).

If a particle hits a certain DNA molecule directly, then besides the above-mentioned local effects it can break both strands of the DNA double helix at the same site (if it has sufficient energy) and so cause complete loss of the genetic information carried by the DNA molecule. This does not usually lead to carcinogenesis. On the other hand, particle radiation can also cause collective excitations (plasmons) in DNA. This was actually measured[30] and some calculations for the first collective excited state of DNA (which corresponds to a collective excitation of the π-electrons of the nucleotide bases) were performed using a simple dispersion relation.[31] The results showed good agreement with experiment. The energy of the collective excitation dissipates after a very short time to a number of localized single excitations or ionizations. These can again create excitons or, if these excitations or ionizations are accompanied by geometrical changes, different conformational solitons. In this way a direct hit of particle radiation on DNA first causes a delocalized state, followed by the creation of several localized changes in the electronic structure that can again cause different delocalized phenomena. In other words, a long-range effect causes a number of short-range effects which, in turn, can again create other kinds of long-range effects. These may initiate the activation or overactivation of oncogenes, which may lead to a malignant transformation of the cell. We shall have to wait until appropriate experiments can be performed on ultrapure and well-characterized DNA samples to prove or disprove this rather complicated mechanism for the initiation of carcinogenesis through direct hits on DNA by particle radiations.

REFERENCES

1. J. D. H. Watson and F. H. C. Crick, *Nature* **171**, 737 (1953); F. H. C. Crick and J. D. H. Watson, *Proc. R. Soc. London, Ser. A* **223**, 171, 738 (1953).
2. P.-O. Löwdin, *Rev. Mod. Phys.* **35**, 724 (1963); *Biopolymers Symp.* **1**, 161 (1964); *Adv. Quantum Chem.* **2**, 213 (1965).
3. P.-O. Löwdin and J. Ladik (unpublished results).
4. J. Ladik, Preprint *QB 8*, Quantum Chemistry Group, Uppsala University (1964); R. Rein and F. E. Harris, *J. Chem. Phys.* **41**, 3393 (1964); *J. Chem. Phys.* **42**, 2177 (1965); *J. Chem. Phys.* **43**, 4415 (1965); G. Biczó, J. Ladik, and J. Gergely, *Acta Phys. Acad. Sci. Hung.* **20**, 11 (1966); S. Lunell and G. Sperber, *J. Chem. Phys.* **46**, 2119 (1967).
5. R. Rein and J. Ladik, *J. Chem. Phys.* **40**, 2466 (1964).
6. J. Ladik, *J. Theor. Biol.* **6**, 201 (1964).
7. E. Santos, S. R. Tronick, S. A. Aaronson, S. Pulciani, and M. Barbacid, *Nature* **298**, 343 (1982); C. J. Tabin, S. M. Bradley, C. T. Bargmann, R. H. Weinberg, A. G. Papageorge, E. M. Scolnick, R. D. Dhar, D. R. Lowy, and E. H. Chang, *Nature* **300**, 143 (1982).
8. J. Ladik and J. Čížek, *Int. J. Quantum Chem., Quantum Biol. Symp.* **26**, 955 (1984).
9. See, for instance, H. P. Yockey, in: *Symposium on Information Theory in Biology* (H. P. Yockey, ed.), pp. 50 and 297, Pergamon Press, London–Paris–New York–Los Angeles (1958).
10. W. Forbes (personal communication).
11. M. R. Pincus, J. van Ranswoude, J. B. Harford, E. H. Chang, and R. D. Klausner, *Proc. Natl. Acad. Sci. U.S.A.* **180**, 5253 (1983).
12. Y. Yuasa, S. K. Srivasta, C. Y. Dunn, J. S. Thim, P. Reddy, and S. A. Aaronson, *Nature* **303**, 775 (1983).
13. See, for instance, E. H. Chang, M. E. Furth, E. M. Scolnick, and P. R. Lowy, *Nature* **294**, 479 (1982).
14. See, for instance, E. Rechavi, D. Givot, and E. Canaani, *Nature* **300**, 607 (1982).
15. I. B. Weinstein (personal communication).
16. J. Ladik, S. Suhai, and M. Seel, *Int. J. Quantum Chem., Quantum Biol. Symp.* **5**, 35 (1978).
17. F. A. Cotton, V. W. Day, E. E. Hazen, Jr., and S. Larsen, *J. Am. Chem. Soc.* **95**, 4834 (1973).
18. R. S. Day, F. Martino, and J. Ladik, *J. Theor. Biol.* **84**, 651 (1980).
19. A. K. Bakhshi, J. Ladik, M. Seel, and P. Otto, *Chem. Phys.* **108**, 233 (1986).
20. K. Laki and J. Ladik, *Int. J. Quantum Chem., Quantum Biol. Symp.* **3**, 51 (1976).
21. J. N. Murrell, M. Randić, and O. R. Williams, *Proc. R. Soc. London, Ser. A* **284**, 566 (1965).
22. I. B. Weinstein (personal communication).

23. See, for instance, A. S. Davydov and N. F. Kisluha, *Phys. Status Solidi* **59**, 463 (1973); A. S. Davydov, *Phys. Scr.* **20**, 387 (1979).
24. W. P. Su, J. R. Schrieffer, and A. J. Heeger, *Phys. Rev. B* **4**, 2099 (1980).
25. J. A. Krumhansl and D. M. Alexander, in: *Structure and Dynamics: Nucleic Acids and Proteins* (E. Clementi and R. H. Sarma, eds.), p. 61, Academic Press, New York (1983).
26. W. P. Su and A. J. Heeger, *Proc. Natl. Acad. Sci. U.S.A.* **77**, 5626 (1980).
27. W. P. Su, in: Proc. Int. Conf. on Low-Dimensional Conductors, Boulder, Colorado [*Mol. Cryst. Liq. Cryst.* **83**, 114 (1982)].
28. W. Förner, C. L. Wang, F. Martino, and J. Ladik, *Phys. REv. B* (submitted).
29. D. Hoffmann, W. Förner, and J. Ladik, *Phys. Rev. B* (submitted).
30. N. Swansson and C. I. Powell, *J. Chem. Phys.* **39**, 630 (1963).
31. J. Jäger and J. Ladik, *Phys. Lett.* **281**, 328 (1969).

Index